Numerical Python

Scientific Computing and Data Science Applications with Numpy, SciPy and Matplotlib

Second Edition

Robert Johansson

Apress®

Numerical Python: Scientific Computing and Data Science Applications with Numpy, SciPy and Matplotlib

Robert Johansson
Urayasu-shi, Chiba, Japan

ISBN-13 (pbk): 978-1-4842-4245-2
https://doi.org/10.1007/978-1-4842-4246-9

ISBN-13 (electronic): 978-1-4842-4246-9

Library of Congress Control Number: 2018966798

Managing Director, Apress Media LLC: Welmoed Spahr
Acquisitions Editor: Todd Green
Development Editor: James Markham
Coordinating Editor: Jill Balzano

Cover designed by eStudioCalamar

Cover image designed by Freepik (www.freepik.com)

Distributed to the book trade worldwide by Springer Science+Business Media New York, 233 Spring Street, 6th Floor, New York, NY 10013. Phone 1-800-SPRINGER, fax (201) 348-4505, e-mail orders-ny@springer-sbm.com, or visit www.springeronline.com. Apress Media, LLC is a California LLC and the sole member (owner) is Springer Science + Business Media Finance Inc (SSBM Finance Inc). SSBM Finance Inc is a **Delaware** corporation.

For information on translations, please e-mail editorial@apress.com; for reprint, paperback, or audio rights, please email bookpermissions@springernature.com.

Apress titles may be purchased in bulk for academic, corporate, or promotional use. eBook versions and licenses are also available for most titles. For more information, reference our Print and eBook Bulk Sales web page at http://www.apress.com/bulk-sales.

Any source code or other supplementary material referenced by the author in this book is available to readers on GitHub via the book's product page, located at www.apress.com/9781484242452. For more detailed information, please visit http://www.apress.com/source-code.

Printed on acid-free paper

To Mika and Erika.

Table of Contents

About the Author

Robert Johansson is an experienced Python programmer and computational scientist, with a Ph.D. in Theoretical Physics from Chalmers University of Technology, Sweden. He has worked with scientific computing in academia and industry for over 10 years, and he has participated in both open source development and proprietary research projects. His open source contributions include work on QuTiP, a popular Python framework for simulating the dynamics of quantum systems; and he has also contributed to several other popular Python libraries in the scientific computing landscape. Robert is passionate about scientific computing and software development and about teaching and communicating best practices for bringing these fields together with optimal outcome: novel, reproducible, and extensible computational results. Robert's background includes 5 years of postdoctoral research in theoretical and computational physics, and he is now working as a data scientist in the IT industry.

About the Technical Reviewers

 Massimo Nardone has more than 24 years of experiences in security, web/mobile development, cloud, and IT architecture. His true IT passions are security and Android.

He has been programming and teaching how to program with Android, Perl, PHP, Java, VB, Python, C/C++, and MySQL for more than 20 years.

He holds an M.Sc. degree in computing science from the University of Salerno, Italy.

He has worked as a project manager, software engineer, research engineer, chief security architect, information security manager, PCI/SCADA auditor, and senior lead IT security/cloud/SCADA architect for many years.

His technical skills include security, Android, cloud, Java, MySQL, Drupal, Cobol, Perl, web and mobile development, MongoDB, D3, Joomla!, Couchbase, C/C++, WebGL, Python, Pro Rails, Django CMS, Jekyll, Scratch, etc.

He worked as visiting lecturer and supervisor for exercises at the Networking Laboratory of the Helsinki University of Technology (Aalto University). He holds four international patents (PKI, SIP, SAML, and Proxy areas).

He currently works as chief information security officer (CISO) for Cargotec Oyj, and he is a member of the ISACA Finland Chapter Board.

Massimo has reviewed more than 45 IT books for different publishers and has coauthored *Pro JPA 2 in Java EE 8* (Apress, 2018), *Beginning EJB in Java EE 8* (Apress, 2018), and *Pro Android Games* (Apress, 2015).

Chinmaya Patnayak is an embedded software developer at NVIDIA and is skilled in C++, CUDA, deep learning, Linux, and file systems. He has been a speaker and instructor for deep learning at various major technology events across India. Chinmaya holds an M.Sc. degree in physics and B.E. in electrical and electronics engineering from BITS Pilani. He has previously worked with Defence Research and Development Organization (DRDO) on encryption algorithms for video streams. His current interest lies in neural networks for image segmentation and applications in biomedical research and self-driving cars. Find more about him at http://chinmayapatnayak.github.io.

Michael Thomas has worked in software development for more than 20 years as an individual contributor, team lead, program manager, and vice president of engineering. Michael has more than 10 years of experience working with mobile devices. His current focus is in the medical sector, using mobile devices to accelerate information transfer between patients and health-care providers.

David Stansby is a Ph.D. student at Imperial College London and an active Python developer. He is on the core development team of Matplotlib, Python's most popular plotting library, and the creator of HelioPy, a Python package for space science data analysis.

Jason Whitehorn is an experienced entrepreneur and software developer and has helped many oil and gas companies automate and enhance their oilfield solutions through field data capture, SCADA, and machine learning. Jason obtained his B.SC. in computer science from Arkansas State University, but he traces his passion for development back many years before then, having first taught himself to program BASIC on his family's computer while still in middle school.

When he's not mentoring and helping his team at work, writing, or pursuing one of his many side projects, Jason enjoys spending time with his wife and four children and living in the Tulsa, Oklahoma region. More information about Jason can be found on his web site: `https://jason.whitehorn.us`.

Introduction

Scientific and numerical computing is a booming field in research, engineering, and analytics. The revolution in the computer industry over the last several decades has provided new and powerful tools for computational practitioners. This has enabled computational undertakings of previously unprecedented scale and complexity. Entire fields and industries have sprung up as a result. This development is still ongoing, and it is creating new opportunities as hardware, software, and algorithms keep improving. Ultimately the enabling technology for this movement is the powerful computing hardware that has been developed in recent decades. However, for a computational practitioner, the software environment used for computational work is as important as, if not more important than, the hardware on which the computations are carried out. This book is about one popular and fast-growing environment for numerical computing: the Python programming language and its vibrant ecosystem of libraries and extensions for computational work.

Computing is an interdisciplinary activity that requires experience and expertise in both theoretical and practical subjects: a firm understanding of mathematics and scientific thinking is a fundamental requirement for effective computational work. Equally important is solid training in computer programming and computer science. The role of this book is to bridge these two subjects by introducing how scientific computing can be done using the Python programming language and the computing environment that has appeared around this language. In this book the reader is assumed to have some previous training in mathematics and numerical methods and basic knowledge about Python programming. The focus of the book is to give a practical introduction to computational problem-solving with Python. Brief introductions to the theory of the covered topics are given in each chapter, to introduce notation and remind readers of the basic methods and algorithms. However, this book is not a self-consistent treatment of numerical methods. To assist readers that are not previously familiar with some of the topics of this book, references for further reading are given at the end of each chapter. Likewise, readers without experience in Python programming will probably find it useful to read this book together with a book that focuses on the Python programming language itself.

How This Book Is Organized

The first chapter in this book introduces general principles for scientific computing and the main development environments that are available for work with computing in Python: the focus is on IPython and its interactive Python prompt, the excellent Jupyter Notebook application, and the Spyder IDE.

In Chapter 2, an introduction to the NumPy library is given, and here we also discuss more generally array-based computing and its virtues. In Chapter 3, we turn our attention to symbolic computing – which in many respects complements array-based computing – using the SymPy library. In Chapter 4, we cover plotting and visualization using the Matplotlib library. Together, Chapters 2 to 4 provide the basic computational tools that will be used for domain-specific problems throughout the rest of the book: numerics, symbolics, and visualization.

In Chapter 5, the topic of study is equation solving, which we explore with both numerical and symbolic methods, using the SciPy and SymPy libraries. In Chapter 6, we explore optimization, which is a natural extension of equation solving. Here we mainly work with the SciPy library and briefly with the cvxopt library. Chapter 7 deals with interpolation, which is another basic mathematical method with many applications of its own, and important roles in higher-level algorithms and methods. In Chapter 8, we cover numerical and symbolic integration. Chapters 5 to 8 cover core computational techniques that are pervasive in all types of computational work. Most of the methods from these chapters are found in the SciPy library.

In Chapter 9, we proceed to cover ordinary differential equations. Chapter 10 is a detour into sparse matrices and graph methods, which helps prepare the field for the following chapter. In Chapter 11, we discuss partial differential equations, which conceptually are closely related to ordinary differential equations, but require a different set of techniques that necessitates the introduction of sparse matrices, the topic of Chapter 10.

Starting with Chapter 12, we make a change of direction and begin exploring data analysis and statistics. In Chapter 12, we introduce the Pandas library and its excellent data analysis framework. In Chapter 13, we cover basic statistical analysis and methods from the SciPy `stats` package. In Chapter 14, we move on to statistical modeling, using the statsmodels library. In Chapter 15, the theme of statistics and data analysis is continued with a discussion of machine learning, using the scikit-learn library. In Chapter 16, we wrap up the statistics-related chapters with a discussion of Bayesian statistics and the PyMC library. Together, Chapters 12 to 16 provide an introduction to the broad field of statistics and data analytics: a field that has been developing rapidly within and outside of the scientific Python community in recent years.

In Chapter 17, we briefly return to a core subject in scientific computing: signal processing. In Chapter 18, we discuss data input and output, and several methods for reading and writing numerical data to files, which is a basic topic that is required for most types of computational work. In Chapter 19, the final regular chapter in this book, two methods for speeding up Python code are introduced, using the Numba and Cython libraries.

The Appendix covers the installation of the software used in this book. To install the required software (mostly Python libraries), we use the conda package manager. Conda can also be used to create virtual and isolated Python environments, which is an important topic for creating stable and reproducible computational environments. The Appendix also discusses how to work with such environments using the conda package manager.

Source Code Listings

Each chapter in this book has an accompanying Jupyter Notebook that contains the chapter's source code listings. These notebooks, and the data files required to run them, can be downloaded by clicking the **Download Source Code** button located at www.apress.com/9781484242452.

CHAPTER 1

Introduction to Computing with Python

This book is about using Python for numerical computing. Python is a high-level, general-purpose interpreted programming language that is widely used in scientific computing and engineering. As a general-purpose language, Python was not specifically designed for numerical computing, but many of its characteristics make it well suited for this task. First and foremost, Python is well known for its clean and easy-to-read code syntax. Good code readability improves maintainability, which in general results in fewer bugs and better applications overall, but it also enables rapid code development. This readability and expressiveness are essential in exploratory and interactive computing, which requires fast turnaround for testing various ideas and models.

In computational problem-solving, it is, of course, important to consider the performance of algorithms and their implementations. It is natural to strive for efficient high-performance code, and optimal performance is indeed crucial for many computational problems. In such cases it may be necessary to use a low-level program language, such as C or Fortran, to obtain the best performance out of the hardware that runs the code. However, it is not always the case that optimal runtime performance is the most suitable objective. It is also important to consider the development time required to implement a solution to a problem in a given programming language or environment. While the best possible runtime performance can be achieved in a low-level programming language, working in a high-level language such as Python usually reduces the development time and often results in more flexible and extensible code.

These conflicting objectives present a trade-off between high performance and long development time and lower performance but shorter development time. See Figure 1-1 for a schematic visualization of this concept. When choosing a computational environment for solving a particular problem, it is important to consider this trade-off and to decide whether man-hours spent on the development or CPU-hours spent on

R. Johansson, *Numerical Python*, https://doi.org/10.1007/978-1-4842-4246-9_1

running the computations is more valuable. It is worth noting that CPU-hours are cheap already and are getting even cheaper, but man-hours are expensive. In particular, your own time is of course a very valuable resource. This makes a strong case for minimizing development time rather than the runtime of a computation by using a high-level programming language and environment such as Python and its scientific computing libraries.

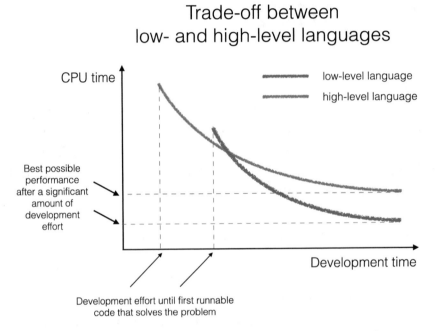

Figure 1-1. *Trade-off between low- and high-level programming languages. While a low-level language typically gives the best performance when a significant amount of development time is invested in the implementation of a solution to a problem, the development time required to obtain a first runnable code that solves the problem is typically shorter in a high-level language such as Python.*

A solution that partially avoids the trade-off between high- and low-level languages is to use a multilanguage model, where a high-level language is used to interface libraries and software packages written in low-level languages. In a high-level scientific computing environment, this type of interoperability with software packages written in low-level languages (e.g., Fortran, C, or C++) is an important requirement. Python excels at this type of integration, and as a result, Python has become a popular "glue language" used as an interface for setting up and controlling computations that use code written in low-level programming languages for time-consuming number crunching. This is an important reason for why Python is a popular language for numerical computing. The multilanguage model enables rapid code development in a high-level language while retaining most of the performance of low-level languages.

As a consequence of the multilanguage model, scientific and technical computing with Python involves much more than just the Python language itself. In fact, the Python language is only a piece of an entire ecosystem of software and solutions that provide a complete environment for scientific and technical computing. This ecosystem includes development tools and interactive programming environments, such as Spyder and IPython, which are designed particularly with scientific computing in mind. It also includes a vast collection of Python packages for scientific computing. This ecosystem of scientifically oriented libraries ranges from generic core libraries – such as NumPy, SciPy, and Matplotlib – to more specific libraries for particular problem domains. Another crucial layer in the scientific Python stack exists below the various Python modules: many scientific Python library interface, in one way or another; low-level high-performance scientific software packages, such as for example optimized LAPACK and BLAS libraries[1] for low-level vector, matrix, and linear algebra routines; or other specialized libraries for specific computational tasks. These libraries are typically implemented in a compiled low-level language and can therefore be optimized and efficient. Without the foundation that such libraries provide, scientific computing with Python would not be practical. See Figure 1-2 for an overview of the various layers of the software stack for computing with Python.

---

[1]For example, MKL, the Math Kernel Library from Intel, `https://software.intel.com/en-us/intel-mkl`; openBLAS, `https://www.openblas.net`; or ATLAS, the Automatically Tuned Linear Algebra Software, available at `http://math-atlas.sourceforge.net`

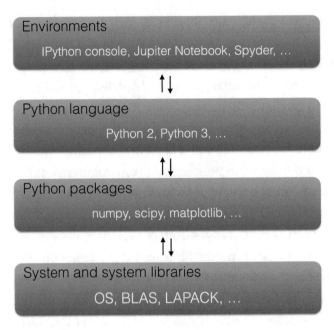

Figure 1-2. *An overview of the components and layers in the scientific computing environment for Python, from a user's perspective from top to bottom. Users typically only interact with the top three layers, but the bottom layer constitutes a very important part of the software stack.*

Tip   The SciPy organization and its web site `www.scipy.org` provide a centralized resource for information about the core packages in the scientific Python ecosystem, and lists of additional specialized packages, as well as documentation and tutorials. As such, it is a valuable resource when working with scientific and technical computing in Python. Another great resource is the *Numeric and Scientific* page on the official Python Wiki: `http://wiki.python.org/moin/NumericAndScientific`.

Apart from the technical reasons for why Python provides a good environment for computational work, it is also significant that Python and its scientific computing libraries are free and open source. This eliminates economic constraints on when and how applications developed with the environment can be deployed and distributed by its users. Equally significant, it makes it possible for a dedicated user to obtain complete insight on how the language and the domain-specific packages are implemented and what methods are used. For academic work where transparency and reproducibility are hallmarks, this is increasingly recognized as an important requirement on software used in research. For commercial use, it provides freedom on how the environment is used and integrated into products and how such solutions are distributed to customers. All users benefit from the relief of not having to pay license fees, which may otherwise inhibit deployments on large computing environments, such as clusters and cloud computing platforms.

The social component of the scientific computing ecosystem for Python is another important aspect of its success. Vibrant user communities have emerged around the core packages and many of the domain-specific projects. Project-specific mailing lists, Stack Overflow groups, and issue trackers (e.g., on Github, `www.github.com`) are typically very active and provide forums for discussing problems and obtaining help, as well as a way of getting involved in the development of these tools. The Python computing community also organizes yearly conferences and meet-ups at many venues around the world, such as the SciPy (`http://conference.scipy.org`) and PyData (`http://pydata.org`) conference series.

Environments for Computing with Python

There are a number of different environments that are suitable for working with Python for scientific and technical computing. This diversity has both advantages and disadvantages compared to a single endorsed environment that is common in proprietary computing products: diversity provides flexibility and dynamism that lends itself to specialization for particular use-cases, but on the other hand, it can also be confusing and distracting for new users, and it can be more complicated to set up a full productive environment. Here I give an orientation of common environments for scientific computing, so that their benefits can be weighed against each other and an informed decision can be reached regarding which one to use in different situations and for different purposes. The three environments discussed here are

- The Python interpreter or the IPython console to run code interactively. Together with a text editor for writing code, this provides a lightweight development environment.

- The Jupyter Notebook, which is a web application in which Python code can be written and executed through a web browser. This environment is great for numerical computing, analysis, and problem-solving, because it allows one to collect the code, the output produced by the code, related technical documentation, and the analysis and interpretation, all in one document.

- The Spyder Integrated Development Environment, which can be
  used to write and interactively run Python code. An IDE such as
  Spyder is a great tool for developing libraries and reusable Python
  modules.

All of these environments have justified use-cases, and it is largely a matter of
personal preference which one to use. However, I do in particular recommend exploring
the Jupyter Notebook environment, because it is highly suitable for interactive and
exploratory computing and data analysis, where data, code, documentation, and results
are tightly connected. For development of Python modules and packages, I recommend
using the Spyder IDE, because of its integration with code analysis tools and the Python
debugger.

Python, and the rest of the software stack required for scientific computing with
Python, can be installed and configured in a large number of ways, and in general the
installation details also vary from system to system. In Appendix 1, we go through one
popular cross-platform method to install the tools and libraries that are required for
this book.

Python

The Python programming language and the standard implementation of the Python
interpreter are frequently updated and made available through new releases.[2] Currently,
there are two active versions of Python available for production use: Python 2 and
Python 3. In this book we will work with Python 3, which by now has practically
superseded Python 2. However, for some legacy applications, using Python 2 may still be
the only option, if it contains libraries that have not been made compatible with Python
3. It is also sometimes the case that only Python 2 is the available in institutionally
provided environments, such as on high-performance clusters or universities' computer
systems. When developing Python code for such environments, it might be necessary
to use Python 2, but otherwise, I strongly recommend using Python 3 in new projects. It
should also be noted that support for Python 2 has now been dropped by many major

---

[2]The Python language and the default Python interpreter are managed and maintained by the
Python Software Foundation: `http://www.python.org`.

Python libraries, and the vast majority of computing-oriented libraries for Python now support Python 3. For the purpose of this book, we require version 2.7 or greater for the Python 2 series or Python 3.2 or greater for the preferred Python 3 series.

Interpreter

The standard way to execute Python code is to run the program directly through the Python interpreter. On most systems, the Python interpreter is invoked using the python command. When a Python source file is passed as an argument to this command, the Python code in the file is executed.

```
$ python hello.py
Hello from Python!
```

Here the file hello.py contains the single line:

```
print("Hello from Python!")
```

To see which version of Python is installed, one can invoke the python command with the --version argument:

```
$ python --version
Python 3.6.5
```

It is common to have more than one version of Python installed on the same system. Each version of Python maintains its own set of libraries and provides its own interpreter command (so each Python environment can have different libraries installed). On many systems, specific versions of the Python interpreter are available through the commands such as, for example, python2.7 and python3.6. It is also possible to set up *virtual* python environments that are independent of the system-provided environments. This has many advantages and I strongly recommend to become familiar with this way of working with Python. Appendix A provides details of how to set up and work with these kinds of environments.

In addition to executing Python script files, a Python interpreter can also be used as an interactive console (also known as a REPL: Read–Evaluate–Print–Loop). Entering python at the command prompt (without any Python files as argument) launches the Python interpreter in an interactive mode. When doing so, you are presented with a prompt:

```
$ python
Python 3.6.1 |Continuum Analytics, Inc.| (default, May 11 2017, 13:04:09)
[GCC 4.2.1 Compatible Apple LLVM 6.0 (clang-600.0.57)] on darwin
Type "help", "copyright", "credits" or "license" for more information.
>>>
```

From here Python code can be entered, and for each statement, the interpreter evaluates the code and prints the result to the screen. The Python interpreter itself already provides a very useful environment for interactively exploring Python code, especially since the release of Python 3.4, which includes basic facilities such as a command history and basic autocompletion (not available by default in Python 2).

IPython Console

Although the interactive command-line interface provided by the standard Python interpreter has been greatly improved in recent versions of Python 3, it is still in certain aspects rudimentary, and it does not by itself provide a satisfactory environment for interactive computing. IPython[3] is an enhanced command-line REPL environment for Python, with additional features for interactive and exploratory computing. For example, IPython provides improved command history browsing (also between sessions), an input and output caching system, improved autocompletion, more verbose and helpful exception tracebacks, and much more. In fact, IPython is now much more than an enhanced Python command-line interface, which we will explore in more detail later in this chapter and throughout the book. For instance, under the hood IPython is a

---

[3]See the IPython project web page, `http://ipython.org`, for more information and its official documentation.

client-server application, which separates the frontend (user interface) from the backend (kernel) that executes the Python code. This allows multiple types of user interfaces to communicate and work with the same kernel, and a user-interface application can connect multiple kernels using IPython's powerful framework for parallel computing.

Running the `ipython` command launches the IPython command prompt:

```
$ ipython
Python 3.6.1 |Continuum Analytics, Inc.| (default, May 11 2017, 13:04:09)
Type 'copyright', 'credits' or 'license' for more information
IPython 6.4.0 -- An enhanced Interactive Python. Type '?' for help.

In [1]:
```

Caution   Note that each IPython installation corresponds to a specific version of Python, and if you have several versions of Python available on your system, you may also have several versions of IPython as well. On many systems, IPython for Python 2 is invoked with the command `ipython2` and for Python 3 with `ipython3`, although the exact setup varies from system to system. Note that here the "2" and "3" refer to the Python version, which is different from the version of IPython itself (which at the time of writing is 6.4.0).

In the following sections, I give a brief overview of some of the IPython features that are most relevant to interactive computing. It is worth noting that IPython is used in many different contexts in scientific computing with Python, for example, as a kernel in the Jupyter Notebook application and in the Spyder IDE, which are covered in more detail later in this chapter. It is time well spent to get familiar with the tricks and techniques that IPython offers to improve your productivity when working with interactive computing.

Input and Output Caching

In the IPython console, the input prompt is denoted as In [1]: and the corresponding output is denoted as Out [1]:, where the numbers within the square brackets are incremented for each new input and output. These inputs and outputs are called *cells* in IPython. Both the input and the output of previous cells can later be accessed through the In and Out variables that are automatically created by IPython. The In and Out variables are a list and a dictionary, respectively, that can be indexed with a cell number. For instance, consider the following IPython session:

```
In [1]: 3 * 3
Out[1]: 9
In [2]: In[1]
Out[2]: '3 * 3'
In [3]: Out[1]
Out[3]: 9
In [4]: In
Out[4]: ['', '3 * 3', 'In[1]', 'Out[1]', 'In']
In [5]: Out
Out[5]: {1: 9, 2: '3 * 3', 3: 9, 4: ['', '3 * 3', 'In[1]', 'Out[1]', 'In', 'Out']}
```

Here, the first input was 3 * 3 and the result was 9, which later is available as In[1] and Out[1]. A single underscore _ is a shorthand notation for referring to the most recent output, and a double underscore __ refers to the output that preceded the most recent output. Input and output caching is often useful in interactive and exploratory computing, since the result of a computation can be accessed even if it was not explicitly assigned to a variable.

Note that when a cell is executed, the value of the last statement in an input cell is by default displayed in the corresponding output cell, unless the statement is an assignment or if the value is Python null value None. The output can be suppressed by ending the statement with a semicolon:

```
In [6]: 1 + 2
Out[6]: 3
In [7]: 1 + 2;     # output suppressed by the semicolon
In [8]: x = 1      # no output for assignments
In [9]: x = 2; x   # these are two statements. The value of 'x' is shown in
                     the output
Out[9]: 2
```

Autocompletion and Object Introspection

In IPython, pressing the TAB key activates autocompletion, which displays a list of symbols (variables, functions, classes, etc.) with names that are valid completions of what has already been typed. The autocompletion in IPython is contextual, and it will look for matching variables and functions in the current namespace or among the attributes and methods of a class when invoked after the name of a class instance. For example, os.<TAB> produces a list of the variables, functions, and classes in the os module, and pressing TAB after having typed os.w results in a list of symbols in the os module that starts with w:

```
In [10]: import os
In [11]: os.w<TAB>
os.wait   os.wait3   os.wait4   os.waitpid   os.walk   os.write   os.writev
```

This feature is called object introspection, and it is a powerful tool for interactively exploring the properties of Python objects. Object introspection works on modules, classes, and their attributes and methods and on functions and their arguments.

Documentation

Object introspection is convenient for exploring the API of a module and its member classes and functions, and together with the documentation strings, or "docstrings", that are commonly provided in Python code, it provides a built-in dynamic reference manual for almost any Python module that is installed and can be imported. A Python object followed by a question mark displays the documentation string for the object. This is similar to the Python function help. An object can also be followed by two question marks, in which case IPython tries to display more detailed documentation, including the Python source code if available. For example, to display help for the cos function in the math library:

```
In [12]: import math
In [13]: math.cos?
Type:         builtin_function_or_method
String form: <built-in function cos>
```

```
Docstring:
cos(x)
```

Return the cosine of x (measured in radians).

Docstrings can be specified for Python modules, functions, classes, and their attributes and methods. A well-documented module therefore includes a full API documentation in the code itself. From a developer's point of view, it is convenient to be able to document a code together with the implementation. This encourages writing and maintaining documentation, and Python modules tend to be well documented.

Interaction with the System Shell

IPython also provides extensions to the Python language that makes it convenient to interact with the underlying system. Anything that follows an exclamation mark is evaluated using the system shell (such as bash shell). For example, on a UNIX-like system, such as Linux or Mac OS X, listing files in the current directory can be done using

```
In[14]: !ls
file1.py    file2.py    file3.py
```

On Microsoft Windows, the equivalent command would be !dir. This method for interacting with the OS is a very powerful feature that makes it easy to navigate the file system and to use the IPython console as a system shell. The output generated by a command following an exclamation mark can easily be captured in a Python variable. For example, a file listing produced by !ls can be stored in a Python list using

```
In[15]: files = !ls
In[16]: len(files)
3
In[17] : files
['file1.py', 'file2.py', 'file3.py']
```

Likewise, we can pass the values of Python variables to shell commands by prefixing the variable name with a $ sign:

```
In[18]: file = "file1.py"
In[19]: !ls -l $file
-rw-r--r--  1 rob  staff 131 Oct 22 16:38 file1.py
```

This two-way communication with the IPython console and the system shell can be very convenient when, for example, processing data files.

IPython Extensions

IPython provides extension commands that are called magic functions in IPython terminology. These commands all start with one or two % signs.[4] A single % sign is used for one-line commands, and two % signs are used for commands that operate on cells (multiple lines). For a complete list of available extension commands, type %lsmagic, and the documentation for each command can be obtained by typing the magic command followed by a question mark:

```
In[20]: %lsmagic?
Type:           Magic function
String form:    <bound method BasicMagics.lsmagic of <IPython.core.magics.
                basic.BasicMagics object at 0x10e3d28d0>>
Namespace:      IPython internal
File:           /usr/local/lib/python3.6/site-packages/IPython/core/magics/
                basic.py
Definition:     %lsmagic(self, parameter_s=")
Docstring:      List currently available magic functions.
```

File System Navigation

In addition to the interaction with the system shell described in the previous section, IPython provides commands for navigating and exploring the file system. The commands will be familiar to UNIX shell users: %ls (list files), %pwd (return current working directory), %cd (change working directory), %cp (copy file), %less (show the content of a file in the pager), and %%writefile filename (write content of a cell to the file filename). Note that autocomplete in IPython also works with the files in the current

---

[4] When %automagic is activated (type %automagic at the IPython prompt to toggle this feature), the % sign that precedes the IPython commands can be omitted, unless there is a name conflict with a Python variable or function. However, for clarity, the % signs are explicitly shown here.

working directory, which makes IPython as convenient to explore the file system as is the system shell. It is worth noting that these IPython commands are system independent and can therefore be used on both UNIX-like operating systems and on Windows.

Running Scripts from the IPython Console

The command %run is an important and useful extension, perhaps one of the most important features of the IPython console. With this command, an external Python source code file can be executed within an interactive IPython session. Keeping a session active between multiple runs of a script makes it possible to explore the variables and functions defined in a script interactively after the execution of the script has finished. To demonstrate this functionality, consider a script file fib.py that contains the following code:

```
def fib(n):
    """
    Return a list of the first n Fibonacci numbers.
    """
    f0, f1 = 0, 1
    f = [1] * n
    for i in range(1, n):
        f[i] = f0 + f1
        f0, f1 = f1, f[i]

    return f

print(fib(10))
```

It defines a function that generates a sequence of n Fibonacci numbers and prints the result for $n = 10$ to the standard output. It can be run from the system terminal using the standard Python interpreter:

```
$ python fib.py
[1, 1, 2, 3, 5, 8, 13, 21, 34, 55]
```

It can also be run from an interactive IPython session, which produces the same output, but also adds the symbols defined in the file to the local namespace, so that the `fib` function is available in the interactive session after the %run command has been issued.

```
In [21]: %run fib.py
Out[22]: [1, 1, 2, 3, 5, 8, 13, 21, 34, 55]

In [23]: %who
fib

In [23]: fib(6)
Out[23]: [1, 1, 2, 3, 5, 8]
```

In the preceding example, we also made use of the %who command, which lists all defined symbols (variables and functions).[5] The %whos command is similar, but also gives more detailed information about the type and value of each symbol, when applicable.

Debugger

IPython includes a handy debugger mode, which can be invoked postmortem after a Python exception (error) has been raised. After the traceback of an unintercepted exception has been printed to the IPython console, it is possible to step directly into the Python debugger using the IPython command %debug. This possibility can eliminate the need to rerun the program from the beginning using the debugger or after having employed the common debugging method of sprinkling print statements into the code. If the exception was unexpected and happened late in a time-consuming computation, this can be a big time-saver.

To see how the %debug command can be used, consider the following incorrect invocation of the `fib` function defined earlier. It is incorrect because a float is passed to the function while the function is implemented with the assumption that the argument passed to it is an integer. On line 7 the code runs into a type error, and the Python interpreter raises an exception of the type `TypeError`. IPython catches the exception and

---

[5]The Python function `dir` provides a similar feature.

prints out a useful traceback of the call sequence on the console. If we are clueless as to why the code on line 7 contains an error, it could be useful to enter the debugger by typing %debug in the IPython console. We then get access to the local namespace at the source of the exception, which can allow us to explore in more detail why the exception was raised.

```
In [24]: fib(1.0)
---------------------------------------------------------------------------
TypeError                                 Traceback (most recent call last)
<ipython-input-24-874ca58a3dfb> in <module>()
 ----> 1 fib.fib(1.0)

/Users/rob/code/fib.py in fib(n)
       5        """
       6        f0, f1 = 0, 1
 ----> 7      f = [1] * n
       8        for i in range(1, n):
       9            f[n] = f0 + f1

TypeError: can't multiply sequence by non-int of type 'float'

In [25]: %debug
> /Users/rob/code/fib.py(7)fib()
       6    f0, f1 = 0, 1
 ----> 7    f = [1] * n
       8    for i in range(1, n):

ipdb> print(n)
1.0
```

> **Tip**   Type a question mark at the debugger prompt to show a help menu that lists available commands:
>
> ```
> ipdb> ?
> ```
>
> More information about the Python debugger and its features is also available in the Python Standard Library documentation: http://docs.python.org/3/library/pdb.html.

Reset

Resetting the namespace of an IPython session is often useful to ensure that a program is run in a pristine environment, uncluttered by existing variables and functions. The %reset command provides this functionality (use the flag –f to force the reset). Using this command can often eliminate the need for otherwise common exit-restart cycles of the console. Although it is necessary to reimport modules after the %reset command has been used, it is important to know that even if the modules have changed since the last import, a new import after a %reset will not import the new module but rather reenable a cached version of the module from the previous import. When developing Python modules, this is usually not the desired behavior. In that case, a reimport of a previously imported (and since updated) module can often be achieved by using the reload function from IPython.lib.deepreload. However, this method does not always work, as some libraries run code at import time that is only intended to run once. In this case, the only option might be to terminate and restart the IPython interpreter.

Timing and Profiling Code

The %timeit and %time commands provide simple benchmarking facilities that are useful when looking for bottlenecks and attempting to optimize code. The %timeit command runs a Python statement a number of times and gives an estimate of the runtime (use %%timeit to do the same for a multiline cell). The exact number of times the statement is ran is determined heuristically, unless explicitly set using the –n and –r flags. See %timeit? for details. The %timeit command does not return the resulting value of the expression. If the result of the computation is required, the %time or %%time (for a multiline cell) commands can be used instead, but %time and %%time only run the statement once and therefore give a less accurate estimate of the average runtime.

The following example demonstrates a typical usage of the %timeit and %time commands:

```
In [26]: %timeit fib(100)
100000 loops, best of 3: 16.9 µs per loop
In [27]: result = %time fib(100)
CPU times: user 33 µs, sys: 0 ns, total: 33 µs
Wall time: 48.2
```

While the %timeit and %time commands are useful for measuring the elapsed runtime of a computation, they do not give any detailed information about what part of the computation takes more time. Such analyses require a more sophisticated code profiler, such as the one provided by Python standard library module cProfile.[6] The Python profiler is accessible in IPython through the commands %prun (for statements) and %run with the flag –p (for running external script files). The output from the profiler is rather verbose and can be customized using optional flags to the %prun and %run –p commands (see %prun? for a detailed description of the available options).

As an example, consider a function that simulates N random walkers each taking M steps and then calculates the furthest distance from the starting point achieved by any of the random walkers:

```
In [28]: import numpy as np
In [29]: def random_walker_max_distance(M, N):
    ...:         """
    ...:         Simulate N random walkers taking M steps, and return the
    ...:         largest distance
    ...:         from the starting point achieved by any of the random walkers.
    ...:         """
    ...:         trajectories = [np.random.randn(M).cumsum() for _ in range(N)]
    ...:         return np.max(np.abs(trajectories))
```

Calling this function using the profiler with %prun results in the following output, which includes information about how many times each function was called and a breakdown of the total and cumulative time spent in each function. From this information we can conclude that in this simple example, the calls to the function np.random.randn consume the bulk of the elapsed computation time.

```
In [30]: %prun random_walker_max_distance(400, 10000)

    20008 function calls in 0.254 seconds

Ordered by: internal time
```

---

[6]Which can, for example, be used with the standard Python interpreter to profile scripts by running python -m cProfile script.py

18

ncalls	tottime	percall	cumtime	percall	filename:lineno(function)
10000	0.169	0.000	0.169	0.000	{method 'randn' of 'mtrand. RandomState' objects}
10000	0.036	0.000	0.036	0.000	{method 'cumsum' of 'numpy. ndarray' objects}
1	0.030	0.030	0.249	0.249	<ipython-input-30>:18(random_ walker_max_distance)
1	0.012	0.012	0.217	0.217	<ipython-input-30>: 19(<listcomp>)
1	0.005	0.005	0.254	0.254	<string>:1(<module>)
1	0.002	0.002	0.002	0.002	{method 'reduce' of 'numpy. ufunc' objects}
1	0.000	0.000	0.254	0.254	{built-in method exec}
1	0.000	0.000	0.002	0.002	_methods.py:25(_amax)
1	0.000	0.000	0.002	0.002	fromnumeric.py:2050(amax)
1	0.000	0.000	0.000	0.000	{method 'disable' of '_ lsprof.Profiler' objects}

Interpreter and Text Editor as Development Environment

In principle, the Python or the IPython interpreter and a good text editor are all that are required for a full productive Python development environment. This simple setup is, in fact, the preferred development environment for many experienced programmers. However, in the following sections, we will look into the Jupyter Notebook and the integrated development environment Spyder. These environments provide richer features that improve productivity when working with interactive and exploratory computing applications.

Jupyter

The Jupyter project[7] is a spin-off from the IPython project that includes the Python independent frontends – most notably the notebook application which we discuss in more detail in the following section – and the communication framework that enables

---

[7]For more information about Jupyter, see http://jupyter.org.

the separation of the frontend from the computational backends, known as kernels. Prior to the creation of the Jupyter project, the notebook application and its underlying framework were a part of the IPython project. However, because the notebook frontend is language agnostic – it can also be used with a large number of other languages, such as R and Julia – it was spun off a separate project to better cater to the wider computational community and to avoid a perceived bias toward Python. Now, the remaining role of IPython is to focus on Python-specific applications, such as the interactive Python console, and to provide a Python kernel for the Jupyter environment.

In the Jupyter framework, the frontend talks to computational backends known as kernels. The frontend can have multiple kernels registered, for example, for different programming languages, for different versions of Python, or for different Python environments. The kernel maintains the state of the interpreter and performs the actual computations, while the frontend manages how code is entered and organized and how the results of calculations are visualized to the user.

In this section, we will discuss the Jupyter QtConsole and Notebook frontends and give a brief introduction to some of their rich display and interactivity features, as well as the workflow organization that the notebook provides. The Jupyter Notebook is the Python environment for computation work that I generally recommend in this book, and the code listings in the rest of this book are understood to be read as if they are cells in a notebook.

The Jupyter QtConsole

The Jupyter QtConsole is an enhanced console application that can serve as a substitute to the standard IPython console. The QtConsole is launched by passing the `qtconsole` argument to the `jupyter` command:

```
$ jupyter qtconsole
```

This opens up a new IPython application in a console that is capable of displaying rich media objects such as images, figures, and mathematical equations. The Jupyter QtConsole also provides a menu-based mechanism for displaying autocompletion results, and it shows docstrings for functions in a pop-up window when typing the opening parenthesis of a function or a method call. A screenshot of the Jupyter Qtconsole is shown in Figure 1-3.

```
                              Jupyter QtConsole
Jupyter QtConsole 4.3.1
Python 3.6.1 |Continuum Analytics, Inc.| (default, May 11 2017, 13:04:09)
Type 'copyright', 'credits' or 'license' for more information
IPython 6.4.0 -- An enhanced Interactive Python. Type '?' for help.

In [1]: import sympy

In [2]: sympy.init_printing()

In [3]: x = sympy.symbols("x")

In [4]: i = sympy.Integral(x**2, (x, 0, 1)); i
Out[4]:
```

$$\int_0^1 x^2\ dx$$

```
In [5]: |
```

Figure 1-3. *A screenshot of the Jupyter QtConsole application*

The Jupyter Notebook

In addition to the interactive console, Jupyter also provides the web-based notebook application that has made it famous. The notebook offers many advantages over a traditional development environment when working with data analysis and computational problem-solving. In particular, the notebook environment allows to write and to run code, to display the output produced by the code, and to document and interpret the code and the results: all in one document. This means that the entire analysis workflow is captured in one file, which can be saved, restored, and reused later on. In contrast, when working with a text editor or an IDE, the code, the corresponding data files and figures, and the documentation are spread out over multiple files in the file system, and it takes a significant effort and discipline to keep such a workflow organized.

The Jupyter Notebook features a rich display system that can show media such as equations, figures, and videos as embedded objects in the notebook. It is also possible to create user interface (UI) elements with HTML and JavaScript, using Jupyter's widget system. These widgets can be used in interactive applications that connect the web application with Python code that is executed in the IPython kernel (on the server side). These and many other features of the Jupyter Notebook make it a great environment for interactive and literate computing, as we will see examples of throughout this book.

To launch the Jupyter Notebook environment, the `notebook` argument is passed to the `jupyter` command-line application.

```
$ jupyter notebook
```

This launches a notebook kernel and a web application that, by default, will serve up a web server on port 8888 on localhost, which is accessed using the local address `http://localhost:8888/` in a web browser.[8] By default, running `jupyter notebook` will open a dashboard web page in the default web browser. The dashboard lists all notebooks that are available in the directory from where the Jupyter Notebook was launched, as well as a simple directory browser that can be used to navigate subdirectories, relative to the location where the notebook server was launched, and to open notebooks from therein. Figure 1-4 shows a screenshot of a web browser and the Jupyter Notebook dashboard page.

---

[8]This web application is by default only accessible locally from the system where the notebook application was launched.

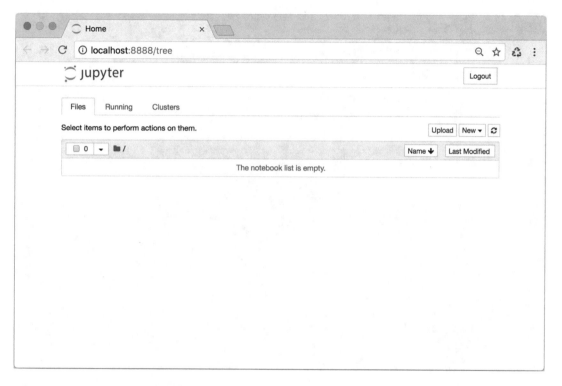

Figure 1-4. *A screenshot of the Jupyter Notebook dashboard page*

Clicking the "New" button creates a new notebook and opens it in a new page in the browser (see Figure 1-5). A newly created notebook is named Untitled, or Untitled1, etc., depending on the availability of unused filenames. A notebook can be renamed by clicking the title field on the top of the notebook page. The Jupyter Notebook files are stored in a JSON file format using the filename extension ipynb. A Jupyter Notebook file is not pure Python code, but if necessary the Python code in a notebook can easily be extracted using either "File ➤ Download as ➤ Python" or the Jupyter utility nbconvert (see in the following section).

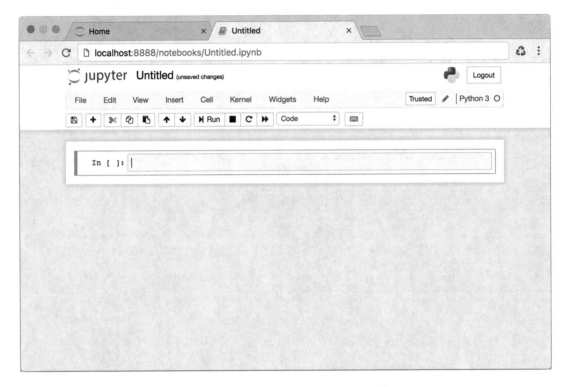

Figure 1-5. *A newly created and empty Jupyter Notebook*

Jupyter Lab

Jupyter Lab is a new alternative development environment from the Jupyter project. It combines the Jupyter Notebook interface with a file browser, text editor, shell, and IPython consoles, in a web-based IDE-like environment; see Figure 1-6.

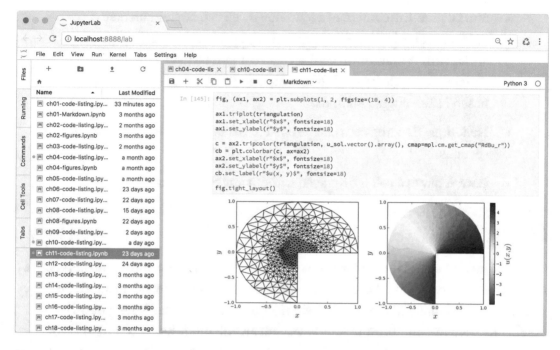

Figure 1-6. *The Jupyter Lab interface, which includes a file browser (left) and multitab notebook editor (right). The notebook displayed here shows code and output from an example in Chapter 11.*

The Jupyter Lab environment consolidates the many advantages of the notebook environment and the strengths of traditional IDEs. Having access to shell consoles and text editors all within the same web frontend is also convenient when working on a Jupyter server that runs on a remote system, such as a computing cluster or in the cloud.

Cell Types

The main content of a notebook, below the menu bar and the toolbar, is organized as input and output cells. The cells can be of several types, and the type of the selected cell can be changed using the cell-type drop-down menu in the toolbar (which initially displays "Code"). The most important types are

- **Code**: A code cell can contain an arbitrary amount of multiline Python code. Pressing Shift-Enter sends the code in the cell to the kernel process, where the kernel evaluates it using the Python interpreter. The result is sent back to the browser and displayed in the corresponding output cell.

- **Markdown**: The content of a Markdown cell can contain marked-up plain text, which is interpreted using the Markdown language and HTML. A Markdown cell can also contain LaTeX formatted equations, which are rendered in the notebook using the JavaScript-based LaTeX engine MathJax.

- **Headings**: Heading cells can be used to structure a notebook into sections.

- **Raw**: A raw text cell is displayed without any processing.

Editing Cells

Using the menu bar and the toolbar, cells can be added, removed, moved up and down, cut and pasted, and so on. These functions are also mapped to keyboard shortcuts, which are convenient and time-saving when working with Jupyter Notebooks. The notebook uses a two-mode input interface, with an edit mode and a command mode. The edit mode can be entered by clicking a cell or by pressing the Enter key on the keyboard when a cell is in focus. Once in edit mode, the content of the input cell can be edited. Leaving the edit mode is done by pressing the ESC key or by using Shift-Enter to execute the cell. When in command mode, the up and down arrows can be used to move focus between cells, and a number of keyboard shortcuts are mapped to the basic cell manipulation actions that are available through the toolbar and the menu bar. Table 1-1 summarizes the most important Jupyter Notebook keyboard shortcuts for the command mode.

Table 1-1. *A Summary of Keyboard Shortcuts in the Jupyter Notebook Command Mode*

Keyboard Shortcut	Description
b	Create a new cell *below* the currently selected cell.
a	Create a new cell *above* the currently selected cell.
d-d	Delete the currently selected cell.
1 to 6	Heading cell of level 1 to 6.
x	Cut currently selected cell.
c	Copy currently selected cell.
v	Paste cell from the clipboard.
m	Convert a cell to a Markdown cell.
y	Convert a cell to a code cell.
Up	Select previous cell.
Down	Select next cell.
Enter	Enter edit mode.
Escape	Exit edit mode.
Shift-Enter	Run the cell.
h	Display a help window with a list of all available keyboard shortcuts.
0-0	Restart the kernel.
i-i	Interrupt an executing cell.
s	Save the notebook.

While a notebook cell is being executed, the input prompt number is represented with an asterisk, In[*], and an indicator in the upper right corner of the page signals that the IPython kernel is busy. The execution of a cell can be interrupted using the menu option "Kernel ➤ Interrupt" or by typing i-i in the command mode (i.e., press the i key twice in a row).

Markdown Cells

One of the key features of the Jupyter Notebook is that code cells and output cells can be complemented with documentation contained in text cells. Text input cells are called Markdown cells. The input text is interpreted and reformatted using the Markdown markup language. The Markdown language is designed to be a lightweight typesetting system that allows text with simple markup rules to be converted to HTML and other formats for richer display. The markup rules are designed to be user-friendly and readable as is in plain-text format. For example, a piece of text can be made italics by surrounding it with asterisks, *text*, and it can be made bold by surrounding it with double asterisks, **text**. Markdown also allows creating enumerated and bulleted lists, tables, and hyper-references. An extension to Markdown supported by Jupyter is that mathematical expressions can be typeset in LaTeX, using the JavaScript LaTeX library MathJax. Taking full advantage of what Jupyter Notebooks offer includes generously documenting the code and resulting output using Markdown cells and the many rich display options they provide. Table 1-2 introduces basic Markdown and equation formatting features that can be used in a Jupyter Notebook Markdown cell.

Table 1-2. *Summary of Markdown Syntax for Jupyter Notebook Markdown Cells*

Function	Syntax by Example
Italics	`*text*`
Bold	`**text**`
Strike-through	`~~text~~`
Fixed-width font	`` `text` ``
URL	`[URL text](http://www.example.com)`
New paragraph	Separate the text of two paragraphs with an empty line.
Verbatim	Lines that start with four blank spaces are displayed as is, without any further processing, using a fixed-width font. This is useful for code-like text segments. ␣␣␣␣def func(x): ␣␣␣␣ return x ** 2

(continued)

Table 1-2. (*continued*)

Function	Syntax by Example

Table	` | A | B | C | `
	` |---|---|---| `
	` | 1 | 2 | 3 | `
	` | 4 | 5 | 6 | `

Horizontal line	A line containing three dashes is rendered as a horizontal line separator: `---`

Heading	`# Level 1 heading`
	`## Level 2 heading`
	`### Level 3 heading`
	…

Block quote	Lines that start with a ">" are rendered as a block quote.
	`> Text here is indented and offset`
	`> from the main text body.`

Unordered list	`* Item one`
	`* Item two`
	`* Item three`

Ordered list	`1. Item one`
	`2. Item two`
	`3. Item three`

Image	`![Alternative text](image-file.png)`[9] or `![Alternative text](http://www.example.com/image.png)`
Inline LaTeX equation	`$\LaTeX$`
Displayed LaTeX equation (centered and on a new line)	`$$\LaTeX$$` or `\begin{env}...\end{env}` where env can be a LaTeX environment such as `equation`, `eqnarray`, `align`, etc.

---

[9]The path/filename is relative to the notebook directory.

Markdown cells can also contain HTML code, and the Jupyter Notebook interface will display it as rendered HTML. This is a very powerful feature for the Jupyter Notebook, but its disadvantage is that such HTML code cannot be converted to other formats, such as PDF, using the nbconvert tool (see later section in this chapter). Therefore, it is in general better to use Markdown formatting when possible and resort to HTML only when absolutely necessary.

More information about MathJax and Markdown is available at the projects web pages at www.mathjax.com and http://daringfireball.net/projects/markdown, respectively.

Rich Output Display

The result produced by the last statement in a notebook cell is normally displayed in the corresponding output cell, just like in the standard Python interpreter or the IPython console. The default output cell formatting is a string representation of the object, generated, for example, by the __repr__ method. However, the notebook environment enables a much richer output formatting, as it in principle allows displaying arbitrary HTML in the output cell area. The IPython.display module provides several classes and functions that make it easy to programmatically render formatted output in a notebook. For example, the Image class provides a way to display images from the local file system or online resources in a notebook, as shown in Figure 1-7. Other useful classes from the same module are HTML, for rendering HTML code, and Math, for rendering LaTeX expressions. The display function can be used to explicitly request an object to be rendered and displayed in the output area.

```
In [1]:  from IPython.display import display, Image, HTML, Math

In [2]:  Image(url='http://python.org/images/python-logo.gif')

Out[2]:
```

Figure 1-7. *An example of rich Jupyter Notebook output cell formatting, where an image has been displayed in the cell output area using the Image class*

An example of how HTML code can be rendered in the notebook using the HTML class is shown in Figure 1-8. Here we first construct a string containing HTML code for a table with version information for a list of Python libraries. This HTML code is then rendered in the output cell area by creating an instance of the HTML class, and since this statement is the last (and only) statement in the corresponding input cell, Jupyter will render the representation of this object in the output cell area.

```
In [3]: import scipy, numpy, matplotlib
        modules = [numpy, matplotlib, scipy]
        row = "<tr> <td>%s</td> <td>%s</td> </tr>"
        rows = "\n".join([row % (module.__name__, module.__version__) for module in modules])
        s = "<table> <tr><th>Library</th><th>Version</th> </tr> %s</table>" % rows
```

```
In [4]: s
```

```
Out[4]: '<table> <tr><th>Library</th><th>Version</th> </tr> <tr> <td>numpy</td> <td>1.13.3</td> </tr>
        \n<tr> <td>matplotlib</td> <td>2.0.2</td> </tr>\n<tr> <td>scipy</td> <td>1.1.0</td> </tr></ta
        ble>'
```

```
In [5]: HTML(s)
```

Out[5]:

Library	Version
numpy	1.13.3
matplotlib	2.0.2
scipy	1.1.0

Figure 1-8. *Another example of rich Jupyter Notebook output cell formatting, where an HTML table containing module version information has been rendered and displayed using the HTML class*

For an object to be displayed in an HTML formatted representation, all we need to do is to add a method called _repr_hmtl_ to the class definition. For example, we can easily implement our own primitive version of the HTML class and use it to render the same HTML code as in the previous example, as demonstrated in Figure 1-9.

```
In [8]: class HTMLDisplayer(object):
            def __init__(self, code):
                self.code = code

            def _repr_html_(self):
                return self.code
```

```
In [9]: HTMLDisplayer(s)
```

Out[9]:

Library	Version
numpy	1.13.3
matplotlib	2.0.2
scipy	1.1.0

Figure 1-9. *Another example of how to render HTML code in the Jupyter Notebook, using a class that implements the _repr_hmtl_ method*

Jupyter supports a large number of representations in addition to the _repr_hmtl_ shown in the preceding text, for example, _repr_png_, _repr_svg_, and _repr_latex_, to mention a few. The former two can be used to generate and display graphics in the notebook output cell, as used by, for example, the Matplotlib library (see the following interactive example and Chapter 4). The Math class, which uses the _repr_latex_ method, can be used to render mathematical formulas in the Jupyter Notebook. This is often useful in scientific and technical applications. Examples of how formulas can be rendered using the Math class and the _repr_latex_ method are shown in Figure 1-10.

```
In [8]: Math(r'\hat{H} = -\frac{1}{2}\epsilon \hat{\sigma}_z-\frac{1}{2}\delta \hat{\sigma}_x')
```

Out[8]: $\hat{H} = -\frac{1}{2}\epsilon\hat{\sigma}_z - \frac{1}{2}\delta\hat{\sigma}_x$

```
In [9]: class QubitHamiltonian(object):
            def __init__(self, epsilon, delta):
                self.epsilon = epsilon
                self.delta = delta

            def _repr_latex_(self):
                return "$\hat{H} = -%.2f\hat{\sigma}_z-%.2f\hat{\sigma}_x$" % \
                    (self.epsilon/2, self.delta/2)
```

```
In [10]: QubitHamiltonian(0.5, 0.25)
```

Out[10]: $\hat{H} = -0.25\hat{\sigma}_z - 0.12\hat{\sigma}_x$

Figure 1-10. *An example of how a LaTeX formula is rendered using the Math class and how the _repr_latex_ method can be used to generate a LaTeX formatted representation of an object*

Using the various representation methods recognized by Jupyter, or the convenience classes in the IPython.display module, we have great flexibility in shaping how results are visualized in the Jupyter Notebook. However, the possibilities do not stop there: an exciting feature of the Jupyter Notebook is that interactive applications, with two-way communication between the frontend and the backend kernel, can be created using, for example, a library of widgets (UI components) or directly with Javascript and HTML. For example, using the interact function from the ipywidgets library, we can very easily create an interactive graph that takes an input parameter that is determined from a UI slider, as shown in Figure 1-11.

```
In [11]: import matplotlib.pyplot as plt
         import numpy as np
         from scipy import stats

         def f(mu):
             X = stats.norm(loc=mu, scale=np.sqrt(mu))
             N = stats.poisson(mu)
             x = np.linspace(0, X.ppf(0.999))
             n = np.arange(0, x[-1])

             fig, ax = plt.subplots()
             ax.plot(x, X.pdf(x), color='black', lw=2, label="Normal($\mu=%d, \sigma^2=%d$)" % (mu, mu))
             ax.bar(n, N.pmf(n), align='edge', label=r"Poisson($\lambda=%d$)" % mu)
             ax.set_ylim(0, X.pdf(x).max() * 1.25)
             ax.legend(loc=2, ncol=2)
             plt.close(fig)
             return fig
```

```
In [12]: from ipywidgets import interact
         import ipywidgets as widgets
```

```
In [13]: interact(f, mu=widgets.FloatSlider(min=1.0, max=20.0, step=1.0));
```

Figure 1-11. *An example of interactive application created using the IPython widget* `interact`*. The interact widget provides a slider UI element which allows the value of an input parameter to be changed. When the slider is dragged, the provided function is reevaluated, which in this case renders a new graph.*

In the example in Figure 1-11, we plot the distribution functions for the Normal distribution and the Poisson distribution, where the mean and the variance of the distributions are taken as an input from the UI object created by the `interact` function. By moving the slider back and forth, we can see how the Normal and Poisson distributions (with equal variance) approach each other as the distribution mean is increased and how they behave very differently for small values of the mean. Interactive graphs like this are a great tool for building intuition and for exploring computation problems, and the Jupyter Notebook is a fantastic enabler for this kind of investigations.[10]

nbconvert

Jupyter Notebooks can be converted to a number of different read-only formats using the nbconvert application, which is invoked by passing nbconvert as the first argument to the jupyter command line. Supported formats include, among others, PDF and HTML. Converting Jupyter Notebooks to PDF or HTML is useful when sharing notebooks with colleagues or when publishing them online, when the reader does not necessarily need to run the code, but primarily view the results contained in the notebooks.

HTML

In the notebook web interface, the menu option "File ➤ Download as ➤ HTML" can be used to generate an HTML document representing a static view of a notebook. An HTML document can also be generated from the command prompt using the nbconvert application. For example, a notebook called Notebook.ipynb can be converted to HTML using the command:

```
$ jupyter nbconvert --to html Notebook.ipynb
```

This generates an HTML page that is self-contained in terms of style sheets and JavaScript resources (which are loaded from public CDN servers), and it can be published as is online. However, image resources that are using Markdown or HTML tags are not included and must be distributed together with the resulting HTML file.

For public online publishing of Jupyter Notebooks, the Jupyter project provides a convenient web service called nbviewer, available at http://nbviewer.jupyter.org. By feeding it a URL to a public notebook file, the nbviewer application automatically

---

[10]For more information about how to create interactive applications using Jupyter and IPython widgets, see the documentation for the ipywidgets library https://ipywidgets.readthedocs. io/en/latest.

converts the notebook to HTML and displays the result. One of the many benefits of this method of publishing Jupyter Notebooks is that the notebook author only needs to maintain one file – the notebook file itself – and when it is updated and uploaded to its online location, the static view of the notebook provided by `nbviewer` is automatically updated as well. However, it requires publishing the source notebook at a publicly accessible URL, so it can only be used for public sharing.

Tip   The Jupyter project maintains a Wiki page that indexes many interesting Jupyter Notebooks that are published online at `http://github.com/jupyter/jupyter/wiki/A-gallery-of-interesting-Jupyter-Notebooks`. These notebooks demonstrate many of IPython's and Jupyter's more advanced features and can be a great resource for learning more about Jupyter Notebooks as well as the many topics covered by those notebooks.

PDF

Converting a Jupyter Notebook to PDF format requires first converting the notebook to LaTeX and then compiling the LaTeX document to PDF format. To be able to do the LaTeX to PDF conversion, a LaTeX environment must be available on the system (see Appendix A for pointers on how to install these tools). The `nbconvert` application can do both the notebook-to-LaTeX and the LaTeX-to-PDF conversions in one go, using the `--to pdf` argument (the `--to latex` argument can be used to obtain the intermediate LaTeX source):

```
$ jupyter nbconvert --to pdf Notebook.ipynb
```

The style of the resulting document can be specified using the `--template name` argument, where built-in templates include `base`, `article`, and `report` (these templates can be found in the `nbconvert/templates/latex` directory where Jupyter is installed). By extending one of the existing templates,[11] it is easy to customize the appearance of

---

[11]The IPython nbconvert application uses the jinja2 template engine. See `http://jinja.pocoo.org` for more information and documentation of its syntax.

the generated document. For example, in LaTeX it is common to include additional information about the document that is not available in Jupyter Notebooks, such as a document title (if different from the notebook filename) and the author of the document. This information can be added to a LaTeX document that is generated by the nbconvert application by creating a custom template. For example, the following template extends the built-in template article and overrides the title and author blocks:

```
((*- extends 'article.tplx' -*))
```

```
((* block title *)) \title{Document title} ((* endblock title *))
((* block author *)) \author{Author's Name} ((* endblock author *))
```

Assuming that this template is stored in a file called custom_template.tplx, the following command can be used to convert a notebook to PDF format using this customized template:

```
$ jupyter nbconvert --to pdf --template custom_template.tplx Notebook.ipynb
```

The result is LaTeX and PDF documents where the title and author fields are set as requested in the template.

Python

A Jupyter Notebook in its JSON-based file format can be converted to a pure Python code using the nbconvert application and the python format:

```
$ jupyter nbconvert --to python Notebook.ipynb
```

This generates the file Notebook.py, which only contains executable Python code (or if IPython extensions were used in the notebook; a file that is executable with ipython). The noncode content of the notebook is also included in the resulting Python code file in the form of comments that do not prevent the file from being interpreted by the Python interpreter. Converting a notebook to pure Python code is useful, for example, when using the Jupyter Notebooks to develop functions and classes that need to be imported in other Python files or notebooks.

Spyder: An Integrated Development Environment

An integrated development environment is an enhanced text editor that also provides features such as integrated code execution, documentation, and debugging. Many free and commercial IDE environments have good support for Python-based projects. Spyder[12] is an excellent free IDE that is particularly well suited for computing and data analysis using Python. The rest of this section focuses on Spyder and explores its features in more detail. However, there are also many other suitable IDEs. For example, Eclipse[13] is a popular and powerful multilanguage IDE, and the PyDev[14] extension to Eclipse provides a good Python environment. PyCharm[15] is another powerful Python IDE that has gained a significant popularity among Python developers recently, and the Atom IDE[16] is yet another great option. For readers with previous experience with any of these tools, they could be a productive and familiar environment also for computational work.

However, the Spyder IDE was specifically created for Python programming and in particular for scientific computing with Python. As such it has features that are useful for interactive and exploratory computing: most notably, integration with the IPython console directly in the IDE. The Spyder user interface consists of several optional panes, which can be arranged in different ways within the IDE application. The most important panes are

- Source code editor

- Consoles for the Python and the IPython interpreters and the system shell

- Object inspector, for showing documentation for Python objects

- Variable explorer

- File explorer

- Command history

- Profiler

---

[12]http://code.google.com/p/spyderlib
[13]http://www.eclipse.org
[14]http://pydev.org
[15]http://www.jetbrains.com/pycharm
[16]https://atom.io

Each pane can be configured to be shown or hidden, depending on the user's preferences and needs, using the "View ➤ Panes" menu option. Furthermore, panes can be organized together in tabbed groups. In the default layout, three pane groups are displayed: The left pane group contains the source code editor. The top-right pane group contains the variable explorer, the file explorer, and the object inspector. The bottom right pane group contains Python and IPython consoles.

Running the command `spyder` at the shell prompt launches the Spyder IDE. See Figure 1-12 for a screenshot of the default layout of the Spyder application.

Figure 1-12. *A screenshot of the Spyder IDE application. The code editor is shown in the left panel, the top-right panel shows the object inspector (help viewer), and the bottom right panel shows an IPython console.*

Source Code Editor

The source code editor in Spyder supports code highlighting, intelligent autocompletion, working with multiple open files simultaneously, parenthesis matching, indentation guidance, and many other features that one would expect from a modern source code editor. The added benefit from using an IDE is that code in the editor can be run – as a whole (shortcut F5) or a selection (shortcut F9) – in attached Python or IPython consoles with persistent sessions between successive runs.

In addition, the Spyder editor has very useful support for static code checking with pylint,[17] pyflakes,[18] and pep8,[19] which are external tools that analyze Python source code and report errors such as undefined symbols, syntax errors, coding style violations, and more. Such warnings and errors are shown on a line-by-line basis as a yellow triangle with an exclamation mark in the left margin of the editor, next to the line number. Static code checking is extremely important in Python programming. Since Python is an interpreted and lazily evaluated language, simple bugs like undefined symbols may not be discovered until the offending code line is reached at runtime, and for rarely used code paths, sometimes such bugs can be very hard to discover. Real-time static code checking and coding style checks in the Spyder editor can be activated and deactivated in the "Editor" section of the preference windows (Python ➤ Preferences, in the menu on OS X, and Tools ➤ Preferences on Linux and Windows). In the Editor section, I recommend checking the "Code analysis" and "Style analysis" boxes in the "Code Introspection/Analysis" tab.

Tip   The Python language is versatile, and equivalent Python source code can be written in a vast variety of styles and manners. However, a Python coding style standard, PEP8, has been put forward to encourage a uniform appearance of Python code. I strongly recommend studying the PEP8 coding style standard and complying to it in your code. The PEP8 is described at `www.python.org/dev/peps/pep-0008`.

---

[17]`http://www.pylint.org`
[18]`http://github.com/pyflakes/pyflakes`
[19]`http://pep8.readthedocs.org`

Consoles in Spyder

The integrated Python and IPython consoles can be used to execute a file that is being edited in the text editor window, or for running interactively typed Python code. When executing Python source code files from the editor, the namespace variables created in the script are retained in the IPython or Python session in the console. This is an important feature that makes Spyder an interactive-computing environment, in addition to a traditional IDE application, since it allows exploring the values of variables after a script has finished executing. Spyder supports having multiple Python and IPython consoles opened simultaneously, and, for example, a new IPython console can be launched through the "Consoles ➤ Open an IPython console" menu. When running a script from the editor, by pressing F5 or pressing the run button in the toolbar, the script is by default ran in the most recently activated console. This makes it possible to maintain different consoles, with independent namespaces, for different scripts or projects.

When possible, use the %reset command and the reload function to clear a namespace and reload updated modules. If that is insufficient, it is possible to restart the IPython kernel corresponding to an IPython console, or the Python interpreter, via the drop-down menu for the top-right icon in the console panel. Finally, a practical feature is that IPython console sessions can be exported as an HTML file by right-clicking the console window and selecting "Save as HTML/XML" in the pop-up menu.

Object Inspector

The object inspector (Help pane) is a great aid when writing Python code. It can display richly formatted documentation strings for objects defined in source code created with the editor and for symbols defined in library modules that are installed on the system. The object text field at the top of the object inspector panel can be used to type the name of a module, function, or class for which to display the documentation string. Modules and symbols do not need to be imported into the local namespace to be able to display their docstrings using the object inspector. The documentation for an object in the editor or the console can also be opened in the object inspector by selecting the object with the cursor and using the shortcut Ctrl-i (Cmd-i on OS X). It is even possible to automatically display docstrings for callable objects when its opening left parenthesis is typed. This gives an immediate reminder of the arguments and their order for the callable object, which can be a great productivity booster. To activate this feature, navigate to the "Help" page in the "Preferences" window and check the boxes in the "Automatic connections" section.

Summary

In this chapter we introduced the Python environment for scientific and technical computing. This environment is, in fact, an entire ecosystem of libraries and tools for computing, which includes not only Python software but everything from low-level number crunching libraries up to graphical user interface applications and web applications. In this multilanguage ecosystem, Python is the language that ties it all together into a coherent and productive environment for computing. IPython is a core component of Python's computing environment, and we briefly surveyed some of its most important features before covering the higher-level user environments provided by the Jupyter Notebook and the Spyder IDE. These are the tools in which the majority of exploratory and interactive computing is carried out. In the rest of this book, we focus on computing using Python libraries, assuming that we are working within one of the environments provided by IPython, the Jupyter Notebook, or Spyder.

Further Reading

The Jupyter Notebook is a particularly rich platform for interactive computing, and it is also a very actively developed software. One of the most recent developments within the Jupyter Notebook is its widget system, which are user-interface components that can be used to create interactive interfaces within the browser that is displaying the notebook. In this book we just briefly touch upon Jupyter widgets, but it is a very interesting and rapidly developing part of the Jupyter project, and I do recommend exploring their potential applications for interactive computing. The Jupyter Notebook widgets, and many other parts of Jupyter, are documented through examples in Jupyter Notebook form that are available here: `http://nbviewer.ipython.org/github/ipython/ipython/tree/master/examples`. There are also two interesting books on this topic (Rossant, *Learning IPython for Interactive Computing and Data Visualization*, 2013; Rossant, *IPython Interactive Computing and Visualization Cookbook*, 2014) that I highly recommend.

References

Rossant, C. (2014). IPython Interactive Computing and Visualization Cookbook. Mumbai: Packt.

Rossant, C. (2013). Learning IPython for Interactive Computing and Data Visualization. Mumbai: Packt.

CHAPTER 2

Vectors, Matrices, and Multidimensional Arrays

Vectors, matrices, and arrays of higher dimensions are essential tools in numerical computing. When a computation must be repeated for a set of input values, it is natural and advantageous to represent the data as arrays and the computation in terms of array operations. Computations that are formulated this way are said to be vectorized.[1] Vectorized computing eliminates the need for many explicit loops over the array elements by applying batch operations on the array data. The result is concise and more maintainable code, and it enables delegating the implementation of (e.g., elementwise) array operations to more efficient low-level libraries. Vectorized computations can therefore be significantly faster than sequential element-by-element computations. This is particularly important in an interpreted language such as Python, where looping over arrays element by element entails a significant performance overhead.

In Python's scientific computing environment, efficient data structures for working with arrays are provided by the NumPy library. The core of NumPy is implemented in C and provides efficient functions for manipulating and processing arrays. At a first glance, NumPy arrays bear some resemblance to Python's list data structure. But an important difference is that while Python lists are generic containers of objects, NumPy arrays are homogenous and typed arrays of fixed size. Homogenous means that all elements in the array have the same data type. Fixed size means that an array cannot be resized (without creating a new array). For these and other reasons, operations and functions acting on NumPy arrays can be much more efficient than those using Python lists. In addition to

---

[1]Many modern processors provide instructions that operate on arrays. These are also known as vectorized operations, but here vectorized refers to high-level array-based operations, regardless of how they are implemented at the processor level.

© Robert Johansson 2019
R. Johansson, *Numerical Python*, https://doi.org/10.1007/978-1-4842-4246-9_2

the data structures for arrays, NumPy also provides a large collection of basic operators and functions that act on these data structures, as well as submodules with higher-level algorithms such as linear algebra and fast Fourier transform.

In this chapter we first look at the basic NumPy data structure for arrays and various methods to create such NumPy arrays. Next we look at operations for manipulating arrays and for doing computations with arrays. The multidimensional data array provided by NumPy is a foundation for nearly all numerical libraries for Python. Spending time on getting familiar with NumPy and developing an understanding of how NumPy works is therefore important.

NumPy   The NumPy library provides data structures for representing a rich variety of arrays and methods and functions for operating on such arrays. NumPy provides the numerical backend for nearly every scientific or technical library for Python. It is therefore a very important part of the scientific Python ecosystem. At the time of writing, the latest version of NumPy is 1.14.2. More information about NumPy is available at `www.numpy.org`.

Importing the Modules

In order to use the NumPy library, we need to import it in our program. By convention, the numPy module imported under the alias np, like so:

```
In [1]: import numpy as np
```

After this, we can access functions and classes in the numpy module using the np namespace. Throughout this book, we assume that the NumPy module is imported in this way.

The NumPy Array Object

The core of the NumPy library is the data structures for representing multidimensional arrays of homogeneous data. Homogeneous refers to all elements in an array having the same data type.[2] The main data structure for multidimensional arrays in NumPy

---

[2]This does not necessarily need to be the case for Python lists, which therefore can be heterogenous.

is the ndarray class. In addition to the data stored in the array, this data structure also contains important metadata about the array, such as its shape, size, data type, and other attributes. See Table 2-1 for a more detailed description of these attributes. A full list of attributes with descriptions is available in the ndarray docstring, which can be accessed by calling help(np.ndarray) in the Python interpreter or np.ndarray? in an IPython console.

Table 2-1. *Basic Attributes of the ndarray Class*

Attribute	Description
Shape	A tuple that contains the number of elements (i.e., the length) for each dimension (axis) of the array.
Size	The total number elements in the array.
Ndim	Number of dimensions (axes).
nbytes	Number of bytes used to store the data.
dtype	The data type of the elements in the array.

The following example demonstrates how these attributes are accessed for an instance data of the class ndarray:

```
In [2]: data = np.array([[1, 2], [3, 4], [5, 6]])
In [3]: type(data)
Out[3]: <class 'numpy.ndarray'>
In [4]: data
Out[4]: array([[1, 2],
               [3, 4],
               [5, 6]])
In [5]: data.ndim
Out[5]: 2
In [6]: data.shape
Out[6]: (3, 2)
In [7]: data.size
Out[7]: 6
```

```
In [8]: data.dtype
Out[8]: dtype('int64')
In [9]: data.nbytes
Out[9]: 48
```

Here the ndarray instance data is created from a nested Python list using the function np.array. More ways to create ndarray instances from data and from rules of various kinds are introduced later in this chapter. In the preceding example, the data is a two-dimensional array (data.ndim) of shape 3 × 2, as indicated by data.shape, and in total it contains six elements (data.size) of type int64 (data.dtype), which amounts to a total size of 48 bytes (data.nbytes).

Data Types

In the previous section, we encountered the dtype attribute of the ndarray object. This attribute describes the data type of each element in the array (remember, since NumPy arrays are homogeneous, all elements have the same data type). The basic numerical data types supported in NumPy are shown in Table 2-2. Nonnumerical data types, such as strings, objects, and user-defined compound types, are also supported.

Table 2-2. *Basic Numerical Data Types Available in NumPy*

dtype	Variants	Description
int	int8, int16, int32, int64	Integers
uint	uint8, uint16, uint32, uint64	Unsigned (nonnegative) integers
bool	Bool	Boolean (True or False)
float	float16, float32, float64, float128	Floating-point numbers
complex	complex64, complex128, complex256	Complex-valued floating-point numbers

For numerical work the most important data types are int (for integers), float (for floating-point numbers), and complex (for complex floating-point numbers). Each of these data types comes in different sizes, such as int32 for 32-bit integers, int64 for 64-bit integers, etc. This offers more fine-grained control over data types than the standard Python types, which only provides one type for integers and one type for floats.

It is usually not necessary to explicitly choose the bit size of the data type to work with, but it is often necessary to explicitly choose whether to use arrays of integers, floating-point numbers, or complex values.

The following example demonstrates how to use the dtype attribute to generate arrays of integer-, float-, and complex-valued elements:

```
In [10]: np.array([1, 2, 3], dtype=np.int)
Out[10]: array([1, 2, 3])
In [11]: np.array([1, 2, 3], dtype=np.float)
Out[11]: array([ 1.,   2.,   3.])
In [12]: np.array([1, 2, 3], dtype=np.complex)
Out[12]: array([ 1.+0.j,   2.+0.j,   3.+0.j])
```

Once a NumPy array is created, its dtype cannot be changed, other than by creating a new copy with type-casted array values. Typecasting an array is straightforward and can be done using either the np.array function:

```
In [13]: data = np.array([1, 2, 3], dtype=np.float)
In [14]: data
Out[14]: array([ 1.,   2.,   3.])
In [15]: data.dtype
Out[15]: dtype('float64')
In [16]: data = np.array(data, dtype=np.int)
In [17]: data.dtype
Out[17]: dtype('int64')
In [18]: data
Out[18]: array([1, 2, 3])
```

or by using the astype method of the ndarray class:

```
In [19]: data = np.array([1, 2, 3], dtype=np.float)
In [20]: data
Out[20]: array([ 1.,   2.,   3.])
In [21]: data.astype(np.int)
Out[21]: array([1, 2, 3])
```

When computing with NumPy arrays, the data type might get promoted from one type to another, if required by the operation. For example, adding float-valued and complex-valued arrays, the resulting array is a complex-valued array:

```
In [22]: d1 = np.array([1, 2, 3], dtype=float)
In [23]: d2 = np.array([1, 2, 3], dtype=complex)
In [24]: d1 + d2
Out[24]: array([ 2.+0.j,  4.+0.j,  6.+0.j])
In [25]: (d1 + d2).dtype
Out[25]: dtype('complex128')
```

In some cases, depending on the application and its requirements, it is essential to create arrays with data type appropriately set to, for example, int or complex. The default type is float. Consider the following example:

```
In [26]: np.sqrt(np.array([-1, 0, 1]))
Out[26]: RuntimeWarning: invalid value encountered in sqrt
         array([ nan,   0.,   1.])
In [27]: np.sqrt(np.array([-1, 0, 1], dtype=complex))
Out[27]: array([ 0.+1.j,  0.+0.j,  1.+0.j])
```

Here, using the np.sqrt function to compute the square root of each element in an array gives different results depending on the data type of the array. Only when the data type of the array is complex is the square root of −1 resulting in the imaginary unit (denoted as 1j in Python).

Real and Imaginary Parts

Regardless of the value of the dtype attribute, all NumPy array instances have the attributes real and imag for extracting the real and imaginary parts of the array, respectively:

```
In [28]: data = np.array([1, 2, 3], dtype=complex)
In [29]: data
Out[29]: array([ 1.+0.j,  2.+0.j,  3.+0.j])
In [30]: data.real
Out[30]: array([ 1.,  2.,  3.])
In [31]: data.imag
Out[31]: array([ 0.,  0.,  0.])
```

The same functionality is also provided by the functions np.real and np.imag, which also can be applied to other array-like objects, such as Python lists. Note that Python itself has support of complex numbers, and the imag and real attributes are also available for Python scalars.

Order of Array Data in Memory

Multidimensional arrays are stored as contiguous data in memory. There is a freedom of choice in how to arrange the array elements in this memory segment. Consider the case of a two-dimensional array, containing rows and columns: one possible way to store this array as a consecutive sequence of values is to store the rows after each other, and another equally valid approach is to store the columns one after another. The former is called row-major format and the latter is column-major format. Whether to use row-major or column-major is a matter of conventions, and row-major format is used, for example, in the C programming language, and Fortran uses the column-major format. A NumPy array can be specified to be stored in row-major format, using the keyword argument order= 'C', and column-major format, using the keyword argument order= 'F', when the array is created or reshaped. The default format is row-major. The 'C' or 'F' ordering of NumPy array is particularly relevant when NumPy arrays are used in interfaces with software written in C and Fortran, which is often required when working with numerical computing with Python.

Row-major and column-major ordering are special cases of strategies for mapping the index used to address an element, to the offset for the element in the array's memory segment. In general, the NumPy array attribute ndarray.strides defines exactly how this mapping is done. The strides attribute is a tuple of the same length as the number of axes (dimensions) of the array. Each value in strides is the factor by which the index for the corresponding axis is multiplied when calculating the memory offset (in bytes) for a given index expression.

For example, consider a C-order array A with shape (2, 3), which corresponds to a two-dimensional array with two and three elements along the first and the second dimensions, respectively. If the data type is int32, then each element uses 4 bytes, and the total memory buffer for the array therefore uses $2 \times 3 \times 4 = 24$ bytes. The strides attribute of this array is therefore $(4 \times 3, 4 \times 1) = (12, 4)$, because each increment of m in A[n, m] increases the memory offset with one item, or 4 bytes. Likewise, each increment of n increases the memory offset with three items or 12 bytes (because the second dimension of the array has length 3). If, on the other hand, the same array were stored in 'F' order, the strides would instead be (4, 8). Using strides to describe the mapping of array index to array memory offset is clever because it can be used to describe different mapping strategies, and many common operations on arrays, such as for example the transpose, can be implemented by simply changing the strides attribute, which can eliminate the need for moving data around in the memory. Operations that only require changing the strides attribute result in new ndarray objects that refer to the same data as the original array. Such arrays are called views. For efficiency, NumPy strives to create views rather than copies when applying operations on arrays. This is generally a good thing, but it is important to be aware of that some array operations result in views rather than new independent arrays, because modifying their data also modifies the data of the original array. Later in this chapter, we will see several examples of this behavior.

Creating Arrays

In the previous section, we looked at NumPy's basic data structure for representing arrays, the ndarray class, and we looked at the basic attributes of this class. In this section we focus on functions from the NumPy library that can be used to create ndarray instances.

Arrays can be generated in a number of ways, depending on their properties and the applications they are used for. For example, as we saw in the previous section, one way to initialize an ndarray instance is to use the np.array function on a Python list, which, for example, can be explicitly defined. However, this method is obviously limited to small arrays. In many situations it is necessary to generate arrays with elements that follow some given rule, such as filled with constant values, increasing integers, uniformly spaced numbers, random numbers, etc. In other cases we might need to create arrays from data stored in a file. The requirements are many and varied, and the NumPy library provides a comprehensive set of functions for generating arrays of various types. In this section we look in more detail at many of these functions. For a complete list, see the NumPy reference manual or the docstrings that are available by typing help(np) or using the autocompletion np.<TAB>. A summary of frequently used array-generating functions is given in Table 2-3.

Table 2-3. *Summary of NumPy Functions for Generating Arrays*

Function Name	Type of Array
np.array	Creates an array for which the elements are given by an array-like object, which, for example, can be a (nested) Python list, a tuple, an iterable sequence, or another ndarray instance.
np.zeros	Creates an array with the specified dimensions and data type that is filled with zeros.
np.ones	Creates an array with the specified dimensions and data type that is filled with ones.
np.diag	Creates a diagonal array with specified values along the diagonal and zeros elsewhere.
np.arange	Creates an array with evenly spaced values between the specified start, end, and increment values.
np.linspace	Creates an array with evenly spaced values between specified start and end values, using a specified number of elements.
np.logspace	Creates an array with values that are logarithmically spaced between the given start and end values.
np.meshgrid	Generates coordinate matrices (and higher-dimensional coordinate arrays) from one-dimensional coordinate vectors.
np.fromfunction	Creates an array and fills it with values specified by a given function, which is evaluated for each combination of indices for the given array size.
np.fromfile	Creates an array with the data from a binary (or text) file. NumPy also provides a corresponding function np.tofile with which NumPy arrays can be stored to disk and later read back using np.fromfile.
np.genfromtxt,np. loadtxt	Create an array from data read from a text file, for example, a comma-separated value (CSV) file. The function np.genfromtxt also supports data files with missing values.
np.random.rand	Generates an array with random numbers that are uniformly distributed between 0 and 1. Other types of distributions are also available in the np. random module.

Arrays Created from Lists and Other Array-Like Objects

Using the np.array function, NumPy arrays can be constructed from explicit Python lists, iterable expressions, and other array-like objects (such as other ndarray instances). For example, to create a one-dimensional array from a Python list, we simply pass the Python list as an argument to the np.array function:

```
In [32]: np.array([1, 2, 3, 4])
Out[32]: array([ 1,   2,   3, 4])
In [33]: data.ndim
Out[33]: 1
In [34]: data.shape
Out[34]: (4,)
```

To create a two-dimensional array with the same data as in the previous example, we can use a nested Python list:

```
In [35]: np.array([[1, 2], [3, 4]])
Out[35]: array([[1,   2],
                [3, 4]])
In [36]: data.ndim
Out[36]: 2
In [37]: data.shape
Out[37]: (2, 2)
```

Arrays Filled with Constant Values

The functions np.zeros and np.ones create and return arrays filled with zeros and ones, respectively. They take, as first argument, an integer or a tuple that describes the number of elements along each dimension of the array. For example, to create a 2 × 3 array filled with zeros, and an array of length 4 filled with ones, we can use

```
In [38]: np.zeros((2, 3))
Out[38]: array([[ 0.,   0.,   0.],
                [ 0.,   0.,   0.]])
In [39]: np.ones(4)
Out[39]: array([ 1.,   1.,   1., 1.])
```

Like other array-generating functions, the np.zeros and np.ones functions also accept an optional keyword argument that specifies the data type for the elements in the array. By default, the data type is float64, and it can be changed to the required type by explicitly specifying the dtype argument.

```
In [40]: data = np.ones(4)
In [41]: data.dtype
Out[41]: dtype('float64')
In [42]: data = np.ones(4, dtype=np.int64)
In [43]: data.dtype
Out[43]: dtype('int64')
```

An array filled with an arbitrary constant value can be generated by first creating an array filled with ones and then multiplying the array with the desired fill value. However, NumPy also provides the function np.full that does exactly this in one step. The following two ways of constructing arrays with ten elements, which are initialized to the numerical value 5.4 in this example, produces the same results, but using np.full is slightly more efficient since it avoids the multiplication.

```
In [44]: x1 = 5.4 * np.ones(10)
In [45]: x2 = np.full(10, 5.4)
```

An already created array can also be filled with constant values using the np.fill function, which takes an array and a value as arguments, and set all elements in the array to the given value. The following two methods to create an array therefore give the same results:

```
In [46]: x1 = np.empty(5)
In [47]: x1.fill(3.0)
In [48]: x1
Out[48]: array([ 3.,   3.,   3.,   3.,   3.])
In [49]: x2 = np.full(5, 3.0)
In [50]: x2
Out[50]: array([ 3.,   3.,   3.,   3.,   3.])
```

In this last example, we also used the np.empty function, which generates an array with uninitialized values, of the given size. This function should only be used when the initialization of all elements can be guaranteed by other means, such as an explicit loop over the array elements or another explicit assignment. This function is described in more detail later in this chapter.

Arrays Filled with Incremental Sequences

In numerical computing it is very common to require arrays with evenly spaced values between a starting value and ending value. NumPy provides two similar functions to create such arrays: np.arange and np.linspace. Both functions take three arguments, where the first two arguments are the start and end values. The third argument of np.arange is the increment, while for np.linspace it is the total number of points in the array.

For example, to generate arrays with values between 1 and 10, with increment 1, we could use either of the following:

```
In [51]: np.arange(0.0, 10, 1)
Out[51]: array([ 0.,  1.,  2.,  3.,  4.,  5.,  6.,  7.,  8.,  9.])
In [52]: np.linspace(0, 10, 11)
Out[52]: array([ 0.,  1.,  2.,  3.,  4.,  5.,  6.,  7.,  8.,  9.,  10.])
```

However, note that np.arange does not include the end value (10), while by default np.linspace does (although this behavior can be changed using the optional endpoint keyword argument). Whether to use np.arange or np.linspace is mostly a matter of personal preference, but it is generally recommended to use np.linspace whenever the increment is a noninteger.

Arrays Filled with Logarithmic Sequences

The function np.logspace is similar to np.linspace, but the increments between the elements in the array are logarithmically distributed, and the first two arguments, for the start and end values, are the powers of the optional base keyword argument (which defaults to 10). For example, to generate an array with logarithmically distributed values between 1 and 100, we can use

```
In [53]: np.logspace(0, 2, 5)  # 5 data points between 10**0=1 to 10**2=100
Out[53]: array([ 1. , 3.16227766, 10. , 31.6227766 , 100.])
```

Meshgrid Arrays

Multidimensional coordinate grids can be generated using the function np.meshgrid. Given two one-dimensional coordinate arrays (i.e., arrays containing a set of coordinates along a given dimension), we can generate two-dimensional coordinate arrays using the np.meshgrid function. An illustration of this is given in the following example:

```
In [54]: x = np.array([-1, 0, 1])
In [55]: y = np.array([-2, 0, 2])
In [56]: X, Y = np.meshgrid(x, y)
In [57]: X
Out[57]: array([[-1,  0,  1],
                [-1,  0,  1],
                [-1,  0,  1]])
In [58]: Y
Out[58]: array([[-2, -2, -2],
                [ 0,  0,  0],
                [ 2,  2,  2]])
```

A common use-case of the two-dimensional coordinate arrays, like X and Y in this example, is to evaluate functions over two variables x and y. This can be used when plotting functions over two variables, as colormap plots and contour plots. For example, to evaluate the expression $(x+y)^2$ at all combinations of values from the x and y arrays in the preceding section, we can use the two-dimensional coordinate arrays X and Y:

```
In [59]: Z = (X + Y) ** 2
In [60]: Z
Out[60]: array([[9, 4, 1],
                [1, 0, 1],
                [1, 4, 9]])
```

It is also possible to generate higher-dimensional coordinate arrays by passing more arrays as argument to the np.meshgrid function. Alternatively, the functions np.mgrid and np.ogrid can also be used to generate coordinate arrays, using a slightly different syntax based on indexing and slice objects. See their docstrings or the NumPy documentation for details.

Creating Uninitialized Arrays

To create an array of specific size and data type, but without initializing the elements in the array to any particular values, we can use the function np.empty. The advantage of using this function, for example, instead of np.zeros, which creates an array initialized with zero-valued elements, is that we can avoid the initiation step. If all elements are guaranteed to be initialized later in the code, this can save a little bit of time, especially when working with large arrays. To illustrate the use of the np.empty function, consider the following example:

```
In [61]: np.empty(3, dtype=np.float)
Out[61]: array([  1.28822975e-231,   1.28822975e-231,   2.13677905e-314])
```

Here we generated a new array with three elements of type float. There is no guarantee that the elements have any particular values, and the actual values will vary from time to time. For this reason it is important that all values are explicitly assigned before the array is used; otherwise unpredictable errors are likely to arise. Often the np.zeros function is a safer alternative to np.empty, and if the performance gain is not essential, it is better to use np.zeros, to minimize the likelihood of subtle and hard-to-reproduce bugs due to uninitialized values in the array returned by np.empty.

Creating Arrays with Properties of Other Arrays

It is often necessary to create new arrays that share properties, such as shape and data type, with another array. NumPy provides a family of functions for this purpose: np.ones_like, np.zeros_like, np.full_like, and np.empty_like. A typical use-case is a function that takes arrays of unspecified type and size as arguments and requires working arrays of the same size and type. For example, a boilerplate example of this situation is given in the following function:

```
def f(x):
    y = np.ones_like(x)
    # compute with x and y
    return y
```

At the first line of the body of this function, a new array y is created using np.ones_like, which results in an array of the same size and data type as x, and filled with ones.

Creating Matrix Arrays

Matrices, or two-dimensional arrays, are an important case for numerical computing. NumPy provides functions for generating commonly used matrices. In particular, the function np.identity generates a square matrix with ones on the diagonal and zeros elsewhere:

```
In [62]: np.identity(4)
Out[62]: array([[ 1.,  0.,  0.,  0.],
                [ 0.,  1.,  0.,  0.],
                [ 0.,  0.,  1.,  0.],
                [ 0.,  0.,  0.,  1.]])
```

The similar function numpy.eye generates matrices with ones on a diagonal (optionally offset). This is illustrated in the following example, which produces matrices with nonzero diagonals above and below the diagonal, respectively:

```
In [63]: np.eye(3, k=1)
Out[63]: array([[ 0.,  1.,  0.],
                [ 0.,  0.,  1.],
                [ 0.,  0.,  0.]])

In [64]: np.eye(3, k=-1)
Out[64]: array([[ 0.,  0.,  0.],
                [ 1.,  0.,  0.],
                [ 0.,  1.,  0.]])
```

To construct a matrix with an arbitrary one-dimensional array on the diagonal, we can use the np.diag function (which also takes the optional keyword argument k to specify an offset from the diagonal), as demonstrated here:

```
In [65]: np.diag(np.arange(0, 20, 5))
Out[65]: array([[0,  0,  0,  0],
                [0,  5,  0,  0],
                [0,  0, 10,  0],
                [0,  0,  0, 15]])
```

Here we gave a third argument to the `np.arange` function, which specifies the step size in the enumeration of elements in the array returned by the function. The resulting array therefore contains the values $[0, 5, 10, 15]$, which is inserted on the diagonal of a two-dimensional matrix by the `np.diag` function.

Indexing and Slicing

Elements and subarrays of NumPy arrays are accessed using the standard square bracket notation that is also used with Python lists. Within the square bracket, a variety of different index formats are used for different types of element selection. In general, the expression within the bracket is a tuple, where each item in the tuple is a specification of which elements to select from each axis (dimension) of the array.

One-Dimensional Arrays

Along a single axis, integers are used to select single elements, and so-called slices are used to select ranges and sequences of elements. Positive integers are used to index elements from the beginning of the array (index starts at 0), and negative integers are used to index elements from the end of the array, where the last element is indexed with –1, the second to last element with –2, and so on.

Slices are specified using the : notation that is also used for Python lists. In this notation, a range of elements can be selected using an expression like m:n, which selects elements starting with m and ending with $n - 1$ (note that the nth element is not included). The slice m:n can also be written more explicitly as m : n : 1, where the number 1 specifies that every element between m and n should be selected. To select every second element between m and n, use m : n : 2, and to select every p elements, use m : n : p, and so on. If p is negative, elements are returned in reversed order starting from m to $n+1$ (which implies that m has to be larger than n in this case). See Table 2-4 for a summary of indexing and slicing operations for NumPy arrays.

Table 2-4. *Examples of Array Indexing and Slicing Expressions*

Expression	Description
a[m]	Select element at index *m*, where *m* is an integer (start counting form 0).
a[-m]	Select the *n*th element from the end of the list, where *n* is an integer. The last element in the list is addressed as –1, the second to last element as –2, and so on.
a[m:n]	Select elements with index starting at *m* and ending at *n* – 1 (*m* and *n* are integers).
a[:] or a[0:-1]	Select all elements in the given axis.
a[:n]	Select elements starting with index 0 and going up to index *n* – 1 (integer).
a[m:] or a[m:-1]	Select elements starting with index *m* (integer) and going up to the last element in the array.
a[m:n:p]	Select elements with index *m* through *n* (exclusive), with increment *p*.
a[::-1]	Select all the elements, in reverse order.

The following examples demonstrate index and slicing operations for NumPy arrays. To begin with, consider an array with a single axis (dimension) that contains a sequence of integers between 0 and 10:

```
In [66]: a = np.arange(0, 11)
In [67]: a
Out[67]: array([ 0,  1,  2,  3,  4,  5,  6,  7,  8,  9, 10])
```

Note that the end value 11 is not included in the array. To select specific elements from this array, for example, the first, the last, and the 5th element, we can use integer indexing:

```
In [68]: a[0]  # the first element
Out[68]: 0
In [69]: a[-1] # the last element
Out[69]: 10
In [70]: a[4]  # the fifth element, at index 4
Out[70]: 4
```

To select a range of element, say from the second to the second-to-last element, selecting every element and every second element, respectively, we can use index slices:

```
In [71]: a[1:-1]
Out[71]: array([1, 2, 3, 4, 5, 6, 7, 8, 9])
In [72]: a[1:-1:2]
Out[72]: array([1, 3, 5, 7, 9])
```

To select the first five and the last five elements from an array, we can use the slices :5 and –5:, since if m or n is omitted in m:n, the defaults are the beginning and the end of the array, respectively.

```
In [73]: a[:5]
Out[73]: array([0, 1, 2, 3, 4])
In [74]: a[-5:]
Out[74]: array([6, 7, 8, 9, 10])
```

To reverse the array and select only every second value, we can use the slice ::-2, as shown in the following example:

```
In [75]: a[::-2]
Out[75]: array([10,  8,  6,  4,  2,  0])
```

Multidimensional Arrays

With multidimensional arrays, element selections like those introduced in the previous section can be applied on each axis (dimension). The result is a reduced array where each element matches the given selection rules. As a specific example, consider the following two-dimensional array:

```
In [76]: f = lambda m, n: n + 10 * m
In [77]: A = np.fromfunction(f, (6, 6), dtype=int)
In [78]: A
Out[78]: array([[ 0,  1,  2,  3,  4,  5],
                [10, 11, 12, 13, 14, 15],
                [20, 21, 22, 23, 24, 25],
                [30, 31, 32, 33, 34, 35],
                [40, 41, 42, 43, 44, 45],
                [50, 51, 52, 53, 54, 55]])
```

We can extract columns and rows from this two-dimensional array using a combination of slice and integer indexing:

```
In [79]: A[:, 1]  # the second column
Out[79]: array([ 1, 11, 21, 31, 41, 51])
In [80]: A[1, :]  # the second row
Out[80]: array([10, 11, 12, 13, 14, 15])
```

By applying a slice on each of the array axes, we can extract subarrays (submatrices in this two-dimensional example):

```
In [81]: A[:3, :3]  # upper half diagonal block matrix
Out[81]: array([[ 0,  1,  2],
                [10, 11, 12],
                [20, 21, 22]])
In [82]: A[3:, :3]  # lower left off-diagonal block matrix
Out[82]: array([[30, 31, 32],
                [40, 41, 42],
                [50, 51, 52]])
```

With element spacing other that 1, submatrices made up from nonconsecutive elements can be extracted:

```
In [83]: A[::2, ::2]  # every second element starting from 0, 0
Out[83]: array([[ 0,  2,  4],
                [20, 22, 24],
                [40, 42, 44]])
In [84]: A[1::2, 1::3]  # every second and third element starting from 1, 1
Out[84]: array([[11, 14],
                [31, 34],
                [51, 54]])
```

This ability to extract subsets of data from a multidimensional array is a simple but very powerful feature with many data processing applications.

Views

Subarrays that are extracted from arrays using slice operations are alternative *views* of the same underlying array data. That is, they are arrays that refer to the same data in the memory as the original array, but with a different `strides` configuration. When elements in a view are assigned new values, the values of the original array are therefore also updated. For example,

```
In [85]: B = A[1:5, 1:5]
In [86]: B
Out[86]: array([[11, 12, 13, 14],
                [21, 22, 23, 24],
                [31, 32, 33, 34],
                [41, 42, 43, 44]])
In [87]: B[:, :] = 0
In [88]: A
Out[88]: array([[ 0,  1,  2,  3,  4,  5],
                [10,  0,  0,  0,  0, 15],
                [20,  0,  0,  0,  0, 25],
                [30,  0,  0,  0,  0, 35],
                [40,  0,  0,  0,  0, 45],
                [50, 51, 52, 53, 54, 55]])
```

Here, assigning new values to the elements in an array B, which is created from the array A, also modifies the values in A (since both arrays refer to the same data in the memory). The fact that extracting subarrays results in views rather than new independent arrays eliminates the need for copying data and improves performance. When a copy rather than a view is needed, the view can be copied explicitly by using the copy method of the ndarray instance.

```
In [89]: C = B[1:3, 1:3].copy()
In [90]: C
Out[90]: array([[0, 0],
                [0, 0]])
In [91]: C[:, :] = 1  # this does not affect B since C is a copy of the
view B[1:3, 1:3]
In [92]: C
```

```
Out[92]: array([[1, 1],
                 [1, 1]])
In [93]: B
Out[93]: array([[0, 0, 0, 0],
                [0, 0, 0, 0],
                [0, 0, 0, 0],
                [0, 0, 0, 0]])
```

In addition to the copy attribute of the ndarray class, an array can also be copied using the function np.copy or, equivalently, using the np.array function with the keyword argument copy=True.

Fancy Indexing and Boolean-Valued Indexing

In the previous section, we looked at indexing NumPy arrays with integers and slices, to extract individual elements or ranges of elements. NumPy provides another convenient method to index arrays, called fancy indexing. With fancy indexing, an array can be indexed with another NumPy array, a Python list, or a sequence of integers, whose values select elements in the indexed array. To clarify this concept, consider the following example: we first create a NumPy array with 11 floating-point numbers, and then index the array with another NumPy array (and Python list), to extract element numbers 0, 2, and 4 from the original array:

```
In [94]: A = np.linspace(0, 1, 11)
Out[94]: array([ 0. ,  0.1,  0.2,  0.3,  0.4,  0.5,  0.6,  0.7,  0.8,  0.9,  1. ])
In [95]: A[np.array([0, 2, 4])]
Out[95]: array([ 0. ,  0.2,  0.4])
In [96]: A[[0, 2, 4]]   # The same thing can be accomplished by indexing with a
Python list
Out[96]: array([ 0. ,  0.2,  0.4])
```

This method of indexing can be used along each axis (dimension) of a multidimensional NumPy array. It requires that the elements in the array or list used for indexing are integers.

Another variant of indexing NumPy arrays is to use Boolean-valued index arrays. In this case, each element (with values True or False) indicates whether or not to select the element from the list with the corresponding index. That is, if element n in the indexing array of Boolean values is True, then element n is selected from the indexed array. If the value is False, then element n is not selected. This index method is handy when filtering out elements from an array. For example, to select all the elements from the array A (as defined in the preceding section) that exceed the value 0.5, we can use the following combination of the comparison operator applied to a NumPy array and indexing using a Boolean-valued array:

```
In [97]: A > 0.5
Out[97]: array([False, False, False, False, False, False, True, True, True,
          True, True], dtype=bool)
In [98]: A[A > 0.5]
Out[98]: array([ 0.6,  0.7,  0.8,  0.9,  1. ])
```

Unlike arrays created by using slices, the arrays returned using fancy indexing and Boolean-valued indexing are not views but rather new independent arrays. Nonetheless, it is possible to assign values to elements selected using fancy indexing:

```
In [99]: A = np.arange(10)
In [100]: indices = [2, 4, 6]
In [101]: B = A[indices]
In [102]: B[0] = -1  # this does not affect A
In [103]: A
Out[103]: array([0, 1, 2, 3, 4, 5, 6, 7, 8, 9])
In [104]: A[indices] = -1  # this alters A
In [105]: A
Out[105]: array([ 0,  1, -1,  3, -1,  5, -1,  7,  8,  9])
```

and likewise for Boolean-valued indexing:

```
In [106]: A = np.arange(10)
In [107]: B = A[A > 5]
In [108]: B[0] = -1  # this does not affect A
In [109]: A
Out[109]: array([0, 1, 2, 3, 4, 5, 6, 7, 8, 9])
```

```
In [110]: A[A > 5] = -1  # this alters A
In [111]: A
Out[111]: array([ 0,  1,  2,  3,  4,  5, -1, -1, -1, -1])
```

A visual summary of different methods to index NumPy arrays is given in Figure 2-1. Note that each type of indexing we have discussed here can be independently applied to each dimension of an array.

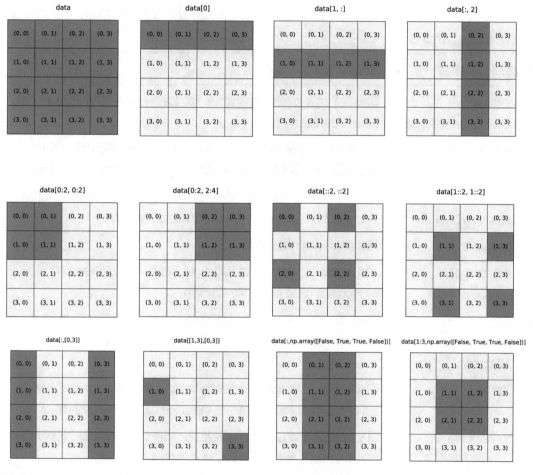

Figure 2-1. *Visual summary of indexing methods for NumPy arrays. These diagrams represent NumPy arrays of shape (4, 4), and the highlighted elements are those that are selected using the indexing expression shown above the block representations of the arrays.*

Reshaping and Resizing

When working with data in array form, it is often useful to rearrange arrays and alter the way they are interpreted. For example, an $N \times N$ matrix array could be rearranged into a vector of length N^2, or a set of one-dimensional arrays could be concatenated together or stacked next to each other to form a matrix. NumPy provides a rich set of functions of this type of manipulation. See Table 2-5 for a summary of a selection of these functions.

Table 2-5. *Summary of NumPy Functions for Manipulating the Dimensions and the Shape of Arrays*

Function/Method	Description
np.reshape, np.ndarray.reshape	Reshape an N-dimensional array. The total number of elements must remain the same.
np.ndarray.flatten	Creates a copy of an N-dimensional array, and reinterpret it as a one-dimensional array (i.e., all dimensions are collapsed into one).
np.ravel, np.ndarray.ravel	Create a view (if possible, otherwise a copy) of an N-dimensional array in which it is interpreted as a one-dimensional array.
np.squeeze	Removes axes with length 1.
np.expand_dims, np.newaxis	Add a new axis (dimension) of length 1 to an array, where np.newaxis is used with array indexing.
np.transpose, np.ndarray.transpose, np.ndarray.T	Transpose the array. The transpose operation corresponds to reversing (or more generally, permuting) the axes of the array.
np.hstack	Stacks a list of arrays horizontally (along axis 1): for example, given a list of column vectors, appends the columns to form a matrix.
np.vstack	Stacks a list of arrays vertically (along axis 0): for example, given a list of row vectors, appends the rows to form a matrix.
np.dstack	Stacks arrays depth-wise (along axis 2).
np.concatenate	Creates a new array by appending arrays after each other, along a given axis.

(continued)

Table 2-5. (*continued*)

Function/Method	Description
np.resize	Resizes an array. Creates a new copy of the original array, with the requested size. If necessary, the original array will be repeated to fill up the new array.
np.append	Appends an element to an array. Creates a new copy of the array.
np.insert	Inserts a new element at a given position. Creates a new copy of the array.
np.delete	Deletes an element at a given position. Creates a new copy of the array.

Reshaping an array does not require modifying the underlying array data; it only changes in how the data is interpreted, by redefining the array's strides attribute. An example of this type of operation is a 2 × 2 array (matrix) that is reinterpreted as a 1 × 4 array (vector). In NumPy, the function np.reshape, or the ndarray class method reshape, can be used to reconfigure how the underlying data is interpreted. It takes an array and the new shape of the array as arguments:

```
In [112]: data = np.array([[1, 2], [3, 4]])
In [113]: np.reshape(data, (1, 4))
Out[113]: array([[1, 2, 3, 4]])
In [114]: data.reshape(4)
Out[114]: array([1, 2, 3, 4])
```

It is necessary that the requested new shape of the array match the number of elements in the original size. However, the number of axes (dimensions) does not need to be conserved, as illustrated in the previous example, where in the first case, the new array has dimension 2 and shape (1, 4), while in the second case, the new array has dimension 1 and shape (4,). This example also demonstrates two different ways of invoking the reshape operation: using the function np.reshape and the ndarray method reshape. Note that reshaping an array produces a view of the array, and if an independent copy of the array is needed, the view has to be copied explicitly (e.g., using np.copy).

The np.ravel (and its corresponding ndarray method) is a special case of reshape, which collapses all dimensions of an array and returns a flattened one-dimensional array with a length that corresponds to the total number of elements in the original array. The ndarray method flatten performs the same function but returns a copy instead of a view.

```
In [115]: data = np.array([[1, 2], [3, 4]])
In [116]: data
Out[116]: array([[1, 2],
                  [3, 4]])
In [117]: data.flatten()
Out[117]: array([ 1,  2,  3,  4])
In [118]: data.flatten().shape
Out[118]: (4,)
```

While np.ravel and np.flatten collapse the axes of an array into a one-dimensional array, it is also possible to introduce new axes into an array, either by using np.reshape or, when adding new empty axes, using indexing notation and the np.newaxis keyword at the place of a new axis. In the following example, the array data has one axis, so it should normally be indexed with a tuple with one element. However, if it is indexed with a tuple with more than one element, and if the extra indices in the tuple have the value np.newaxis, then the corresponding new axes are added:

```
In [119]: data = np.arange(0, 5)
In [120]: column = data[:, np.newaxis]
In [121]: column
Out[121]: array([[0],
                  [1],
                  [2],
                  [3],
                  [4]])
In [122]: row = data[np.newaxis, :]
In [123]: row
Out[123]: array([[0, 1, 2, 3, 4]])
```

The function np.expand_dims can also be used to add new dimensions to an array, and in the preceding example, the expression data[:, np.newaxis] is equivalent to np.expand_dims(data, axis=1), and data[np.newaxis, :] is equivalent to np.expand_dims(data, axis=0). Here the axis argument specifies the location relative to the existing axes where the new axis is to be inserted.

We have up to now looked at methods to rearrange arrays in ways that do not affect the underlying data. Earlier in this chapter, we also looked at how to extract subarrays using various indexing techniques. In addition to reshaping and selecting subarrays, it is often necessary to merge arrays into bigger arrays, for example, when joining separately computed or measured data series into a higher-dimensional array, such as a matrix. For this task, NumPy provides the functions np.vstack, for vertical stacking of, for example, rows into a matrix, and np.hstack for horizontal stacking of, for example, columns into a matrix. The function np.concatenate provides similar functionality, but it takes a keyword argument axis that specifies the axis along which the arrays are to be concatenated.

The shape of the arrays passed to np.hstack, np.vstack, and np.concatenate is important to achieve the desired type of array joining. For example, consider the following cases: say we have one-dimensional arrays of data, and we want to stack them vertically to obtain a matrix where the rows are made up of the one-dimensional arrays. We can use np.vstack to achieve this

```
In [124]: data = np.arange(5)
In [125]: data
Out[125]: array([0, 1, 2, 3, 4])
In [126]: np.vstack((data, data, data))
Out[126]: array([[0, 1, 2, 3, 4],
                  [0, 1, 2, 3, 4],
                  [0, 1, 2, 3, 4]])
```

If we instead want to stack the arrays horizontally, to obtain a matrix where the arrays are the column vectors, we might first attempt something similar using np.hstack:

```
In [127]: data = np.arange(5)
In [128]: data
Out[128]: array([0, 1, 2, 3, 4])
In [129]: np.hstack((data, data, data))
Out[129]: array([0, 1, 2, 3, 4, 0, 1, 2, 3, 4, 0, 1, 2, 3, 4])
```

This indeed stacks the arrays horizontally, but not in the way intended here. To make `np.hstack` treat the input arrays as columns and stack them accordingly, we need to make the input arrays two-dimensional arrays of shape $(1, 5)$ rather than one-dimensional arrays of shape $(5,)$. As discussed earlier, we can insert a new axis by indexing with `np.newaxis`:

```
In [130]: data = data[:, np.newaxis]
In [131]: np.hstack((data, data, data))
Out[131]: array([[0, 0, 0],
                 [1, 1, 1],
                 [2, 2, 2],
                 [3, 3, 3],
                 [4, 4, 4]])
```

The behavior of the functions for horizontal and vertical stacking, as well as concatenating arrays using `np.concatenate`, is clearest when the stacked arrays have the same number of dimensions as the final array and when the input arrays are stacked along an axis for which they have length 1.

The number of elements in a NumPy array cannot be changed once the array has been created. To insert, append, and remove elements from a NumPy array, for example, using the function `np.append`, `np.insert`, and `np.delete`, a new array must be created and the data copied to it. It may sometimes be tempting to use these functions to grow or shrink the size of a NumPy array, but due to the overhead of creating new arrays and copying the data, it is usually a good idea to preallocate arrays with size such that they do not later need to be resized.

Vectorized Expressions

The purpose of storing numerical data in arrays is to be able to process the data with concise vectorized expressions that represent batch operations that are applied to all elements in the arrays. Efficient use of vectorized expressions eliminates the need of many explicit for loops. This results in less verbose code, better maintainability, and higher-performing code. NumPy implements functions and vectorized operations corresponding to most fundamental mathematical functions and operators. Many of these functions and operations act on arrays on an elementwise basis, and binary operations require all arrays in an expression to be of compatible size. The meaning of compatible size is normally that the variables in an expression represent either scalars or arrays of the same size and shape. More generally, a binary operation involving two arrays is well defined if the arrays can be *broadcasted* into the same shape and size.

In the case of an operation between a scalar and an array, broadcasting refers to the scalar being distributed and the operation applied to each element in the array. When an expression contains arrays of unequal sizes, the operations may still be well defined if the smaller of the array can be broadcasted ("effectively expanded") to match the larger array according to NumPy's broadcasting rule: an array can be broadcasted over another array if their axes on a one-by-one basis either have the same length or if either of them have length 1. If the number of axes of the two arrays is not equal, the array with fewer axes is padded with new axes of length 1 from the left until the numbers of dimensions of the two arrays agree.

Two simple examples that illustrate array broadcasting are shown in Figure 2-2: a 3×3 matrix is added to a 1×3 row vector and a 3×1 column vector, respectively, and in both cases the result is a 3×3 matrix. However, the elements in the two resulting matrices are different, because the way the elements of the row and column vectors are broadcasted to the shape of the larger array is different depending on the shape of the arrays, according to NumPy's broadcasting rule.

Figure 2-2. *Visualization of broadcasting of row and column vectors into the shape of a matrix. The highlighted elements represent true elements of the arrays, while the light gray-shaded elements describe the broadcasting of the elements of the array of smaller size.*

Arithmetic Operations

The standard arithmetic operations with NumPy arrays perform elementwise operations. Consider, for example, the addition, subtraction, multiplication, and division of equal-sized arrays:

```
In [132]: x = np.array([[1, 2], [3, 4]])
In [133]: y = np.array([[5, 6], [7, 8]])
In [134]: x + y
Out[134]: array([[ 6,  8],
                 [10, 12]])
```

```
In [135]: y - x
Out[135]: array([[4, 4],
                 [4, 4]])
In [136]: x * y
Out[136]: array([[ 5, 12],
                 [21, 32]])
In [137]: y / x
Out[137]: array([[ 5.        ,  3.        ],
                 [ 2.33333333,  2.        ]])
```

In operations between scalars and arrays, the scalar value is applied to each element in the array, as one could expect:

```
In [138]: x * 2
Out[138]: array([[2, 4],
                 [6, 8]])
In [139]: 2 ** x
Out[139]: array([[ 2,  4],
                 [ 8, 16]])
In [140]: y / 2
Out[140]: array([[ 2.5,  3. ],
                 [ 3.5,  4. ]])
In [141]: (y / 2).dtype
Out[141]: dtype('float64')
```

Note that the dtype of the resulting array for an expression can be promoted if the computation requires it, as shown in the preceding example with division between an integer array and an integer scalar, which in that case resulted in an array with a dtype that is np.float64.

If an arithmetic operation is performed on arrays with incompatible size or shape, a ValueError exception is raised:

```
In [142]: x = np.array([1, 2, 3, 4]).reshape(2, 2)
In [143]: z = np.array([1, 2, 3, 4])
In [144]: x / z
```

```
------------------------------------------------------------------------
ValueError                                Traceback (most recent call last)
<ipython-input-144-b88ced08eb6a> in <module>()
----> 1 x / z
ValueError: operands could not be broadcast together with shapes (2,2) (4,)
```

Here the array x has shape $(2, 2)$ and the array z has shape $(4,)$, which cannot be broadcasted into a form that is compatible with $(2, 2)$. If, on the other hand, z has shape $(2,)$, $(2, 1)$, or $(1, 2)$, then it can broadcasted to the shape $(2, 2)$ by effectively repeating the array z along the axis with length 1. Let's first consider an example with an array z of shape $(1, 2)$, where the first axis (axis 0) has length 1:

```
In [145]: z = np.array([[2, 4]])
In [146]: z.shape
Out[146]: (1, 2)
```

Dividing the array x with array z is equivalent to dividing x with an array zz that is constructed by repeating (here using np.concatenate) the row vector z to obtain an array zz that has the same dimensions as x:

```
In [147]: x / z
Out[147]: array([[ 0.5,  0.5],
                 [ 1.5,  1. ]])
In [148]: zz = np.concatenate([z, z], axis=0)
In [149]: zz
Out[149]: array([[2, 4],
                 [2, 4]])
In [150]: x / zz
Out[150]: array([[ 0.5,  0.5],
                 [ 1.5,  1. ]])
```

Let's also consider the example in which the array z has shape $(2, 1)$ and where the second axis (axis 1) has length 1:

```
In [151]: z = np.array([[2], [4]])
In [152]: z.shape
Out[152]: (2, 1)
```

In this case, dividing x with z is equivalent to dividing x with an array zz that is constructed by repeating the column vector z until a matrix with the same dimension as x is obtained.

```
In [153]: x / z
Out[153]: array([[ 0.5 ,  1.  ],
                 [ 0.75,  1.  ]])
In [154]: zz = np.concatenate([z, z], axis=1)
In [155]: zz
Out[155]: array([[2, 2],
                 [4, 4]])
In [156]: x / zz
Out[156]: array([[ 0.5 ,  1.  ],
                 [ 0.75,  1.  ]])
```

In summary, these examples show how arrays with shape (1, 2) and (2, 1) are broadcasted to the shape (2, 2) of the array x when the operation x / z is performed. In both cases, the result of the operation x / z is the same as first repeating the smaller array z along its axis of length 1 to obtain a new array zz with the same shape as x and then performing the equal-sized array operation x / zz. However, the implementation of the broadcasting does not explicitly perform this expansion and the corresponding memory copies, but it can be helpful to think of the array broadcasting in these terms.

A summary of the operators for arithmetic operations with NumPy arrays is given in Table 2-6. These operators use the standard symbols used in Python. The result of an arithmetic operation with one or two arrays is a new independent array, with its own data in the memory. Evaluating complicated arithmetic expression might therefore trigger many memory allocation and copy operations, and when working with large arrays, this can lead to a large memory footprint and impact the performance negatively. In such cases, using inplace operation (see Table 2-6) can reduce the memory footprint and improve performance. As an example of inplace operators, consider the following two statements, which have the same effect:

```
In [157]: x = x + y
In [158]: x += y
```

Table 2-6. *Operators for Elementwise Arithmetic Operation on NumPy Arrays*

Operator	Operation
+, +=	Addition
-, -=	Subtraction
*, *=	Multiplication
/, /=	Division
//, //=	Integer division
**, **=	Exponentiation

The two expressions have the same effect, but in the first case, x is reassigned to a new array, while in the second case, the values of array x are updated inplace. Extensive use of inplace operators tends to impair code readability, and inplace operators should therefore be used only when necessary.

Elementwise Functions

In addition to arithmetic expressions using operators, NumPy provides vectorized functions for elementwise evaluation of many elementary mathematical functions and operations. Table 2-7 gives a summary of elementary mathematical functions in NumPy.[3] Each of these functions takes a single array (of arbitrary dimension) as input and returns a new array of the same shape, where for each element the function has been applied to the corresponding element in the input array. The data type of the output array is not necessarily the same as that of the input array.

---

[3]Note that this is not a complete list of the available elementwise functions in NumPy. See the NumPy reference documentations for comprehensive lists.

Table 2-7. *Selection of NumPy Functions for Elementwise Elementary Mathematical Functions*

NumPy Function	Description
`np.cos, np.sin, np.tan`	Trigonometric functions.
`np.arccos, np.arcsin, np.arctan`	Inverse trigonometric functions.
`np.cosh, np.sinh, np.tanh`	Hyperbolic trigonometric functions.
`np.arccosh, np.arcsinh, np.arctanh`	Inverse hyperbolic trigonometric functions.
`np.sqrt`	Square root.
`np.exp`	Exponential.
`np.log, np.log2, np.log10`	Logarithms of base e, 2, and 10, respectively.

For example, the `np.sin` function (which takes only one argument) is used to compute the sine function for all values in the array:

```
In [159]: x = np.linspace(-1, 1, 11)
In [160]: x
Out[160]: array([-1. , -0.8, -0.6, -0.4, -0.2,  0. ,  0.2,  0.4,  0.6,  0.8,  1.])
In [161]: y = np.sin(np.pi * x)
In [162]: np.round(y, decimals=4)
Out[162]: array([-0., -0.5878, -0.9511, -0.9511, -0.5878, 0., 0.5878, 0.9511,
          0.9511, 0.5878, 0.])
```

Here we also used the constant `np.pi` and the function `np.round` to round the values of y to four decimals. Like the `np.sin` function, many of the elementary math functions take one input array and produce one output array. In contrast, many of the mathematical operator functions (Table 2-8) operates on two input arrays returns one array:

```
In [163]: np.add(np.sin(x) ** 2, np.cos(x) ** 2)
Out[163]: array([ 1.,  1.,  1.,  1.,  1.,  1.,  1.,  1.,  1.,  1.,  1.])
In [164]: np.sin(x) ** 2 + np.cos(x) ** 2
Out[164]: array([ 1.,  1.,  1.,  1.,  1.,  1.,  1.,  1.,  1.,  1.,  1.])
```

Table 2-8. *Summary of NumPy Functions for Elementwise Mathematical Operations*

NumPy Function	Description
np.add, np.subtract, np.multiply, np.divide	Addition, subtraction, multiplication, and division of two NumPy arrays.
np.power	Raises first input argument to the power of the second input argument (applied elementwise).
np.remainder	The remainder of division.
np.reciprocal	The reciprocal (inverse) of each element.
np.real, np.imag, np.conj	The real part, imaginary part, and the complex conjugate of the elements in the input arrays.
np.sign, np.abs	The sign and the absolute value.
np.floor, np.ceil, np.rint	Convert to integer values.
np.round	Rounds to a given number of decimals.

Note that in this example, np.add and the operator + are equivalent, and for normal use the operator should be used.

Occasionally it is necessary to define new functions that operate on NumPy arrays on an element-by-element basis. A good way to implement such functions is to express it in terms of already existing NumPy operators and expressions, but in cases when this is not possible, the np.vectorize function can be a convenient tool. This function takes a nonvectorized function and returns a vectorized function. For example, consider the following implementation of the Heaviside step function, which works for scalar input:

```
In [165]: def heaviside(x):
     ...:     return 1 if x > 0 else 0
In [166]: heaviside(-1)
Out[166]: 0
In [167]: heaviside(1.5)
Out[167]: 1
```

However, unfortunately this function does not work for NumPy array input:

```
In [168]: x = np.linspace(-5, 5, 11)
In [169]: heaviside(x)
...
ValueError: The truth value of an array with more than one element is
ambiguous. Use a.any() or a.all()
```

Using np.vectorize the scalar Heaviside function can be converted into a vectorized function that works with NumPy arrays as input:

```
In [170]: heaviside = np.vectorize(heaviside)
In [171]: heaviside(x)
Out[171]: array([0, 0, 0, 0, 0, 0, 1, 1, 1, 1, 1])
```

Although the function returned by np.vectorize works with arrays, it will be relatively slow since the original function must be called for each element in the array. There are much better ways to implementing this particular function using arithmetic with Boolean-valued arrays, as discussed later in this chapter:

```
In [172]: def heaviside(x):
    ...:        return 1.0 * (x > 0)
```

Nonetheless, np.vectorize can often be a quick and convenient way to vectorize a function written for scalar input.

In addition to NumPy's functions for elementary mathematical function, as summarized in Table 2-7, there are also numerous functions in NumPy for mathematical operations. A summary of a selection of these functions is given in Table 2-8.

Aggregate Functions

NumPy provides another set of functions for calculating aggregates for NumPy arrays, which take an array as input and by default return a scalar as output. For example, statistics such as averages, standard deviations, and variances of the values in the input array, and functions for calculating the sum and the product of elements in an array, are all aggregate functions.

A summary of aggregate functions is given in Table 2-9. All of these functions are also available as methods in the ndarray class. For example, np.mean(data) and data.mean() in the following example are equivalent:

```
In [173]: data = np.random.normal(size=(15,15))
In [174]: np.mean(data)
Out[174]: -0.032423651106794522
In [175]: data.mean()
Out[175]: -0.032423651106794522
```

Table 2-9. *NumPy Functions for Calculating Aggregates of NumPy Arrays*

NumPy Function	Description
np.mean	The average of all values in the array.
np.std	Standard deviation.
np.var	Variance.
np.sum	Sum of all elements.
np.prod	Product of all elements.
np.cumsum	Cumulative sum of all elements.
np.cumprod	Cumulative product of all elements.
np.min, np.max	The minimum/maximum value in an array.
np.argmin, np.argmax	The index of the minimum/maximum value in an array.
np.all	Returns True if all elements in the argument array are nonzero.
np.any	Returns True if any of the elements in the argument array is nonzero.

By default, the functions in Table 2-9 aggregate over the entire input array. Using the axis keyword argument with these functions, and their corresponding ndarray methods, it is possible to control over which axis in the array aggregation is carried out. The axis argument can be an integer, which specifies the axis to aggregate values over. In many cases the axis argument can also be a tuple of integers, which specifies multiple axes to aggregate over. The following example demonstrates how calling the aggregate function `np.sum` on the array of shape (5, 10, 15) reduces the dimensionality of the array depending on the values of the `axis` argument:

```
In [176]: data = np.random.normal(size=(5, 10, 15))
In [177]: data.sum(axis=0).shape
Out[177]: (10, 15)
In [178]: data.sum(axis=(0, 2)).shape
Out[178]: (10,)
In [179]: data.sum()
Out[179]: -31.983793284860798
```

A visual illustration of how aggregation over all elements, over the first axis, and over the second axis of a 3 × 3 array is shown in Figure 2-3. In this example, the `data` array is filled with integers between 1 and 9:

```
In [180]: data = np.arange(1,10).reshape(3,3)
In [181]: data
Out[181]: array([[1, 2, 3],
                 [4, 5, 6],
                 [7, 8, 9]])
```

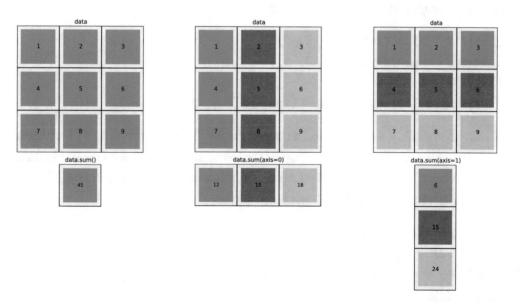

Figure 2-3. *Illustration of array aggregation functions along all axes (left), the first axis (center), and the second axis (right) of a two-dimensional array of shape 3 × 3*

and we compute the aggregate sum of the entire array, over the axis 0, and over axis 1, respectively:

```
In [182]: data.sum()
Out[182]: 45
In [183]: data.sum(axis=0)
Out[183]: array([12, 15, 18])
In [184]: data.sum(axis=1)
Out[184]: array([ 6, 15, 24])
```

Boolean Arrays and Conditional Expressions

When computing with NumPy arrays, there is often a need to compare elements in different arrays and perform conditional computations based on the results of such comparisons. Like with arithmetic operators, NumPy arrays can be used with the usual comparison operators, for example, >, <, >=, <=, ==, and !=, and the comparisons are made on an element-by-element basis. The broadcasting rules also apply to comparison operators, and if two operators have compatible shapes and sizes, the result of the comparison is a new array with Boolean values (with dtype as np.bool) that gives the result of the comparison for each element:

```
In [185]: a = np.array([1, 2, 3, 4])
In [186]: b = np.array([4, 3, 2, 1])
In [187]: a < b
Out[187]: array([ True,  True, False, False], dtype=bool)
```

To use the result of a comparison between arrays in, for example, an if statement, we need to aggregate the Boolean values of the resulting arrays in some suitable fashion, to obtain a single True or False value. A common use-case is to apply the np.all or np. any aggregation functions, depending on the situation at hand:

```
In [188]: np.all(a < b)
Out[188]: False
In [189]: np.any(a < b)
Out[189]: True
In [190]: if np.all(a < b):
```

```
    ...: print("All elements in a are smaller than their corresponding
         element in b")
    ...: elif np.any(a < b):
    ...: print("Some elements in a are smaller than their corresponding
         element in b")
    ...: else:
    ...: print("All elements in b are smaller than their corresponding
         element in a")
Some elements in a are smaller than their corresponding element in b
```

The advantage of Boolean-valued arrays, however, is that they often make it possible to avoid conditional if statements altogether. By using Boolean-valued arrays in arithmetic expressions, it is possible to write conditional computations in vectorized form. When appearing in an arithmetic expression together with a scalar number, or another NumPy array with a numerical data type, a Boolean array is converted to a numerical-valued array with values 0 and 1 inplace of False and True, respectively.

```
In [191]: x = np.array([-2, -1, 0, 1, 2])
In [192]: x > 0
Out[192]: array([False, False, False,  True,  True], dtype=bool)
In [193]: 1 * (x > 0)
Out[193]: array([0, 0, 0, 1, 1])
In [194]: x * (x > 0)
Out[194]: array([0, 0, 0, 1, 2])
```

This is a useful property for conditional computing, such as when defining piecewise functions. For example, if we need to define a function describing a pulse of a given height, width, and position, we can implement this function by multiplying the height (a scalar variable) with two Boolean-valued arrays for the spatial extension of the pulse:

```
In [195]: def pulse(x, position, height, width):
    ...:     return height * (x >= position) * (x <= (position + width))
In [196]: x = np.linspace(-5, 5, 11)
In [197]: pulse(x, position=-2, height=1, width=5)
Out[197]: array([0, 0, 0, 1, 1, 1, 1, 1, 1, 0, 0])
In [198]: pulse(x, position=1, height=1, width=5)
Out[198]: array([0, 0, 0, 0, 0, 0, 1, 1, 1, 1, 1])
```

In this example, the expression (x >= position) * (x <= (position + width)) is a multiplication of two Boolean-valued arrays, and for this case the multiplication operator acts as an elementwise AND operator. The function pulse could also be implemented using NumPy's function for elementwise AND operations, np.logical_and:

```
In [199]: def pulse(x, position, height, width):
    ...:        return height * np.logical_and(x >= position, x <= (position +
           width))
```

There are also functions for other logical operations, such as NOT, OR, and XOR, and functions for selectively picking values from different arrays depending on a given condition np.where, a list of conditions np.select, and an array of indices np.choose. See Table 2-10 for a summary of such functions, and the following examples demonstrate the basic usage of some of these functions. The np.where function selects elements from two arrays (second and third arguments), given a Boolean-valued array condition (the first argument). For elements where the condition is True, the corresponding values from the array given as second argument are selected, and if the condition is False, elements from the third argument array are selected:

```
In [200]: x = np.linspace(-4, 4, 9)
In [201]: np.where(x < 0, x**2, x**3)
Out[201]: array([ 16.,    9.,    4.,    1.,    0.,    1.,    8.,   27.,   64.])
```

Table 2-10. *NumPy Functions for Conditional and Logical Expressions*

Function	Description
np.where	Chooses values from two arrays depending on the value of a condition array.
np.choose	Chooses values from a list of arrays depending on the values of a given index array.
np.select	Chooses values from a list of arrays depending on a list of conditions.
np.nonzero	Returns an array with indices of nonzero elements.
np.logical_and	Performs an elementwise AND operation.
np.logical_or, np.logical_xor	Elementwise OR/XOR operations.
np.logical_not	Elementwise NOT operation (inverting).

The np.select function works similarly, but instead of a Boolean-valued condition array, it expects a list of Boolean-valued condition arrays and a corresponding list of value arrays:

```
In [202]: np.select([x < -1, x < 2, x >= 2],
     ...:            [x**2  , x**3 , x**4])
Out[202]: array([  16.,     9.,     4.,    -1.,     0.,     1.,    16.,
                  81.,   256.])
```

The np.choose takes as a first argument a list or an array with indices that determine from which array in a given list of arrays an element is picked from:

```
In [203]: np.choose([0, 0, 0, 1, 1, 1, 2, 2, 2],
     ...:           [x**2,  x**3,  x**4])
Out[203]: array([  16.,     9.,     4.,    -1.,     0.,     1.,    16.,
                  81.,   256.])
```

The function np.nonzero returns a tuple of indices that can be used to index the array (e.g., the one that the condition was based on). This has the same results as indexing the array directly with abs(x) > 2, but it uses fancy indexing with the indices returned by np.nonzero rather than Boolean-valued array indexing.

```
In [204]: np.nonzero(abs(x) > 2)
Out[204]: (array([0, 1, 7, 8]),)
In [205]: x[np.nonzero(abs(x) > 2)]
Out[205]: array([-4., -3.,  3.,  4.])
In [206]: x[abs(x) > 2]
Out[206]: array([-4., -3.,  3.,  4.])
```

Set Operations

The Python language provides a convenient *set* data structure for managing unordered collections of unique objects. The NumPy array class ndarray can also be used to describe such sets, and NumPy contains functions for operating on sets stored as NumPy arrays. These functions are summarized in Table 2-11. Using NumPy arrays to describe and operate on sets allows expressing certain operations in vectorized form. For example, testing if the values in a NumPy array are included in a set can be done using the np.in1d function, which tests for the existence of each element of its first argument in the array passed as the second argument. To see how this works, consider the following example: first, to ensure that a NumPy array is a proper set, we can use the np.unique function, which returns a new array with unique values:

```
In [207]: a = np.unique([1, 2, 3, 3])
In [208]: b = np.unique([2, 3, 4, 4, 5, 6, 5])
In [209]: np.in1d(a, b)
Out[209]: array([False,  True,  True], dtype=bool)
```

Table 2-11. *NumPy Functions for Operating on Sets*

Function	Description
np.unique	Creates a new array with unique elements, where each value only appears once.
np.in1d	Tests for the existence of an array of elements in another array.
np.intersect1d	Returns an array with elements that are contained in two given arrays.
np.setdiff1d	Returns an array with elements that are contained in one, but not the other, of two given arrays.
np.union1d	Returns an array with elements that are contained in either, or both, of two given arrays.

Here, the existence of each element in a in the set b was tested, and the result is a Boolean-valued array. Note that we can use the in keyword to test for the existence of single elements in a set represented as NumPy array:

```
In [210]: 1 in a
Out[210]: True
In [211]: 1 in b
Out[211]: False
```

To test if a is a subset of b, we can use the np.in1d, as in the previous example, together with the aggregation function np.all (or the corresponding ndarray method):

```
In [212]: np.all(np.in1d(a, b))
Out[212]: False
```

The standard set operations union (the set of elements included in either or both sets), intersection (elements included in both sets), and difference (elements included in one of the sets but not the other) are provided by np.union1d, np.intersect1d, and np. setdiff1d, respectively:

```
In [213]: np.union1d(a, b)
Out[213]: array([1, 2, 3, 4, 5, 6])
In [214]: np.intersect1d(a, b)
Out[214]: array([2, 3])
In [215]: np.setdiff1d(a, b)
Out[215]: array([1])
In [216]: np.setdiff1d(b, a)
Out[216]: array([4, 5, 6])
```

Operations on Arrays

In addition to elementwise and aggregation functions, some operations act on arrays as a whole and produce a transformed array of the same size. An example of this type of operation is the transpose, which flips the order of the axes of an array. For the special case of a two-dimensional array, i.e., a matrix, the transpose simply exchanges rows and columns:

```
In [217]: data = np.arange(9).reshape(3, 3)
In [218]: data
Out[218]: array([[0, 1, 2],
                 [3, 4, 5],
                 [6, 7, 8]])
In [219]: np.transpose(data)
Out[219]: array([[0, 3, 6],
                 [1, 4, 7],
                 [2, 5, 8]])
```

The transpose function np.transpose also exists as a method in ndarray and as the special method name ndarray.T. For an arbitrary N-dimensional array, the transpose operation reverses all the axes, as can be seen from the following example (note that the shape attribute is used here to display the number of values along each axis of the array) :

```
In [220]: data = np.random.randn(1, 2, 3, 4, 5)
In [221]: data.shape
Out[221]: (1, 2, 3, 4, 5)
In [222]: data.T.shape
Out[222]: (5, 4, 3, 2, 1)
```

The np.fliplr (flip left-right) and np.flipud (flip up-down) functions perform operations that are similar to the transpose: they reshuffle the elements of an array so that the elements in rows (np.fliplr) or columns (np.flipud) are reversed, and the shape of the output array is the same as the input. The np.rot90 function rotates the elements in the first two axes in an array by 90 degrees, and like the transpose function, it can change the shape of the array. Table 2-12 gives a summary of NumPy functions for common array operations.

Table 2-12. *Summary of NumPy Functions for Array Operations*

Function	Description
np.transpose, np.ndarray.transpose, np.ndarray.T	The transpose (reverse axes) of an array.
np.fliplr/np.flipud	Reverse the elements in each row/column.
np.rot90	Rotates the elements along the first two axes by 90 degrees.
np.sort, np.ndarray.sort	Sort the elements of an array along a given specified axis (which default to the last axis of the array). The np.ndarray method sort performs the sorting in place, modifying the input array.

Matrix and Vector Operations

We have so far discussed general N-dimensional arrays. One of the main applications of such arrays is to represent the mathematical concepts of vectors, matrices, and tensors, and in this use-case, we also frequently need to calculate vector and matrix operations such as scalar (inner) products, dot (matrix) products, and tensor (outer) products. A summary of NumPy's functions for matrix operations is given in Table 2-13.

Table 2-13. *Summary of NumPy Functions for Matrix Operations*

NumPy Function	Description
np.dot	Matrix multiplication (dot product) between two given arrays representing vectors, arrays, or tensors.
np.inner	Scalar multiplication (inner product) between two arrays representing vectors.
np.cross	The cross product between two arrays that represent vectors.
np.tensordot	Dot product along specified axes of multidimensional arrays.
np.outer	Outer product (tensor product of vectors) between two arrays representing vectors.
np.kron	Kronecker product (tensor product of matrices) between arrays representing matrices and higher-dimensional arrays.
np.einsum	Evaluates Einstein's summation convention for multidimensional arrays.

In NumPy, the * operator is used for elementwise multiplication. For two two-dimensional arrays A and B, the expression A * B therefore does not compute a matrix product (in contrast to many other computing environments). Currently there is no operator for denoting matrix multiplication,[4] and instead the NumPy function np.dot is used for this purpose. There is also a corresponding method in the ndarray class. To compute the product of two matrices A and B, of size $N \times M$ and $M \times P$, which results in a matrix of size $N \times P$, we can use:

```
In [223]: A = np.arange(1, 7).reshape(2, 3)
In [224]: A
Out[224]: array([[1, 2, 3],
                 [4, 5, 6]])
In [225]: B = np.arange(1, 7).reshape(3, 2)
In [226]: B
```

---

[4]Python recently adopted the @ symbol for denoting matrix multiplication, and as of Python 3.5, this operator is now available. However, at the time of writing, this operator is still not widely used. See http://legacy.python.org/dev/peps/pep-0465 for details.

```
Out[226]: array([[1, 2],
                  [3, 4],
                  [5, 6]])
In [227]: np.dot(A, B)
Out[227]: array([[22, 28],
                  [49, 64]])
In [228]: np.dot(B, A)
Out[228]: array([[ 9, 12, 15],
                  [19, 26, 33],
                  [29, 40, 51]])
```

The np.dot function can also be used for matrix-vector multiplication (i.e., multiplication of a two-dimensional array, which represents a matrix, with a one-dimensional array representing a vector) . For example,

```
In [229]: A = np.arange(9).reshape(3, 3)
In [230]: A
Out[230]: array([[0, 1, 2],
                  [3, 4, 5],
                  [6, 7, 8]])
In [231]: x = np.arange(3)
In [232]: x
Out[232]: array([0, 1, 2])
In [233]: np.dot(A, x)
Out[233]: array([5, 14, 23])
```

In this example, x can be either a two-dimensional array of shape (1, 3) or a one-dimensional array with shape (3,). In addition to the function np.dot, there is also a corresponding method dot in ndarray, which can be used as in the following example:

```
In [234]: A.dot(x)
Out[234]: array([5, 14, 23])
```

Unfortunately, nontrivial matrix multiplication expressions can often become complex and hard to read when using either np.dot or np.ndarray.dot. For example, even a relatively simple matrix expression like the one for a similarity transform, $A' = BAB^{-1}$, must be represented with relatively cryptic nested expressions,[5] such as either

```
In [235]: A = np.random.rand(3,3)
In [236]: B = np.random.rand(3,3)
In [237]: Ap = np.dot(B, np.dot(A, np.linalg.inv(B)))
```

or

```
In [238]: Ap = B.dot(A.dot(np.linalg.inv(B)))
```

To improve this situation, NumPy provides an alternative data structure to ndarray named matrix, for which expressions like A * B are implemented as matrix multiplication. It also provides some convenient special attributes, like matrix.I for the inverse matrix and matrix.H for the complex conjugate transpose of a matrix. Using instances of this matrix class, one can therefore use the vastly more readable expression:

```
In [239]: A = np.matrix(A)
In [240]: B = np.matrix(B)
In [241]: Ap = B * A * B.I
```

This may seem like a practical compromise, but unfortunately using the matrix class does have a few disadvantages, and its use is therefore often discouraged. The main objection against using matrix is that expression like A * B is then context dependent: that is, it is not immediately clear if A * B denotes elementwise or matrix multiplication, because it depends on the type of A and B, and this creates another code-readability problem. This can be a particularly relevant issue if A and B are user-supplied arguments to a function, in which case it would be necessary to cast all input arrays explicitly to matrix instances, using, for example, np.asmatrix or the function np.matrix (since there would be no guarantee that the user calls the function with arguments of type matrix rather than ndarray). The np.asmatrix function creates a view of the original array in the form of an np.matrix instance. This does not add much in computational costs, but explicitly casting arrays back and forth between ndarray and matrix does

---

[5]With the new infix matrix multiplication operator, this same expression can be expressed as the considerably more readable: Ap = B @ A @ np.linalg.inv(B).

offset much of the benefits of the improved readability of matrix expressions. A related issue is that some functions that operate on arrays and matrices might not respect the type of the input and may return an ndarray even though it was called with an input argument of type matrix. This way, a matrix of type matrix might be unintentionally converted to ndarray, which in turn would change the behavior of expressions like A * B. This type of behavior is not likely to occur when using NumPy's array and matrix functions, but it is not unlikely to happen when using functions from other packages. However, in spite of all the arguments for not using matrix matrices too extensively, personally I think that using matrix class instances for complicated matrix expressions is an important use-case, and in these cases, it might be a good idea to explicitly cast arrays to matrices before the computation and explicitly cast the result back to the ndarray type, following the pattern:

```
In [242]: A = np.asmatrix(A)
In [243]: B = np.asmatrix(B)
In [244]: Ap = B * A * B.I
In [245]: Ap = np.asarray(Ap)
```

The inner product (scalar product) between two arrays representing vectors can be computed using the np.inner function:

```
In [246]: np.inner(x, x)
Out[246]: 5
```

or, equivalently, using np.dot:

```
In [247]: np.dot(x, x)
Out[247]: 5
```

The main difference is that np.inner expects two input arguments with the same dimension, while np.dot can take input vectors of shape $1 \times N$ and $N \times 1$, respectively:

```
In [248]: y = x[:, np.newaxis]
In [249]: y
Out[249]: array([[0],
                 [1],
                 [2]])
In [250]: np.dot(y.T, y)
Out[250]: array([[5]])
```

While the inner product maps two vectors to a scalar, the outer product performs the complementary operation of mapping two vectors to a matrix.

```
In [251]: x = np.array([1, 2, 3])
In [252]: np.outer(x, x)
Out[252]: array([[1, 2, 3],
                 [2, 4, 6],
                 [3, 6, 9]])
```

The outer product can also be calculated using the Kronecker product using the function np.kron, which, however, in contrast to np.outer, produces an output array of shape (M*P, N*Q) if the input arrays have shapes (M, N) and (P, Q), respectively. Thus, for the case of two one-dimensional arrays of length M and P, the resulting array has shape (M*P,):

```
In [253]: np.kron(x, x)
Out[253]: array([1, 2, 3, 2, 4, 6, 3, 6, 9])
```

To obtain the result that corresponds to np.outer(x, x), the input array x must be expanded to shape (N, 1) and (1, N), in the first and second argument to np.kron, respectively:

```
In [254]: np.kron(x[:, np.newaxis], x[np.newaxis, :])
Out[254]: array([[1, 2, 3],
                 [2, 4, 6],
                 [3, 6, 9]])
```

In general, while the np.outer function is primarily intended for vectors as input, the np.kron function can be used for computing tensor products of arrays of arbitrary dimension (but both inputs must have the same number of axes). For example, to compute the tensor product of two 2×2 matrices, we can use:

```
In [255]: np.kron(np.ones((2,2)), np.identity(2))
Out[255]: array([[ 1.,  0.,  1.,  0.],
                 [ 0.,  1.,  0.,  1.],
                 [ 1.,  0.,  1.,  0.],
                 [ 0.,  1.,  0.,  1.]])
```

```
In [256]: np.kron(np.identity(2), np.ones((2,2)))
Out[256]: array([[ 1.,   1.,   0.,   0.],
                 [ 1.,   1.,   0.,   0.],
                 [ 0.,   0.,   1.,   1.],
                 [ 0.,   0.,   1.,   1.]])
```

When working with multidimensional arrays, it is often possible to express common array operations concisely using Einstein's summation convention, in which an implicit summation is assumed over each index that occurs multiple times in an expression. For example, the scalar product between two vectors x and y is compactly expressed as $x_n y_n$, and the matrix multiplication of two matrices A and B is expressed as $A_{mk} B_{kn}$. NumPy provides the function np.einsum for carrying out Einstein summations. Its first argument is an index expression, followed by an arbitrary number of arrays that are included in the expression. The index expression is a string with comma-separated indices, where each comma separates the indices of each array. Each array can have any number of indices. For example, the scalar product expression $x_n y_n$ can be evaluated with np.einsum using the index expression "n,n", that is using np.einsum("n,n", x, y):

```
In [257]: x = np.array([1, 2, 3, 4])
In [258]: y = np.array([5, 6, 7, 8])
In [259]: np.einsum("n,n", x, y)
Out[259]: 70
In [260]: np.inner(x, y)
Out[260]: 70
```

Similarly, the matrix multiplication $A_{mk} B_{kn}$ can be evaluated using np.einsum and the index expression "mk,kn":

```
In [261]: A = np.arange(9).reshape(3, 3)
In [262]: B = A.T
In [263]: np.einsum("mk,kn", A, B)
Out[263]: array([[  5,  14,  23],
                 [ 14,  50,  86],
                 [ 23,  86, 149]])
In [264]: np.alltrue(np.einsum("mk,kn", A, B) == np.dot(A, B))
Out[264]: True
```

The Einstein summation convention can be particularly convenient when dealing with multidimensional arrays, since the index expression that defines the operation makes it explicit which operation is carried out and along which axes it is performed. An equivalent computation using, for example, `np.tensordot` might require giving the axes along which the dot product is to be evaluated.

Summary

In this chapter we have given a brief introduction to array-based programming with the NumPy library that can serve as a reference for the following chapters in this book. NumPy is a core library for computing with Python that provides a foundation for nearly all computational libraries for Python. Familiarity with the NumPy library and its usage patterns is a fundamental skill for using Python for scientific and technical computing. Here we started with introducing NumPy's data structure for N-dimensional arrays – the `ndarray` object – and we continued by discussing functions for creating and manipulating arrays, including indexing and slicing for extracting elements from arrays. We also discussed functions and operators for performing computations with `ndarray` objects, with an emphasis on vectorized expressions and operators for efficient computation with arrays. Throughout the rest of this book, we will see examples of higher-level libraries for specific fields in scientific computing that use the array framework provided by NumPy.

Further Reading

The NumPy library is the topic of several books, including the *Guide to NumPy*, by the creator of the NumPy T. Oliphant, available for free online at `http://web.mit.edu/dvp/Public/numpybook.pdf`, and a series of books by Ivan Idris: *Numpy Beginner's Guide* (2015), *NumPy Cookbook* (2012), and *Learning NumPy Array* (2014). NumPy is also covered in fair detail in McKinney (2013).

References

Idris, I. *Learning NumPy Array*. Mumbai: Packt, 2014.

—. *Numpy Beginner's Guide*. 3rd. Mumbai: Packt, 2015.

—. *NumPy Cookbook*. Mumbai: Packt, 2012.

McKinney, Wes. *Python for Data Analysis*. Sebastopol: O'Reilly, 2013.

CHAPTER 3

Symbolic Computing

Symbolic computing is an entirely different paradigm in computing compared to the numerical array-based computing introduced in the previous chapter. In symbolic computing software, also known as computer algebra systems (CASs), representations of mathematical objects and expressions are manipulated and transformed analytically. Symbolic computing is mainly about using computers to automate analytical computations that can in principle be done by hand with pen and paper. However, by automating the book-keeping and the manipulations of mathematical expressions using a computer algebra system, it is possible to take analytical computing much further than can realistically be done by hand. Symbolic computing is a great tool for checking and debugging analytical calculations that are done by hand, but more importantly it enables carrying out analytical analysis that may not otherwise be possible.

Analytical and symbolic computing is a key part of the scientific and technical computing landscape, and even for problems that can only be solved numerically (which is common, because analytical methods are not feasible in many practical problems), it can make a big difference to push the limits for what can be done analytically before resorting to numerical techniques. This can, for example, reduce the complexity or size of the numerical problem that finally needs to be solved. In other words, instead of tackling a problem in its original form directly using numerical methods, it may be possible to use analytical methods to simplify the problem first.

In the scientific Python environment, the main module for symbolic computing is SymPy (Symbolic Python). SymPy is entirely written in Python and provides tools for a wide range of analytical and symbolic problems. In this chapter we look in detail into how SymPy can be used for symbolic computing with Python.

© Robert Johansson 2019
R. Johansson, *Numerical Python*, https://doi.org/10.1007/978-1-4842-4246-9_3

SymPy   The Symbolic Python (SymPy) library aims to provide a full-featured computer algebra system (CAS). In contrast to many other CASs, SymPy is primarily a library, rather than a full environment. This makes SymPy well suited for integration in applications and computations that also use other Python libraries. At the time of writing, the latest version is 1.1.1. More information about SymPy is available at `www.sympy.org` and `https://github.com/sympy/sympy/wiki/Faq`.

Importing SymPy

The SymPy project provides the Python module named `sympy`. It is common to import all symbols from this module when working with SymPy, using `from sympy import *`, but in the interest of clarity and for avoiding namespace conflicts between functions and variables from SymPy and from other packages such NumPy and SciPy (see later chapters), here we will import the library in its entirety as `sympy`. In the rest of this book, we will assume that SymPy is imported in this way.

```
In [1]: import sympy
In [2]: sympy.init_printing()
```

Here we have also called the `sympy.init_printing` function, which configures SymPy's printing system to display nicely formatted renditions of mathematical expressions, as we will see examples of such later in this chapter. In the Jupyter Notebook, this sets up printing so that the MathJax JavaScript library renders SymPy expressions, and the results are displayed on the browser page of the notebook.

For the sake of convenience and readability of the example codes in this chapter, we will also assume that the following frequently used symbols are explicitly imported from SymPy into the local namespace:

```
In [3]: from sympy import I, pi, oo
```

Caution   Note that NumPy and SymPy, as well as many other libraries, provide many functions and variables with the same name. But these symbols are rarely interchangeable. For example, numpy.pi is a numerical approximation of the mathematical symbol π, while sympy.pi is a symbolic representation of π. It is therefore important to not mix them up and use, for instance, numpy.pi in place of sympy.pi when doing symbolic computations, or vice versa. The same holds true for many fundamental mathematical functions, such as for example numpy.sin vs. sympy.sin. Therefore, when using more than one package in computing with Python, it is important to consistently use namespaces.

Symbols

A core feature in SymPy is to represent mathematical symbols as Python objects. In the SymPy library, for example, the class sympy.Symbol can be used for this purpose. An instance of Symbol has a name and set of attributes describing its properties and methods for querying those properties and for operating on the symbol object. A symbol by itself is not of much practical use, but symbols are used as nodes in expression trees to represent algebraic expressions (see next section). Among the first steps in setting up and analyzing a problem with SymPy is to create symbols for the various mathematical variables and quantities that are required to describe the problem.

The symbol name is a string, which optionally can contain LaTeX-like markup to make the symbol name display well in, for example, IPython's rich display system. The name of a Symbol object is set when it is created. Symbols can be created in a few different ways in SymPy, for example, using sympy.Symbol, sympy.symbols, and sympy.var. Normally it is desirable to associate SymPy symbols with Python variables with the same name or a name that closely corresponds to the symbol name. For example, to create a symbol named x, and binding it to the Python variable with the same name, we can use the constructor of the Symbol class and pass a string containing the symbol name as the first argument:

```
In [4]: x = sympy.Symbol("x")
```

The variable x now represents an abstract mathematical symbol x of which very little information is known by default. At this point, x could represent, for example, a real number, an integer, a complex number, a function, as well as a large number of other possibilities. In many cases it is sufficient to represent a mathematical symbol with this abstract, unspecified Symbol object, but sometimes it is necessary to give the SymPy library more hints about exactly what type of symbol a Symbol object is representing. This may help SymPy to more efficiently manipulate or simplify analytical expressions. We can add on various assumptions that narrow down the possible properties of a symbol by adding optional keyword arguments to the symbol-creating functions, such as Symbol. Table 3-1 summarizes a selection of frequently used assumptions that can be associated with a Symbol class instance. For example, if we have a mathematical variable y that is known to be a real number, we can use the real=True keyword argument when creating the corresponding symbol instance. We can verify that SymPy indeed recognizes that the symbol is real by using the is_real attribute of the Symbol class:

```
In [5]: y = sympy.Symbol("y", real=True)
In [6]: y.is_real
Out[6]: True
```

If, on the other hand, we were to use is_real to query the previously defined symbol x, which was not explicitly specified to real, and therefore can represent both real and nonreal variables, we get None as a result:

```
In [7]: x.is_real is None
Out[7]: True
```

Note that the is_real returns True if the symbol is known to be real, False if the symbol is known to not be real, and None if it is not known if the symbol is real or not. Other attributes (see Table 3-1) for querying assumptions on Symbol objects work in the same way. For an example that demonstrates a symbol for which the is_real attribute is False, consider

```
In [8]: sympy.Symbol("z", imaginary=True).is_real
Out[8]: False
```

Table 3-1. *Selected Assumptions and Their Corresponding Keyword for* Symbol *Objects. For a complete list, see the docstring for* sympy. Symbol

Assumption Keyword Arguments	Attributes	Description
real, imaginary	is_real, is_imaginary	Specify that a symbol represents a real or imaginary number.
positive, negative	is_positive, is_negative	Specify that a symbol is positive or negative.
integer	is_integer	The symbol represents an integer.
odd, even	is_odd, is_even	The symbol represents an odd or even integer.
prime	is_prime	The symbol is a prime number and therefore also an integer.
finite, infinite	is_finite, is_infinite	The symbol represents a quantity that is finite or infinite.

Among the assumptions in Table 3-1, the most important ones to explicitly specify when creating new symbols are real and positive. When applicable, adding these assumptions to symbols can frequently help SymPy to simplify various expressions further than otherwise possible. Consider the following simple example:

```
In  [9]: x = sympy.Symbol("x")
In [10]: y = sympy.Symbol("y", positive=True)
In [11]: sympy.sqrt(x ** 2)
Out[11]: √x²
In [12]: sympy.sqrt(y ** 2)
Out[12]: y
```

Here we have created two symbols, x and y, and computed the square root of the square of that symbol using the SymPy function sympy.sqrt. If nothing is known about the symbol in the computation, then no simplification can be done. If, on the other hand, the symbol is known to be representing a positive number, then obviously $\sqrt{y^2} = y$, and SymPy correctly recognizes this in the latter example.

When working with mathematical symbols that represent integers, rather than say real numbers, it is also useful to explicitly specify this when creating the corresponding SymPy symbols, using, for example, the integer=True, or even=True or odd=True, if applicable. This may also allow SymPy to analytically simplify certain expressions and function evaluations, such as in the following example:

```
In [13]: n1 = sympy.Symbol("n")
In [13]: n2 = sympy.Symbol("n", integer=True)
In [13]: n3 = sympy.Symbol("n", odd=True)
In [14]: sympy.cos(n1 * pi)
Out[14]: cos(πn)
In [15]: sympy.cos(n2 * pi)
Out[15]: (-1)n
In [16]: sympy.cos(n3 * pi)
Out[16]: -1
```

To formulate a nontrivial mathematical problem, it is often necessary to define a large number of symbols. Using Symbol to specify each symbol one-by-one may become tedious, and for convenience, SymPy contains a function sympy.symbols for creating multiple symbols in one function call. This function takes a comma-separated string of symbol names, as well as an arbitrary set of keyword arguments (which apply to all the symbols), and it returns a tuple of newly created symbols. Using Python's tuple unpacking syntax together with a call to sympy.symbols is a convenient way to create symbols:

```
In [17]: a, b, c = sympy.symbols("a, b, c", negative=True)
In [18]: d, e, f = sympy.symbols("d, e, f", positive=True)
```

Numbers

The purpose of representing mathematical symbols as Python objects is to use them in expression trees that represent mathematical expressions. To be able to do this, we also need to represent other mathematical objects, such as numbers, functions, and constants. In this section we look at SymPy's classes for representing number objects. All of these classes have many methods and attributes shared with instances of Symbol, which allows us to treat symbols and numbers on equal footing when representing expressions.

For example, in the previous section, we saw that Symbol instances have attributes for querying properties of symbol objects, such as is_real. We need to be able to use the same attributes for all types of objects, including for example numbers such as integers and floating-point numbers, when manipulating symbolic expressions in SymPy. For this reason, we cannot directly use the built-in Python objects for integers, int, and floating-point numbers, float, and so on. Instead, SymPy provides the classes sympy.Integer and sympy.Float for representing integers and floating-point numbers within the SymPy framework. This distinction is important to be aware of when working with SymPy, but fortunately we rarely need to concern ourselves with creating objects of type sympy. Integer and sympy.Float to representing specific numbers, since SymPy automatically promotes Python numbers to instances of these classes when they occur in SymPy expressions. However, to demonstrate this difference between Python's built-in number types and the corresponding types in SymPy, in the following example, we explicitly create instances of sympy.Integer and sympy.Float and use some of their attributes to query their properties:

```
In [19]: i = sympy.Integer(19)
In [20]: type(i)
Out[20]: sympy.core.numbers.Integer
In [21]: i.is_Integer, i.is_real, i.is_odd
Out[21]: (True, True, True)
In [22]: f = sympy.Float(2.3)
In [23]: type(f)
Out[23]: sympy.core.numbers.Float
In [24]: f.is_Integer, f.is_real, f.is_odd
Out[24]: (False, True, False)
```

Tip   We can cast instances of sympy.Integer and sympy.Float back to Python built-in types using the standard type casting int(i) and float(f).

To create a SymPy representation of a number, or in general, an arbitrary expression, we can also use the `sympy.sympify` function. This function takes a wide range of inputs and derives a SymPy compatible expression, and it eliminates the need for specifying explicitly what types of objects are to be created. For the simple case of number input, we can use

```
In [25]: i, f = sympy.sympify(19), sympy.sympify(2.3)
In [26]: type(i), type(f)
Out[26]: (sympy.core.numbers.Integer, sympy.core.numbers.Float)
```

Integer

In the previous section, we have already used the `Integer` class to represent integers. It's worth pointing out that there is a difference between a `Symbol` instance with the assumption `integer=True` and an instance of `Integer`. While the `Symbol` with `integer=True` represents some integer, the `Integer` instance represents a specific integer. For both cases, the `is_integer` attribute is `True`, but there is also an attribute `is_Integer` (note the capital I), which is only `True` for `Integer` instances. In general, attributes with names in the form `is_Name` indicate if the object is of type `Name`, and attributes with names in the form `is_name` indicate if the object is known to satisfy the condition name. Thus, there is also an attribute `is_Symbol` that is `True` for `Symbol` instances.

```
In [27]: n = sympy.Symbol("n", integer=True)
In [28]: n.is_integer, n.is_Integer, n.is_positive, n.is_Symbol
Out[28]: (True, False, None, True)
In [29]: i = sympy.Integer(19)
In [30]: i.is_integer, i.is_Integer, i.is_positive, i.is_Symbol
Out[30]: (True, True, True, False)
```

Integers in SymPy are arbitrary precision, meaning that they have no fixed lower and upper bounds, which is the case when representing integers with a specific bit size, as, for example, in NumPy. It is therefore possible to work with very large numbers, as shown in the following examples:

```
In [31]: i ** 50
Out[31]: 8663234049605954426644038200675212212900743262211018069459689001
```

```
In [32]: sympy.factorial(100)
Out[32]: 9332621544394415268169923885626670049071596826438162146859296389521
7599993229915608941463976156518286253697920827223758251185210916864000000000
0000000000000000
```

Float

We have also already encountered the type sympy.Float in the previous sections.
Like Integer, Float is arbitrary precision, in contrast to Python's built-in float type
and the float types in NumPy. This means that a Float can represent a float with an
arbitrary number of decimals. When a Float instance is created using its constructor,
there are two arguments: the first argument is a Python float or a string representing a
floating-point number, and the second (optional) argument is the precision (number
of significant decimal digits) of the Float object. For example, it is well known
that the real number 0.3 cannot be represented exactly as a normal fixed bit-size
floating-point number, and when printing 0.3 to 20 significant digits, it is displayed as
0.29999999999999998897698. The SymPy Float object can represent the real number
0.3 without the limitations of floating-point numbers:

```
In [33]: "%.25f" % 0.3   # create a string representation with 25 decimals
Out[33]: '0.299999999999999988897698'
In [34]: sympy.Float(0.3, 25)
Out[34]: 0.299999999999999988897698
In [35]: sympy.Float('0.3', 25)
Out[35]: 0.3
```

However, note that to correctly represent 0.3 as a Float object, it is necessary to
initialize it from a string '0.3' rather than the Python float 0.3, which already contains a
floating-point error.

Rational

A rational number is a fraction p/q of two integers, the numerator p and the
denominator q. SymPy represents this type of numbers using the sympy.Rational class.
Rational numbers can be created explicitly, using sympy.Rational and the numerator
and denominator as arguments:

```
In [36]: sympy.Rational(11, 13)
```
$$Out[36]: \frac{11}{13}$$

or they can be a result of a simplification carried out by SymPy. In either case, arithmetic operations between rational and integers remain rational.

```
In [37]: r1 = sympy.Rational(2, 3)
In [38]: r2 = sympy.Rational(4, 5)
In [39]: r1 * r2
```
$$Out[39]: \frac{8}{15}$$
```
In [40]: r1 / r2
```
$$Out[40]: \frac{5}{6}$$

Constants and Special Symbols

SymPy provides predefined symbols for various mathematical constants and special objects, such as the imaginary unit i and infinity. These are summarized in Table 3-2, together with their corresponding symbols in SymPy. Note in particular that the imaginary unit is written as I in SymPy.

Table 3-2. *Selected Mathematical Constants and Special Symbols and Their Corresponding Symbols in SymPy*

Mathematical Symbol	SymPy Symbol	Description
π	sympy.pi	Ratio of the circumference to the diameter of a circle.
e	sympy.E	The base of the natural logarithm, $e = \exp(1)$.
γ	sympy.EulerGamma	Euler's constant.
i	sympy.I	The imaginary unit.
∞	sympy.oo	Infinity.

Functions

In SymPy, objects that represent functions can be created with sympy.Function. Like Symbol, this Function object takes a name as the first argument. SymPy distinguishes between defined and undefined functions, as well as between applied and unapplied functions. Creating a function with Function results in an undefined (abstract) and unapplied function, which has a name but cannot be evaluated because its expression, or body, is not defined. Such a function can represent an arbitrary function of arbitrary numbers of input variables, since it also has not yet been applied to any particular symbols or input variables. An unapplied function can be applied to a set of input symbols that represent the domain of the function by calling the function instance with those symbols as arguments.[1] The result is still an unevaluated function, but one that has been applied to the specified input variables, and therefore has a set of dependent variables. As an example of these concepts, consider the following code listing where we create an undefined function f, which we apply to the symbol x, and another function g which we directly apply to the set of symbols x, y, z:

```
In [41]: x, y, z = sympy.symbols("x, y, z")
In [42]: f = sympy.Function("f")
In [43]: type(f)
Out[43]: sympy.core.function.UndefinedFunction
In [44]: f(x)
Out[44]: f(x)
In [45]: g = sympy.Function("g")(x, y, z)
In [46]: g
Out[46]: g(x,y,z)
In [47]: g.free_symbols
Out[47]: {x,y,z}
```

Here we have also used the property free_symbols, which returns a set of unique symbols contained in a given expression (in this case the applied undefined function g), to demonstrate that an applied function indeed is associated with a specific set of input symbols. This will be important later in this chapter, for example, when we consider derivatives of abstract functions. One important application of undefined functions is for specifying differential equations or, in other words, when an equation for the function is known, but the function itself is unknown.

In contrast to undefined functions, a defined function is one that has a specific implementation and can be numerically evaluated for all valid input parameters. It is possible to define this type of function, for example, by subclassing sympy.Function,

---

[1]Here it is important to keep in mind the distinction between a Python function, or callable Python object such as sympy.Function, and the symbolic function that a sympy.Function class instance represents.

but in most cases it is sufficient to use the mathematical functions provided by SymPy. Naturally, SymPy has built-in functions for many standard mathematical functions that are available in the global SymPy namespace (see the module documentation for `sympy.functions.elementary`, `sympy.functions.combinatorial`, and `sympy.functions.special` and their subpackages for comprehensive lists of the numerous functions that are available, using the Python `help` function). For example, the SymPy function for the sine function is available as `sympy.sin` (with our import convention). Note that this is not a function in the Python sense of the word (it is, in fact, a subclass of `sympy.Function`), and it represents an unevaluated sin function that can be applied to a numerical value, a symbol, or an expression.

```
In [48]: sympy.sin
Out[48]: sympy.functions.elementary.trigonometric.sin
In [49]: sympy.sin(x)
Out[49]: sin(x)
In [50]: sympy.sin(pi * 1.5)
Out[50]: -1
```

When applied to an abstract symbol, such as x, the `sin` function remains unevaluated, but when possible it is evaluated to a numerical value, for example, when applied to a number or, in some cases, when applied to expressions with certain properties, as in the following example:

```
In [51]: n = sympy.Symbol("n", integer=True)
In [52]: sympy.sin(pi * n)
Out[52]: 0
```

A third type of function in SymPy is lambda functions, or anonymous functions, which do not have names associated with them, but do have a specific function body that can be evaluated. Lambda functions can be created with `sympy.Lambda`:

```
In [53]: h = sympy.Lambda(x, x**2)
In [54]: h
```
$$\text{Out}[54]: \left(x \mapsto x^2 \right)$$

```
In [55]: h(5)
Out[55]: 25
In [56]: h(1 + x)
```
$$\text{Out}[56]: (1 + x)^2$$

Expressions

The various symbols introduced in the previous sections are the fundamental building blocks required to express mathematical expressions. In SymPy, mathematical expressions are represented as trees where leaves are symbols and nodes are class instances that represent mathematical operations. Examples of these classes are Add, Mul, and Pow for basic arithmetic operators and Sum, Product, Integral, and Derivative for analytical mathematical operations. In addition, there are many other classes for mathematical operations, which we will see more examples of later in this chapter.

Consider, for example, the mathematical expression $1+2x^2+3x^3$. To represent this in SymPy, we only need to create the symbol x and then write the expression as Python code:

```
In [54]: x = sympy.Symbol("x")
In [55]: expr = 1 + 2 * x**2 + 3 * x**3
In [56]: expr
Out[56]: 3x³ + 2x² + 1
```

Here expr is an instance of Add, with the subexpressions 1, 2*x**2, and 3*x**3. The entire expression tree for expr is visualized in Figure 3-1. Note that we do not need to explicitly construct the expression tree, since it is automatically built up from the expression with symbols and operators. Nevertheless, to understand how SymPy works, it is important to know how expressions are represented.

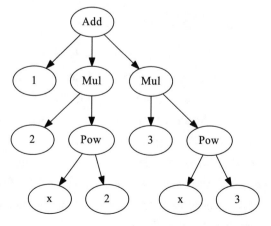

Figure 3-1. *Visualization of the expression tree for 1 + 2*x**2 + 3*x**3*

The expression tree can be traversed explicitly using the `args` attribute, which all SymPy operations and symbols provide. For an operator, the `args` attribute is a tuple of subexpressions that are combined with the rule implemented by the operator class. For symbols, the `args` attribute is an empty tuple, which signifies that it is a leaf in the expression tree. The following example demonstrates how the expression tree can be explicitly accessed:

```
In [57]: expr.args
Out[57]: (1,2x²,3x³)
In [58]: expr.args[1]
Out[58]: 2x²
In [59]: expr.args[1].args[1]
Out[59]: x²
In [60]: expr.args[1].args[1].args[0]
Out[60]: x
In [61]: expr.args[1].args[1].args[0].args
Out[61]: ()
```

In the basic use of SymPy, it is rarely necessary to explicitly manipulate expression trees, but when the methods for manipulating expressions that are introduced in the following section are not sufficient, it is useful to be able to implement functions of your own that traverse and manipulate the expression tree using the `args` attribute.

Manipulating Expressions

Manipulating expression trees is one of the main jobs for SymPy, and numerous functions are provided for different types of transformations. The general idea is that expression trees can be transformed between mathematically equivalent forms using simplification and rewrite functions. These functions generally do not change the expressions that are passed to the functions, but rather create a new expression that corresponds to the modified expression. Expressions in SymPy should thus be considered immutable objects (that cannot be changed). All the functions we consider in this section treat SymPy expressions as immutable objects and return new expression trees rather than modify expressions in place.

Simplification

The most desirable manipulation of a mathematical expression is to simplify it. This is perhaps and also the most ambiguous operation, since it is nontrivial to determine algorithmically if one expression appears simpler than another to a human being, and in general it is also not obvious which methods should be employed to arrive at a simpler expression. Nonetheless, black-box simplification is an important part of any CAS, and SymPy includes the function `sympy.simplify` that attempts to simplify a given expression using a variety of methods and approaches. The simplification function can also be invoked through the method `simplify`, as illustrated in the following example.

```
In [67]: expr = 2 * (x**2 - x) - x * (x + 1)
In [68]: expr
Out[68]: 2x² - x(x+1)−2x
In [69]: sympy.simplify(expr)
Out[69]: x(x−3)
In [70]: expr.simplify()
Out[70]: x(x−3)
In [71]: expr
Out[71]: 2x² - x(x+1)−2x
```

Note that here both `sympy.simplify(expr)` and `expr.simplify()` return new expression trees and leave the expression `expr` untouched, as mentioned earlier. In this example, the expression `expr` can be simplified by expanding the products, canceling terms, and then factoring the expression again. In general, `sympy.simplify` will attempt a variety of different strategies and will also simplify, for example, trigonometric and power expressions, as exemplified here:

```
In [72]: expr = 2 * sympy.cos(x) * sympy.sin(x)
In [73]: expr
Out[73]: 2 sin(x)cos(x)
In [74]: sympy.simplify(expr)
Out[74]: sin(2x)
```

and

```
In [75]: expr = sympy.exp(x) * sympy.exp(y)
In [76]: expr
Out[76]: exp(x)exp(y)
In [77]: sympy.simplify(expr)
Out[77]: exp(x+y)
```

Each specific type of simplification can also be carried out with more specialized functions, such as `sympy.trigsimp` and `sympy.powsimp`, for trigonometric and power simplifications, respectively. These functions only perform the simplification that their names indicate and leave other parts of an expression in its original form. A summary of simplification functions is given in Table 3-3. When the exact simplification steps are known, it is in general better to rely on the more specific simplification functions, since their actions are more well defined and less likely to change in future versions of SymPy. The `sympy.simplify` function, on the other hand, relies on heuristic approaches that may change in the future and, as a consequence, produce different results for a particular input expression.

Table 3-3. *Summary of Selected SymPy Functions for Simplifying Expressions*

Function	Description
sympy.simplify	Attempt various methods and approaches to obtain a simpler form of a given expression.
sympy.trigsimp	Attempt to simplify an expression using trigonometric identities.
sympy.powsimp	Attempt to simplify an expression using laws of powers.
sympy.compsimp	Simplify combinatorial expressions.
sympy.ratsimp	Simplify an expression by writing on a common denominator.

Expand

When the black-box simplification provided by `sympy.simplify` does not produce satisfying results, it is often possible to make progress by manually guiding SymPy using more specific algebraic operations. An important tool in this process is to expand expression in various ways. The function `sympy.expand` performs a variety of expansions, depending on the values of optional keyword arguments. By default the function distributes products over additions, into a fully expanded expression. For example, a product of the type $(x+1)(x+2)$ can be expanded to x^2+3x+2 using

```
In [78]: expr = (x + 1) * (x + 2)
In [79]: sympy.expand(expr)
Out[79]: x² + 3x + 2
```

Some of the available keyword arguments are mul=True for expanding products (as in the preceding example), trig=True for trigonometric expansions,

```
In [80]: sympy.sin(x + y).expand(trig=True)
Out[80]: sin(x)cos(y) + sin(y)cos(x)
```

log=True for expanding logarithms,

```
In [81]: a, b = sympy.symbols("a, b", positive=True)
In [82]: sympy.log(a * b).expand(log=True)
Out[82]: log(a) + log(b)
```

complex=True for separating real and imaginary parts of an expression,

```
In [83]: sympy.exp(I*a + b).expand(complex=True)
Out[83]: ieᵇ sin(a) + eᵇ cos(a)
```

and power_base=True and power_exp=True for expanding the base and the exponent of a power expression, respectively.

```
In [84]: sympy.expand((a * b)**x, power_base=True)
Out[84]: aˣbˣ
In [85]: sympy.exp((a-b)*x).expand(power_exp=True)
Out[85]: eⁱᵃˣe⁻ⁱᵇˣ
```

Calling the sympy.expand function with these keyword arguments set to True is equivalent to calling the more specific functions sympy.expand_mul, sympy.expand_trig, sympy.expand_log, sympy.expand_complex, sympy.expand_power_base, and sympy.expand_power_exp, respectively, but an advantage of the sympy.expand function is that several types of expansions can be performed in a single function call.

Factor, Collect, and Combine

A common use pattern for the sympy.expand function is to expand an expression, let SymPy cancel terms or factors, and then factor or combine the expression again. The sympy.factor function attempts to factor an expression as far as possible and is in some sense the opposite to sympy.expand with mul=True. It can be used to factor algebraic expressions, such as

```
In [86]: sympy.factor(x**2 - 1)
Out[86]: (x - 1)(x + 1)
In [87]: sympy.factor(x * sympy.cos(y) + sympy.sin(z) * x)
Out[87]: x(sin(x) + cos(y))
```

The inverse of the other types of expansions in the previous section can be carried out using sympy.trigsimp, sympy.powsimp, and sympy.logcombine, for example

```
In [90]: sympy.logcombine(sympy.log(a) - sympy.log(b))
```
$$\text{Out}[90]: \log\left(\frac{a}{b}\right)$$

When working with mathematical expressions, it is often necessary to have fine-grained control over factoring. The SymPy function sympy.collect factors terms that contain a given symbol or list of symbols. For example, $x+y+xyz$ cannot be completely factorized, but we can partially factor terms containing x or y:

```
In [89]: expr = x + y + x * y * z
In [90]: expr.collect(x)
Out[90]: x(yz + 1) + y
In [91]: expr.collect(y)
Out[91]: x + y(xz + 1)
```

By passing a list of symbols or expressions to the sympy.collect function or to the corresponding collect method, we can collect multiple symbols in one function call. Also, when using the method collect, which returns the new expression, it is possible to chain multiple method calls in the following way:

```
In [93]: expr = sympy.cos(x + y) + sympy.sin(x - y)
In [94]: expr.expand(trig=True).collect([sympy.cos(x),
    ...:                                  sympy.sin(x)]).collect(sympy.
         cos(y) - sympy.sin(y))
Out[95]: (sin(x) + cos(x))(-sin(y) + cos(y))
```

Apart, Together, and Cancel

The final type of mathematical simplification that we will consider here is the rewriting of fractions. The functions sympy.apart and sympy.together, which, respectively, rewrite a fraction as a partial fraction and combine partial fractions to a single fraction, can be used in the following way:

```
In [95]: sympy.apart(1/(x**2 + 3*x + 2), x)
```
$$Out[95]: -\frac{1}{x+2}+\frac{1}{x+1}$$
```
In [96]: sympy.together(1 / (y * x + y) + 1 / (1+x))
```
$$Out[96]: \frac{y+1}{y(x+1)}$$
```
In [97]: sympy.cancel(y / (y * x + y))
```
$$Out[97]: \frac{1}{x+1}$$

In the first example, we used sympy.apart to rewrite the expression $(x^2+3x+2)^{-1}$ as the partial fraction $-\frac{1}{x+2}+\frac{1}{x+1}$, and we used sympy.together to combine the sum of fractions $1/(yx+y)+1/(1+x)$ into an expression in the form of a single fraction. In this example we also used the function sympy.cancel to cancel shared factors between numerator and the denominator in the expression $y/(yx+y)$.

Substitutions

The previous sections have been concerned with rewriting expressions using various mathematical identities. Another frequently used form of manipulation of mathematical expressions is substitutions of symbols or subexpressions within an expression. For example, we may want to perform a variable substitution and replace the variable x with y or replace a symbol with another expression. In SymPy there are two methods for carrying out substitutions: subs and replace. Usually subs is the most suitable alternative, but in some cases replace provides a more powerful tool, which, for example, can make replacements based on wildcard expressions (see docstring for sympy.Symbol.replace for details).

In the most basic use of subs, the method is called in an expression, and the symbol or expression that is to be replaced (x) is given as the first argument, and the new symbol or the expression (y) is given as the second argument. The result is that all occurrences of x in the expression are replaced with y:

```
In [98]: (x + y).subs(x, y)
Out[98]: 2y
In [99]: sympy.sin(x * sympy.exp(x)).subs(x, y)
Out[99]: sin(ye^y)
```

Instead of chaining multiple subs calls when multiple substitutions are required, we can alternatively pass a dictionary as the first and only argument to subs that maps old symbols or expressions to new symbols or expressions:

```
In [100]: sympy.sin(x * z).subs({z: sympy.exp(y), x: y, sympy.sin: sympy.cos})
Out[100]: cos(ye^y)
```

A typical application of the subs method is to substitute numerical values in place of symbols, for numerical evaluation (see the following section for more details). A convenient way of doing this is to define a dictionary that translates the symbols to numerical values and pass this dictionary as the argument to the subs method. For example, consider

```
In [101]: expr = x * y + z**2 *x
In [102]: values = {x: 1.25, y: 0.4, z: 3.2}
In [103]: expr.subs(values)
Out[103]: 13.3
```

Numerical Evaluation

Even when working with symbolic mathematics, it is almost invariably sooner or later required to evaluate the symbolic expressions numerically, for example, when producing plots or concrete numerical results. A SymPy expression can be evaluated using either the sympy.N function or the evalf method of SymPy expression instances:

```
In [104]: sympy.N(1 + pi)
Out[104]: 4.14159265358979
In [105]: sympy.N(pi, 50)
Out[105]: 3.1415926535897932384626433832795028841971693993751
In [106]: (x + 1/pi).evalf(10)
Out[106]: x + 0.3183098862
```

Both sympy.N and the evalf method take an optional argument that specifies the number of significant digits to which the expression is to be evaluated, as shown in the previous example where SymPy's multiprecision float capabilities were leveraged to evaluate the value of π up to 50 digits.

When we need to evaluate an expression numerically for a range of input values, we could in principle loop over the values and perform successive evalf calls, for example

```
In [114]: expr = sympy.sin(pi * x * sympy.exp(x))
In [115]: [expr.subs(x, xx).evalf(3) for xx in range(0, 10)]
Out[115]: [0,0.774,0.642,0.722,0.944,0.205,0.974,0.977,-0.870,-0.695]
```

However, this method is rather slow, and SymPy provides a more efficient method for doing this operation using the function sympy.lambdify. This function takes a set of free symbols and an expression as arguments and generates a function that efficiently evaluates the numerical value of the expression. The produced function takes the same number of arguments as the number of free symbols passed as the first argument to sympy.lambdify.

```
In [109]: expr_func = sympy.lambdify(x, expr)
In [110]: expr_func(1.0)
Out[110]: 0.773942685266709
```

Note that the function expr_func expects numerical (scalar) values as arguments, so we cannot, for example, pass a symbol as an argument to this function; it is strictly for numerical evaluation. The expr_func created in the previous example is a scalar function and is not directly compatible with vectorized input in the form of NumPy arrays, as discussed in Chapter 2. However, SymPy is also able to generate functions that are NumPy-array aware: by passing the optional argument 'numpy' as the third argument to sympy.lambdify SymPy creates a vectorized function that accepts NumPy arrays as input. This is in general an efficient way to numerically evaluate symbolic expressions[2] for a large number of input parameters. The following code exemplifies how the SymPy expression expr is converted into a NumPy-array aware vectorized function that can be efficiently evaluated:

```
In [111]: expr_func = sympy.lambdify(x, expr, 'numpy')
In [112]: import numpy as np
In [113]: xvalues = np.arange(0, 10)
In [114]: expr_func(xvalues)
Out[114]: array([ 0.        ,  0.77394269,  0.64198244,  0.72163867,
          0.94361635,
                 0.20523391,  0.97398794,  0.97734066, -0.87034418,
          -0.69512687])
```

This method for generating data from SymPy expressions is useful for plotting and many other data-oriented applications.

Calculus

So far we have looked at how to represent mathematical expression in SymPy and how to perform basic simplification and transformation of such expressions. With this framework in place, we are now ready to explore symbolic calculus, or analysis, which is a cornerstone in applied mathematics and has a great number of applications throughout science and engineering. The central concept in calculus is the change of functions as input variables are varied, as quantified with derivatives and differentials, and accumulations of functions over ranges of input, as quantified by integrals. In this section we look at how to compute derivatives and integrals of functions in SymPy.

---

[2]See also the ufuncity from the sympy.utilities.autowrap module and the theano_function from the sympy.printing.theanocode module. These provide similar functionality as sympy.lambdify, but use different computational backends.

Derivatives

The derivative of a function describes its rate of change at a given point. In SymPy we can calculate the derivative of a function using `sympy.diff` or alternatively by using the `diff` method of SymPy expression instances. The argument to these functions is a symbol, or a number of symbols, with respect to which the function or the expression is to be derived. To represent the first-order derivative of an abstract function $f(x)$ with respect to x, we can do

```
In [119]: f = sympy.Function('f')(x)
In [120]: sympy.diff(f, x)                # equivalent to f.diff(x)
```
$$Out[120]: \frac{d}{dx} f(x)$$

and to represent higher-order derivatives, all we need to do is to repeat the symbol x in the argument list in the call to `sympy.diff` or, equivalently, specify an integer as an argument following a symbol, which defines the number of times the expression should be derived with respect to that symbol:

```
In [117]: sympy.diff(f, x, x)
```
$$Out[117]: \frac{d^2}{dx^2} f(x)$$

```
In [118]: sympy.diff(f, x, 3)   # equivalent to sympy.diff(f, x, x, x)
```
$$Out[118]: \frac{d^3}{dx^3} f(x)$$

This method is readily extended to multivariate functions:

```
In [119]: g = sympy.Function('g')(x, y)
In [120]: g.diff(x, y)               # equivalent to sympy.diff(g, x, y)
```
$$Out[120]: \frac{\partial^2}{\partial x \partial y} g(x,y)$$

```
In [121]: g.diff(x, 3, y, 2)    # equivalent to sympy.diff(g, x, x, x, y, y)
```
$$Out[121]: \frac{\partial^5}{\partial x^3 \partial y^2} g(x,y)$$

These examples so far only involve formal derivatives of undefined functions. Naturally, we can also evaluate the derivatives of defined functions and expressions, which result in new expressions that correspond to the evaluated derivatives. For example, using `sympy.diff` we can easily evaluate derivatives of arbitrary mathematical expressions, such as polynomials:

```
In [122]: expr = x**4 + x**3 + x**2 + x + 1
In [123]: expr.diff(x)
Out[123]: 4x³ + 3x² + 2x+1
In [124]: expr.diff(x, x)
Out[124]: 2(6x² + 3x + 1)
In [125]: expr = (x + 1)**3 * y ** 2 * (z - 1)
In [126]: expr.diff(x, y, z)
Out[126]: 6y(x + 1)²
```

as well as trigonometric and other more complicated mathematical expressions:

```
In [127]: expr = sympy.sin(x * y) * sympy.cos(x / 2)
In [128]: expr.diff(x)
```

$$\text{Out[128]: } y\cos\left(\frac{x}{2}\right)\cos(xy) - \frac{1}{2}\sin\left(\frac{x}{2}\right)\sin(xy)$$

```
In [129]: expr = sympy.special.polynomials.hermite(x, 0)
In [130]: expr.diff(x).doit()
```

$$\text{Out[130]: } \frac{2^x \sqrt{\pi}\, \text{polygamma}\left(0, -\frac{x}{2} + \frac{1}{2}\right)}{2\Gamma\left(-\frac{x}{2} + \frac{1}{2}\right)} + \frac{2^x \sqrt{\pi}\, \log(2)}{\Gamma\left(-\frac{x}{2} + \frac{1}{2}\right)}$$

Derivatives are usually relatively easy to compute, and `sympy.diff` should be able to evaluate the derivative of most standard mathematical functions defined in SymPy.

Note that in these examples, calling `sympy.diff` on an expression directly results in a new expression. If we rather want to symbolically represent the derivative of a definite expression, we can create an instance of the class `sympy.Derivative`, passing the expression as the first argument, followed by the symbols with respect to the derivative that is to be computed:

```
In [131]: d = sympy.Derivative(sympy.exp(sympy.cos(x)), x)
In [132]: d
```

Out[132]: $\dfrac{d}{dx}e^{\cos(x)}$

This formal representation of a derivative can then be evaluated by calling the doit method on the sympy.Derivative instance:

```
In [133]: d.doit()
Out[133]: -e^{cos(x)} sin(x)
```

This pattern of delayed evaluation is reoccurring throughout SymPy, and full control of when a formal expression is evaluated to a specific result is useful in many situations, in particular with expressions that can be simplified or manipulated while represented as a formal expression rather than after it has been evaluated.

Integrals

In SymPy, integrals are evaluated using the function sympy.integrate, and formal integrals can be represented using sympy.Integral (which, as in the case with sympy.Derivative, can be explicitly evaluated by calling the doit method). Integrals come in two basic forms: definite and indefinite, where a definite integral has specified integration limits and can be interpreted as an area or volume, while an indefinite integral does not have integration limits and denotes the antiderivative (inverse of the derivative of a function). SymPy handles both indefinite and definite integrals using the sympy.integrate function.

If the sympy.integrate function is called with only an expression as an argument, the indefinite integral is computed. On the other hand, a definite integral is computed if the sympy.integrate function additionally is passed a tuple in the form (x, a, b), where x is the integration variable and a and b are the integration limits. For a single-variable function $f(x)$, the indefinite and definite integrals are therefore computed using

```
In [135]: a, b, x, y = sympy.symbols("a, b, x, y")
     ...: f = sympy.Function("f")(x)
In [136]: sympy.integrate(f)
Out[136]: ∫ f(x)dx
In [137]: sympy.integrate(f, (x, a, b))
```

Out[137]: $\int_{a}^{b} f(x)dx$

and when these methods are applied to explicit functions, the integrals are evaluated accordingly:

```
In [138]: sympy.integrate(sympy.sin(x))
Out[138]: -cos(x)
In [139]: sympy.integrate(sympy.sin(x), (x, a, b))
Out[139]: cos(a) - cos(b)
```

Definite integrals can also include limits that extend from negative infinity, and/or to positive infinite, using SymPy's symbol for infinity oo:

```
In [139]: sympy.integrate(sympy.exp(-x**2), (x, 0, oo))
```
$$Out[139]: \frac{\sqrt{\pi}}{2}$$
```
In [140]: a, b, c = sympy.symbols("a, b, c", positive=True)
In [141]: sympy.integrate(a * sympy.exp(-((x-b)/c)**2), (x, -oo, oo))
```
$Out[141]: \sqrt{\pi}\,ac$

Computing integrals symbolically is in general a difficult problem, and SymPy will not be able to give symbolic results for any integral you can come up with. When SymPy fails to evaluate an integral, an instance of sympy.Integral, representing the formal integral, is returned instead.

```
In [142]: sympy.integrate(sympy.sin(x * sympy.cos(x)))
```
$Out[142]: \int \sin(x\,\cos(x))dx$

Multivariable expressions can also be integrated with sympy.integrate. In the case of an indefinite integral of a multivariable expression, the integration variable has to be specified explicitly:

```
In [140]: expr = sympy.sin(x*sympy.exp(y))
In [141]: sympy.integrate(expr, x)
```
$Out[141]: -e^{-y}\cos(xe^{y})$

```
In [142]: expr = (x + y)**2
In [143]: sympy.integrate(expr, x)
```
$$Out[143]: \frac{x^3}{3}+x^2y+xy^2$$

By passing more than one symbol, or multiple tuples that contain symbols and their integration limits, we can carry out multiple integrations:

In [144]: sympy.integrate(expr, x, y)

Out[144]: $\dfrac{x^3 y}{3} + \dfrac{x^2 y^2}{2} + \dfrac{x y^3}{3}$

In [145]: sympy.integrate(expr, (x, 0, 1), (y, 0, 1))

Out[145]: $\dfrac{7}{6}$

Series

Series expansions are an important tool in many disciplines in computing. With a series expansion, an arbitrary function can be written as a polynomial, with coefficients given by the derivatives of the function at the point around which the series expansion is made. By truncating the series expansion at some order n, the nth order approximation of the function is obtained. In SymPy, the series expansion of a function or an expression can be computed using the function sympy.series or the series method available in SymPy expression instances. The first argument to sympy.series is a function or expression that is to be expanded, followed by a symbol with respect to which the expansion is to be computed (it can be omitted for single-variable expressions and function). In addition, it is also possible to request a particular point around which the series expansions are to be performed (using the x0 keyword argument, with default x0=0), specifying the order of the expansion (using the n keyword argument, with default n=6) and specifying the direction from which the series is computed, i.e., from below or above x0 (using the dir keyword argument, which defaults to dir='+').

For an undefined function f(x), the expansion up to sixth order around x0=0 is computed using

In [147]: x, y = sympy.symbols("x, y")
In [148]: f = sympy.Function("f")(x)
In [149]: sympy.series(f, x)

Out[149]: $f(0) + x \dfrac{d}{dx} f(x) \Big|_{x=0} + \dfrac{x^2}{2} \dfrac{d^2}{dx^2} f(x) \Big|_{x=0} + \dfrac{x^3}{6} \dfrac{d^3}{dx^3} f(x) \Big|_{x=0}$

$+ \dfrac{x^4}{24} \dfrac{d^4}{dx^4} f(x) \Big|_{x=0} + \dfrac{x^5}{120} \dfrac{d^5}{dx^5} f(x) \Big|_{x=0} + \mathcal{O}(x^6)$

To change the point around which the function is expanded, we specify the x0 argument as in the following example:

```
In [147]: x0 = sympy.Symbol("{x_0}")
In [151]: f.series(x, x0, n=2)
```

Out[151]: $f(x_0)+(x-x_0)\dfrac{d}{d\xi_1}f(\xi_1)\Big|_{\xi_1=x_0}+\mathcal{O}\left((x-x_0)^2;x\to x_0\right)$

Here we also specified n=2, to request a series expansion with only terms up to and including the second order. Note that the errors due to the truncated terms are represented by the order object $\mathcal{O}(\ldots)$. The order object is useful for keeping track of the order of an expression when computing with series expansions, such as multiplying or adding different expansions. However, for concrete numerical evolution, it is necessary to remove the order term from the expression, which can be done using the method removeO:

```
In [152]: f.series(x, x0, n=2).removeO()
```

Out[152]: $f(x_0)+(x-x_0)\dfrac{d}{d\xi_1}f(\xi_1)\Big|_{\xi_1=x_0}$

While the expansions shown in the preceding text were computed for an unspecified function f(x), we can naturally also compute the series expansions of specific functions and expressions, and in those cases we obtain specific evaluated results. For example, we can easily generate the well-known expansions of many standard mathematical functions:

```
In [153]: sympy.cos(x).series()
```

Out[153]: $1-\dfrac{x^2}{2}+\dfrac{x^4}{24}+\mathcal{O}\left(x^6\right)$

```
In [154]: sympy.sin(x).series()
```

Out[154]: $x-\dfrac{x^3}{6}+\dfrac{x^5}{120}+\mathcal{O}\left(x^6\right)$

```
In [155]: sympy.exp(x).series()
```

Out[155]: $1+x+\dfrac{x^2}{2}+\dfrac{x^3}{6}+\dfrac{x^4}{24}+\dfrac{x^5}{120}+\mathcal{O}\left(x^6\right)$

```
In [156]: (1/(1+x)).series()
```

Out[156]: $1-x+x^2-x^3+x^4-x^5+\mathcal{O}\left(x^6\right)$

as well as arbitrary expressions of symbols and functions, which in general can also be multivariable functions:

```
In [157]: expr = sympy.cos(x) / (1 + sympy.sin(x * y))
In [158]: expr.series(x, n=4)
```

$$\text{Out}[158]: 1 - xy + x^2\left(y^2 - \frac{1}{2}\right) + x^3\left(-\frac{5y^3}{6} + \frac{y}{2}\right) + \mathcal{O}(x^4)$$

```
In [159]: expr.series(y, n=4)
```

$$\text{Out}[159]: \cos(x) - xy\cos(x) + x^2 y^2 \cos(x) - \frac{5x^3 y^3 \cos(x)}{6} + \mathcal{O}(y^4)$$

Limits

Another important tool in calculus is limits, which denotes the value of a function as one of its dependent variables approaches a specific value or as the value of the variable approaches negative or positive infinity. An example of a limit is one of the definitions of the derivative:

$$\frac{d}{dx} f(x) = \lim_{h \to 0} \frac{f(x+h) - f(x)}{h}.$$

While limits are more of a theoretical tool and do not have as many practical applications as, say, series expansions, it is still useful to be able to compute limits using SymPy. In SymPy, limits can be evaluated using the sympy.limit function, which takes an expression, a symbol it depends on, as well as the value that the symbol approaches in the limit. For example, to compute the limit of the function sin(x)/x, as the variable x goes to zero, that is, $\lim_{x \to 0} \sin(x)/x$, we can use

```
In [161]: sympy.limit(sympy.sin(x) / x, x, 0)
Out[161]: 1
```

Here we obtained the well-known answer 1 for this limit. We can also use sympy.limit to compute symbolic limits, which can be illustrated by computing derivatives using the previous definition (although it is of course more efficient to use sympy.diff):

```
In [162]: f = sympy.Function('f')
     ...: x, h = sympy.symbols("x, h")
In [163]: diff_limit = (f(x + h) - f(x))/h
```

```
In [164]: sympy.limit(diff_limit.subs(f, sympy.cos), h, 0)
Out[164]: -sin(x)
In [165]: sympy.limit(diff_limit.subs(f, sympy.sin), h, 0)
Out[165]: cos(x)
```

A more practical example of using limits is to find the asymptotic behavior as a function, for example, as its dependent variable approaches infinity. As an example, consider the function $f(x) = (x^2 - 3x)/(2x - 2)$, and suppose we are interested in the large-x dependence of this function. It will be in the form $f(x) \rightarrow px+q$, and we can compute p and q using sympy.limit as in the following:

```
In [166]: expr = (x**2 - 3*x) / (2*x - 2)
In [167]: p = sympy.limit(expr/x, x, sympy.oo)
In [168]: q = sympy.limit(expr - p*x, x, sympy.oo)
In [169]: p, q
```
$$\text{Out[169]: } \left(\frac{1}{2}, -1 \right)$$

Thus, the asymptotic behavior of $f(x)$ as x becomes large is the linear function $f(x) \rightarrow x/2 - 1$.

Sums and Products

Sums and products can be symbolically represented using the SymPy classes sympy. Sum and sympy.Product. They both take an expression as their first argument, and as a second argument, they take a tuple of the form (n, n1, n2), where n is a symbol and n1 and n2 are the lower and upper limits for the symbol n, in the sum or product, respectively. After sympy.Sum or sympy.Product objects have been created, they can be evaluated using the doit method:

```
In [171]: n = sympy.symbols("n", integer=True)
In [172]: x = sympy.Sum(1/(n**2), (n, 1, oo))
In [173]: x
```
$$\text{Out[173]: } \sum_{n=1}^{\infty} \frac{1}{n^2}$$

```
In [174]: x.doit()
```
$$\text{Out[174]: } \frac{\pi^2}{6}$$

```
In [175]: x = sympy.Product(n, (n, 1, 7))
In [176]: x
```

$$\text{Out[176]:} \quad \prod_{n=1}^{7} n$$

```
In [177]: x.doit()
Out[177]: 5040
```

Note that the sum in the previous example was specified with an upper limit of infinity. It is therefore clear that this sum was not evaluated by explicit summation, but was rather computed analytically. SymPy can evaluate many summations of this type, including when the summand contains symbolic variables other than the summation index, such as in the following example:

```
In [178]: x = sympy.Symbol("x")
In [179]: sympy.Sum((x)**n/(sympy.factorial(n)), (n, 1, oo)).doit().
          simplify()
Out[179]: e^x - 1
```

Equations

Equation solving is a fundamental part of mathematics with applications in nearly every branch of science and technology, and it is therefore immensely important. SymPy can solve a wide variety of equations symbolically, although many equations cannot be solved analytically even in principle. If an equation, or a system of equations, can be solved analytically, there is a good chance that SymPy is able to find the solution. If not, numerical methods might be the only option.

In its simplest form, equation solving involves a single equation with a single unknown variable, and no additional parameters: for example, finding the value of x that satisfies the second-degree polynomial equation $x^2+2x-3 = 0$. This equation is of course easy to solve, even by hand, but in SymPy we can use the function `sympy.solve` to find the solutions of x that satisfy this equation using

```
In [170]: x = sympy.Symbol("x")
In [171]: sympy.solve(x**2 + 2*x - 3)
Out[171]: [-3,1]
```

That is, the solutions are x=-3 and x=1. The argument to the `sympy.solve` function is an expression that will be solved under the assumption that it equals zero. When this expression contains more than one symbol, the variable that is to be solved for must be given as a second argument. For example,

```
In [172]: a, b, c = sympy.symbols("a, b, c")
In [173]: sympy.solve(a * x**2 + b * x + c, x)
```

$$\text{Out}[173]: \left[\frac{1}{2a}\left(-b+\sqrt{-4ac+b^2}\right), -\frac{1}{2a}\left(b+\sqrt{-4ac+b^2}\right)\right]$$

and in this case the resulting solutions are expressions that depend on the symbols representing the parameters in the equation.

The `sympy.solve` function is also capable of solving other types of equations, including trigonometric expressions:

```
In [174]: sympy.solve(sympy.sin(x) - sympy.cos(x), x)
```

$$\text{Out}[174]: \left[-\frac{3\pi}{4},\right]$$

and equations whose solution can be expressed in terms of special functions:

```
In [180]: sympy.solve(sympy.exp(x) + 2 * x, x)
```

$$\text{Out}[180]: \left[-\text{LambertW}\left(\frac{1}{2}\right)\right]$$

However, when dealing with general equations, even for a univariate case, it is not uncommon to encounter equations that are not solvable algebraically or which SymPy is unable to solve. In these cases SymPy will return a formal solution, which can be evaluated numerically if needed, or raise an error if no method is available for that particular type of equation:

```
In [176]: sympy.solve(x**5 - x**2 + 1, x)
Out[176]: [RootOf(x⁵ - x² + 1,0), RootOf(x⁵ - x² + 1,1), RootOf(x⁵ - x² + 1,2),
          RootOf(x⁵ - x² + 1,3), RootOf(x⁵ - x² + 1,4)]
In [177]: sympy.solve(sympy.tan(x) + x, x)
---------------------------------------------------------------------------
NotImplementedError                       Traceback (most recent call last)
...
NotImplementedError: multiple generators [x, tan(x)] No algorithms are
implemented to solve equation x + tan(x)
```

Solving a system of equations for more than one unknown variable in SymPy is a straightforward generalization of the procedure used for univariate equations. Instead of passing a single expression as the first argument to `sympy.solve`, a list of expressions that represents the system of equations is used, and in this case the second argument should be a list of symbols to solve for. For example, the following two examples demonstrate how to solve two systems that are linear and nonlinear equations in x and y, respectively:

```
In [178]: eq1 = x + 2 * y - 1
     ...: eq2 = x - y + 1
In [179]: sympy.solve([eq1, eq2], [x, y], dict=True)
```

$$\text{Out[179]:} \quad \left[\left\{ x:-\frac{1}{3}, y:\frac{2}{3} \right\} \right]$$

```
In [180]: eq1 = x**2 - y
     ...: eq2 = y**2 - x
In [181]: sols = sympy.solve([eq1, eq2], [x, y], dict=True)
In [182]: sols
Out[182]:
```

$$\left[\{x:0, y:0\}, \{x:1, y:1\}, \left\{ x:-\frac{1}{2}+\frac{\sqrt{3}i}{2}, y:-\frac{1}{2}-\frac{\sqrt{3}i}{2} \right\}, \left\{ x:\frac{\left(1-\sqrt{3}i\right)^2}{4}, y:-\frac{1}{2}+\frac{\sqrt{3}i}{2} \right\} \right]$$

Note that in both these examples, the function `sympy.solve` returns a list where each element represents a solution to the equation system. The optional keyword argument `dict=True` was also used, to request that each solution is returned in dictionary format, which maps the symbols that have been solved for to their values. This dictionary can conveniently be used in, for example, calls to `subs`, as in the following code that checks that each solution indeed satisfies the two equations:

```
In [183]: [eq1.subs(sol).simplify() == 0 and eq2.subs(sol).simplify() == 0
              for sol in sols]
Out[183]: [True, True, True, True]
```

Linear Algebra

Linear algebra is another fundamental branch of mathematics with important applications throughout scientific and technical computing. It concerns vectors, vector spaces, and linear mappings between vector spaces, which can be represented as matrices. In SymPy we can represent vectors and matrices symbolically using the `sympy.Matrix` class, whose elements can in turn be represented by numbers, symbols, or even arbitrary symbolic expressions. To create a matrix with numerical entries, we can, as in the case of NumPy arrays in Chapter 2, pass a Python list to `sympy.Matrix`:

```
In [184]: sympy.Matrix([1, 2])
Out[184]: ⎡1⎤
          ⎣2⎦
In [185]: sympy.Matrix([[1, 2]])
Out[185]: [1 2]
In [186]: sympy.Matrix([[1, 2], [3, 4]])
Out[186]: ⎡1  2⎤
          ⎣3  4⎦
```

As this example demonstrates, a single list generates a column vector, while a matrix requires a nested list of values. Note that unlike the multidimensional arrays in NumPy discussed in Chapter 2, the `sympy.Matrix` object in SymPy is only for up to two-dimensional arrays, i.e., vectors and matrices. Another way of creating new `sympy.Matrix` objects is to pass as arguments the number of rows, the number of columns, and a function that takes the row and column index as arguments and returns the value of the corresponding element:

```
In [187]: sympy.Matrix(3, 4, lambda m, n: 10 * m + n)
Out[187]: ⎡ 0   1   2   3⎤
          ⎢10  11  12  13⎥
          ⎣20  21  22  23⎦
```

The most powerful features of SymPy's matrix objects, which distinguish it from, for example, NumPy arrays, are of course that its elements themselves can be symbolic expressions. For example, an arbitrary 2x2 matrix can be represented with a symbolic variable for each of its elements:

```
In [188]: a, b, c, d = sympy.symbols("a, b, c, d")
In [189]: M = sympy.Matrix([[a, b], [c, d]])
In [190]: M
Out[190]:
```
$$\begin{bmatrix} a & b \\ c & d \end{bmatrix}$$

and such matrices can naturally also be used in computations, which then remains parameterized with the symbolic values of the elements. The usual arithmetic operators are implemented for matrix objects, but note that multiplication operator * in this case denotes matrix multiplication:

```
In [191]: M * M
Out[191]:
```
$$\begin{bmatrix} a^2 + bc & ab + bd \\ ac + cd & bc + d^2 \end{bmatrix}$$
```
In [192]: x = sympy.Matrix(sympy.symbols("x_1, x_2"))
In [194]: M * x
Out[194]:
```
$$\begin{bmatrix} ax_1 + bx_2 \\ cx_1 + dx_2 \end{bmatrix}$$

In addition to arithmetic operations, many standard linear algebra operations on vectors and matrices are also implemented as SymPy functions and methods of the sympy.Matrix class. Table 3-4 gives a summary of the frequently used linear algebra-related functions (see the docstring for sympy.Matrix for a complete list), and SymPy matrices can also be used in an element-oriented fashion using indexing and slicing operations that closely resemble those discussed for NumPy arrays in Chapter 2.

As an example of a problem that can be solved with symbolic linear algebra using SymPy, but which is not directly solvable with purely numerical approaches, consider the following parameterized linear equation system:

$$x + p\,y = b_1,$$

$$q\,x + y = b_2,$$

which we would like to solve for the unknown variables x and y. Here p, q, b_1, and b_2 are unspecified parameters. On matrix form, we can write these two equations as

$$\begin{pmatrix} 1 & p \\ q & 1 \end{pmatrix}\begin{pmatrix} x \\ y \end{pmatrix} = \begin{pmatrix} b_1 \\ b_2 \end{pmatrix}.$$

With purely numerical methods, we would have to choose particular values of the parameters p and q before we could begin to solve this problem, for example, using an LU factorization (or by computing the inverse) of the matrix on the left-hand side of the equation. With a symbolic computing approach, on the other hand, we can directly proceed with computing the solution, as if we carried out the calculation analytically by hand. With SymPy, we can simply define symbols for the unknown variables and parameters and set up the required matrix objects:

```
In [195]: p, q = sympy.symbols("p, q")
In [196]: M = sympy.Matrix([[1, p], [q, 1]])
In [203]: M
```
$$\text{Out}[203]: \begin{bmatrix} 1 & p \\ q & 1 \end{bmatrix}$$
```
In [197]: b = sympy.Matrix(sympy.symbols("b_1, b_2"))
In [198]: b
```
$$\text{Out}[198]: \begin{bmatrix} b_1 & b_2 \end{bmatrix}$$

and then use, for example, the LUsolve method to solve the linear equation system:

```
In [199]: x = M.LUsolve(b)
In [200]: x
```
$$\text{Out}[200]: \begin{bmatrix} b_1 - \dfrac{p(-b_1q + b_2)}{-pq + 1} \\ \dfrac{-b_1q + b_2}{-pq + 1} \end{bmatrix}$$

Alternatively, we could also directly compute the inverse of the matrix M and multiply it with the vector b:

```
In [201]: x = M.inv() * b
In [202]: x
```

$$\text{Out[202]:} \begin{bmatrix} b_1\left(\dfrac{pq}{-pq+1}+1\right)-\dfrac{b_2 p}{-pq+1} \\[4mm] -\dfrac{b_1 q}{-pq+1}+\dfrac{b_2}{-pq+1} \end{bmatrix}$$

Table 3-4. *Selected Functions and Methods for Operating on SymPy Matrices*

Function/Method	Description
transpose/T	Compute the transpose of a matrix.
adjoint/H	Compute the adjoint of a matrix.
trace	Compute the trace (sum of diagonal elements) of a matrix.
det	Compute the determinant of a matrix.
inv	Compute the inverse of a matrix.
LUdecomposition	Compute the LU decomposition of a matrix.
LUsolve	Solve a linear system of equations in the form $Mx = b$, for the unknown vector x, using LU factorization.
QRdecomposition	Compute the QR decomposition of a matrix.
QRsolve	Solve a linear system of equations in the form $Mx = b$, for the unknown vector x, using QR factorization.
diagonalize	Diagonalize a matrix M, such that it can be written in the form $D = P^{-1}MP$, where D is diagonal.
norm	Compute the norm of a matrix.
nullspace	Compute a set of vectors that span the null space of a Matrix.
rank	Compute the rank of a matrix.
singular_values	Compute the singular values of a matrix.
solve	Solve a linear system of equations in the form $Mx = b$.

However, computing the inverse of a matrix is more difficult than performing the LU factorization, so if solving the equation $Mx = b$ is the objective, as it was here, then using LU factorization is more efficient. This becomes particularly noticeable for larger equation systems. With both methods considered here, we obtain a symbolic expression for the solution that is trivial to evaluate for any parameter values, without having to recompute the solution. This is the strength of symbolic computing and an example of how it sometimes can excel over direct numerical computing. The example considered here could of course also be solved easily by hand, but as the number of equations and unspecified parameters grow, analytical treatment by hand quickly becomes prohibitively lengthy and tedious. With the help of a computer algebra system such as SymPy, we can push the limits of which problems that can be treated analytically.

Summary

This chapter introduced computer-assisted symbolic computing using Python and the SymPy library. Although analytical and numerical techniques are often considered separately, it is a fact that analytical methods underpin everything in computing and are essential in developing algorithms and numerical methods. Whether analytical mathematics is carried by hand or using a computer algebra system such as SymPy, it is an essential tool for computational work. The approach that I would like to encourage is therefore the following: Analytical and numerical methods are closely intertwined, and it is often worthwhile to start analyzing a computational problem with analytical and symbolic methods. When such methods turn out to be unfeasible, it is time to resort to numerical methods. However, by directly applying numerical methods to a problem, before analyzing it analytically, it is likely that one ends up solving a more difficult computational problem than is really necessary.

Further Reading

For a quick and short introduction to SymPy, see, for example, Lamy (2013). The official SymPy documentation also provides a great tutorial for getting started with SymPy, which is available at `http://docs.sympy.org/latest/tutorial/index.html`.

Reference

Lamy, R. (2013). *Instant SymPy Starter.* Mumbai: Packt.

CHAPTER 4

Plotting and Visualization

Visualization is a universal tool for investigating and communicating results of computational studies, and it is hardly an exaggeration to say that the end product of nearly all computations – be it numeric or symbolic – is a plot or a graph of some sort. It is when visualized in graphical form that knowledge and insights can be most easily gained from computational results. Visualization is therefore a tremendously important part of the workflow in all fields of computational studies.

In the scientific computing environment for Python, there are a number of high-quality visualization libraries. The most popular general-purpose visualization library is Matplotlib, which mainly focuses on generating static publication-quality 2D and 3D graphs. Many other libraries focus on niche areas of visualization. A few prominent examples are Bokeh (`http://bokeh.pydata.org`) and Plotly (`http://plot.ly`), which both primarily focus on interactivity and web connectivity, Seaborn (`http://stanford.edu/~mwaskom/software/seaborn`) which is a high-level plotting library which targets statistical data analysis and which is based on the Matplotlib library, and the Mayavi library (`http://docs.enthought.com/mayavi/mayavi`) for high-quality 3D visualization, which uses the venerable VTK software (`http://www.vtk.org`) for heavy-duty scientific visualization. It is also worth noting that other VTK-based visualization software, such as ParaView (`www.paraview.org`), is scriptable with Python and can also be controlled from Python applications. In the 3D visualization space, there are also more recent players, such as VisPy (`http://vispy.org`), which is an OpenGL-based 2D and 3D visualization library with great interactivity and connectivity with browser-based environments, such as the Jupyter Notebook.

The visualization landscape in the scientific computing environment for Python is vibrant and diverse, and it provides ample options for various visualization needs. In this chapter we focus on exploring traditional scientific visualization in Python using the Matplotlib library. With traditional visualization, I mean plots and figures that are commonly used to visualize results and data in scientific and technical disciplines, such as line plots, bar plots, contour plots, colormap plots, and 3D surface plots.

© Robert Johansson 2019
R. Johansson, *Numerical Python*, https://doi.org/10.1007/978-1-4842-4246-9_4

Matplotlib   Matplotlib is a Python library for publication-quality 2D and 3D graphics, with support for a variety of different output formats. At the time of writing, the latest version is 2.2.2. More information about Matplotlib is available at the project's web site `www.matplotlib.org`. This web site contains detailed documentation and an extensive gallery that showcases the various types of graphs that can be generated using the Matplotlib library, together with the code for each example. This gallery is a great source of inspiration for visualization ideas, and I highly recommend exploring Matplotlib by browsing this gallery.

There are two common approaches to creating scientific visualizations: using a graphical user interface to manually build up graphs and using a programmatic approach where the graphs are created with code. Both approaches have their advantages and disadvantages. In this chapter we will take the programmatic approach, and we will explore how to use the Matplotlib API to create graphs and control every aspect of their appearance. The programmatic approach is a particularly suitable method for creating graphics for scientific and technical applications and in particular for creating publication-quality figures. An important part of the motivation for this is that programmatically created graphics can guarantee consistency across multiple figures, can be made reproducible, and can easily be revised and adjusted without having to redo potentially lengthy and tedious procedures in a graphical user interface.

Importing Modules

Unlike most Python libraries, Matplotlib actually provides multiple entry points into the library, with different application programming interfaces (APIs). Specifically, it provides a stateful API and an object-oriented API, both provided by the module `matplotlib.pyplot`. I strongly recommend to only use the object-oriented approach, and the remainder of this chapter will solely focus on this part of Matplotlib.[1]

---

[1]Although the stateful API may be convenient and simple for small examples, the readability and maintainability of code written for stateful APIs scale poorly, and the context-dependent nature of such code makes it hard to rearrange or reuse. I therefore recommend to avoid it altogether and to only use the object-oriented API.

To use the object-oriented Matplotlib API, we first need to import its Python modules. In the following we will assume that Matplotlib is imported using the following standard convention:

```
In [1]: %matplotlib inline
In [2]: import matplotlib as mpl
In [3]: import matplotlib.pyplot as plt
In [4]: from mpl_toolkits.mplot3d.axes3d import Axes3D
```

The first line is assuming that we are working in an IPython environment and more specifically in the Jupyter Notebook or the IPython QtConsole. The IPython magic command `%matplotlib inline` configures the Matplotlib to use the "inline" backend, which results in the created figures being displayed directly in, for example, the Jupyter Notebook, rather than in a new window. The statement `import matplotlib as mpl` imports the main Matplotlib module, and the import statement `import matplotlib.pyplot as plt`, is for convenient access to the submodule `matplotlib.pyplot` that provides the functions that we will use to create new Figure instances.

Throughout this chapter we also make frequent use of the NumPy library, and as in Chapter 2, we assume that NumPy is imported using

```
In [5]: import numpy as np
```

and we also use the SymPy library, imported as:

```
In [6]: import sympy
```

Getting Started

Before we delve deeper into the details of how to create graphics with Matplotlib, we begin here with a quick example of how to create a simple but typical graph. We also cover some of the fundamental principles of the Matplotlib library, to build up an understanding for how graphics can be produced with the library.

A graph in Matplotlib is structured in terms of a Figure instance and one or more Axes instances within the figure. The Figure instance provides a canvas area for drawing, and the Axes instances provide coordinate systems that are assigned to fixed regions of the total figure canvas; see Figure 4-1.

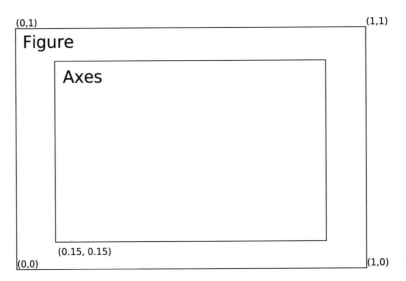

Figure 4-1. *Illustration of the arrangement of a Matplotlib* Figure *instance and an* Axes *instance. The* Axes *instance provides a coordinate system for plotting, and the* Axes *instance itself is assigned to a region within the figure canvas. The figure canvas has a simple coordinate system where (0, 0) is the lower-left corner and (1,1) is the upper-right corner. This coordinate system is only used when placing elements, such as an* Axes, *directly on the figure canvas.*

A Figure can contain multiple Axes instances, for example, to show multiple panels in a figure or to show insets within another Axes instance. An Axes instance can manually be assigned to an arbitrary region of a figure canvas, or, alternatively, Axes instances can be automatically added to a figure canvas using one of several layout managers provided by Matplotlib. The Axes instance provides a coordinate system that can be used to plot data in a variety of plot styles, including line graphs, scatter plots, bar plots, and many other styles. In addition, the Axes instance also determines how the coordinate axes are displayed, for example, with respect to the axis labels, ticks and tick labels, and so on. In fact, when working with Matplotlib's object-oriented API, most functions that are needed to tune the appearance of a graph are methods of the Axes class.

As a simple example for getting started with Matplotlib, say that we would like to graph the function $y(x) = x^3+5x^2+10$, together with its first and second derivatives, over the range $x \in [-5, 2]$. To do this we first create NumPy arrays for the x range and then compute the three functions we want to graph. When the data for the graph is prepared, we need to create Matplotlib Figure and Axes instances, then use the plot method of the Axes instance to plot the data, and set basic graph properties such as x and y axis labels, using the set_xlabel and set_ylabel methods and generating a legend using the legend method. These steps are carried out in the following code, and the resulting graph is shown in Figure 4-2.

```
In [7]: x = np.linspace(-5, 2, 100)
   ...: y1 = x**3 + 5*x**2 + 10
   ...: y2 = 3*x**2 + 10*x
   ...: y3 = 6*x + 10
   ...:
   ...: fig, ax = plt.subplots()
   ...: ax.plot(x, y1, color="blue", label="y(x)")
   ...: ax.plot(x, y2, color="red", label="y'(x)")
   ...: ax.plot(x, y3, color="green", label="y"(x)")
   ...: ax.set_xlabel("x")
   ...: ax.set_ylabel("y")
   ...: ax.legend()
```

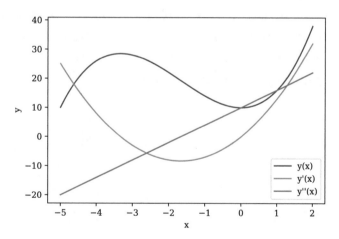

Figure 4-2. *Example of a simple graph created with Matplotlib*

Here we used the plt.subplots function to generate Figure and Axes instances. This function can be used to create grids of Axes instances within a newly created Figure instance, but here it was merely used as a convenient way of creating a Figure and an Axes instance in one function call. Once the Axes instance is available, note that all the remaining steps involve calling methods of this Axes instance. To create the actual graphs, we use ax.plot, which takes as first and second arguments NumPy arrays with numerical data for the x and y values of the graph, and it draws a line connecting these data points. We also used the optional `color` and `label` keyword arguments to specify the color of each line and assign a text label to each line that is used in the legend. These few lines of code are enough to generate the graph we set out to produce, but as a bare minimum, we should also set labels on the x and y axes and if suitable add a legend for the curves we have plotted. The axis labels are set with `ax.set_xlabel` and `ax.set_ylabel` methods, which takes as argument a text string with the corresponding label. The legend is added using the `ax.legend` method, which does not require any arguments in this case since we used the `label` keyword argument when plotting the curves.

These are the typical steps required to create a graph using Matplotlib. While this graph, Figure 4-2, is complete and fully functional, there is certainly room for improvements in many aspects of its appearance. For example, to meet publication or production standards, we may need to change the font and the fontsize of the axis labels, the tick labels, and the legend, and we should probably move the legend to a part of the graph where it does not interfere with the curves we are plotting. We might even want to change the number of axis ticks and label and add annotations and additional help lines to emphasize certain aspects of the graph and so on. With a few changes along these lines, the figure may, for example, appear like in Figure 4-3, which is considerably more presentable. In the remainder of this chapter, we look at how to fully control the appearance of the graphics produced using Matplotlib.

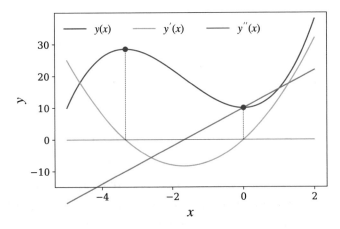

Figure 4-3. *Revised version of Figure 4-2*

Interactive and Noninteractive Modes

The Matplotlib library is designed to work well with many different environments and platforms. As such, the library does not only contain routines for generating graphs, but it also contains support for displaying graphs in different graphical environments. To this end, Matplotlib provides *backends* for generating graphics in different formats (e.g., PNG, PDF, Postscript, and SVG) and for displaying graphics in a graphical user interface using a variety of different widget toolkits (e.g., Qt, GTK, wxWidgets, and Cocoa for Mac OS X) that are suitable for different platforms.

Which backend to use can be selected in that Matplotlib resource file,[2] or using the function mpl.use, which must be called right after importing matplotlib, before importing the matplotlib.pyplot module. For example, to select the Qt4Agg backend, we can use

```
import matplotlib as mpl
mpl.use('qt4agg')
import matplotlib.pyplot as plt
```

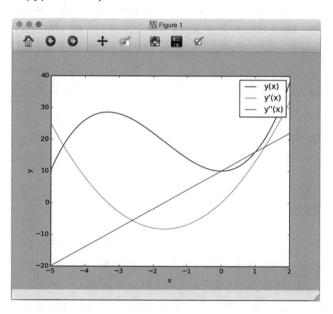

Figure 4-4. *A screenshot of the Matplotlib graphical user interface for displaying figures, using the Qt4 backend on Mac OS X. The detailed appearance varies across platforms and backends, but the basic functionality is the same.*

---

[2]The Matplotlib resource file, matplotlibrc, can be used to set default values of many Matplotlib parameters, including which backend to use. The location of the file is platform dependent. For details, see http://matplotlib.org/users/customizing.html.

The graphical user interface for displaying Matplotlib figures, as shown in Figure 4-4, is useful for interactive use with Python script files or the IPython console, and it allows to interactively explore figures, for example, by zooming and panning. When using an interactive backend, which displays the figure in a graphical user interface, it is necessary to call the function `plt.show` to get the window to appear on the screen. By default, the `plt.show` call will hang until the window is closed. For a more interactive experience, we can activate *interactive mode* by calling the function `plt.ion`. This instructs Matplotlib to take over the GUI event loop and show a window for a figure as soon as it is created, returning the control flow to the Python or IPython interpreter. To have the changes to a figure take effect, we need to issue a redraw command using the function `plt.draw`. We can deactivate the interactive mode using the function `plt.ioff`, and we can use the function `mpl.is_interactive` to check if Matplotlib is in interactive or noninteractive mode.

While the interactive graphical user interfaces have unique advantages, when working the Jupyter Notebook or Qtconsole, it is often more convenient to display Matplotlib-produced graphics embedded directly in the notebook. This behavior is activated using the IPython command `%matplotlib inline`, which activates the "inline backend" provided by IPython. This configures Matplotlib to use a noninteractive backend to generate graphics images, which are then displayed as static images in, for example, the Jupyter Notebook. The IPython "inline backend" for Matplotlib can be fine-tuned using the IPython `%config` command. For example, we can select the output format for the generated graphics using the `InlineBackend.figure_format` option,[3] which, for example, we can set to 'svg' to generate SVG graphics rather than PNG files:

```
In [8]: %matplotlib inline
In [9]: %config InlineBackend.figure_format='svg'
```

With this approach the interactive aspect of the graphical user interface is lost (e.g., zooming and panning), but embedding the graphics directly in the notebook has many other advantages. For example, keeping the code that was used to generate a figure together with the resulting figure in the same document eliminates the need for rerunning the code to display a figure, and interactive nature of the Jupyter Notebook itself replaces some of the interactivity of Matplotlib's graphical user interface.

---

[3]For Max OS X users, `%config InlineBackend.figure_format='retina'` is another useful option, which improves the quality of the Matplotlib graphics when viewed on retina displays.

When using the IPython inline backend, it is not necessary to use `plt.show` and `plt.draw`, since the IPython rich display system is responsible for triggering the rendering and the displaying of the figures. In this book, I will assume that code examples are executed in the Jupyter Notebooks, and the calls to the function `plt.show` are therefore not in the code examples. When using an interactive backend, it is necessary to add this function call at the end of each example.

Figure

As introduced in the previous section, the `Figure` object is used in Matplotlib to represent a graph. In addition to providing a canvas on which, for example, `Axes` instances can be placed, the `Figure` object also provides methods for performing actions on figures, and it has several attributes that can be used to configure the properties of a figure.

A `Figure` object can be created using the function `plt.figure`, which takes several optional keyword arguments for setting figure properties. In particular, it accepts the `figsize` keyword argument, which should be assigned to a tuple on the form (`width`, `height`), specifying the width and height of the figure canvas in inches. It can also be useful to specify the color of the figure canvas by setting the `facecolor` keyword argument.

Once a `Figure` is created, we can use the `add_axes` method to create a new `Axes` instance and assign it to a region on the figure canvas. The `add_axes` takes one mandatory argument: a list containing the coordinates of the lower-left corner and the width and height of the `Axes` in the figure canvas coordinate system, on the format (`left`, `bottom`, `width`, `height`).[4] The coordinates and the width and height of the `Axes` object are expressed as fractions of total canvas width and height; see Figure 4-1. For example, an `Axes` object that completely fills the canvas corresponds to (`0`, `0`, `1`, `1`), but this leaves no space for axis labels and ticks. A more practical size could be (`0.1`, `0.1`, `0.8`, `0.8`), which corresponds to a centered `Axes` instance that covers 80% of the width and height of the canvas. The `add_axes` method takes a large number of keyword arguments for setting properties of the new `Axes` instance. These will be described in more detail later in this chapter, when we discuss the `Axes` object in depth. However,

---

[4]An alternative to passing a coordinate and size tuple to `add_axes` is to pass an already existing `Axes` instance.

one keyword argument that is worth to emphasize here is `facecolor`, with which we can assign a background color for the Axes object. Together with the `facecolor` argument of `plt.figure`, this allows selecting colors of both the canvas and the regions covered by Axes instances.

With the Figure and Axes objects obtained from `plt.figure` and `fig.add_axes`, we have the necessary preparations to start plotting data using the methods of the Axes objects. For more details on this, see the next section of this chapter. However, once the required plots have been created, there are more methods in the Figure objects that are important in the graph creation workflow. For example, to set an overall figure title, we can use `suptitle`, which takes a string with the title as argument. To save a figure to a file, we can use the `savefig` method. This method takes a string with the output filename as first argument, as well as several optional keyword arguments. By default, the output file format will be determined from the file extension of the filename argument, but we can also specify the format explicitly using the `format` argument. The available output formats depend on which Matplotlib backend is used, but commonly available options are PNG, PDF, EPS, and SVG formats. The resolution of the generated image can be set with the `dpi` argument. DPI stands for "dots per inch," and since the figure size is specified in inches using the `figsize` argument, multiplying these numbers gives the output image size in pixels. For example, with `figsize=(8, 6)` and `dpi=100`, the size of the generated image is 800x600 pixels. The `savefig` method also takes some arguments that are similar to those of the `plt.figure` function, such as the `facecolor` argument. Note that even though the `facecolor` argument is used with `plt.figure`, it also needs to be specified with `savefig` for it to apply to the generated image file. Finally, the figure canvas can also be made transparent using the `transparent=True` argument to `savefig`. The following code listing illustrates these techniques, and the result is shown in Figure 4-5.

```
In [10]: fig = plt.figure(figsize=(8, 2.5), facecolor="#f1f1f1")
    ...:
    ...: # axes coordinates as fractions of the canvas width and height
    ...: left, bottom, width, height = 0.1, 0.1, 0.8, 0.8
    ...: ax = fig.add_axes((left, bottom, width, height),
        facecolor="#e1e1e1")
    ...:
    ...: x = np.linspace(-2, 2, 1000)
    ...: y1 = np.cos(40 * x)
```

```
...: y2 = np.exp(-x**2)
...:
...: ax.plot(x, y1 * y2)
...: ax.plot(x, y2, 'g')
...: ax.plot(x, -y2, 'g')
...: ax.set_xlabel("x")
...: ax.set_ylabel("y")
...:
...: fig.savefig("graph.png", dpi=100, facecolor="#f1f1f1")
```

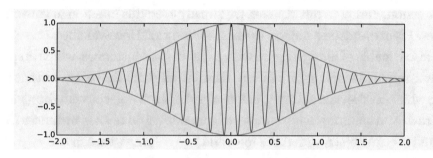

Figure 4-5. *Graph showing the result of setting the size of a figure with* figsize,
adding a new Axes instance with add_axes, *setting the background colors of the*
Figure *and Axes objects using* facecolor, *and finally saving the figure to file using*
savefig

Axes

The Figure object introduced in the previous section provides the backbone of a
Matplotlib graph, but all the interesting content is organized within or around Axes
instances. We have already encountered Axes objects on a few occasions earlier in this
chapter. The Axes object is central to most plotting activities with the Matplotlib library.
It provides the coordinate system in which we can plot data and mathematical functions,
and in addition it contains the axis objects that determine where the axis labels and the
axis ticks are placed. The functions for drawing different types of plots are also methods
of this Axes class. In this section we first explore different types of plots that can be drawn
using Axes methods and how to customize the appearance of the x and y axes and the
coordinate systems used with an Axes object.

We have seen how new Axes instances can be added to a figure explicitly using the add_axes method. This is a flexible and powerful method for placing Axes objects at arbitrary positions, which has several important applications, as we will see later in the chapter. However, for most common use-cases, it is tedious to specify explicitly the coordinates of the Axes instances within the figure canvas. This is especially true when using multiple panels of Axes instances within a figure, for example, in a grid layout. Matplotlib provides several different Axes layout managers, which create and place Axes instances within a figure canvas following different strategies. Later in this chapter, we look into more detail of how to use such layout managers. However, to facilitate the forthcoming examples, we here briefly look at one of these layout managers: the plt. subplots function. Earlier in this chapter, we already used this function to conveniently generate new Figure and Axes objects in one function call. However, the plt.subplots function is also capable of filling a figure with a grid of Axes instances, which is specified using the first and the second arguments, or alternatively with the nrows and ncols arguments, which, as the names imply, create a grid of Axes objects, with the given number of rows and columns. For example, to generate a grid of Axes instances in a newly created Figure object, with three rows and two columns, we can use

```
fig, axes = plt.subplots(nrows=3, ncols=2)
```

Here, the function plt.subplots returns a tuple (fig, axes), where fig is a Figure instance and axes is a NumPy array of size (nrows, ncols), in which each element is an Axes instance that has been appropriately placed in the corresponding figure canvas. At this point we can also specify that columns and/or rows should share x and y axes, using the sharex and sharey arguments, which can be set to True or False.

The plt.subplots function also takes two special keyword arguments fig_kw and subplot_kw, which are dictionaries with keyword arguments that are used when creating the Figure and Axes instances, respectively. This allows us to set and retain full control of the properties of the Figure and Axes objects with plt.subplots in a similar way as when directly using plt.figure and the make_axes method.

Plot Types

Effective scientific and technical visualization of data requires a wide variety of graphing techniques. Matplotlib implements many types of plotting techniques as methods of the Axes object. For example, in the previous examples, we have already used the plot method, which draws curves in the coordinate system provided by the Axes object.

In the following sections, we explore some of Matplotlib's plotting functions in more depth by using these functions in example graphs. A summary of commonly used 2D plot functions is shown in Figure 4-6. Other types of graphs, such as color maps and 3D graphs, are discussed later in this chapter. All plotting functions in Matplotlib expect data as NumPy arrays as input, typically as arrays with x and y coordinates as the first and second arguments. For details, see the docstrings for each method shown in Figure 4-6, using, for example, `help(plt.Axes.bar)`.

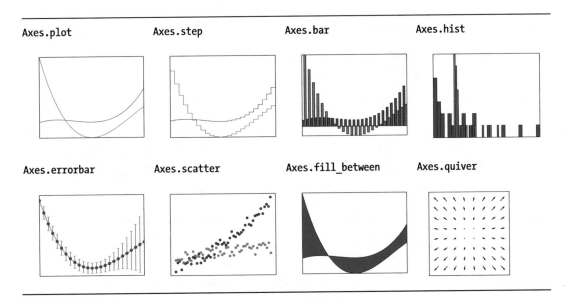

Figure 4-6. *Overview of selected 2D graph types. The name of the* Axes *method for generating each type of graph is shown together with the corresponding graph.*

Line Properties

The most basic type of plot is the simple line plot. It may, for example, be used to depict the graph of a univariate function or to plot data as a function of a control variable. In line plots, we frequently need to configure properties of the lines in the graph, for example, the line width, line color, and line style (solid, dashed, dotted, etc.). In Matplotlib we set these properties with keyword arguments to the plot methods, such as plot, step, and bar. A few of these graph types are shown in Figure 4-6. Many of the plot methods have their own specific arguments, but basic properties such as colors and line width are shared among most plotting methods. These basic properties and the corresponding keyword arguments are summarized in Table 4-1.

Table 4-1. *Basic Line Properties and Their Corresponding Argument Names for Use with the Matplotlib Plotting Methods*

Argument	Example Values	Description
`color`	A color specification can be a string with a color name, such as "red," "blue," etc., or a RGB color code on the form "#aabbcc."	A color specification.
`alpha`	Float number between 0.0 (completely transparent) and 1.0 (completely opaque).	The amount of transparency.
`linewidth, lw`	Float number.	The width of a line.
`linestyle, ls`	"-" – solid "--" – dashed ":" – dotted ".-" – dash-dotted	The style of the line, i.e., whether the line is to be drawn as a solid line or if it should be, for example, dotted or dashed.
`marker`	+, 0, * = cross, circle, star s = square . = small dot 1, 2, 3, 4, ... = triangle-shaped symbols with different angles.	Each data point, whether or not it is connected with adjacent data points, can be represented with a marker symbol as specified with this argument.
`markersize`	Float number.	The marker size.
`markerfacecolor`	Color specification (see in the preceding text).	The fill color for the marker.
`markeredgewidth`	Float number.	The line width of the marker edge.
`markeredgecolor`	Color specification (see above).	The marker edge color.

To illustrate the use of these properties and arguments, consider the following code, which draws horizontal lines with various values of the line width, line style, marker symbol, color, and size. The resulting graph is shown in Figure 4-7.

```
In [11]: x = np.linspace(-5, 5, 5)
    ...: y = np.ones_like(x)
    ...:
    ...: def axes_settings(fig, ax, title, ymax):
    ...:     ax.set_xticks([])
    ...:     ax.set_yticks([])
    ...:     ax.set_ylim(0, ymax+1)
    ...:     ax.set_title(title)
    ...:
    ...: fig, axes = plt.subplots(1, 4, figsize=(16,3))
    ...:
    ...: # Line width
    ...: linewidths = [0.5, 1.0, 2.0, 4.0]
    ...: for n, linewidth in enumerate(linewidths):
    ...:     axes[0].plot(x, y + n, color="blue", linewidth=linewidth)
    ...: axes_settings(fig, axes[0], "linewidth", len(linewidths))
    ...:
    ...: # Line style
    ...: linestyles = ['-', '-.', ':']
    ...: for n, linestyle in enumerate(linestyles):
    ...:     axes[1].plot(x, y + n, color="blue", lw=2, linestyle=linestyle)
    ...: # custom dash style
    ...: line, = axes[1].plot(x, y + 3, color="blue", lw=2)
    ...: length1, gap1, length2, gap2 = 10, 7, 20, 7
    ...: line.set_dashes([length1, gap1, length2, gap2])
    ...: axes_settings(fig, axes[1], "linetypes", len(linestyles) + 1)

    ...: # marker types
    ...: markers = ['+', 'o', '*', 's', '.', '1', '2', '3', '4']
    ...: for n, marker in enumerate(markers):
    ...:     # lw = shorthand for linewidth, ls = shorthand for linestyle
    ...:     axes[2].plot(x, y + n, color="blue", lw=2, ls='*',
    ...:     marker=marker)
    ...: axes_settings(fig, axes[2], "markers", len(markers))
    ...:
```

149

```
...: # marker size and color
...: markersizecolors = [(4, "white"), (8, "red"), (12, "yellow"),
     (16, "lightgreen")]
...: for n, (markersize, markerfacecolor) in enumerate
     (markersizecolors):
...:     axes[3].plot(x, y + n, color="blue", lw=1, ls='-',
...:                  marker='o', markersize=markersize,
...:                  markerfacecolor=markerfacecolor,
                      markeredgewidth=2)
...: axes_settings(fig, axes[3], "marker size/color", len
     (markersizecolors))
```

Figure 4-7. *Graphs showing the result of setting the line properties line width, line style, marker type and marker size, and color*

In practice, using different colors, line widths and line styles are important tools for making a graph easily readable. In a graph with a large number of lines, we can use a combination of colors and line style to make each line uniquely identifiable, for example, via a legend. The line width property is best used to give emphasis to important lines. Consider the following example, where the function $\sin(x)$ is plotted together with its first few series expansions around $x = 0$, as shown in Figure 4-8.

```
In [12]: # a symbolic variable for x, and a numerical array with specific
         values of x
    ...: sym_x = sympy.Symbol("x")
    ...: x = np.linspace(-2 * np.pi, 2 * np.pi, 100)
    ...:
    ...: def sin_expansion(x, n):
    ...:     """
    ...:     Evaluate the nth order Taylor. series expansion
```

```
...:        of sin(x) for the numerical values in the array x.
...:        """
...:        return sympy.lambdify(sym_x, sympy.sin(sym_x).series(n=n+1).
           removeO(), 'numpy')(x)
...:
...: fig, ax = plt.subplots()
...:
...: ax.plot(x, np.sin(x), linewidth=4, color="red", label='exact')
...:
...: colors = ["blue", "black"]
...: linestyles = [':', '-.', '--']
...: for idx, n in enumerate(range(1, 12, 2)):
...:        ax.plot(x, sin_expansion(x, n), color=colors[idx // 3],
...:                linestyle=linestyles[idx % 3], linewidth=3,
...:                label="order %d approx." % (n+1))
...:
...:  ax.set_ylim(-1.1, 1.1)
...: ax.set_xlim(-1.5*np.pi, 1.5*np.pi)
...:
...: # place a legend outsize of the Axes
...: ax.legend(bbox_to_anchor=(1.02, 1), loc=2, borderaxespad=0.0)
...: # make room for the legend to the right of the Axes
...: fig.subplots_adjust(right=.75)
```

Figure 4-8. *Graph for sin(x) together with its Taylor series approximation of the few lowest orders*

Legends

A graph with multiple lines may often benefit from a legend, which displays a label along each line type somewhere within the figure. As we have seen in the previous example, a legend may be added to an Axes instance in a Matplotlib figure using the legend method. Only lines with assigned labels are included in the legend (to assign a label to a line, use the label argument of, for example, Axes.plot). The legend method accepts a large number of optional arguments. See help(plt.legend) for details. Here we emphasize a few of the more useful arguments. In the example in the previous section, we used the loc argument, which allows to specify where in the Axes area the legend is to be added: loc=1 for upper-right corner, loc=2 for upper-left corner, loc=3 for the lower-left corner, and loc=4 for lower-right corner, as shown in Figure 4-9.

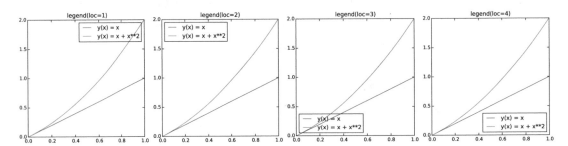

Figure 4-9. *Legend at different positions within an Axes instance, specified using the loc argument of the method legend*

In the example of the previous section, we also used the bbox_to_anchor, with which help the legend can be placed at an arbitrary location within the figure canvas. The bbox_to_anchor argument takes the value of a tuple on the form (x, y), where x and y are the *canvas coordinates* within the Axes object. That is, the point (0, 0) corresponds to the lower-left corner, and (1, 1) corresponds to the upper-right corner. Note that x and y can be smaller than 0 and larger than 1 in this case, which indicates that the legend is to be placed outside the Axes area, as was used in the previous section.

By default all lines in the legend are shown in a vertical arrangement. Using the ncols argument, it is possible to split the legend labels into multiple columns, as illustrated in Figure 4-10.

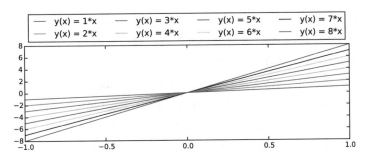

Figure 4-10. *Legend displayed outside the* Axes *object and shown with four columns instead of the single one, here using* ax.legend(ncol=4, loc=3, bbox_ to_anchor=(0, 1))

Text Formatting and Annotations

Text labels, titles, and annotations are important components in most graphs, and having full control of, for example, the font types and fontsizes that are used to render such texts is a basic requirement for producing publication-quality graphs. Matplotlib provides several ways of configuring font properties. The default values can be set in the Matplotlib resource file, and session-wide configuration can be set in the mpl.rcParams dictionary. This dictionary is a cache of the Matplotlib resource file, and changes to parameters within this dictionary are valid until the Python interpreter is restarted and Matplotlib is imported again. Parameters that are relevant to how text is displayed include, for example, 'font.family' and 'font.size'.

Tip   Try print(mpl.rcParams) to get a list of possible configuration parameters and their current values. Updating a parameter is as simple as assigning a new value to the corresponding item in the dictionary mpl.rcParams, for example, mpl.rcParams['savefig.dpi'] = 100. See also the mpl.rc function, which can be used to update the mpl.rcParams dictionary, and mpl.rcdefaults for restoring the default values.

It is also possible to set text properties on a case-to-case basis, by passing a set of standard keyword arguments to functions that create text labels in a graph. Most Matplotlib functions that deal with text labels, in one way or another, accept the keyword arguments summarized in Table 4-2 (this list is an incomplete selection of common arguments; see help(mpl.text.Text) for a complete reference). For example, these arguments can be used with the method `Axes.text`, which create a new text label at a given coordinate. They may also be used with `set_title`, `set_xlabel`, `set_ylabel`, etc. For more information on these methods, see the next section.

In scientific and technical visualization, it is clearly important to be able to render mathematical symbols and expressions in text labels. Matplotlib provides excellent support for this through LaTeX markup within its text labels: any text label in Matplotlib can include LaTeX math by enclosing it within $ signs, for example, `"Regular text:` `$f(x)=1-x^2$"`. By default, Matplotlib uses an internal LaTeX rendering, which supports a subset of LaTeX language. However, by setting the configuration parameter `mpl.` `rcParams["text.usetex"]=True`, it is also possible to use an external full-featured LaTeX engine (if it is available on your system).

When embedding LaTeX code in strings in Python, there is a common stumbling block: Python uses \ as escape character, while in LaTeX it is used to denote the start of commands. To prevent the Python interpreter from escaping characters in strings containing LaTeX expressions, it is convenient to use raw strings, which are literal string expressions that are prepended with and an `r`, for example, `r"$\int f(x) dx$"` and `r'$x_{\rm A}$'`.

The following example demonstrates how to add text labels and annotations to a Matplotlib figure using `ax.text` and `ax.annotate`, as well as how to render a text label that includes an equation that is typeset in LaTeX. The resulting graph is shown in Figure 4-11.

```
In [13]: fig, ax = plt.subplots(figsize=(12, 3))
    ...:
    ...: ax.set_yticks([])
    ...: ax.set_xticks([])
    ...: ax.set_xlim(-0.5, 3.5)
    ...: ax.set_ylim(-0.05, 0.25)
    ...: ax.axhline(0)
    ...:
    ...: # text label
    ...: ax.text(0, 0.1, "Text label", fontsize=14, family="serif")
    ...:
    ...: # annotation
    ...: ax.plot(1, 0, "o")
    ...: ax.annotate("Annotation",
```

```
...:          fontsize=14, family="serif",
...:            xy=(1, 0), xycoords="data",
...:          xytext=(+20, +50), textcoords="offset points",
...:          arrowprops=dict(arrowstyle="->", connectionstyle="arc3,
             rad=.5"))
...:
...: # equation
...: ax.text(2, 0.1, r"Equation: $i\hbar\partial_t \Psi = \hat{H}\
     Psi$", fontsize=14, family="serif")
...:
```

Figure 4-11. *Example demonstrating the result of adding text labels and annotations using* ax.text *and* ax.annotation *and including LaTeX formatted equations in a Matplotlib text label*

Table 4-2. *Summary of Selected Font Properties and the Corresponding Keyword Arguments*

Argument	Description
fontsize	The size of the font, in points.
family or fontname	The font type.
backgroundcolor	Color specification for the background color of the text label.
color	Color specification for the font color.
alpha	Transparency of the font color.
rotation	Rotation angle of the text label.

Axis Properties

After having created Figure and Axes objects, the data or functions are plotted using some of the many plot functions provided by Matplotlib, and the appearance of lines and markers are customized – the last major aspect of a graph that remains to be configured and fine-tuned is the Axis instances. A two-dimensional graph has two axis objects: for the horizontal x axis and the vertical y axis. Each axis can be individually configured with respect to attributes such as the axis labels, the placement of ticks and the tick labels, and the location and appearance of the axis itself. In this section we look into the details of how to control these aspects of a graph.

Axis Labels and Titles

Arguably the most important property of an axis, which needs to be set in nearly all cases, is the axis label. We can set the axis labels using the set_xlabel and set_ylabel methods: they both take a string with the label as first arguments. In addition, the optional labelpad argument specifies the spacing, in units of points, from the axis to the label. This padding is occasionally necessary to avoid overlap between the axis label and the axis tick labels. The set_xlabel and set_ylabel methods also take additional arguments for setting text properties, such as color, fontsize, and fontname, as discussed in detail in the previous section. The following code, which produces Figure 4-12, demonstrates how to use the set_xlabel and set_ylabel methods and the keyword arguments discussed here.

```
In [14]: x = np.linspace(0, 50, 500)
    ...: y = np.sin(x) * np.exp(-x/10)
    ...:
    ...: fig, ax = plt.subplots(figsize=(8, 2), subplot_kw={'facecolor':
        "#ebf5ff"})
    ...:
    ...: ax.plot(x, y, lw=2)
    ...:
    ...: ax.set_xlabel("x", labelpad=5, fontsize=18, fontname='serif',
                    color="blue")
    ...: ax.set_ylabel("f(x)", labelpad=15, fontsize=18, fontname='serif',
                    color="blue")
    ...: ax.set_title("axis labels and title example", fontsize=16,
                    fontname='serif', color="blue")
```

Figure 4-12. *Graph demonstrating the result of using* `set_xlabel` *and* `set_ylabel` *for setting the x and y axis labels*

In addition to labels on the x and y axes, we can also set a title of an `Axes` object, using the `set_title` method. This method takes mostly the same arguments as `set_xlabel` and `set_ylabel`, with the exception of the loc argument, which can be assigned to `'left'`, `'centered'`, to `'right'`, and which dictates that the title is to be left aligned, centered, or right aligned.

Axis Range

By default, the range of the *x* and *y* axes of a Matplotlib is automatically adjusted to the data that is plotted in the `Axes` object. In many cases these default ranges are sufficient, but in some situations, it may be necessary to explicitly set the axis ranges. In such cases, we can use the `set_xlim` and `set_ylim` methods of the `Axes` object. Both these methods take two arguments that specify the lower and upper limit that is to be displayed on the axis, respectively. An alternative to `set_xlim` and `set_ylim` is the `axis` method, which, for example, accepts the string argument `'tight'`, for a coordinate range that tightly fit the lines it contains, and `'equal'`, for a coordinate range where one unit length along each axis corresponds to the same number of pixels (i.e., a ratio preserving coordinate system).

It is also possible to use the `autoscale` method to selectively turn on and off autoscaling, by passing `True` and `False` as first argument, for the *x* and/or *y* axis by setting its `axis` argument to `'x'`, `'y'`, or `'both'`. The example below shows how to use these methods to control axis ranges. The resulting graphs are shown in Figure 4-13.

```
In [15]: x = np.linspace(0, 30, 500)
    ...: y = np.sin(x) * np.exp(-x/10)
    ...:
    ...:
    ...: fig, axes = plt.subplots(1, 3, figsize=(9, 3), subplot_
         kw={'facecolor': "#ebf5ff"})
```

```
...:
...: axes[0].plot(x, y, lw=2)
...: axes[0].set_xlim(-5, 35)
...: axes[0].set_ylim(-1, 1)
...: axes[0].set_title("set_xlim / set_y_lim")
...:
...: axes[1].plot(x, y, lw=2)
...: axes[1].axis('tight')
...: axes[1].set_title("axis('tight')")
...:
...: axes[2].plot(x, y, lw=2)
...: axes[2].axis('equal')
...: axes[2].set_title("axis('equal')")
```

Figure 4-13. *Graphs that show the result of using the* set_xlim, set_ylim, *and* axis *methods for setting the axis ranges that are shown in a graph*

Axis Ticks, Tick Labels, and Grids

The final basic properties of the axis that remain to be configured are the placement of axis ticks and the placement and the formatting of the corresponding tick labels. The axis ticks are an important part of the overall appearance of a graph, and when preparing publication and production-quality graphs, it is often necessary to have detailed control over the axis ticks. Matplotlib module mpl.ticker provides a general and extensible tick management system that gives full control of the tick placement. Matplotlib distinguishes between major ticks and minor ticks. By default, every major tick has a corresponding label, and the distances between major ticks may be further marked with minor ticks that do not have labels, although this feature must be explicitly turned on. See Figure 4-14 for an illustration of major and minor ticks.

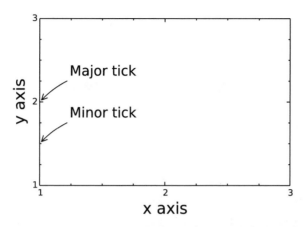

Figure 4-14. *The difference between major and minor ticks*

When approaching the configuration of ticks, the most common design target is to determine where the major tick with labels should be placed along the coordinate axis. The mpl.ticker module provides classes for different tick placement strategies. For example, the mpl.ticker.MaxNLocator can be used to set the maximum number ticks (at unspecified locations), the mpl.ticker.MultipleLocator can be used for setting ticks at multiples of a given base, and the mpl.ticker.FixedLocator can be used to place ticks at explicitly specified coordinates. To change ticker strategy, we can use the set_major_locator and the set_minor_locator methods in Axes.xaxis and Axes. yaxis. These methods accept an instance of a ticker class defined in mpl.ticker or a custom class that is derived from one of those classes.

When explicitly specifying tick locations, we can also use the methods set_xticks and set_yticks, which accept a list of coordinates for where to place major ticks. In this case, it is also possible to set custom labels for each tick using the set_xticklabels and set_yticklabels, which expects lists of strings to use as labels for the corresponding ticks. If possible, it is a good idea to use generic tick placement strategies, for example, mpl.ticker.MaxNLocator, because they dynamically adjust if the coordinate range is changed, whereas explicit tick placement using set_xticks and set_yticks then would require manual code changes. However, when the exact placement of ticks must be controlled, then set_xticks and set_yticks are convenient methods.

The following code demonstrates how to change the default tick placement using combinations of the methods discussed in the previous paragraphs, and the resulting graphs are shown in Figure 4-15.

```
In [16]: x = np.linspace(-2 * np.pi, 2 * np.pi, 500)
    ...: y = np.sin(x) * np.exp(-x**2/20)
    ...:
    ...: fig, axes = plt.subplots(1, 4, figsize=(12, 3))
    ...:
    ...: axes[0].plot(x, y, lw=2)
    ...: axes[0].set_title("default ticks")

    ...: axes[1].plot(x, y, lw=2)
    ...: axes[1].set_title("set_xticks")
    ...: axes[1].set_yticks([-1, 0, 1])
    ...: axes[1].set_xticks([-5, 0, 5])
    ...:
    ...: axes[2].plot(x, y, lw=2)
    ...: axes[2].set_title("set_major_locator")
    ...: axes[2].xaxis.set_major_locator(mpl.ticker.MaxNLocator(4))
    ...: axes[2].yaxis.set_major_locator(mpl.ticker.FixedLocator([-1, 0, 1]))
    ...: axes[2].xaxis.set_minor_locator(mpl.ticker.MaxNLocator(8))
    ...: axes[2].yaxis.set_minor_locator(mpl.ticker.MaxNLocator(8))
    ...:
    ...: axes[3].plot(x, y, lw=2)
    ...: axes[3].set_title("set_xticklabels")
    ...: axes[3].set_yticks([-1, 0, 1])
    ...: axes[3].set_xticks([-2 * np.pi, -np.pi, 0, np.pi, 2 * np.pi])
    ...: axes[3].set_xticklabels([r'$-2\pi$', r'$-\pi$', 0, r'$\pi$',
        r'$2\pi$'])
    ...: x_minor_ticker = mpl.ticker.FixedLocator([-3 * np.pi / 2,
        -np.pi / 2, 0,
    ...:                                           np.pi / 2, 3 * np.pi / 2])
    ...: axes[3].xaxis.set_minor_locator(x_minor_ticker)
    ...: axes[3].yaxis.set_minor_locator(mpl.ticker.MaxNLocator(4))
```

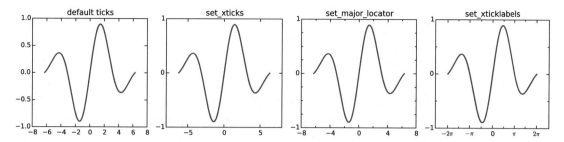

Figure 4-15. *Graphs that demonstrate different ways of controlling the placement and appearance of major and minor ticks along the x axis and the y axis*

A commonly used design element in graphs is grid lines, which are intended as a visual guide when reading values from the graph. Grids and grid lines are closely related to axis ticks, since they are drawn at the same coordinate values, and are therefore essentially extensions of the ticks that span across the graph. In Matplotlib, we can turn on axis grids using the grid method of an axes object. The grid method takes optional keyword arguments that are used to control the appearance of the grid. For example, like many of the plot functions in Matplotlib, the grid method accepts the arguments color, linestyle, and linewidth, for specifying the properties of the grid lines. In addition, it takes argument which and axis that can be assigned values 'major', 'minor', or 'both', and 'x', 'y', or 'both', respectively. These arguments are used to indicate which ticks along which axis the given style is to be applied to. If several different styles for the grid lines are required, multiple calls to grid can be used, with different values of which and axis. For an example of how to add grid lines and how to style them in different ways, see the following example, which produces the graphs shown in Figure 4-16.

```
In [17]: fig, axes = plt.subplots(1, 3, figsize=(12, 4))
    ...: x_major_ticker = mpl.ticker.MultipleLocator(4)
    ...: x_minor_ticker = mpl.ticker.MultipleLocator(1)
    ...: y_major_ticker = mpl.ticker.MultipleLocator(0.5)
    ...: y_minor_ticker = mpl.ticker.MultipleLocator(0.25)
    ...:
    ...: for ax in axes:
    ...:     ax.plot(x, y, lw=2)
    ...:     ax.xaxis.set_major_locator(x_major_ticker)
    ...:     ax.yaxis.set_major_locator(y_major_ticker)
    ...:     ax.xaxis.set_minor_locator(x_minor_ticker)
    ...:     ax.yaxis.set_minor_locator(y_minor_ticker)
```

```
...:
...: axes[0].set_title("default grid")
...: axes[0].grid()
...:
...: axes[1].set_title("major/minor grid")
...: axes[1].grid(color="blue", which="both", linestyle=':',
    linewidth=0.5)
...:
...: axes[2].set_title("individual x/y major/minor grid")
...: axes[2].grid(color="grey", which="major", axis='x', linestyle='-',
    linewidth=0.5)
...: axes[2].grid(color="grey", which="minor", axis='x', linestyle=':',
    linewidth=0.25)
...: axes[2].grid(color="grey", which="major", axis='y', linestyle='-',
    linewidth=0.5)
```

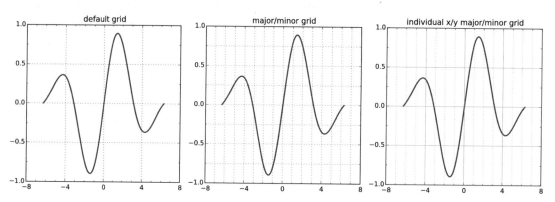

Figure 4-16. *Graphs demonstrating the result of using grid lines*

In addition to controlling the tick placements, the Matplotlib mpl.ticker module also provides classes for customizing the tick labels. For example, the ScalarFormatter from the mpl.ticker module can be used to set several useful properties related to displaying tick labels with scientific notation, for displaying axis labels for large numerical values. If scientific notation is activated using the set_scientific method, we can control the threshold for when scientific notation is used with the set_ powerlimits method (by default, tick labels for small numbers are not displayed using the scientific notation), and we can use the useMathText=True argument when creating the ScalarFormatter instance in order to have the exponents shown in math style rather than using code style exponents (e.g., 1e10). See the following code for an example of using scientific notation in tick labels. The resulting graphs are shown in Figure 4-17.

```
In [19]: fig, axes = plt.subplots(1, 2, figsize=(8, 3))
    ...:
    ...: x = np.linspace(0, 1e5, 100)
    ...: y = x ** 2
    ...:
    ...: axes[0].plot(x, y, 'b.')
    ...: axes[0].set_title("default labels", loc='right')
    ...:
    ...: axes[1].plot(x, y, 'b')
    ...: axes[1].set_title("scientific notation labels", loc='right')
    ...:
    ...: formatter = mpl.ticker.ScalarFormatter(useMathText=True)
    ...: formatter.set_scientific(True)
    ...: formatter.set_powerlimits((-1,1))
    ...: axes[1].xaxis.set_major_formatter(formatter)
    ...: axes[1].yaxis.set_major_formatter(formatter)
```

Figure 4-17. *Graphs with tick labels in scientific notation. The left panel uses the default label formatting, while the right panel uses tick labels in scientific notation, rendered as math text.*

Log Plots

In visualization of data that spans several orders of magnitude, it is useful to work with logarithmic coordinate systems. In Matplotlib, there are several plot functions for graphing functions in such coordinate systems, for example, `loglog`, `semilogx`, and `semilogy`, which use logarithmic scales for both the x and y axes, for only the x axis, and for only the y axis, respectively. Apart from the logarithmic axis scales, these functions behave similarly to the standard plot method. An alternative approach is to use the standard plot method and to separately configure the axis scales to be logarithmic using the set_xscale and/or set_yscale method with 'log' as first argument. These methods of producing log-scale plots are exemplified in the following section, and the resulting graphs are shown in Figure 4-18.

```
In [20]: fig, axes = plt.subplots(1, 3, figsize=(12, 3))
    ...:
    ...: x = np.linspace(0, 1e3, 100)
    ...: y1, y2 = x**3, x**4
    ...:
    ...: axes[0].set_title('loglog')
    ...: axes[0].loglog(x, y1, 'b', x, y2, 'r')
    ...:
    ...: axes[1].set_title('semilogy')
    ...: axes[1].semilogy(x, y1, 'b', x, y2, 'r')
    ...:
    ...: axes[2].set_title('plot / set_xscale / set_yscale')
    ...: axes[2].plot(x, y1, 'b', x, y2, 'r')
    ...: axes[2].set_xscale('log')
    ...: axes[2].set_yscale('log')
```

Figure 4-18. *Examples of log-scale plots*

Twin Axes

An interesting trick with axes that Matplotlib provides is the twin axis feature, which allows displaying two independent axes overlaid on each other. This is useful when plotting two different quantities, for example, with different units, within the same graph. A simple example that demonstrates this feature is shown as follows, and the resulting graph is shown in Figure 4-19. Here we use the twinx method (there is also a twiny method) to produce second Axes instance with shared x axis and a new independent y axis, which is displayed on the right side of the graph.

```
In [21]: fig, ax1 = plt.subplots(figsize=(8, 4))
    ...:
    ...: r = np.linspace(0, 5, 100)
    ...: a = 4 * np.pi * r ** 2   # area
    ...: v = (4 * np.pi / 3) * r ** 3   # volume
    ...:
    ...: ax1.set_title("surface area and volume of a sphere", fontsize=16)
    ...: ax1.set_xlabel("radius [m]", fontsize=16)
    ...:
    ...: ax1.plot(r, a, lw=2, color="blue")
    ...: ax1.set_ylabel(r"surface area ($m^2$)", fontsize=16, color="blue")
    ...: for label in ax1.get_yticklabels():
    ...:     label.set_color("blue")
    ...:
    ...: ax2 = ax1.twinx()
    ...: ax2.plot(r, v, lw=2, color="red")
    ...: ax2.set_ylabel(r"volume ($m^3$)", fontsize=16, color="red")
    ...: for label in ax2.get_yticklabels():
    ...:     label.set_color("red")
```

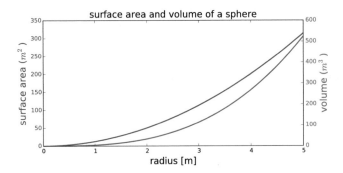

Figure 4-19. *Example of graphs with twin axes*

Spines

In all graphs generated so far, we have always had a box surrounding the Axes region. This is indeed a common style for scientific and technical graphs, but in some cases, for example, when representing schematic graphs, moving these coordinate lines may be desired. The lines that make up the surrounding box are called axis spines in Matplotlib, and we can use the Axes.spines attribute to change their properties. For example, we might want to remove the top and the right spines and move the spines to coincide with the origin of the coordinate systems.

The spines attribute of the Axes object is a dictionary with the keys right, left, top, and bottom that can be used to access each spine individually. We can use the set_color method to set the color to 'None' to indicate that a particular spine should not be displayed, and in this case, we also need to remove the ticks associated with that spine, using the set_ticks_position method of Axes.xaxis and Axes.yaxis (which accepts the arguments 'both', 'top', or 'bottom' and 'both', 'left', or 'right', respectively). With these methods we can transform the surrounding box to x and y coordinate axes, as demonstrated in the following example. The resulting graph is shown in Figure 4-20.

```
In [22]: x = np.linspace(-10, 10, 500)
    ...: y = np.sin(x) / x
    ...:
    ...: fig, ax = plt.subplots(figsize=(8, 4))
    ...:
    ...: ax.plot(x, y, linewidth=2)
    ...:
    ...: # remove top and right spines
```

```
...: ax.spines['right'].set_color('none')
...: ax.spines['top'].set_color('none')
...:
...: # remove top and right spine ticks
...: ax.xaxis.set_ticks_position('bottom')
...: ax.yaxis.set_ticks_position('left')
...:
...: # move bottom and left spine to x = 0 and y = 0
...: ax.spines['bottom'].set_position(('data', 0))
...: ax.spines['left'].set_position(('data', 0))
...:
...: ax.set_xticks([-10, -5, 5, 10])
...: ax.set_yticks([0.5, 1])
...:
...: # give each label a solid background of white, to not overlap with
        the plot line
...: for label in ax.get_xticklabels() + ax.get_yticklabels():
...:     label.set_bbox({'facecolor': 'white',
...:                     'edgecolor': 'white'})
```

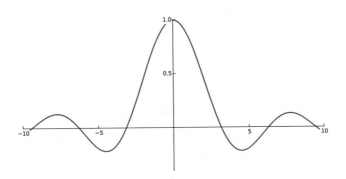

Figure 4-20. *Example of a graph with axis spines*

Advanced Axes Layouts

So far, we have repeatedly used `plt.figure`, `Figure.make_axes,` and `plt.subplots` to create new Figure and Axes instances, which we then used for producing graphs. In scientific and technical visualization, it is common to pack together multiple figures in different panels, for example, in a grid layout. In Matplotlib there are functions for automatically creating Axes objects and placing them on a figure canvas, using a variety of different layout strategies. We have already used the `plt.subplots` function, which is capable of generating a uniform grid of Axes objects. In this section we explore additional features of the `plt.subplots` function and introduce the `subplot2grid` and `GridSpec` layout managers, which are more flexible in how the Axes objects are distributed within a figure canvas.

Insets

Before diving into the details of how to use more advanced Axes layout managers, it is worth taking a step back and considering an important use-case of the very first approach we used to add Axes instances to a figure canvas: the `Figure.add_axes` method. This approach is well suited for creating so-called inset, which is a smaller graph that is displayed within the region of another graph. Insets are, for example, frequently used for displaying a magnified region of special interest in the larger graph or for displaying some related graphs of secondary importance.

In Matplotlib we can place additional Axes objects at arbitrary locations within a figure canvas, even if they overlap with existing Axes objects. To create an inset, we therefore simply add a new Axes object with `Figure.make_axes` and with the (figure canvas) coordinates for where the inset should be placed. A typical example of a graph with an inset is produced by the following code, and the graph that this code generates is shown in Figure 4-21. When creating the Axes object for the inset, it may be useful to use the argument `facecolor='none'`, which indicates that there should be no background color, that is, that the Axes background of the inset should be transparent.

```
In [23]: fig = plt.figure(figsize=(8, 4))
    ...:
    ...: def f(x):
    ...:     return 1/(1 + x**2) + 0.1/(1 + ((3 - x)/0.1)**2)
    ...:
```

```
...: def plot_and_format_axes(ax, x, f, fontsize):
...:     ax.plot(x, f(x), linewidth=2)
...:     ax.xaxis.set_major_locator(mpl.ticker.MaxNLocator(5))
...:     ax.yaxis.set_major_locator(mpl.ticker.MaxNLocator(4))
...:     ax.set_xlabel(r"$x$", fontsize=fontsize)
...:     ax.set_ylabel(r"$f(x)$", fontsize=fontsize)
...:
...: # main graph
...: ax = fig.add_axes([0.1, 0.15, 0.8, 0.8], facecolor="#f5f5f5")
...: x = np.linspace(-4, 14, 1000)
...: plot_and_format_axes(ax, x, f, 18)
...:
...: # inset
...: x0, x1 = 2.5, 3.5
...: ax.axvline(x0, ymax=0.3, color="grey", linestyle=":")
...: ax.axvline(x1, ymax=0.3, color="grey", linestyle=":")
...:
...: ax_insert = fig.add_axes([0.5, 0.5, 0.38, 0.42], facecolor='none')
...: x = np.linspace(x0, x1, 1000)
...: plot_and_format_axes(ax_insert, x, f, 14)
```

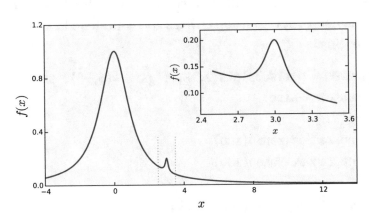

Figure 4-21. *Example of a graph with an inset*

Subplots

We have already used `plt.subplots` extensively, and we have noted that it returns a tuple with a `Figure` instance and a NumPy array with the `Axes` objects for each row and column that was requested in the function call. It is often the case when plotting grids of subplots that either the *x* or the *y* axis, or both, is shared among the subplots. Using the `sharex` and `sharey` arguments to `plt.subplots` can be useful in such situations, since it prevents the same axis labels to be repeated across multiple `Axes`.

It is also worth noting that the dimension of the NumPy array with `Axes` instances that is returned by `plt.subplots` is "squeezed" by default: that is, the dimensions with length 1 are removed from the array. If both the requested numbers of column and row are greater than one, then a two-dimensional array is returned, but if either (or both) the number of columns or rows is one, then a one-dimensional (or scalar, i.e., the only `Axes` object itself) is returned. We can turn off the squeezing of the dimensions of the NumPy arrays by passing the argument `squeeze=False` to the `plt.subplots` function. In this case the `axes` variable in `fig, axes = plt.subplots(nrows, ncols)` is always a two-dimensional array.

A final touch of configurability can be achieved using the `plt.subplots_adjust` function, which allows to explicitly set the left, right, bottom, and top coordinates of the overall `Axes` grid, as well as the width (`wspace`) and height spacing (`hspace`) between `Axes` instances in the grid. See the following code, and the corresponding Figure 4-22, for a step-by-step example of how to set up an `Axes` grid with shared *x* and *y* axes and with adjusted `Axes` spacing.

```
In [24]: fig, axes = plt.subplots(2, 2, figsize=(6, 6), sharex=True,
sharey=True, squeeze=False)
   ...:
   ...: x1 = np.random.randn(100)
   ...: x2 = np.random.randn(100)
   ...:
   ...: axes[0, 0].set_title("Uncorrelated")
   ...: axes[0, 0].scatter(x1, x2)
   ...:
   ...: axes[0, 1].set_title("Weakly positively correlated")
   ...: axes[0, 1].scatter(x1, x1 + x2)
   ...:
```

```
...: axes[1, 0].set_title("Weakly negatively correlated")
...: axes[1, 0].scatter(x1, -x1 + x2)
...:
...: axes[1, 1].set_title("Strongly correlated")
...: axes[1, 1].scatter(x1, x1 + 0.15 * x2)
...:
...: axes[1, 1].set_xlabel("x")
...: axes[1, 0].set_xlabel("x")
...: axes[0, 0].set_ylabel("y")
...: axes[1, 0].set_ylabel("y")
...:
...: plt.subplots_adjust(left=0.1, right=0.95, bottom=0.1, top=0.95,
     wspace=0.1, hspace=0.2)
```

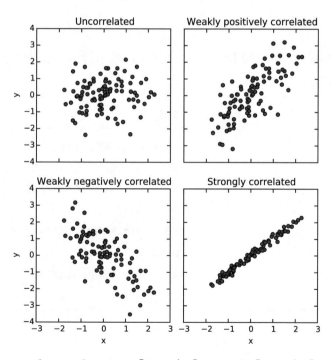

Figure 4-22. *Example graph using* `plt.subplot` *and* `plt.subplot_adjust`

Subplot2grid

The `plt.subplot2grid` function is an intermediary between `plt.subplots` and `gridspec` (see the next section) that provides a more flexible Axes layout management than `plt.subplots` while at the same time being simpler to use than `gridspec`. In particular, `plt.subplot2grid` is able to create grids with Axes instances that span multiple rows and/or columns. The `plt.subplot2grid` takes two mandatory arguments: the first argument is the shape of the Axes grid, in the form of a tuple (`nrows, ncols`), and the second argument is a tuple (`row, col`) that specifies the starting position within the grid. The two optional keyword arguments `colspan` and `rowspan` can be used to indicate how many rows and columns the new Axes instance should span. An example of how to use the `plt.subplot2grid` function is given in Table 4-3. Note that each call to the `plt.subplot2grid` function results in one new Axes instance, in contrast to `plt.subplots` which creates all Axes instances in one function call and returns them in a NumPy array.

Table 4-3. *Example of a Grid Layout Created with* `plt.subplot2grid` *and the Corresponding Code*

Axes Grid Layout	Code
| | ```
ax0 = plt.subplot2grid((3, 3), (0, 0))
ax1 = plt.subplot2grid((3, 3), (0, 1))
ax2 = plt.subplot2grid((3, 3), (1, 0),
colspan=2)
ax3 = plt.subplot2grid((3, 3), (2, 0),
colspan=3)
ax4 = plt.subplot2grid((3, 3), (0, 2),
rowspan=2)
``` |
```

GridSpec

The final grid layout manager that we cover here is GridSpec from the mpl.gridspec module. This is the most general grid layout manager in Matplotlib, and in particular it allows creating grids where not all rows and columns have equal width and height, which is not easily achieved with the grid layout managers we have used earlier in this chapter.

A GridSpec object is only used to specify the grid layout, and by itself it does not create any Axes objects. When creating a new instance of the GridSpec class, we must specify the number of rows and columns in the grid. Like for other grid layout managers, we can also set the position of the grid using the keyword arguments left, bottom, right, and top, and we can set the width and height spacing between subplots using wspace and hspace. Additionally, GricSpec allows specifying the relative width and heights of columns and rows using the width_ratios and height_ratios arguments. These should both be lists with relative weights for the size of each column and row in the grid. For example, to generate a grid with two rows and two columns, where the first row and column is twice as big as the second row and column, we could use mpl.gridspec.GridSpec(2, 2, width_ratios=[2, 1], height_ratios=[2, 1]).

Once a GridSpec instance has been created, we can use the Figure.add_subplot method to create Axes objects and place them on a figure canvas. As argument to add_subplot, we need to pass an mpl.gridspec.SubplotSpec instance, which we can generate from the GridSpec object using an array-like indexing: for example, given a GridSpec instance gs, we obtain a SubplotSpec instance for the upper-left grid element using gs[0, 0] and for a SubplotSpec instance that covers the first row we use gs[:, 0] and so on. See Table 4-4 for concrete examples of how to use GridSpec and add_subplot to create Axes instance.

Table 4-4. *Examples of How to Use the Subplot Grid Manager* `mpl.gridspec.` `GridSpec`

Axes Grid Layout	Code

ax0 ax4 ax1 ax6 ax5 ax2 ax8 ax7 ax9 ax3	```fig = plt.figure(figsize=(6, 4))```
	```gs = mpl.gridspec.GridSpec(4, 4)```
	```ax0 = fig.add_subplot(gs[0, 0])```
	```ax1 = fig.add_subplot(gs[1, 1])```
	```ax2 = fig.add_subplot(gs[2, 2])```
	```ax3 = fig.add_subplot(gs[3, 3])```
	```ax4 = fig.add_subplot(gs[0, 1:])```
	```ax5 = fig.add_subplot(gs[1:, 0])```
	```ax6 = fig.add_subplot(gs[1, 2:])```
	```ax7 = fig.add_subplot(gs[2:, 1])```
	```ax8 = fig.add_subplot(gs[2, 3])```
	```ax9 = fig.add_subplot(gs[3, 2])```

ax1 ax0 ax2	```fig = plt.figure(figsize=(4, 4))```
	```gs = mpl.gridspec.GridSpec(```
	``` 2, 2,```
	``` width_ratios=[4, 1],```
	``` height_ratios=[1, 4],```
	``` wspace=0.05, hspace=0.05)```
	```ax0 = fig.add_subplot(gs[1, 0])```
	```ax1 = fig.add_subplot(gs[0, 0])```
	```ax2 = fig.add_subplot(gs[1, 1])```

Colormap Plots

We have so far only considered graphs of univariate functions or, equivalently, two-dimensional data in x-y format. The two-dimensional Axes objects that we have used for this purpose can also be used to visualize bivariate functions, or three-dimensional data on x-y-z format, using so-called color maps (or heat maps), where each pixel in the Axes area is colored according to the *z* value corresponding to that point in the coordinate system. Matplotlib provides the functions pcolor and imshow for these types of plots, and the contour and contourf functions graph data on the same format by drawing contour lines rather than color maps. Examples of graphs generated with these functions are shown in Figure 4-23.

Figure 4-23. *Example graphs generated with* pcolor, imshow, contour, *and* contourf

To produce a colormap graph, for example, using pcolor, we first need to prepare the data in the appropriate format. While standard two-dimensional graphs expect one-dimensional coordinate arrays with *x* and *y* values, in the present case, we need to use two-dimensional coordinate arrays, as, for example, generated using the NumPy meshgrid function. To plot a bivariate function or data with two dependent variables, we start by defining one-dimensional coordinate arrays, x and y, that span the desired coordinate range or correspond to the values for which data is available. The x and y arrays can then be passed to the np.meshgrid function, which produces the required two-dimensional coordinate arrays X and Y. If necessary, we can use NumPy array computations with X and Y to evaluate bivariate functions to obtain a data array Z, as done in lines 1 to 3 in In [25] (see in the following section).

Once the two-dimensional coordinate and data arrays are prepared, they are easily visualized using, for example, pcolor, contour, or contourf, by passing the X, Y, and Z arrays as the first three arguments. The imshow method works similarly but only expects the data array Z as argument, and the relevant coordinate ranges must instead be set using the extent argument, which should be set to a list on the format [xmin, xmax, ymin, ymax]. Additional keyword arguments that are important for controlling the appearance of colormap graphs are vmin, vmax, norm, and cmap: the vmin and vmax can be used to set the range of values that are mapped to the color axis. This can equivalently be achieved by setting norm=mpl.colors.Normalize(vmin, vmax). The cmap argument specifies a color map for mapping the data values to colors in the graph. This argument can either be a string with a predefined colormap name or a colormap instance. The predefined color maps in Matplotlib are available in mpl.cm. Try help(mpl.cm) or try to autocomplete in IPython on the mpl.cm module for a full list of available color maps.[5]

The last piece required for a complete colormap plot is the colorbar element, which gives the viewer of the graph a way to read off the numerical values that different colors correspond to. In Matplotlib we can use the plt.colorbar function to attach a colorbar to an already plotted colormap graph. It takes a handle to the plot as first argument, and it takes two optional arguments ax and cax, which can be used to control where in the graph the colorbar is to appear. If ax is given, the space will be taken from this Axes object for the new colorbar. If, on the other hand, cax is given, then the colorbar will draw on this Axes object. A colorbar instance cb has its own axis object, and the standard methods for setting axis attributes can be used on the cb.ax object, and we can use, for example, the set_label, set_ticks, and set_ticklabels method in the same manner as for x and y axes.

The steps outlined in the previous paragraphs are shown in the following code, and the resulting graph is shown in Figure 4-24. The functions imshow, contour, and contourf can be used in a nearly similar manner, although these functions take additional arguments for controlling their characteristic properties. For example, the contour and contourf functions additionally take an argument N that specifies the number of contour lines to draw.

```
In [25]: x = y = np.linspace(-10, 10, 150)
    ...: X, Y = np.meshgrid(x, y)
    ...: Z = np.cos(X) * np.cos(Y) * np.exp(-(X/5)**2-(Y/5)**2)
    ...:
    ...: fig, ax = plt.subplots(figsize=(6, 5))
    ...:
    ...: norm = mpl.colors.Normalize(-abs(Z).max(), abs(Z).max())
    ...: p = ax.pcolor(X, Y, Z, norm=norm, cmap=mpl.cm.bwr)
    ...:
    ...: ax.axis('tight')
    ...: ax.set_xlabel(r"$x$", fontsize=18)
```

---

[5]A nice visualization of all the available color maps is available at http://wiki.scipy.org/Cookbook/Matplotlib/Show_colormaps. This page also describes how to create new color maps.

```
...: ax.set_ylabel(r"$y$", fontsize=18)
...: ax.xaxis.set_major_locator(mpl.ticker.MaxNLocator(4))
...: ax.yaxis.set_major_locator(mpl.ticker.MaxNLocator(4))
...:
...: cb = fig.colorbar(p, ax=ax)
...: cb.set_label(r"$z$", fontsize=18)
...: cb.set_ticks([-1, -.5, 0, .5, 1])
```

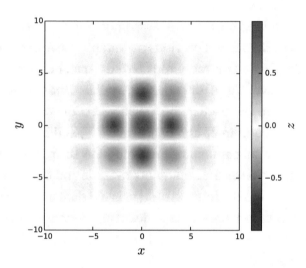

Figure 4-24. *Example using* pcolor *to produce a colormap graph*

3D Plots

The colormap graphs discussed in the previous section were used to visualize data with two dependent variables by color-coding data in 2D graphs. Another way of visualizing the same type of data is to use 3D graphs, where a third axis z is introduced and the graph is displayed in a perspective on the screen. In Matplotlib, drawing 3D graphs requires using a different axes object, namely, the Axes3D object that is available from the mpl_toolkits.mplot3d module. We can create a 3D-aware Axes instance explicitly using the constructor of the Axes3D class, by passing a Figure instance as argument: ax = Axes3D(fig). Alternatively, we can use the add_subplot function with the projection='3d' argument:

```
ax = ax = fig.add_subplot(1, 1, 1, projection='3d')
```

or use `plt.subplots` with the `subplot_kw={'projection': '3d'}` argument:

```
fig, ax = plt.subplots(1, 1, figsize=(8, 6), subplot_kw={'projection': '3d'})
```

In this way, we can use all of the axes layout approaches we have previously used for 2D graphs, if only we specify the `projection` argument in the appropriate manner. Note that using `add_subplot`, it is possible to mix axes objects with 2D and 3D projections within the same figure, but when using `plt.subplots`, the `subplot_kw` argument applies to all the subplots added to a figure.

Having created and added 3D-aware Axes instances to a figure, for example, using one of the methods described in the previous paragraph, the `Axes3D` class methods – such as `plot_surface`, `plot_wireframe`, and `contour` – can be used to plot data as surfaces in a 3D perspective. These functions are used in a manner that is nearly the same as how the color map was used in the previous section: these 3D plotting functions all take two-dimensional coordinate and data arrays X, Y, and Z as first arguments. Each function also takes additional parameters for tuning specific properties. For example, the `plot_surface` function takes the arguments `rstride` and `cstride` (row and column stride) for selecting data from the input arrays (to avoid data points that are too dense). The `contour` and `contourf` functions take optional arguments `zdir` and `offset`, which is used to select a projection direction (the allowed values are "x," "y," and "z") and the plane to display the projection on.

In addition to the methods for 3D surface plotting, there are also straightforward generalizations of the line and scatter plot functions that are available for 2D axes, for example, `plot`, `scatter`, `bar`, and `bar3d`, which in the version that is available in the Axes3D class takes an additional argument for the z coordinates. Like their 2D relatives, these functions expect one-dimensional data arrays rather than the two-dimensional coordinate arrays that are used for surface plots.

When it comes to axes titles, labels, ticks, and tick labels, all the methods used for 2D graphs, as described in detail earlier in this chapter, are straightforwardly generalized to 3D graphs. For example, there are new methods `set_zlabel`, `set_zticks`, and `set_zticklabels` for manipulating the attributes of the new z axis. The `Axes3D` object also provides new class methods for 3D specific actions and attributes. In particular, the `view_init` method can be used to change the angle from which the graph is viewed, and it takes the elevation and the azimuth, in degrees, as first and second arguments.

Examples of how to use these 3D plotting functions are given in the following section, and the produced graphs are shown in Figure 4-25.

```
In [26]: fig, axes = plt.subplots(1, 3, figsize=(14, 4), subplot_
         kw={'projection': '3d'})
    ...:
    ...: def title_and_labels(ax, title):
    ...:     ax.set_title(title)
    ...:     ax.set_xlabel("$x$", fontsize=16)
    ...:     ax.set_ylabel("$y$", fontsize=16)
    ...:     ax.set_zlabel("$z$", fontsize=16)
    ...:
    ...: x = y = np.linspace(-3, 3, 74)
    ...: X, Y = np.meshgrid(x, y)
    ...:
    ...: R = np.sqrt(X**2 + Y**2)
    ...: Z = np.sin(4 * R) / R
    ...:
    ...: norm = mpl.colors.Normalize(-abs(Z).max(), abs(Z).max())
    ...:
    ...: p = axes[0].plot_surface(X, Y, Z, rstride=1, cstride=1,
            linewidth=0, antialiased=False, norm=norm, cmap=mpl.cm.Blues)
    ...:
    ...: cb = fig.colorbar(p, ax=axes[0], shrink=0.6)
    ...: title_and_labels(axes[0], "plot_surface")
    ...:
    ...: p = axes[1].plot_wireframe(X, Y, Z, rstride=2, cstride=2,
         color="darkgrey")
    ...: title_and_labels(axes[1], "plot_wireframe")
    ...:
    ...: cset = axes[2].contour(X, Y, Z, zdir='z', offset=0, norm=norm,
         cmap=mpl.cm.Blues)
    ...: cset = axes[2].contour(X, Y, Z, zdir='y', offset=3, norm=norm,
         cmap=mpl.cm.Blues)
    ...: title_and_labels(axes[2], "contour")
```

Figure 4-25. *3D surface and contour graphs generated by using* `plot_surface,` `plot_wireframe,` *and* `contour`

Summary

In this chapter, we have covered the basics of how to produce 2D and 3D graphics using Matplotlib. Visualization is one of the most important tools for computational scientists and engineers, both as an analysis tool while working on computational problems and for presenting and communicating computational results. Visualization is therefore an integral part of the computational workflow, and it is equally important to be able to quickly visualize and explore data and to be able to produce picture-perfect publication-quality graphs, with detailed control over every graphical element. Matplotlib is a great general-purpose tool for both exploratory visualization and for producing publication-quality graphics. However, there are limitations to what can be achieved with Matplotlib, especially with respect to interactivity and high-quality 3D graphics. For more specialized use-cases, I therefore recommend to also explore some of the other graphic libraries that are available in the scientific Python ecosystem, some of which was briefly mentioned at the beginning of this chapter.

Further Reading

The Matplotlib is treated in books dedicated to the library, such as Tosi (2009) and Devert (2014), and in several books with a wider scope, for example, Milovanovi (2013) and McKinney (2013). For interesting discussions on data visualization and style guides and good practices in visualization, see, for example, Yau (2011) and J. Steele (2010).

References

Devert, A. (2014). *matplotlib Plotting Cookbook.* Mumbai: Packt.

J. Steele, N. I. (2010). *Beautiful Visualization.* Sebastopol: O'Reilly.

McKinney, W. (2013). *Python for Data Analysis.* Sebastopol: O'Reilly.

Milovanovi, I. (2013). *Python Data Visualization Cookbook.* Mumbai: Packt.

Tosi, S. (2009). *Matplotlib for Python Developers.* Mumbai: Packt.

Yau, N. (2011). *Visualize this.* Indianapolis: Wiley.

CHAPTER 5

Equation Solving

In the previous chapters, we have discussed general methodologies and techniques, namely, array-based numerical computing, symbolic computing, and visualization. These methods are the cornerstones of scientific computing that make up a fundamental toolset we have at our disposal when attacking computational problems.

Starting from this chapter, we begin to explore how to solve problems from different domains of applied mathematics and computational sciences, using the basic techniques introduced in the previous chapters. The topic of this chapter is algebraic equation solving. This is a broad topic that requires the application of theory and approaches from multiple fields of mathematics. In particular, when discussing equation solving, we have to distinguish between univariate and multivariate equations (i.e., equations that contain one unknown variable or many unknown variables). In addition, we need to distinguish between linear and nonlinear equations. This classification is useful because solving equations of these different types requires applying different mathematical methods and approaches.

We begin with linear equation systems, which are tremendously useful and have important applications in every field of science. The reason for this universality is that linear algebra theory allows us to straightforwardly solve linear equations, while nonlinear equations are difficult to solve in general and typically require more complicated and computationally demanding methods. Because linear systems are readily solvable, they are also an important tool for local approximations of nonlinear systems. For example, by considering small variations from an expansion point, a nonlinear system can often be approximated by a linear system in the local vicinity of the expansion point. However, a linearization can only describe local properties, and for global analysis of nonlinear problems, other techniques are required. Such methods typically employ iterative approaches for gradually constructing an increasingly accurate estimate of the solution.

© Robert Johansson 2019
R. Johansson, *Numerical Python*, https://doi.org/10.1007/978-1-4842-4246-9_5

In this chapter, we use SymPy for solving equations symbolically, when possible, and use the linear algebra module from the SciPy library for numerically solving linear equation systems. For tackling nonlinear problems, we will use the root-finding functions in the `optimize` module of SciPy.

SciPy   SciPy is a Python library, the collective name of the scientific computing environment for Python, and the umbrella organization for many of the core libraries for scientific computing with Python. The library, `scipy`, is in fact rather a collection of libraries for high-level scientific computing, which are more or less independent of each other. The SciPy library is built on top of NumPy, which provides the basic array data structures and fundamental operations on such arrays. The modules in SciPy provide domain-specific high-level computation methods, such as routines for linear algebra, optimization, interpolation, integration, and much more. At the time of writing, the most recent version of SciPy is 1.1.0. See `www.scipy.org` for more information.

Importing Modules

The SciPy module `scipy` should be considered a collection of modules that are selectively imported when required. In this chapter we will use the `scipy.linalg` module, for solving linear systems of equations, and the `scipy.optimize` module, for solving nonlinear equations. In this chapter we assume that these modules are imported as:

```
In [1]: from scipy import linalg as la
In [2]: from scipy import optimize
```

In this chapter we also use the NumPy, SymPy, and Matplotlib libraries introduced in the previous chapters, and we assume that those libraries are imported following the previously introduced convention:

```
In [3]: import sympy
In [4]: sympy.init_printing()
In [5]: import numpy as np
In [6]: import matplotlib.pyplot as plt
```

To get the same behavior in both Python 2 and Python 3 with respect to integer division, we also include the following statement (which is only necessary in Python 2):

```
In [7]: from __future__ import division
```

Linear Equation Systems

An important application of linear algebra is solving systems of linear equations. We have already encountered linear algebra functionality in the SymPy library, in Chapter 3. There are also linear algebra modules in the NumPy and SciPy libraries, numpy.linalg and scipy.linalg, which together provide linear algebra routines for numerical problems, that is, for problems that are completely specified in terms of numerical factors and parameters.

In general, a linear equation system can be written on the form

$$a_{11}x_1 + a_{12}x_2 + \ldots + a_{1n}x_n = b_1,$$

$$a_{21}x_1 + a_{22}x_2 + \ldots + a_{2n}x_n = b_2,$$

$$\ldots$$

$$a_{m1}x_1 + a_{m2}x_2 + \ldots + a_{mn}x_n = b_m.$$

This is a linear system of m equations in n unknown variables $\{x_1, x_2, \ldots, x_n\}$, where a_{mn} and b_m are known parameters or constant values. When working with linear equation systems, it is convenient to write the equations in matrix form:

$$\begin{pmatrix} a_{11} & a_{12} & \cdots & a_{1n} \\ a_{21} & a_{22} & \cdots & a_{2n} \\ \vdots & \vdots & \ddots & \vdots \\ a_{m1} & a_{m2} & \cdots & a_{mn} \end{pmatrix} \begin{pmatrix} x_1 \\ x_2 \\ \vdots \\ x_n \end{pmatrix} = \begin{pmatrix} b_1 \\ b_2 \\ \vdots \\ b_m \end{pmatrix},$$

or simply $Ax = b$, where A is a $m \times n$ matrix, b is a $m \times 1$ matrix (or m-vector), and x is the unknown $n \times 1$ solution matrix (or n-vector). Depending on the properties of the matrix A, the solution vector x may or may not exist, and if a solution does exist, it is not necessarily unique. However, if a solution exists, then it can be interpreted as an expression for the vector b as a linear combination of the columns of the matrix A, where the coefficients are given by the elements in the solution vector x.

A system for which $n < m$ is said to be underdetermined, because it has fewer equations than unknown and therefore cannot completely determine a unique solution. If, on the other hand, $m > n$, then the equations are said to be overdetermined. This will in general lead to conflicting constraints, resulting in that a solution does not exist.

Square Systems

Square systems with $m = n$ is an important special case. It corresponds to the situation where the number of equations equals the number of unknown variables, and it can therefore potentially have a unique solution. In order for a unique solution to exist, the matrix A must be *nonsingular*, in which case the inverse of A exists, and the solution can be written as $x = A^{-1}b$. If the matrix A is singular, that is, the rank of the matrix is less than n, $\text{rank}(A) < n$, or, equivalently, if its determinant is zero, $\det A = 0$, then the equation $Ax = b$ can either have no solution or infinitely many solutions, depending on the right-hand side vector b. For a matrix with rank deficiency, $\text{rank}(A) < n$, there are columns or rows that can be expressed as linear combinations of other columns or vectors, and they therefore correspond to equations that do not contain any new constraints, and the system is really underdetermined. Computing the rank of the matrix A that defines a linear equation system is therefore a useful method that can tell us whether the matrix is singular or not and therefore whether there exists a solution or not.

When A has full rank, the solution is guaranteed to exist. However, it may or may not be possible to accurately compute the solution. The *condition number* of the matrix, $\text{cond}(A)$, gives a measure of how well or poorly conditioned a linear equation system is. If the conditioning number is close to 1, if the system is said to be *well conditioned* (a condition number 1 is ideal), and if the condition number is large, the system is said to be *ill-conditioned*. The solution to an equation system that is ill-conditioned can have large errors. An intuitive interpretation of the condition number can be obtained from a simple error analysis. Assume that we have a linear equation system on the form $Ax = b$, where x is the solution vector. Now consider a small variation of b, say δb, which gives a corresponding change in the solution, δx, given by $A(x+\delta x) = b+\delta b$. Because of linearity of the equation, we have $A\delta x = \delta b$. An important question to consider now is: how large is the relative change in x compared to the relative change in b? Mathematically we can formulate this question in terms of the ratios of the norms of these vectors. Specifically, we are interested in comparing $\|\delta x\|/\|x\|$ and $\|\delta b\|/\|b\|$, where $\|x\|$ denotes the norm of x. Using the matrix norm relation $\|Ax\| \le \|A\| \cdot \|x\|$, we can write

$$\frac{\|\delta x\|}{\|x\|} = \frac{\|A^{-1}\delta b\|}{\|x\|} \leq \frac{\|A^{-1}\|\cdot\|\delta b\|}{\|x\|} = \frac{\|A^{-1}\|\cdot\|b\|}{\|x\|}\cdot\frac{\|\delta b\|}{\|b\|} \leq \|A^{-1}\|\cdot\|A\|\cdot\frac{\|\delta b\|}{\|b\|}$$

A bound of the relative error in the solution x, given a relative error in the b vector, is therefore given by $\text{cond}(A) \equiv \|A^{-1}\| \cdot \|A\|$, which by definition is the condition number of the matrix A. This means that for linear equation systems characterized by a matrix A that is ill-conditioned, even a small perturbation in the b vector can give large errors in the solution vector x. This is particularly relevant in numerical solution using floating-point numbers, which are only approximations to real numbers. When solving a system of linear equations, it is therefore important to look at the condition number to estimate the accuracy of the solution.

The rank, condition number, and norm of a symbolic matrix can be computed in SymPy using the `Matrix` methods `rank`, `condition_number`, and `norm`, and for numerical problems, we can use the NumPy functions `np.linalg.matrix_rank`, `np.linalg. cond`, and `np.linalg.norm`. For example, consider the following system of two linear equations:

$$2x_1 + 3x_2 = 4$$
$$5x_1 + 4x_2 = 3$$

These two equations correspond to lines in the (x_1,x_2) plane, and their intersection is the solution to the equation system. As can be seen in Figure 5-1, which graphs the lines corresponding to the two equations, the lines intersect at $(-1,2)$.

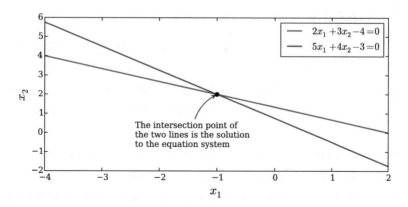

Figure 5-1. *Graphical representation of a system of two linear equations*

We can define this problem in SymPy by creating matrix objects for A and b and computing the rank, condition number, and norm of the matrix A using

```
In [8]: A = sympy.Matrix([[2, 3], [5, 4]])
In [9]: b = sympy.Matrix([4, 3])
In [10]: A.rank()
Out[10]: 2
In [11]: A.condition_number()
```

$$\text{Out[11]: } \frac{\sqrt{27+2\sqrt{170}}}{\sqrt{27-2\sqrt{170}}}$$

```
In [12]: sympy.N(_)
Out[12]: 7.58240137440151
In [13]: A.norm()
```

$\text{Out[13]: } 3\sqrt{6}$

We can do the same thing in NumPy/SciPy using NumPy arrays for A and b and functions from the `np.linalg` and `scipy.linalg` modules:

```
In [14]: A = np.array([[2, 3], [5, 4]])
In [15]: b = np.array([4, 3])
In [16]: np.linalg.matrix_rank(A)
Out[16]: 2
In [17]: np.linalg.cond(A)
Out[17]: 7.5824013744
In [18]: np.linalg.norm(A)
Out[18]: 7.34846922835
```

A direct approach to solving the linear problem is to compute the inverse of the matrix A and multiply it with the vector b, as used, for example, in the previous analytical discussions. However, this is not the most efficient computational method to find the solution vector x. A better method is LU factorization of the matrix A, such that $A = LU$ and where L is a lower triangular matrix and U is an upper triangular matrix. Given L and U, the solution vector x can be efficiently constructed by first solving $Ly = b$ with forward substitution and then solving $Ux = y$ with backward substitution. Owning to the fact that L and U are triangular matrices, these two procedures are computationally efficient.

In SymPy we can perform a symbolic LU factorization by using the LUdecomposition method of the sympy.Matrix class. This method returns new Matrix objects for the L and U matrices, as well as a row swap matrix. When we are interested in solving an equation system $Ax = b$, we do not explicitly need to calculate the L and U matrices, but rather we can use the LUsolve method, which performs the LU factorization internally and solves the equation system using those factors. Returning to the previous example, we can compute the L and U factors and solve the equation system using

```
In [19]: A = sympy.Matrix([[2, 3], [5, 4]])
In [20]: b = sympy.Matrix([4, 3])
In [21]: L, U, _ = A.LUdecomposition()
In [22]: L
```

$$
Out[22]: \begin{bmatrix} 1 & 0 \\ 5/2 & 1 \end{bmatrix}
$$

```
In [23]: U
```

$$
Out[23]: \begin{bmatrix} 2 & 3 \\ 0 & -7/2 \end{bmatrix}
$$

```
In [24]: L * U
```

$$
Out[24]: \begin{bmatrix} 2 & 3 \\ 5 & 4 \end{bmatrix}
$$

```
In [25]: x = A.solve(b); x   # equivalent to A.LUsolve(b)
```

$$
Out[25]: \begin{bmatrix} -1 \\ 2 \end{bmatrix}
$$

For numerical problems we can use the la.lu function form SciPy's linear algebra module. It returns a permutation matrix P and the L and U matrices, such that $A = PLU$. Like with SymPy, we can solve the linear system $Ax = b$ without explicitly calculating the L and U matrices by using the la.solve function, which takes the A matrix and the b vector as arguments. This is in general the preferred method for solving numerical linear equation systems using SciPy.

```
In [26]: A = np.array([[2, 3], [5, 4]])
In [27]: b = np.array([4, 3])
In [28]: P, L, U = la.lu(A)
In [29]: L
Out[29]: array([[ 1. ,  0. ],
                [ 0.4,  1. ]])
```

```
In [30]: U
Out[30]: array([[ 5. ,   4. ],
                [ 0. ,   1.4]])
In [31]: P.dot(L.dot(U))
Out[31]: array([[ 2., 3.],
                [ 5., 4.]])
In [32]: la.solve(A, b)
Out[32]: array([-1.,   2.])
```

The advantage of using SymPy is of course that we may obtain exact results and we can also include symbolic variables in the matrices. However, not all problems are solvable symbolically, or it may give exceedingly lengthy results. The advantage of using a numerical approach with NumPy/SciPy, on the other hand, is that we are guaranteed to obtain a result, although it will be an approximate solution due to floating-point errors. See the following code (In [38]) for an example that illustrates the differences between the symbolic and numerical approaches and for an example that show numerical approaches can be sensitive for equation systems with large condition number. In this example we solve the equation system

$$\begin{pmatrix} 1 & \sqrt{p} \\ 1 & \dfrac{1}{\sqrt{p}} \end{pmatrix}\begin{pmatrix} x_1 \\ x_2 \end{pmatrix} = \begin{pmatrix} 1 \\ 2 \end{pmatrix}$$

which for $p = 1$ is singular and for p in the vicinity of one is ill-conditioned. Using SymPy, the solution is easily found to be

```
In [33]: p = sympy.symbols("p", positive=True)
In [34]: A = sympy.Matrix([[1, sympy.sqrt(p)], [1, 1/sympy.sqrt(p)]])
In [35]: b = sympy.Matrix([1, 2])
In [36]: x = A.solve(b)
In [37]: x
Out[37]:
```
$$\begin{pmatrix} \dfrac{2p-1}{p-1} \\ -\dfrac{\sqrt{p}}{p-1} \end{pmatrix}$$

A comparison between this symbolic solution and the numerical solution is shown in Figure 5-2. Here the errors in the numerical solution are due to numerical floating-point errors, and the numerical errors are significantly larger in the vicinity of $p = 1$, where the system has a large condition number. Also, if there are other sources of errors in either A or b, the corresponding errors in x can be even more severe.

```
In [38]: # Symbolic problem specification
    ...: p = sympy.symbols("p", positive=True)
    ...: A = sympy.Matrix([[1, sympy.sqrt(p)], [1, 1/sympy.sqrt(p)]])
    ...: b = sympy.Matrix([1, 2])
    ...:
    ...: # Solve symbolically
    ...: x_sym_sol = A.solve(b)
    ...: Acond = A.condition_number().simplify()
    ...:
    ...: # Numerical problem specification
    ...: AA = lambda p: np.array([[1, np.sqrt(p)], [1, 1/np.sqrt(p)]])
    ...: bb = np.array([1, 2])
    ...: x_num_sol = lambda p: np.linalg.solve(AA(p), bb)
    ...:
    ...: # Graph the difference between the symbolic (exact) and numerical
            results.
    ...: fig, axes = plt.subplots(1, 2, figsize=(12, 4))
    ...:
    ...: p_vec = np.linspace(0.9, 1.1, 200)
    ...: for n in range(2):
    ...:     x_sym = np.array([x_sym_sol[n].subs(p, pp).evalf() for pp in
                p_vec])
    ...:     x_num = np.array([x_num_sol(pp)[n] for pp in p_vec])
    ...:     axes[0].plot(p_vec, (x_num - x_sym)/x_sym, 'k')
    ...: axes[0].set_title("Error in solution\n(numerical - symbolic)/
            symbolic")
    ...: axes[0].set_xlabel(r'$p$', fontsize=18)
    ...:
    ...: axes[1].plot(p_vec, [Acond.subs(p, pp).evalf() for pp in p_vec])
    ...: axes[1].set_title("Condition number")
    ...: axes[1].set_xlabel(r'$p$', fontsize=18)
```

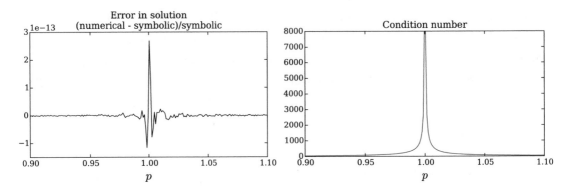

Figure 5-2. *Graph of the relative numerical errors (left) and condition number (right) as a function of the parameter p*

Rectangular Systems

Rectangular systems, with $m \neq n$, can be either underdetermined or overdetermined. Underdetermined systems have more variables than equations, so the solution cannot be fully determined. Therefore, for such a system, the solution must be given in terms of the remaining free variables. This makes it difficult to treat this type of problem numerically, but a symbolic approach can often be used instead.

For example, consider the underdetermined linear equation system

$$\begin{pmatrix} 1 & 2 & 3 \\ 4 & 5 & 6 \end{pmatrix} \begin{pmatrix} x_1 \\ x_2 \\ x_3 \end{pmatrix} = \begin{pmatrix} 7 \\ 8 \end{pmatrix}.$$

Here we have three unknown variables, but only two equations impose constraints on the relations between these variables. By writing this equation as $Ax - b = 0$, we can use the SymPy sympy.solve function to obtain a solution for x_1 and x_2 parameterized by the remaining free variable x_3:

```
In [39]: x_vars = sympy.symbols("x_1, x_2, x_3")
In [40]: A = sympy.Matrix([[1, 2, 3], [4, 5, 6]])
In [41]: x = sympy.Matrix(x_vars)
In [42]: b = sympy.Matrix([7, 8])
In [43]: sympy.solve(A*x - b, x_vars)
Out[43]: {x_1 = x_3 - 19/3, x_2 = -2x_3 + 20/3}
```

Here we obtained the symbolic solution $x_1 = x_3 - 19/3$ and $x_2 = -2x_3 + 20/3$, which defines a line in the three-dimensional space spanned by $\{x_1, x_2, x_3\}$. Any point on this line therefore satisfies this underdetermined equation system.

On the other hand, if the system is overdetermined and has more equations than unknown variables, $m > n$, then we have more constraints than degrees of freedom, and in general there is no exact solution to such a system. However, it is often interesting to find an approximate solution to an overdetermined system. An example of when this situation arises is data fitting: say we have a model where a variable y is a quadratic polynomial in the variable x, so that $y = A + Bx + Cx^2$, and that we would like to fit this model to experimental data. Here y is nonlinear in x, but y is linear in the three unknown coefficients A, B, and C, and this fact can be used to write the model as a linear equation system. If we collect data for m pairs $\{(x_i, y_i)\}_{i=1}^{m}$ of the variables x and y, we can write the model as an $m \times 3$ equation system:

$$\begin{pmatrix} 1 & x_1 & x_1^2 \\ \vdots & \vdots & \vdots \\ 1 & x_m & x_m^2 \end{pmatrix} \begin{pmatrix} A \\ B \\ C \end{pmatrix} = \begin{pmatrix} y_1 \\ \vdots \\ y_m \end{pmatrix}.$$

If $m = 3$, we can solve for the unknown model parameters A, B, and C, assuming the system matrix is nonsingular. However, it is intuitively clear that if the data is noisy and if we were to use more than three data points, we should be able to get a more accurate estimate of the model parameters.

However, for $m > 3$, there is in general no exact solution, and we need to introduce an approximate solution that gives the *best fit* for the overdetermined system $Ax \approx b$. A natural definition of the best fit for this system is to minimize the sum of square errors, $\min_x \sum_{i=1}^{m}(r_i)^2$, where $r = b - Ax$ is the residual vector. This leads to the *least square* solution of the problem $Ax \approx b$, which minimizes the distances between the data points and the linear solution. In SymPy we can solve for the least square solution of an overdetermined system using the solve_least_squares method, and for numerical problems, we can use the SciPy function la.lstsq.

The following code demonstrates how the SciPy la.lstsq method can be used to fit the example model considered in the preceding section, and the result is shown in Figure 5-3. We first define the true parameters of the model, and then we simulate measured data by adding random noise to the true model relation. The least square problem is then solved using the `la.lstsq` function, which in addition to the solution vector x also returns the total sum of square errors (the residual r), the rank rank and the singular values sv of the matrix A. However, in the following example, we only use the solution vector x.

```
In [44]: # define true model parameters
    ...: x = np.linspace(-1, 1, 100)
    ...: a, b, c = 1, 2, 3
    ...: y_exact = a + b * x + c * x**2
    ...:
    ...: # simulate noisy data
    ...: m = 100
    ...: X = 1 - 2 * np.random.rand(m)
    ...: Y = a + b * X + c * X**2 + np.random.randn(m)
    ...:
    ...: # fit the data to the model using linear least square
    ...: A = np.vstack([X**0, X**1, X**2])  # see np.vander for alternative
    ...: sol, r, rank, sv = la.lstsq(A.T, Y)
    ...:
    ...: y_fit = sol[0] + sol[1] * x + sol[2] * x**2
    ...: fig, ax = plt.subplots(figsize=(12, 4))
    ...:
    ...: ax.plot(X, Y, 'go', alpha=0.5, label='Simulated data')
    ...: ax.plot(x, y_exact, 'k', lw=2, label='True value $y = 1 + 2x +
        3x^2$')
    ...: ax.plot(x, y_fit, 'b', lw=2, label='Least square fit')
    ...: ax.set_xlabel(r"$x$", fontsize=18)
    ...: ax.set_ylabel(r"$y$", fontsize=18)
    ...: ax.legend(loc=2)
```

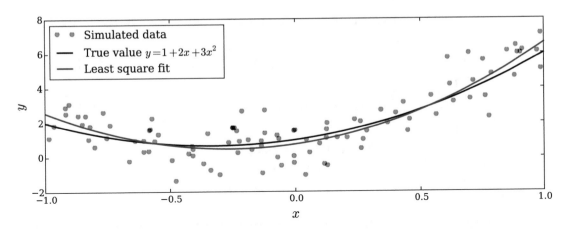

Figure 5-3. *Linear least square fit*

A good fit of data to a model obviously requires that the model used to describe the data correspond well to the underlying process that produced the data. In the following example (In [45]), and in Figure 5-4, we fit the same data used in the previous example to linear model and to a higher-order polynomial model (up to order 15). The former case corresponds to underfitting, where we have used a too simple model for the data, and the latter case corresponds to overfitting, where we have used a too complex model for the data, and thus fit the model not only to the underlying trend but also to the measurement noise. Using an appropriate model is an important and delicate aspect of data fitting.

```
In [45]: # fit the data to the model using linear least square:
    ...: # 1st order polynomial
    ...: A = np.vstack([X**n for n in range(2)])
    ...: sol, r, rank, sv = la.lstsq(A.T, Y)
    ...: y_fit1 = sum([s * x**n for n, s in enumerate(sol)])
    ...:
    ...: # 15th order polynomial
    ...: A = np.vstack([X**n for n in range(16)])
    ...: sol, r, rank, sv = la.lstsq(A.T, Y)
    ...: y_fit15 = sum([s * x**n for n, s in enumerate(sol)])
    ...:
    ...: fig, ax = plt.subplots(figsize=(12, 4))
    ...: ax.plot(X, Y, 'go', alpha=0.5, label='Simulated data')
```

```
...: ax.plot(x, y_exact, 'k', lw=2, label='True value $y = 1 + 2x +
     3x^2$')
...: ax.plot(x, y_fit1, 'b', lw=2, label='Least square fit [1st
     order]')
...: ax.plot(x, y_fit15, 'm', lw=2, label='Least square fit [15th
     order]')
...: ax.set_xlabel(r"$x$", fontsize=18)
...: ax.set_ylabel(r"$y$", fontsize=18)
...: ax.legend(loc=2)
```

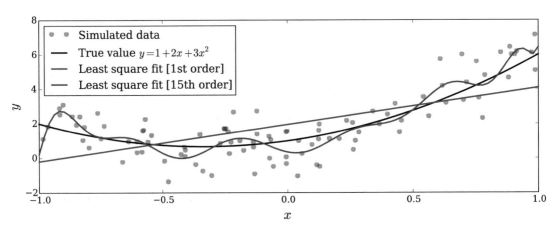

Figure 5-4. *Graph demonstrating underfitting and overfitting of data using the linear least square method*

Eigenvalue Problems

A special system of equations of great theoretical and practical importance is the eigenvalue equation $Ax = \lambda x$, where A is a $N \times N$ square matrix, x is an unknown vector, and λ is an unknown scalar. Here x is an eigenvector and λ an eigenvalue of the matrix A. The eigenvalue equation $Ax = \lambda x$ closely resembles the linear equation system $Ax = b$, but note that here both x and λ are unknown, so we cannot directly apply the same techniques to solve this equation. A standard approach to solve this eigenvalue problem is to rewrite the equation as $(A - I\lambda)x = 0$ and note that for there to exist a nontrivial solution, $x \neq 0$, the matrix $A - I\lambda$ must be singular, and its determinant must be zero, $\det(A - I\lambda) = 0$. This gives a polynomial equation (the characteristic polynomial) of Nth order whose N roots give the N eigenvalues $\{\lambda_n\}_{n=1}^{N}$. Once the eigenvalues are known, the equation $(A - I\lambda_n)x_n = 0$ can be solved for the nth eigenvector x_n using standard forward substitution.

Both SymPy and the linear algebra package in SciPy contain solvers for eigenvalue problems. In SymPy, we can use the `eigenvals` and `eigenvects` methods of the `Matrix` class, which are able to compute the eigenvalues and eigenvectors of some matrices with elements that are symbolic expressions. For example, to compute the eigenvalues and eigenvectors of symmetric 2×2 matrix with symbolic elements, we can use

```
In [46]: eps, delta = sympy.symbols("epsilon, Delta")
In [47]: H = sympy.Matrix([[eps, delta], [delta, -eps]])
In [48]: H
```

$$\text{Out[48]:} \quad \begin{pmatrix} \varepsilon & \Delta \\ \Delta & -\varepsilon \end{pmatrix}$$

```
In [49]: H.eigenvals()
```

$$\text{Out[49]:} \quad \left\{ -\sqrt{\varepsilon^2 + \Delta^2} : 1, \sqrt{\varepsilon^2 + \Delta^2} : 1 \right\}$$

```
In [50]: H.eigenvects()
```

$$\text{Out[50]:} \quad \left[\left(-\sqrt{\varepsilon^2 + \Delta^2}, 1, \left[\begin{bmatrix} -\dfrac{\Delta}{\varepsilon + \sqrt{\varepsilon^2 + \Delta^2}} \\ 1 \end{bmatrix} \right] \right), \left(\sqrt{\varepsilon^2 + \Delta^2}, 1, \left[\begin{bmatrix} -\dfrac{\Delta}{\varepsilon - \sqrt{\varepsilon^2 + \Delta^2}} \\ 1 \end{bmatrix} \right] \right) \right]$$

The return value of the `eigenvals` method is a dictionary where each eigenvalue is a key, and the corresponding value is the multiplicity of that particular eigenvalue. Here the eigenvalues are $-\sqrt{\varepsilon^2 + \Delta^2}$ and $\sqrt{\varepsilon^2 + \Delta^2}$, each with multiplicity one. The return value of `eigenvects` is a bit more involved: a list is returned where each element is a tuple containing an eigenvalue, the multiplicity of the eigenvalue, and a list of eigenvectors. The number of eigenvectors for each eigenvalue equals the multiplicity. For the current example, we can unpack the value returned by eigenvects and verify that the two eigenvectors are orthogonal using, for example,

```
In [51]: (eval1, _, evec1), (eval2, _, evec2) = H.eigenvects()
In [52]: sympy.simplify(evec1[0].T * evec2[0])
Out[52]: [0]
```

Obtaining analytical expressions for eigenvalues and eigenvectors using these methods is often very desirable indeed, but unfortunately it only works for small matrices. For anything larger than a 3×3, the analytical expression typically becomes extremely lengthy and cumbersome to work with even using a computer algebra system such as SymPy. Therefore, for larger systems we must resort to a fully numerical approach. For this we can use the `la.eigvals` and `la.eig` functions in the SciPy linear algebra package. Matrices that are either Hermitian or real symmetric have real-valued eigenvalues, and for such matrices, it is advantageous to instead use the functions `la.eigvalsh` and `la.eigh`, which guarantees that the eigenvalues returned by the function are stored in a NumPy array with real values. For example, to solve a numerical eigenvalue problem with `la.eig`, we can use

```
In [53]: A = np.array([[1, 3, 5], [3, 5, 3], [5, 3, 9]])
In [54]: evals, evecs = la.eig(A)
In [55]: evals
Out[55]: array([ 13.35310908+0.j,  -1.75902942+0.j,   3.40592034+0.j])
In [56]: evecs
Out[56]: array([[ 0.42663918,  0.90353276, -0.04009445],
                [ 0.43751227, -0.24498225, -0.8651975 ],
                [ 0.79155671, -0.35158534,  0.49982569]])
In [57]: la.eigvalsh(A)
Out[57]: array([ -1.75902942,   3.40592034,  13.35310908])
```

Since the matrix in this example is symmetric, we could use `la.eigh` and `la.eigvalsh`, giving real-valued eigenvalue arrays, as shown in the cell `Out[57]` in the preceding code listing.

Nonlinear Equations

In this section we consider *nonlinear* equations. Systems of linear equations, as considered in the previous sections of this chapter, are of fundamental importance in scientific computing because they are easily solved and can be used as important building blocks in many computational methods and techniques. However, in natural sciences and in engineering disciplines, many, if not most, systems are intrinsically nonlinear.

A linear function $f(x)$ by definition satisfies additivity $f(x+y) = f(x)+f(y)$ and homogeneity $f(\alpha x) = \alpha f(x)$, which can be written together as the superposition principle $f(\alpha x+\beta y) = \alpha f(x)+\beta f(y)$. This gives a precise definition of linearity. A *nonlinear* function, in contrast, is a function that does not satisfy these conditions. Nonlinearity is therefore a much broader concept, and a function can be nonlinear in many different ways. However, in general, an expression that contains a variable with a power greater that one is nonlinear. For example, x^2+x+1 is nonlinear because of the x^2 term.

A nonlinear equation can always be written on the form $f(x) = 0$, where $f(x)$ is a nonlinear function and we seek the value of x (which can be a scalar or a vector) such that $f(x)$ is zero. This x is called the root of the function $f(x)$, and equation solving is therefore often referred to as *root finding*. In contrast to the previous section of this chapter, in this section we need to distinguish between univariate equation solving and multivariate equations, in addition to single equations and system of equations.

Univariate Equations

A univariate function is a function that depends only on a single variable $f(x)$, where x is a scalar, and the corresponding univariate equation is on the form $f(x) = 0$. Typical examples of this type of equation are polynomials, such as $x^2 - x + 1 = 0$, and expressions containing elementary functions, such as $x^3 - 3 \sin(x) = 0$ and $\exp(x) - 2 = 0$. Unlike for linear systems, there are no general methods for determining if a nonlinear equation has a solution, or multiple solutions, or if a given solution is unique. This can be understood intuitively from the fact that graphs of nonlinear functions correspond to curves that can intersect $x = 0$ in an arbitrary number of ways.

Because of the vast number of possible situations, it is difficult to develop a completely automatic approach to solving nonlinear equations. Analytically, only equations on special forms can be solved exactly. For example, polynomials of up to 4th order, and in some special cases also higher orders, can be solved analytically, and some equations containing trigonometric and other elementary functions may be solvable analytically. In SymPy we can solve many analytically solvable univariate and nonlinear equations using the `sympy.solve` function. For example, to solve the standard quadratic equation $a + bx + cx^2 = 0$, we define an expression for the equation and pass it to the `sympy.solve` function:

```
In [58]: x, a, b, c = sympy.symbols("x, a, b, c")
In [59]: sympy.solve(a + b*x + c*x**2, x)
Out[59]: [(-b + sqrt(-4*a*c + b**2))/(2*c), -(b + sqrt(-4*a*c + b**2))/(2*c)]
```

The solution is indeed the well-known formula for the solution of this equation. The same method can be used to solve some trigonometric equations:

```
In [60]: sympy.solve(a * sympy.cos(x) - b * sympy.sin(x), x)
Out[60]: [-2*atan((b - sqrt(a**2 + b**2))/a), -2*atan((b +
         sqrt(a**2 + b**2))/a)]
```

However, in general nonlinear equations are typically not solvable analytically. For example, equations that contain both polynomial expressions and elementary functions, such as $\sin x = x$, are often transcendental and do not have an algebraic solution. If we attempt to solve such an equation using SymPy, we obtain an error in the form of an exception:

```
In [61]: sympy.solve(sympy.sin(x)-x, x)
...
NotImplementedError: multiple generators [x, sin(x)]
No algorithms are implemented to solve equation -x + sin(x)
```

In this type of situation, we need to resort to various numerical techniques. As a first step, it is often very useful to graph the function. This can give important clues about the number of solutions to the equation and their approximate locations. This information is often necessary when applying numerical techniques to find good approximations to the roots of the equations. For example, consider the following example (In [62]), which plots four examples of nonlinear functions, as shown in Figure 5-5. From these graphs, we can immediately conclude that the plotted functions, from left to right, have two, three, one, and a large number of roots (at least within the interval that is being graphed).

```
In [62]: x = np.linspace(-2, 2, 1000)
    ...: # four examples of nonlinear functions
    ...: f1 = x**2 - x - 1
    ...: f2 = x**3 - 3 * np.sin(x)
    ...: f3 = np.exp(x) - 2
    ...: f4 = 1 - x**2 + np.sin(50 / (1 + x**2))
    ...:
    ...: # plot each function
    ...: fig, axes = plt.subplots(1, 4, figsize=(12, 3), sharey=True)
    ...:
    ...: for n, f in enumerate([f1, f2, f3, f4]):
    ...:     axes[n].plot(x, f, lw=1.5)
    ...:     axes[n].axhline(0, ls=':', color='k')
    ...:     axes[n].set_ylim(-5, 5)
    ...:     axes[n].set_xticks([-2, -1, 0, 1, 2])
    ...:     axes[n].set_xlabel(r'$x$', fontsize=18)
    ...:
```

```
...: axes[0].set_ylabel(r'$f(x)$', fontsize=18)
...:
...: titles = [r'$f(x)=x^2-x-1$', r'$f(x)=x^3-3\sin(x)$',
...:           r'$f(x)=\exp(x)-2$', r'$f(x)=\sin\left(50/(1+x^2)\
            right)+1-x^2$']
...: for n, title in enumerate(titles):
...:     axes[n].set_title(title)
```

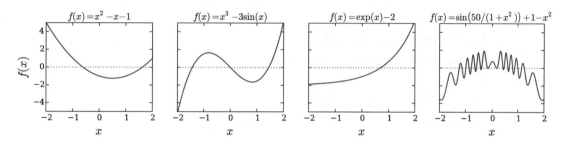

Figure 5-5. *Graphs of four examples of nonlinear functions*

To find the approximate location of a root to an equation, we can apply one of the many techniques for numerical root finding, which typically applies an iterative scheme where the function is evaluated at successive points until the algorithm has narrowed in on the solution, to the desired accuracy. Two standard methods that illustrate the basic idea of how many numerical root-finding methods work are the bisection method and the Newton method.

The bisection method requires a starting interval [*a*, *b*] such that $f(a)$ and $f(b)$ have different signs. This guarantees that there is at least one root within this interval. In each iteration, the function is evaluated in the middle point *m* between *a* and *b*, and sign of the function is different at *a* and *m*, and then the new interval [*a*, *b* = *m*] is chosen for the next iteration. Otherwise the interval [*a* = *m*, *b*] is chosen for the next iteration. This guarantees that in each iteration, the function has a different sign at the two endpoints of the interval, and in each iteration the interval is halved and therefore converges toward a root of the equation. The following code example demonstrates a simple implementation of the bisection method with a graphical visualization of each step, as shown in Figure 5-6.

```
In [63]: # define a function, desired tolerance and starting interval [a, b]
    ...: f = lambda x: np.exp(x) - 2
    ...: tol = 0.1
    ...: a, b = -2, 2
    ...: x = np.linspace(-2.1, 2.1, 1000)
    ...:
    ...: # graph the function f
    ...: fig, ax = plt.subplots(1, 1, figsize=(12, 4))
    ...:
    ...: ax.plot(x, f(x), lw=1.5)
    ...: ax.axhline(0, ls=':', color='k')
    ...: ax.set_xticks([-2, -1, 0, 1, 2])
    ...: ax.set_xlabel(r'$x$', fontsize=18)
    ...: ax.set_ylabel(r'$f(x)$', fontsize=18)
    ...:
    ...: # find the root using the bisection method and visualize
    ...: # the steps in the method in the graph
    ...: fa, fb = f(a), f(b)
    ...:
    ...: ax.plot(a, fa, 'ko')
    ...: ax.plot(b, fb, 'ko')
    ...: ax.text(a, fa + 0.5, r"$a$", ha='center', fontsize=18)
    ...: ax.text(b, fb + 0.5, r"$b$", ha='center', fontsize=18)
    ...:
    ...: n = 1
    ...: while b - a > tol:
    ...:     m = a + (b - a)/2
    ...:     fm = f(m)
    ...:
    ...:     ax.plot(m, fm, 'ko')
    ...:     ax.text(m, fm - 0.5, r"$m_%d$" % n, ha='center')
    ...:     n += 1
    ...:
    ...:     if np.sign(fa) == np.sign(fm):
```

```
...:             a, fa = m, fm
...:         else:
...:             b, fb = m, fm
...:
...: ax.plot(m, fm, 'r*', markersize=10)
...: ax.annotate("Root approximately at %.3f" % m,
...:             fontsize=14, family="serif",
...:             xy=(a, fm), xycoords='data',
...:             xytext=(-150, +50), textcoords='offset points',
...:             arrowprops=dict(arrowstyle="->",
...:             connectionstyle="arc3, rad=-.5"))
...:
...: ax.set_title("Bisection method")
```

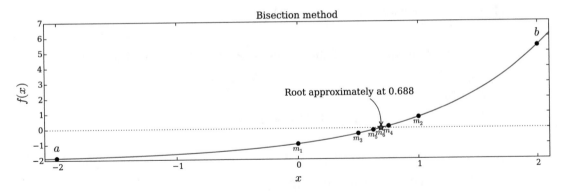

Figure 5-6. *Graphical visualization of how the bisection method works*

Another standard method for root finding is Newton's method, which converges faster than the bisection method discussed in the previous paragraph. While the bisection method only uses the sign of the function at each point, Newton's method uses the actual function values to obtain a more accurate approximation of the nonlinear function. In particular, it approximates the function $f(x)$ with its first-order Taylor expansion $f(x+dx) = f(x)+dx\,f'(x)$, which is a linear function whose root is easily found to be $x - f(x)/f'(x)$. Of course, this does not need to be a root of the function f(x), but in many cases it is a good approximation for getting closer to a root of $f(x)$. By iterating this scheme, $x_{k+1} = x_k - f(x_k)/f'(x_k)$, we may approach the root of the function. A potential problem with this method is that it fails if $f'(x_k)$ is zero at some point x_k. This special case would have to be dealt with in a real implementation of this method. The following example (In [64]) demonstrates how this method can be used to solve for the root of the equation $\exp(x) - 2 = 0$, using SymPy to evaluate the derivative of the function $f(x)$, and Figure 5-7 visualizes the steps in this root-finding process.

```
In [64]: # define a function, desired tolerance and starting point xk
    ...: tol = 0.01
    ...: xk = 2
    ...:
    ...: s_x = sympy.symbols("x")
    ...: s_f = sympy.exp(s_x) - 2
    ...:
    ...: f = lambda x: sympy.lambdify(s_x, s_f, 'numpy')(x)
    ...: fp = lambda x: sympy.lambdify(s_x, sympy.diff(s_f, s_x), 'numpy')(x)
    ...:
    ...: x = np.linspace(-1, 2.1, 1000)
    ...:
    ...: # setup a graph for visualizing the root finding steps
    ...: fig, ax = plt.subplots(1, 1, figsize=(12, 4))
    ...: ax.plot(x, f(x))
    ...: ax.axhline(0, ls=':', color='k')
    ...:
    ...: # iterate Newton's method until convergence to the desired
           tolerance has been reached
    ...: n = 0
    ...: while f(xk) > tol:
    ...:     xk_new = xk - f(xk) / fp(xk)
    ...:
    ...:     ax.plot([xk, xk], [0, f(xk)], color='k', ls=':')
    ...:     ax.plot(xk, f(xk), 'ko')
    ...:     ax.text(xk, -.5, r'$x_%d$' % n, ha='center')
    ...:     ax.plot([xk, xk_new], [f(xk), 0], 'k-')
    ...:
    ...:     xk = xk_new
    ...:     n += 1
    ...:
    ...: ax.plot(xk, f(xk), 'r*', markersize=15)
```

```
...: ax.annotate("Root approximately at %.3f" % xk,
...:             fontsize=14, family="serif",
...:             xy=(xk, f(xk)), xycoords='data',
...:             xytext=(-150, +50), textcoords='offset points',
...:             arrowprops=dict(arrowstyle="->",
...:             connectionstyle="arc3, rad=-.5"))

...:
...: ax.set_title("Newtown's method")
...: ax.set_xticks([-1, 0, 1, 2])
```

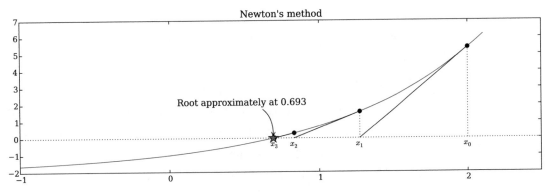

Figure 5-7. *Visualization of the root-finding steps in Newton's method for the equation exp(x) − 2 = 0*

A potential issue with Newton's method is that it requires both the function values and the values of the derivative of the function in each iteration. In the previous example, we used SymPy to symbolically compute the derivatives. In an all-numerical implementation, this is of course not possible, and a numerical approximation of the derivative would be necessary, which would in turn require further function evaluations. A variant of Newton's method that bypasses the requirement to evaluate function derivatives is the secant method, which uses two previous function evaluations to obtain a linear approximation of the function, which can be used to compute a new estimate of the root. The iteration formula for the secant method is $x_{k+1} = x_k - f(x_k)\dfrac{x_k - x_{k-1}}{f(x_k) - f(x_{k-1})}$.

This is only one example of the many variants and possible refinements on the basic idea of Newton's method. State-of-the-art implementations of numerical root-finding functions typically use the basic idea of either the bisection method of Newton's method or a combination of both but additionally use various refinement strategies, such as higher-order interpolations of the function to achieve faster convergence.

The SciPy optimize module provides multiple functions for numerical root finding. The optimize.bisect and optimize.newton functions implement variants of bisection and Newton methods. The optimize.bisect takes three arguments: first a Python function (e.g., a lambda function) that represents the mathematical function for the equation for which a root is to be calculated and the second and third arguments are the lower and upper values of the interval for which to perform the bisection method. Note that the sign of the function has to be different at the points a and b for the bisection method to work, as discussed earlier. Using the optimize.bisect function, we can calculate the root of the equation $\exp(x) - 2 = 0$ that we used in the previous examples, using

```
In [65]: optimize.bisect(lambda x: np.exp(x) - 2, -2, 2)
Out[65]: 0.6931471805592082
```

As long as $f(a)$ and $f(b)$ indeed have different signs, this is guaranteed to give a root within the interval $[a, b]$. In contrast, the function optimize.newton for Newton's method takes a function as the first argument and an initial guess for the root of the function as the second argument. Optionally, it also takes an argument for specifying the derivative of the function, using the fprime keyword argument. If fprime is given, Newton's method is used; otherwise the secant method is used instead. To find the root of the equation $\exp x - 2 = 0$, with and without specifying its derivative, we can use

```
In [66]: x_root_guess = 2
In [67]: f = lambda x: np.exp(x) - 2
In [68]: fprime = lambda x: np.exp(x)
In [69]: optimize.newton(f, x_root_guess)
Out[69]: 0.69314718056
In [70]: optimize.newton(f, x_root_guess, fprime=fprime)
Out[70]: 0.69314718056
```

Note that with this method, we have less control over which root is being computed, if the function has multiple roots. For instance, there is no guarantee that the root the function returns is the closest one to the initial guess; we cannot know in advance if the root is larger or smaller than the initial guess.

The SciPy optimize module provides additional functions for root finding. In particular, the optimize.brentq and optimize.brenth functions, which are variants of the bisection method, also work on an interval where the function changes sign. The optimize.brentq function is generally considered the preferred all-around root-finding function in SciPy. To find a root of the same equation that we considered previously, using optimize.brentq and optimize.brenth functions, we can use

```
In [71]: optimize.brentq(lambda x: np.exp(x) - 2, -2, 2)
Out[71]: 0.6931471805599453
In [72]: optimize.brenth(lambda x: np.exp(x) - 2, -2, 2)
Out[72]: 0.6931471805599381
```

Note that these two functions take a Python function for the equation as the first argument and the lower and upper values of the sign-changing interval as the second and third arguments.

Systems of Nonlinear Equations

In contrast to a linear system of equations, we cannot in general write a system of nonlinear equations as a matrix-vector multiplication. Instead we represent a system of multivariate nonlinear equations as a vector-valued function, for example, $f: \mathbb{R}^N \to \mathbb{R}^N$, that takes an N-dimensional vector and maps it to another N-dimensional vector. Multivariate systems of equations are much more complicated to solve than univariate equations, partly because there are so many more possible behaviors. As a consequence, there is no method that strictly guarantees convergence to a solution, such as the bisection method for a univariate nonlinear equation, and the methods that do exist are much more computationally demanding than the univariate case, especially as the number of variables increases.

Not all methods that we previously discussed for univariate equation solving can be generalized to the multivariate case. For example, the bisection method cannot be directly generalized to a multivariate equation system. On the other hand, Newton's method can be used for multivariate problems, and in this case its iteration formula is $x_{k+1} = x_k - J_f(x_k)^{-1} f(x_k)$, where $J_f(x_k)$ is the Jacobian matrix of the function $f(x)$, with elements $[J_f(x_k)]_{ij} = \partial f_i(x_k)/\partial x_j$. Instead of inverting the Jacobian matrix, it is sufficient to solve the linear equation system $J_f(x_k)\delta x_k = -f(x_k)$ and update x_k using $x_{k+1} = x_k + \delta x_k$. Like the secant variants of the Newton method for univariate equation systems, there are also variants of the multivariate method that avoid computing the Jacobian by estimating it from previous function evaluations. Broyden's method is a popular example of this type of secant updating method for multivariate equation systems. In the SciPy optimize module, `broyden1` and `broyden2` provide two implementations of Broyden's method using different approximations of the Jacobian, and the function `optimize.fsolve` provides an implementation of a Newton-like method, where optionally the Jacobian can be specified, if available. The functions all have a similar function signature: The first argument is a Python function that represents the equation to be solved, and it should take a NumPy array as the first argument and return an array of the same shape. The second argument is an initial guess for the solution, as a NumPy array. The `optimize.fsolve` function also takes an optional keyword argument `fprime`, which can be used to provide a function that returns the Jacobian of the function $f(x)$. In addition, all these functions take numerous optional keyword arguments for tuning their behavior (see the docstrings for details).

For example, consider the following system of two multivariate and nonlinear equations

$$\begin{cases} y - x^3 - 2x^2 + 1 = 0 \\ \quad\quad y + x^2 - 1 = 0 \end{cases},$$

which can be represented by the vector-valued function $f([x_1, x_2]) = [x_2 - x_1^3 - 2x_1^2 + 1, x_2 + x_1^2 - 1]$. To solve this equation system using SciPy, we need to define a Python function for $f([x_1, x_2])$ and call, for example, the `optimize.fsolve` using the function and an initial guess for the solution vector:

```
In [73]: def f(x):
    ...:     return [x[1] - x[0]**3 - 2 * x[0]**2 + 1, x[1] + x[0]**2 - 1]
In [74]: optimize.fsolve(f, [1, 1])
Out[74]: array([ 0.73205081,  0.46410162])
```

The `optimize.broyden1` and `optimize.broyden2` can be used in a similar manner. To specify a Jacobian for `optimize.fsolve` to use, we need to define a function that evaluates the Jacobian for a given input vector. This requires that we first derive the Jacobian by hand or, for example, using SymPy

```
In [75]: x, y = sympy.symbols("x, y")
In [76]: f_mat = sympy.Matrix([y - x**3 -2*x**2 + 1, y + x**2 - 1])
In [77]: f_mat.jacobian(sympy.Matrix([x, y]))
```

$$\text{Out[77]:} \quad \begin{pmatrix} -3x^2 - 4x & 1 \\ 2x & 1 \end{pmatrix}$$

which we can then easily implement as a Python function that can be passed to the `optimize.fsolve` function:

```
In [78]: def f_jacobian(x):
    ...:     return [[-3*x[0]**2-4*x[0], 1], [2*x[0], 1]]
In [79]: optimize.fsolve(f, [1, 1], fprime=f_jacobian)
Out[79]: array([ 0.73205081,  0.46410162])
```

As with the Newton's method for a univariate nonlinear equation system, the initial guess for the solution is important, and different initial guesses may result in different solutions that are found for the equations. There is no guarantee that any particular solution is found, although the proximity of the initial guess to the true solution often is correlated with convergence to that particular solution. When possible, it is often a good approach to graph the equations that are being solved, to give a visual indication of the number of solutions and their locations. For example, the following code demonstrates how three different solutions can be found to the equation systems we are considering here, by using different initial guesses with the `optimize.fsolve` function. The result is shown in Figure 5-8.

```
In [80]: def f(x):
    ...:     return [x[1] - x[0]**3 - 2 * x[0]**2 + 1,
    ...:             x[1] + x[0]**2 - 1]
    ...:
    ...: x = np.linspace(-3, 2, 5000)
    ...: y1 = x**3 + 2 * x**2 -1
    ...: y2 = -x**2 + 1
    ...:
    ...: fig, ax = plt.subplots(figsize=(8, 4))
    ...:
    ...: ax.plot(x, y1, 'b', lw=1.5, label=r'$y = x^3 + 2x^2 - 1$')
    ...: ax.plot(x, y2, 'g', lw=1.5, label=r'$y = -x^2 + 1$')
    ...:
    ...: x_guesses = [[-2, 2], [1, -1], [-2, -5]]
    ...: for x_guess in x_guesses:
    ...:     sol = optimize.fsolve(f, x_guess)
    ...:     ax.plot(sol[0], sol[1], 'r*', markersize=15)
    ...:
```

```
...:         ax.plot(x_guess[0], x_guess[1], 'ko')
...:         ax.annotate("", xy=(sol[0], sol[1]), xytext=(x_guess[0],
                   x_guess[1]),
...:                 arrowprops=dict(arrowstyle="->", linewidth=2.5))
...:
...: ax.legend(loc=0)
...: ax.set_xlabel(r'$x$', fontsize=18)
```

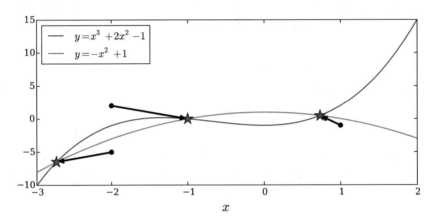

Figure 5-8. *Graph of a system of two nonlinear equations. The solutions are indicated with red stars and the initial guess with a black dot and an arrow to the solution each initial guess eventually converged to.*

By systematically solving the equation systems with different initial guesses, we can build a visualization of how different initial guesses converge to different solutions. This is done in the following code example, and the result is shown in Figure 5-9. This example demonstrates that even for this relatively simple example, the regions of initial guesses that converge to different solutions are highly nontrivial, and there are also missing dots which corresponds to initial guesses for which the algorithm fails to converge to any solution. Nonlinear equation solving is a complex task, and visualizations of different types can often be a valuable tool when building an understanding of the characteristics of a particular problem.

```
In [81]: fig, ax = plt.subplots(figsize=(8, 4))
    ...:
    ...: ax.plot(x, y1, 'k', lw=1.5)
    ...: ax.plot(x, y2, 'k', lw=1.5)
```

```
   ...:
   ...: sol1 = optimize.fsolve(f, [-2,  2])
   ...: sol2 = optimize.fsolve(f, [ 1, -1])
   ...: sol3 = optimize.fsolve(f, [-2, -5])
   ...: sols = [sol1, sol2, sol3]
   ...: for idx, s in enumerate(sols):
   ...:     ax.plot(s[0], s[1], colors[idx]+'*', markersize=15)
   ...:
   ...: colors = ['r', 'b', 'g']
   ...: for m in np.linspace(-4, 3, 80):
   ...:     for n in np.linspace(-15, 15, 40):
   ...:         x_guess = [m, n]
   ...:         sol = optimize.fsolve(f, x_guess)
   ...:         idx = (abs(sols - sol)**2).sum(axis=1).argmin()
   ...:         ax.plot(x_guess[0], x_guess[1], colors[idx]+'.')
   ...:
   ...: ax.set_xlabel(r'$x$', fontsize=18)
```

Figure 5-9. *Visualization of the convergence of different initial guesses to different solutions. Each dot represents an initial guess, and its color encodes which solution it eventually converges to. The solutions are marked with correspondingly color-coded stars.*

Summary

In this chapter we have explored methods for solving algebraic equations using the SymPy and SciPy libraries. Equation solving is one of the most elementary mathematical tools for computational sciences, and it is both an important component in many algorithms and methods and has direct applications in many problem-solving situations. In some cases, analytical algebraic solutions exist, especially for equations that are polynomials or contain certain combinations of elementary functions, and such equations can often be handled symbolically with SymPy. For equations with no algebraic solution, and for larger systems of equations, numerical methods are usually the only feasible approach. Linear equation systems can always be systematically solved, and for this reason there is an abundance of important applications for linear equation systems, be it for originally linear systems or as approximations to originally nonlinear systems. Nonlinear equation solving requires a different set of methods, and it is in general much more complex and computationally demanding compared to linear equation systems. In fact, solving linear equation systems is an important step in the iterative methods employed in many of the methods that exist to solve nonlinear equation systems. For numerical equation solving, we can use the linear algebra and optimization modules in SciPy, which provide efficient and well-tested methods for numerical root finding and equation solving of both linear and nonlinear systems.

Further Reading

Equation solving is a basic numerical technique whose methods are covered in most introductory numerical analysis texts. A good example of a book that covers these topics is Heath (2001), and W.H. Press (2007) gives a practical introduction with implementation details.

References

Heath, M. (2001). *Scientific Computing*. Boston: McGraw-Hill.

W.H. Press, S. T. (2007). *Numerical Recipes: The Art of Scientific Computing* (3rd ed.). Cambridge: Cambridge University Press.

CHAPTER 6

Optimization

In this chapter, we will build on Chapter 5 about equation solving and explore the related topic of solving optimization problems. In general, optimization is the process of finding and selecting the optimal element from a set of feasible candidates. In mathematical optimization, this problem is usually formulated as determining the extreme value of a function on a given domain. An extreme value, or an optimal value, can refer to either the minimum or maximum of the function, depending on the application and the specific problem. In this chapter we are concerned with the optimization of real-valued functions of one or several variables, which optionally can be subject to a set of constraints that restricts the domain of the function.

The applications of mathematical optimization are many and varied, and so are the methods and algorithms that must be employed to solve optimization problems. Since optimization is a universally important mathematical tool, it has been developed and adapted for use in many fields of science and engineering, and the terminology used to describe optimization problems varies between fields. For example, the mathematical function that is optimized may be called a cost function, loss function, energy function, or objective function, to mention a few. Here we use the generic term objective function.

Optimization is closely related to equation solving because at an optimal value of a function, its derivative, or gradient in the multivariate case, is zero. The converse, however, is not necessarily true, but a method to solve optimization problems is to solve for the zeros of the derivative or the gradient and test the resulting candidates for optimality. This approach is not always feasible though, and often it is required to take other numerical approaches, many of which are closely related to the numerical methods for root finding that was covered in Chapter 5.

In this chapter we discuss using SciPy's optimization module `optimize` for nonlinear optimization problems, and we will briefly explore using the convex optimization library `cvxopt` for linear optimization problems with linear constraints. This library also has powerful solvers for quadratic programming problems.

© Robert Johansson 2019
R. Johansson, *Numerical Python*, https://doi.org/10.1007/978-1-4842-4246-9_6

---

cvxopt   The convex optimization library cvxopt provides solvers for linear and quadratic optimization problems. At the time of writing, the latest version is 1.1.9. For more information, see the project's web site `http://cvxopt.org`. Here we use this library for constrained linear optimization.

Importing Modules

Like in the previous chapter, here we use the `optimize` module from the SciPy library. Here we assume that this module is imported in the following manner:

```
In [1]: from scipy import optimize
```

In the later part of this chapter, we also look at linear programming using the `cvxopt` library, which we assume to be imported in its entirety without any alias:

```
In [2]: import cvxopt
```

For basic numerics, symbolics, and plotting, here we also use the NumPy, SymPy, and Matplotlib libraries, which are imported and initialized using the conventions introduced in earlier chapters:

```
In [3]: import matplotlib.pyplot as plt
In [4]: import numpy as np
In [5]: import sympy
In [6]: sympy.init_printing()
```

Classification of Optimization Problems

Here we restrict our attention to mathematical optimization of real-valued functions, with one or more dependent variables. Many mathematical optimization problems can be formulated in this way, but a notable exception is optimization of functions over discrete variables, for example, integers, which is beyond the scope of this book.

A general optimization problem of the type considered here can be formulated as a minimization problem, $\min_x f(x)$, subject to sets of m equality constraints $g(x) = 0$ and p inequality constraints $h(x) \leq 0$. Here $f(x)$ is a real-valued function of x, which can be a scalar or a vector $x = (x_0, x_1, ..., x_n)^T$, while $g(x)$ and $h(x)$ can be vector-valued functions: $f: \mathbb{R}^n \longrightarrow \mathbb{R}$, $g: \mathbb{R}^n \longrightarrow \mathbb{R}^m$ and $h: \mathbb{R}^n \longrightarrow \mathbb{R}^p$. Note that maximizing $f(x)$ is equivalent to minimizing $-f(x)$, so without loss of generality, it is sufficient to consider only minimization problems.

Depending on the properties of the objective function $f(x)$ and the equality and inequality constraints $g(x)$ and $h(x)$, this formulation includes a rich variety of problems. A general mathematical optimization on this form is difficult to solve, and there are no efficient methods for solving completely generic optimization problems. However, there are efficient methods for many important special cases, and in optimization it is therefore important to know as much as possible about the objective functions and the constraints in order to be able to solve a problem.

Optimization problems are classified depending on the properties of the functions $f(x)$, $g(x)$, and $h(x)$. First and foremost, the problem is *univariate* or *one dimensional* if x is a scalar, $x \in \mathbb{R}$, and *multivariate* or *multidimensional* if x is a vector, $x \in \mathbb{R}^n$. For high-dimensional objective functions, with larger n, the optimization problem is harder and more computationally demanding to solve. If the objective function and the constraints all are linear, the problem is a linear optimization problem, or *linear programming* problem.[1] If either the objective function or the constraints are nonlinear, it is a nonlinear optimization problem, or a *nonlinear programming* problem. With respect to constraints, important subclasses of optimization are unconstrained problems, and those with linear and nonlinear constraints. Finally, handling equality and inequality constraints requires different approaches.

As usual, nonlinear problems are much harder to solve than linear problems, because they have a wider variety of possible behaviors. A general nonlinear problem can have both local and global minima, which turns out to make it very difficult to find the global minima: iterative solvers may often converge to local minima rather than the global minima or may even fail to converge altogether if there are both local and global minima. However, an important subclass of nonlinear problems that can be solved efficiently is *convex problems*, which is directly related to the absence of strictly local minima and the existence of a unique global minimum. By definition, a function is convex on an interval $[a, b]$ if the values of the function on this interval lie below the line through the endpoints $(a, f(a))$ and $(b, f(b))$. This condition, which can be readily generalized to the multivariate case, implies a number of important properties, such as

---

[1] For historical reasons optimization problems are often referred to as *programming* problems, which are not related to computer programming.

the existence of a unique minimum on the interval. Because of strong properties like this one, convex problems can be solved efficiently even though they are nonlinear. The concepts of local and global minima, and convex and nonconvex functions, are illustrated in Figure 6-1.

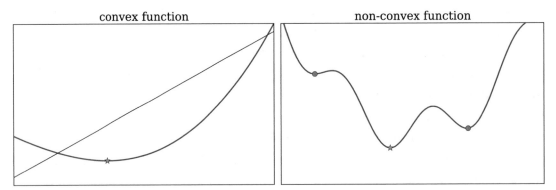

Figure 6-1. *Illustration of a convex function (left) and a nonconvex function (right) with a global minimum and two local minima*

Whether the objective function $f(x)$ and the constraints $g(x)$ and $h(x)$ are continuous and smooth are properties that have very important implications for the methods and techniques that can be used to solve an optimization problem. Discontinuities in these functions, or their derivatives or gradients, cause difficulties for many of the available methods of solving optimization problems, and in the following, we assume that these functions are indeed continuous and smooth. On a related note, if the function itself is not known exactly, but contains noise due to measurements or for other reasons, many of the methods discussed in the following may not be suitable.

Optimization of continuous and smooth functions are closely related to nonlinear equation solving, because extremal values of a function $f(x)$ correspond to points where its derivative, or gradient, is zero. Finding candidates for the optimal value of $f(x)$ is therefore equivalent to solving the (in general nonlinear) equation system $\nabla f(x) = 0$. However, a solution to $\nabla f(x) = 0$, which is known as a stationary point, does not necessarily correspond to a minimum of $f(x)$; it can also be maximum or a saddle point; see Figure 6-2. Candidates obtained by solving $\nabla f(x) = 0$ should therefore be tested for optimality. For unconstrained objective functions, the higher-order derivatives, or Hessian matrix

$$\left\{H_f\left(x\right)\right\}_{ij} = \frac{\partial^2 f\left(x\right)}{\partial x_i \partial x_j},$$

for the multivariate case, can be used to determine if a stationary point is a local minimum or not. In particular if the second-order derivative is positive, or the Hessian positive definite, when evaluated at stationary point x^*, then x^* is a local minimum. Negative second-order derivative, or negative definite Hessian, corresponds to a local maximum, and a zero second-order derivative, or an indefinite Hessian, corresponds to saddle point.

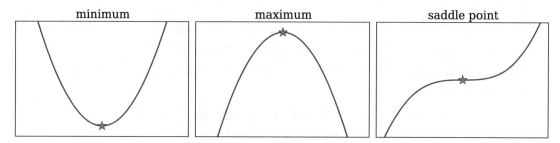

Figure 6-2. *Illustration of different stationary points of a one-dimensional function*

Algebraically solving the equation system $\nabla f(x) = 0$ and testing the candidate solutions for optimality is therefore one possible strategy for solving an optimization problem. However, it is not always a feasible method. In particular, we may not have an analytical expression for $f(x)$ from which we can compute the derivatives, and the resulting nonlinear equation system may not be easy to solve, especially not to find all of its roots. For such cases, there are alternative numerical optimization approaches, some of which have analogs among the root-finding methods discussed in Chapter 5. In the remaining part of this chapter, we explore the various classes of optimization problems and how such problems can be solved in practice using available optimization libraries for Python.

Univariate Optimization

Optimization of a function that only depends on a single variable is relatively easy. In addition to the analytical approach of seeking the roots of the derivative of the function, we can employ techniques that are similar to the root-finding methods for univariate functions, namely, bracketing methods and Newton's method. Like the bisection method for univariate root finding, it is possible to use bracketing and iteratively refine an interval using function evaluations alone. Refining an interval $[a, b]$ that contains a minimum can be achieved by evaluating the function at two interior points x_1 and

x_2, $x_1 < x_2$, and selecting $[x_1, b]$ as new interval if $f(x_1) > f(x_2)$, and $[a, x_2]$ otherwise. This idea is used in the *golden section search* method, which additionally uses the trick of choosing x_1 and x_2 such that their relative positions in the $[a, b]$ interval satisfy the golden ratio. This has the advantage of allowing to reuse one function evaluation from the previous iteration and thus only requiring one new function evaluation in each iteration but still reducing the interval with a constant factor in each iteration. For functions with a unique minimum on the given interval, this approach is guaranteed to converge to an optimal point, but this is unfortunately not guaranteed for more complicated functions. It is therefore important to carefully select the initial interval, ideally relatively close to an optimal point. In the SciPy `optimize` module, the function `golden` implements the golden search method.

As the bisection method for root finding, the golden search method is a (relatively) safe but a slowly converging method. Methods with better convergence can be constructed if the values of the function evaluations are used, rather than only comparing the values to each other (which is similar to using only the sign of the functions, as in the bisection method). The function values can be used to fit a polynomial, for example, a quadratic polynomial, which can be interpolated to find a new approximation for the minimum, giving a candidate for a new function evaluation, after which the process can be iterated. This approach can converge faster but is riskier than bracketing and may not converge at all or may converge to local minima outside the given bracket interval.

Newton's method for root finding is an example of a quadratic approximation method that can be applied to find a function minimum, by applying the method to the derivative rather than the function itself. This yields the iteration formula $x_{k+1} = x_k - f'(x_k)/f''(x_k)$, which can converge quickly if started close to an optimal point but may not converge at all if started too far from the optimal value. This formula also requires evaluating both the derivative and the second-order derivative in each iteration. If analytical expressions for these derivatives are available, this can be a good method. If only function evaluations are available, the derivatives may be approximated using an analog of the secant method for root finding.

A combination of the two previous methods is typically used in practical implementations of univariate optimization routines, giving both stability and fast convergence. In SciPy's `optimize` module, the `brent` function is such a hybrid method, and it is generally the preferred method for optimization of univariate functions with SciPy. This method is a variant of the golden section search method that uses inverse parabolic interpolation to obtain faster convergence.

Instead of calling the `optimize.golden` and `optimize.brent` functions directly, it is convenient to use the unified interface function `optimize.minimize_scalar`, which dispatches to the `optimize.golden` and `optimize.brent` functions depending on the value of the `method` keyword argument, where the currently allowed options are `'Golden'`, `'Brent'`, or `'Bounded'`. The last option dispatches to `optimize.fminbound`, which performs optimization on a bounded interval, which corresponds to an optimization problem with inequality constraints that limit the domain of objective function $f(x)$. Note that the `optimize.golden` and `optimize.brent` functions may converge to a local minimum outside the initial bracket interval, but `optimize.fminbound` would in such circumstances return the value at the end of the allowed range.

As an example for illustrating these techniques, consider the following classic optimization problem: Minimize the area of a cylinder with unit volume. Here, suitable variables are the radius r and height h of the cylinder, and the objective function is $f([r, h]) = 2\pi r^2 + 2\pi rh$, subject to the equality constraint $g([r, h]) = \pi r^2 h - 1 = 0$. As this problem is formulated here, it is a two-dimensional optimization problem with an equality constraint. However, we can algebraically solve the constraint equation for one of the dependent variables, for example, $h = 1/\pi r^2$, and substitute this into the objective function to obtain an unconstrained one-dimensional optimization problem: $f(r) = 2\pi r^2 + 2/r$. To begin with, we can solve this problem symbolically using SymPy, using the method of equating the derivative of $f(r)$ to zero:

```
In [7]: r, h = sympy.symbols("r, h")
In [8]: Area = 2 * sympy.pi * r**2 + 2 * sympy.pi * r * h
In [9]: Volume = sympy.pi * r**2 * h
In [10]: h_r = sympy.solve(Volume - 1)[0]
In [11]: Area_r = Area.subs(h_r)
In [12]: rsol = sympy.solve(Area_r.diff(r))[0]
In [13]: rsol
```
$$\text{Out[13]:} \quad \frac{2^{2/3}}{2\sqrt[3]{\pi}}$$
```
In [14]: _.evalf()
Out[14]: 0.541926070139289
```

Now verify that the second derivative is positive and that `rsol` corresponds to a minimum:

```
In [15]: Area_r.diff(r, 2).subs(r, rsol)
Out[15]: 12π
In [16]: Area_r.subs(r, rsol)
Out[16]: 3∛2π

In [17]: _.evalf()
Out[17]: 5.53581044593209
```

For simple problems this approach is often feasible, but for more realistic problems, we typically need to resort to numerical techniques. To solve this problem using SciPy's numerical optimization functions, we first define a Python function f that implements the objective function. To solve the optimization problem, we then pass this function to, for example, `optimize.brent`. Optionally we can use the `brack` keyword argument to specify a starting interval for the algorithm:

```
In [18]: def f(r):
    ...:        return 2 * np.pi * r**2 + 2 / r
In [19]: r_min = optimize.brent(f, brack=(0.1, 4))
In [20]: r_min
Out[20]: 0.541926077256
In [21]: f(r_min)
Out[21]: 5.53581044593
```

Instead of calling `optimize.brent` directly, we could use the generic interface for scalar minimization problems `optimize.minimize_scalar`. Note that to specify a starting interval in this case, we must use the `bracket` keyword argument:

```
In [22]: optimize.minimize_scalar(f, bracket=(0.1, 4))
Out[22]:   nit: 13
           fun: 5.5358104459320856
             x: 0.54192606489766715
          nfev: 14
```

All these methods give that the radius that minimizes the area of the cylinder is approximately 0.54 (the exact result from the symbolic calculation is $2^{2/3}/2\sqrt[3]{\pi}$) and a minimum area of approximately 5.54 (the exact result is $3\sqrt[3]{2\pi}$). The objective function that we minimized in this example is plotted in Figure 6-3, where the minimum is marked with a red star. When possible, it is a good idea to visualize the objective function before attempting a numerical optimization, because it can help in identifying a suitable initial interval or a starting point for the numerical optimization routine.

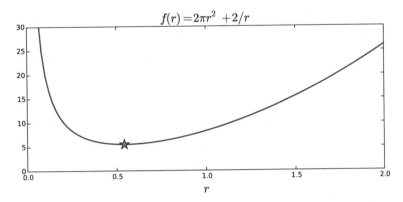

Figure 6-3. *The surface area of a cylinder with unit volume as a function of the radius r*

Unconstrained Multivariate Optimization

Multivariate optimization is significantly harder than the univariate optimization discussed in the previous section. In particular, the analytical approach of solving the nonlinear equations for roots of the gradient is rarely feasible in the multivariate case, and the bracketing scheme used in the golden search method is also not directly applicable. Instead we must resort to techniques that start at some point in the coordinate space and use different strategies to move toward a better approximation of the minimum point. The most basic approach of this type is to consider the gradient $\nabla f(x)$ of the objective function $f(x)$ at a given point x. In general, the negative of the gradient, $-\nabla f(x)$, always points in the direction in which the function $f(x)$ decreases the most. As minimization strategy, it is therefore sensible to move along this direction for some distance α_k and then iterate this scheme at the new point. This method is known as the *steepest descent method,* and it gives the iteration formula $x_{k+1} = x_k - \alpha_k \nabla f(x_k)$, where α_k is a free parameter known as the *line search parameter* that describes how far along the given direction to move in each iteration. An appropriate α_k can, for example, be selected by solving the one-dimensional optimization problem $\min_{a_k} f\left(x_k - \alpha_k \nabla f\left(x_k\right)\right)$. This method is guaranteed to make progress and eventually converge to a minimum of the function, but the convergence can be quite slow because this method tends to overshoot along the direction of the gradient, giving a zigzag approach to the minimum. Nonetheless, the steepest descent method is the conceptual basis for many multivariate optimization algorithms, and with suitable modifications, the convergence can be speed up.

Newton's method for multivariate optimization is a modification of the steepest descent method that can improve convergence. As in the univariate case, Newton's method can be viewed as a local quadratic approximation of the function, which when minimized gives an iteration scheme. In the multivariate case, the iteration formula is $x_{k+1} = x_k - H_f^{-1}(x_k)\nabla f(x_k)$, where compared to the steepest descent method, the gradient has been replaced with the gradient multiplied from the left with the inverse of Hessian matrix for the function.[2] In general this alters both the direction and the length of the step, so this method is not strictly a steepest descent method and may not converge if started too far from a minimum. However, when close to a minimum, it converges quickly. As usual there is a trade-off between convergence rate and stability. As it is formulated here, Newton's method requires both the gradient and the Hessian of the function.

In SciPy, Newton's method is implemented in the function `optimize.fmin_ncg`. This function takes the following arguments: a Python function for the objective function, a starting point, a Python function for evaluating the gradient, and (optionally) a Python function for evaluating the Hessian. To see how this method can be used to solve an optimization problem, we consider the following problem: $\min_x f(x)$ where the objective function is $f(x) = (x_1 - 1)^4 + 5(x_2 - 1)^2 - 2x_1 x_2$. To apply Newton's method, we need to calculate the gradient and the Hessian. For this particular case, this can easily be done by hand. However, for the sake of generality, in the following we use SymPy to compute symbolic expressions for the gradient and the Hessian. To this end, we begin by defining symbols and a symbolic expression for the objective function, and then use the `sympy.diff` function for each variable to obtain the gradient and Hessian in symbolic form:

```
In [23]: x1, x2 = sympy.symbols("x_1, x_2")
In [24]: f_sym = (x1-1)**4 + 5 * (x2-1)**2 - 2*x1*x2
In [25]: fprime_sym = [f_sym.diff(x_) for x_ in (x1, x2)]
```

---

[2]In practice, the inverse of the Hessian does not need to be computed, and instead we can solve the linear equation system $H_f(x_k)y_k = -\nabla f(x_k)$ and use the integration formula $x_{k+1} = x_k + y_k$.

```
In [26]: # Gradient
    ...: sympy.Matrix(fprime_sym)
```

$$\text{Out}[26]: \begin{bmatrix} -2x_2 + 4(x_1 - 1)^3 \\ -2x_1 + 10x_2 - 10 \end{bmatrix}$$

```
In [27]: fhess_sym = [[f_sym.diff(x1_, x2_) for x1_ in (x1, x2)] for x2_ in
         (x1, x2)]
In [28]: # Hessian
    ...: sympy.Matrix(fhess_sym)
```

$$\text{Out}[28]: \begin{bmatrix} 12(x_1 - 1)^2 & -2 \\ -2 & 10 \end{bmatrix}$$

Now that we have a symbolic expression for the gradient and the Hessian, we can create vectorized functions for these expressions using sympy.lambdify.

```
In [29]: f_lmbda = sympy.lambdify((x1, x2), f_sym, 'numpy')
In [30]: fprime_lmbda = sympy.lambdify((x1, x2), fprime_sym, 'numpy')
In [31]: fhess_lmbda = sympy.lambdify((x1, x2), fhess_sym, 'numpy')
```

However, the functions produced by sympy.lambdify take one argument for each variable in the corresponding expression, and the SciPy optimization functions expect a vectorized function where all coordinates are packed into one array. To obtain functions that are compatible with the SciPy optimization routines, we wrap each of the functions generated by sympy.lambdify with a Python function that rearranges the arguments:

```
In [32]: def func_XY_to_X_Y(f):
    ...:     """
    ...:     Wrapper for f(X) -> f(X[0], X[1])
    ...:     """
    ...:     return lambda X: np.array(f(X[0], X[1]))
In [33]: f = func_XY_to_X_Y(f_lmbda)
In [34]: fprime = func_XY_to_X_Y(fprime_lmbda)
In [35]: fhess = func_XY_to_X_Y(fhess_lmbda)
```

Now the functions f, fprime, and fhess are vectorized Python functions on the form that, for example, optimize.fmin_ncg expects, and we can proceed with a numerical optimization of the problem at hand by calling this function. In addition to the functions that we have prepared from SymPy expressions, we also need to give a starting point for the Newton method. Here we use (0, 0) as the starting point.

```
In [36]: x_opt = optimize.fmin_ncg(f, (0, 0), fprime=fprime, fhess=fhess)
         Optimization terminated successfully.
                 Current function value: -3.867223
                 Iterations: 8
                 Function evaluations: 10
                 Gradient evaluations: 17
                 Hessian evaluations: 8
In [37]: x_opt
Out[37]: array([ 1.88292613,  1.37658523])
```

The routine found a minimum point at $(x_1, x_2) = (1.88292613, 1.37658523)$, and diagnostic information about the solution was also printed to standard output, including the number of iterations and the number of function, gradient, and Hessian evaluations that were required to arrive at the solution. As usual it is illustrative to visualize the objective function and the solution (see Figure 6-4):

```
In [38]: fig, ax = plt.subplots(figsize=(6, 4))
    ...: x_ = y_ = np.linspace(-1, 4, 100)
    ...: X, Y = np.meshgrid(x_, y_)
    ...: c = ax.contour(X, Y, f_lmbda(X, Y), 50)
    ...: ax.plot(x_opt[0], x_opt[1], 'r*', markersize=15)
    ...: ax.set_xlabel(r"$x_1$", fontsize=18)
    ...: ax.set_ylabel(r"$x_2$", fontsize=18)
    ...: plt.colorbar(c, ax=ax)
```

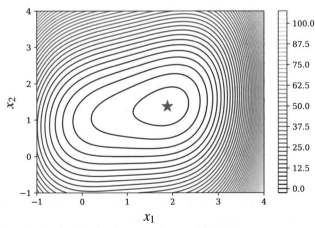

Figure 6-4. *Contour plot of the objective function* $f(x) = (x_1 - 1)^4 + 5(x_2 - 1)^2 - 2x_1x_2.$ *The minimum point is marked by a red star.*

In practice, it may not always be possible to provide functions for evaluating both the gradient and the Hessian of the objective function, and often it is convenient with a solver that only requires function evaluations. For such cases, several methods exist to numerically estimate the gradient or the Hessian or both. Methods that approximate the Hessian are known as quasi-Newton methods, and there are also alternative iterative methods that completely avoid using the Hessian. Two popular methods are the Broyden-Fletcher-Goldfarb-Shanno (BFGS) and the conjugate gradient methods, which are implemented in SciPy as the functions `optimize.fmin_bfgs` and `optimize.fmin_cg`. The BFGS method is a quasi-Newton method that can gradually build up numerical estimates of the Hessian, and also the gradient, if necessary. The conjugate gradient method is a variant of the steepest descent method and does not use the Hessian, and it can be used with numerical estimates of the gradient obtained from only function evaluations. With these methods, the number of function evaluations that are required to solve a problem is much larger than for Newton's method, which on the other hand also evaluates the gradient and the Hessian. Both `optimize.fmin_bfgs` and `optimize. fmin_cg` can optionally accept a function for evaluating the gradient, but if not provided, the gradient is estimated from function evaluations.

The preceding problem given, which was solved with the Newton method, can also be solved using the `optimize.fmin_bfgs` and `optimize.fmin_cg`, without providing a function for the Hessian:

```
In [39]: x_opt = optimize.fmin_bfgs(f, (0, 0), fprime=fprime)
         Optimization terminated successfully.
             Current function value: -3.867223
             Iterations: 10
             Function evaluations: 14
             Gradient evaluations: 14
In [40]: x_opt
Out[40]: array([ 1.88292605,  1.37658523])

In [41]: x_opt = optimize.fmin_cg(f, (0, 0), fprime=fprime)
         Optimization terminated successfully.
             Current function value: -3.867223
             Iterations: 7
             Function evaluations: 17
             Gradient evaluations: 17
```

```
In [42]: x_top
Out[42]: array([ 1.88292613,  1.37658522])
```

Note that here, as shown in the diagnostic output from the optimization solvers in the preceding section, the number of function and gradient evaluations is larger than for Newton's method. As already mentioned, both of these methods can also be used without providing a function for the gradient as well, as shown in the following example using the optimize.fmin_bfgs solver:

```
In [43]: x_opt = optimize.fmin_bfgs(f, (0, 0))
         Optimization terminated successfully.
             Current function value: -3.867223
             Iterations: 10
             Function evaluations: 56
             Gradient evaluations: 14
In [44]: x_opt
Out[44]: array([ 1.88292604,  1.37658522])
```

In this case the number of function evaluations is even larger, but it is clearly convenient to not have to implement functions for the gradient and the Hessian.

In general, the BFGS method is often a good first approach to try, in particular if neither the gradient nor the Hessian is known. If only the gradient is known, then the BFGS method is still the generally recommended method to use, although the conjugate gradient method is often a competitive alternative to the BFGS method. If both the gradient and the Hessian are known, then Newton's method is the method with the fastest convergence in general. However, it should be noted that although the BFGS and the conjugate gradient methods theoretically have slower convergence than Newton's method, they can sometimes offer improved stability and can therefore be preferable. Each iteration can also be more computationally demanding with Newton's method compared to quasi-Newton methods and the conjugate gradient method, and especially for large problems, these methods can be faster in spite of requiring more iterations.

The methods for multivariate optimization that we have discussed so far all converge to a local minimum in general. For problems with many local minima, this can easily lead to a situation when the solver easily gets stuck in a local minimum, even if a global minimum exists. Although there is no complete and general solution to this problem, a practical approach that can partially alleviate this problem is to use a brute force search over a coordinate grid to find a suitable starting point for an iterative solver. At least this gives a systematic approach to find a global minimum within given coordinate ranges. In SciPy, the function `optimize.brute` can carry out such a systematic search. To illustrate this method, consider the problem of minimizing the function $4 \sin x\pi + 6 \sin y\pi + (x - 1)^2 + (y - 1)^2$, which has a large number of local minima. This can make it tricky to pick a suitable initial point for an iterative solver. To solve this optimization problem with SciPy, we first define a Python function for the objective function:

```
In [45]: def f(X):
   ...:       x, y = X
   ...:       return (4 * np.sin(np.pi * x) + 6 * np.sin(np.pi * y)) +
            (x - 1)**2 + (y - 1)**2
```

To systematically search for the minimum over a coordinate grid, we call `optimize.brute` with the objective function f as the first parameter and a tuple of `slice` objects as the second argument, one for each coordinate. The `slice` objects specify the coordinate grid over which to search for a minimum value. Here we also set the keyword argument `finish=None`, which prevents the `optimize.brute` from automatically refining the best candidate.

```
In [46]: x_start = optimize.brute(f, (slice(-3, 5, 0.5),
           slice(-3, 5, 0.5)), finish=None)
In [47]: x_start
Out[47]: array([ 1.5,   1.5])
In [48]: f(x_start)
Out[48]: −9.5
```

On the coordinate grid specified by the given tuple of `slice` objects, the optimal point is $(x_1, x_2) = (1.5, 1.5)$, with corresponding objective function minimum −9.5. This is now a good starting point for a more sophisticated iterative solver, such as `optimize.fmin_bfgs`:

```
In [49]: x_opt = optimize.fmin_bfgs(f, x_start)
           Optimization terminated successfully.
                 Current function value: -9.520229
                 Iterations: 4
                 Function evaluations: 28
                 Gradient evaluations: 7
```

```
In [50]: x_opt
Out[50]: array([ 1.47586906,  1.48365788])
In [51]: f(x_opt)
Out[51]: -9.52022927306
```

Here the BFGS method gave the final minimum point $(x_1, x_2) = (1.47586906, 1.48365788)$, with the minimum value of the objective function -9.52022927306. For this type of problem, guessing the initial starting point easily results in that the iterative solver converges to a local minimum, and the systematic approach that `optimize.brute` provides is frequently useful.

As always, it is important to visualize the objective function and the solution when possible. The following two code cells plot a contour graph of the current objective function and mark the obtained solution with a red star (see Figure 6-5). As in the previous example, we need a wrapper function for reshuffling the parameters of the objective function because of the different conventions of how the coordinated vectors are passed to the function (separate arrays and packed into one array, respectively).

```
In [52]: def func_X_Y_to_XY(f, X, Y):
    ...:     """
    ...:     Wrapper for f(X, Y) -> f([X, Y])
    ...:     """
    ...:     s = np.shape(X)
    ...:     return f(np.vstack([X.ravel(), Y.ravel()])).reshape(*s)
In [53]: fig, ax = plt.subplots(figsize=(6, 4))
    ...: x_ = y_ = np.linspace(-3, 5, 100)
    ...: X, Y = np.meshgrid(x_, y_)
    ...: c = ax.contour(X, Y, func_X_Y_to_XY(f, X, Y), 25)
    ...: ax.plot(x_opt[0], x_opt[1], 'r*', markersize=15)
    ...: ax.set_xlabel(r"$x_1$", fontsize=18)
    ...: ax.set_ylabel(r"$x_2$", fontsize=18)
    ...: plt.colorbar(c, ax=ax)
```

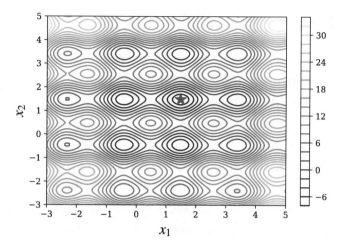

Figure 6-5. *Contour plot of the objective function f(x) = 4 sin xπ+6 sin yπ+(x − 1)²+ (y − 1)². The minimum is marked with a red star.*

In this section, we have explicitly called functions for specific solvers, for example, `optimize.fmin_bfgs`. However, like for scalar optimization, SciPy also provides a unified interface for all multivariate optimization solver with the function `optimize.minimize`, which dispatches out to the solver-specific functions depending on the value of the `method` keyword argument (remember, the univariate minimization function that provides a unified interface is `optimize.scalar_minimize`). For clarity, here we have favored explicitly calling functions for specific solvers, but in general it is a good idea to use `optimize.minimize`, as this makes it easier to switch between different solvers. For example, in the previous example, where we used `optimize.fmin_bfgs` in the following way,

```
In [54]: x_opt = optimize.fmin_bfgs(f, x_start)
```

we could just as well have used

```
In [55]: result = optimize.minimize(f, x_start, method= 'BFGS')
In [56]: x_opt = result.x
```

The `optimize.minimize` function returns an instance of `optimize.OptimizeResult` that represents the result of the optimization. In particular, the solution is available via the `x` attribute of this class.

Nonlinear Least Square Problems

In Chapter 5 we encountered linear least square problems and explored how they can be solved with linear algebra methods. In general, a least square problem can be viewed as an optimization problem with the objective function $g(\beta) = \sum_{i=0}^{m} r_i(\beta)^2 = \|r(\beta)\|^2$, where $r(\beta)$ is a vector with the residuals $r_i(\beta) = y_i - f(x_i, \beta)$ for a set of m observations (x_i, y_i). Here β is a vector with unknown parameters that specifies the function $f(x, \beta)$. If this problem is nonlinear in the parameters β, it is known as a nonlinear least square problem, and since it is nonlinear, it cannot be solved with the linear algebra techniques discussed in Chapter 5. Instead, we can use the multivariate optimization techniques described in the previous section, such as Newton's method or a quasi-Newton method. However, this nonlinear least square optimization problem has a specific structure, and several methods that are tailored to solve this particular optimization problem have been developed. One example is the Levenberg-Marquardt method, which is based on the idea of successive linearizations of the problem in each iteration.

In SciPy, the function `optimize.leastsq` provides a nonlinear least square solver that uses the Levenberg-Marquardt method. To illustrate how this function can be used, consider a nonlinear model on the form $f(x, \beta) = \beta_0 + \beta_1 \exp(-\beta_2 x^2)$ and a set of observations (x_i, y_i). In the following example, we simulate the observations with random noise added to the true values, and we solve the minimization problem that gives the best least square estimates of the parameters β. To begin with, we define a tuple with the true values of the parameter vector β and a Python function for the model function. This function, which should return the y value corresponding to a given x value, takes as first argument the variable x, and the following arguments are the unknown function parameters:

```
In [57]: beta = (0.25, 0.75, 0.5)
In [58]: def f(x, b0, b1, b2):
    ...:     return b0 + b1 * np.exp(-b2 * x**2)
```

Once the model function is defined, we generate randomized data points that simulate the observations.

```
In [59]: xdata = np.linspace(0, 5, 50)
In [60]: y = f(xdata, *beta)
In [61]: ydata = y + 0.05 * np.random.randn(len(xdata))
```

With the model function and observation data prepared, we are ready to start solving the nonlinear least square problem. The first step is to define a function for the residuals given the data and the model function, which is specified in terms of the yet-to-be determined model parameters β.

```
In [62]: def g(beta):
    ...:        return ydata - f(xdata, *beta)
```

Next we define an initial guess for the parameter vector and let the optimize. leastsq function solve for the best least square fit for the parameter vector:

```
In [63]: beta_start = (1, 1, 1)
In [64]: beta_opt, beta_cov = optimize.leastsq(g, beta_start)
In [65]: beta_opt
Out[65]: array([ 0.25733353,  0.76867338,  0.54478761])
```

Here the best fit is quite close to the true parameter values $(0.25, 0.75, 0.5)$, as defined earlier. By plotting the observation data and the model function for the true and fitted function parameters, we can visually confirm that the fitted model seems to describe the data well (see Figure 6-6).

```
In [66]: fig, ax = plt.subplots()
    ...: ax.scatter(xdata, ydata, label='samples')
    ...: ax.plot(xdata, y, 'r', lw=2, label='true model')
    ...: ax.plot(xdata, f(xdata, *beta_opt), 'b', lw=2, label='fitted model')
    ...: ax.set_xlim(0, 5)
    ...: ax.set_xlabel(r"$x$", fontsize=18)
    ...: ax.set_ylabel(r"$f(x, \beta)$", fontsize=18)
    ...: ax.legend()
```

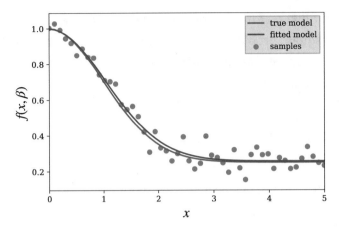

Figure 6-6. *Nonlinear least square fit to the function* $f(x, \beta) = \beta_0 + \beta_1 \exp\left(-\beta_2 x^2\right)$ *with* $\beta = (0.25, 0.75, 0.5)$

The SciPy `optimize` module also provides an alternative interface to nonlinear least square fitting, through the function `optimize.curve_fit`. This is a convenience wrapper around `optimize.leastsq`, which eliminates the need to explicitly define the residual function for the least square problem. The previous problem could therefore be solved more concisely using the following:

```
In [67]: beta_opt, beta_cov = optimize.curve_fit(f, xdata, ydata)
In [68]: beta_opt
Out[68]: array([ 0.25733353,  0.76867338,  0.54478761])
```

Constrained Optimization

Constraints add another level of complexity to optimization problems, and they require a classification of their own. A simple form of constrained optimization is the optimization where the coordinate variables are subject to some bounds. For example: $\min_x f(x)$ subject to $0 \leq x \leq 1$. The constraint $0 \leq x \leq 1$ is simple because it only restricts the range of the coordinate without dependencies on the other variables. This type of problems can be solved using the L-BFGS-B method in SciPy, which is a variant of the BFGS method we used earlier. This solver is available through the function `optimize.fmin_l_bgfs_b` or via `optimize.minimize` with the `method` argument set to `'L-BFGS-B'`. To define the coordinate boundaries, the bound keyword argument must be used, and its value should be a list of tuples that contain the minimum and maximum value of each constrained variable. If the minimum or maximum value is set to `None`, it is interpreted as an unbounded.

As an example of solving a bounded optimization problem with the L-BFGS-B solver, consider minimizing the objective function $f(x) = (x_1 - 1)^2 - (x_2 - 1)^2$ subject to the constraints $2 \leq x_1 \leq 3$ and $0 \leq x_2 \leq 2$. To solve this problem, we first define a Python function for the objective functions and tuples with the boundaries for each of the two variables in this problem, according to the given constraints. For comparison, in the following code, we also solve the unconstrained optimization problem with the same objective function, and we plot a contour graph of the objective function where the unconstrained and constrained minimum values are marked with blue and red stars, respectively (see Figure 6-7).

```
In [69]: def f(X):
    ...:     x, y = X
    ...:     return (x - 1)**2 + (y - 1)**2
In [70]: x_opt = optimize.minimize(f, [1, 1], method='BFGS').x
In [71]: bnd_x1, bnd_x2 = (2, 3), (0, 2)
In [72]: x_cons_opt = optimize.minimize(f, [1, 1], method='L-BFGS-B',
    ...:                                 bounds=[bnd_x1, bnd_x2]).x
In [73]: fig, ax = plt.subplots(figsize=(6, 4))
    ...: x_ = y_ = np.linspace(-1, 3, 100)
    ...: X, Y = np.meshgrid(x_, y_)
    ...: c = ax.contour(X, Y, func_X_Y_to_XY(f, X, Y), 50)
    ...: ax.plot(x_opt[0], x_opt[1], 'b*', markersize=15)
    ...: ax.plot(x_cons_opt[0], x_cons_opt[1], 'r*', markersize=15)
    ...: bound_rect = plt.Rectangle((bnd_x1[0], bnd_x2[0]),
    ...:                            bnd_x1[1] - bnd_x1[0], bnd_x2[1] -
    ...:                            bnd_x2[0], facecolor="grey")
    ...: ax.add_patch(bound_rect)
    ...: ax.set_xlabel(r"$x_1$", fontsize=18)
    ...: ax.set_ylabel(r"$x_2$", fontsize=18)
    ...: plt.colorbar(c, ax=ax)
```

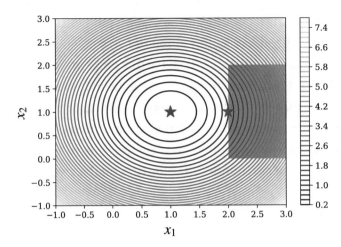

Figure 6-7. *Contours of the objective function f(x), with the unconstrained (blue star) and constrained minima (red star). The feasible region of the constrained problem is shaded in gray.*

Constraints that are defined by equalities or inequalities that include more than one variable are somewhat more complicated to deal with. However, there are general techniques also for this type of problems. For example, using the Lagrange multipliers, it is possible to convert a constrained optimization problem to an unconstrained problem by introducing additional variables. For example, consider the optimization problem $\min_x f(x)$ subject to the equality constraint $g(x) = 0$. In an unconstrained optimization problem, the gradient of $f(x)$ vanishes at the optimal points, $\nabla f(x) = 0$. It can be shown that the corresponding condition for constrained problems is that the negative gradient lies in the space supported by the constraint normal, i.e., $-\nabla f(x) = \lambda J_g^T(x)$. Here $J_g(x)$ is the Jacobian matrix of the constraint function $g(x)$ and λ is the vector of Lagrange multipliers (new variables). This condition arises from equating to zero the gradient of the function $\Lambda(x, \lambda) = f(x) + \lambda^T g(x)$, which is known as the Lagrangian function. Therefore, if both $f(x)$ and $g(x)$ are continuous and smooth, a stationary point (x_0, λ_0) of the function $\Lambda(x, \lambda)$ corresponds to an x_0 that is an optimum of the original constrained optimization problem. Note that if $g(x)$ is a scalar function (i.e., there is only one constraint), then the Jacobian $J_g(x)$ reduces to the gradient $\nabla g(x)$.

To illustrate this technique, consider the problem of maximizing the volume of a rectangle with sides of length x_1, x_2, and x_3, subject to the constraint that the total surface area should be unity: $g(x) = 2x_1x_2 + 2x_0x_2 + 2x_1x_0 - 1 = 0$. To solve this optimization problem using Lagrange multipliers, we form the Lagrangian $\Lambda(x) = f(x) + \lambda g(x)$ and seek the stationary points for $\nabla \Lambda(x) = 0$. With SymPy, we can carry out this task by first defining the symbols for the variables in the problem, then constructing expressions for $f(x)$, $g(x)$, and $\Lambda(x)$,

```
In [74]: x = x0, x1, x2, l = sympy.symbols("x_0, x_1, x_2, lambda")
In [75]: f = x0 * x1 * x2
In [76]: g = 2 * (x0 * x1 + x1 * x2 + x2 * x0) - 1
In [77]: L = f + l * g
```

and finally computing $\nabla \Lambda(x)$ using `sympy.diff` and solving the equation $\nabla \Lambda(x) = 0$ using `sympy.solve`:

```
In [78]: grad_L = [sympy.diff(L, x_) for x_ in x]
In [79]: sols = sympy.solve(grad_L)
In [80]: sols
```
$$\text{Out[80]: } \left[\left\{ \lambda : -\frac{\sqrt{6}}{24}, x_0 : \frac{\sqrt{6}}{6}, x_1 : \frac{\sqrt{6}}{6}, x_2 : \frac{\sqrt{6}}{6} \right\}, \left\{ \lambda : \frac{\sqrt{6}}{24}, x_0 : -\frac{\sqrt{6}}{6}, x_1 : -\frac{\sqrt{6}}{6}, x_2 : -\frac{\sqrt{6}}{6} \right\} \right]$$

This procedure gives two stationary points. We could determine which one corresponds to the optimal solution by evaluating the objective function for each case. However, here only one of the stationary points corresponds to a physically acceptable solution: since x_i is the length of a rectangle side in this problem, it must be positive. We can therefore immediately identify the interesting solution, which corresponds to the intuitive result $x_0 = x_1 = x_2 = \frac{\sqrt{6}}{6}$ (a cube). As a final verification, we evaluate the constraint function and the objective function using the obtained solution:

```
In [81]: g.subs(sols[0])
Out[81]: 0
In [82]: f.subs(sols[0])
```
$$\text{Out[82]: } \frac{\sqrt{6}}{36}$$

This method can be extended to handle inequality constraints as well, and there exist various numerical methods of applying this approach. One example is the method known as sequential least square programming, abbreviated as SLSQP, which is available in the SciPy as the `optimize.slsqp` function and via `optimize.minimize` with method='SLSQP'. The optimize.minimize function takes the keyword argument constraints, which should be a list of dictionaries that each specifies a constraint. The allowed keys (values) in this dictionary are type ('eq' or 'ineq'), fun (constraint function), jac (Jacobian of the constraint function), and args (additional arguments to constraint function and the function for evaluating its Jacobian). For example, the constraint dictionary describing the constraint in the previous problem would be dict(type='eq', fun=g).

To solve the full problem numerically using SciPy's SLSQP solver, we need to define Python functions for the objective function and the constraint function:

```
In [83]: def f(X):
   ...:        return -X[0] * X[1] * X[2]
In [84]: def g(X):
   ...:        return 2 * (X[0]*X[1] + X[1] * X[2] + X[2] * X[0]) - 1
```

Note that since the SciPy optimization functions solve minimization problems, and here we are interested in maximization, the function f is here the negative of the original objective function. Next we define the constraint dictionary for $g(x) = 0$ and finally call the optimize.minimize function

```
In [85]: constraint = dict(type='eq', fun=g)
In [86]: result = optimize.minimize(f, [0.5, 1, 1.5], method='SLSQP',
         constraints=[constraint])
In [87]: result
Out[87]:  status: 0
          success: True
             njev: 18
             nfev: 95
              fun: -0.068041368623352985
                x: array([ 0.40824187,  0.40825127,  0.40825165])
          message: 'Optimization terminated successfully.'
              jac: array([-0.16666925, -0.16666542, -0.16666527,  0.])
              nit: 18
In [88]: result.x
Out[88]: array([ 0.40824187,  0.40825127,  0.40825165])
```

As expected, the solution agrees well with the analytical result obtained from the symbolic calculation using Lagrange multipliers.

To solve problems with inequality constraints, all we need to do is to set `type='ineq'` in the constraint dictionary and provide the corresponding inequality function. To demonstrate minimization of a nonlinear objective function with a nonlinear inequality constraint, we return to the quadratic problem considered previously but in this case with inequality constraint $g(x) = x_1 - 1.75 - (x_0 - 0.75)^4 \geq 0$. As usual, we begin by defining the objective function and the constraint function, as well as the constraint dictionary:

```
In [89]: def f(X):
    ...:     return (X[0] - 1)**2 + (X[1] - 1)**2
In [90]: def g(X):
    ...:     return X[1] - 1.75 - (X[0] - 0.75)**4
In [91]: constraints = [dict(type='ineq', fun=g)]
```

Next, we are ready to solve the optimization problem by calling the `optimize.minimize` function. For comparison, here we also solve the corresponding unconstrained problem.

```
In [92]: x_opt = optimize.minimize(f, (0, 0), method='BFGS').x
In [93]: x_cons_opt = optimize.minimize(f, (0, 0), method='SLSQP',
             constraints=constraints).x
```

To verify the soundness of the obtained solution, we plot the contours of the objective function together with a shaded area representing the feasible region (where the inequality constraint is satisfied). The constrained and unconstrained solutions are marked with a red and a blue star, respectively (see Figure 6-8).

```
In [94]: fig, ax = plt.subplots(figsize=(6, 4))
In [95]: x_ = y_ = np.linspace(-1, 3, 100)
    ...: X, Y = np.meshgrid(x_, y_)
    ...: c = ax.contour(X, Y, func_X_Y_to_XY(f, X, Y), 50)
    ...: ax.plot(x_opt[0], x_opt[1], 'b*', markersize=15)
    ...: ax.plot(x_, 1.75 + (x_-0.75)**4, 'k-', markersize=15)
    ...: ax.fill_between(x_, 1.75 + (x_-0.75)**4, 3, color='grey')
    ...: ax.plot(x_cons_opt[0], x_cons_opt[1], 'r*', markersize=15)
    ...:
```

```
...: ax.set_ylim(-1, 3)
...: ax.set_xlabel(r"$x_0$", fontsize=18)
...: ax.set_ylabel(r"$x_1$", fontsize=18)
...: plt.colorbar(c, ax=ax)
```

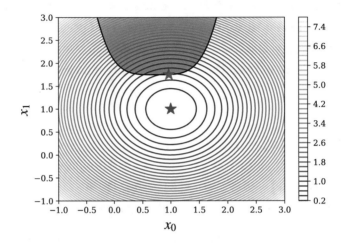

Figure 6-8. *Contour plot of the objective function with the feasible region of the constrained problem shaded gray. The red and blue stars are the optimal points in the constrained and unconstrained problems, respectively.*

For optimization problems with *only* inequality constraints, SciPy provides an alternative solver using the constrained optimization by linear approximation (COBYLA) method. This solver is accessible either through `optimize.fmin_cobyla` or `optimize.minimize` with `method='COBYLA'`. The previous example could just as well have been solved with this solver, by replacing `method='SLSQP'` with `method='COBYLA'`.

Linear Programming

In the previous section, we considered methods for very general optimization problems, where the objective function and constraint functions all can be nonlinear. However, at this point it is worth taking a step back and considering a much more restricted type of optimization problem, namely, *linear programming*, where the objective function is linear and all constraints are linear equality or inequality constraints. The class of problems is clearly much less general, but it turns out that linear programming has many important applications, and they can be solved vastly more efficiently than general nonlinear problems. The reason for this is that linear problems have properties that enable completely different methods to be used. In particular, the solution to a linear

238

optimization problem must necessarily lie on a constraint boundary, so it is sufficient to search the vertices of the intersections of the linear constraint functions. This can be done efficiently in practice. A popular algorithm for this type of problems is known as *simplex*, which systematically moves from one vertex to another until the optimal vertex has been reached. There are also more recent interior point methods that efficiently solve linear programming problems. With these methods, linear programming problems with thousands of variables and constraints are readily solvable.

Linear programming problems are typically written in the so-called standard form: $\min_x c^T x$ where $Ax \leq b$ and $x \geq 0$. Here c and x are vectors of length n, and A is a $m \times n$ matrix and b a m-vector. For example, consider the problem of minimizing the function $f(x) = -x_0 + 2x_1 - 3x_2$, subject to the three inequality constraints $x_0 + x_1 \leq 1$, $-x_0 + 3x_1 \leq 2$, and $-x_1 + x_2 \leq 3$. On the standard form, we have $c = (-1, 2, -3)$, $b = (1, 2, 3)$, and

$$A = \begin{pmatrix} 1 & 1 & 0 \\ -1 & 3 & 0 \\ 0 & -1 & 1 \end{pmatrix}.$$

To solve this problem, here we use the cvxopt library, which provides the linear programming solver with the cvxopt.solvers.lp function. This solver expects as arguments the c, A, and b vectors and matrix used in the standard form introduced in the preceding text, in the given order. The cvxopt library uses its own classes for representing matrices and vectors, but fortunately they are interoperable with NumPy arrays via the array interface[3] and can therefore be cast from one form to another using the cvxopt.matrix and np.array functions. Since NumPy array is the de facto standard array format in the scientific Python environment, it is sensible to use NumPy array as far as possible and only convert to cvxopt matrices when necessary, i.e., before calling one of the solvers in cvxopt.solvers.

To solve the stated example problem using the cvxopt library, we therefore first create NumPy arrays for the A matrix and the c and b vectors and convert them to cvxopt matrices using the cvxpot.matrix function:

```
In [96]: c = np.array([-1.0, 2.0, -3.0])
In [97]: A = np.array([[ 1.0, 1.0, 0.0],
                       [-1.0, 3.0, 0.0],
                       [ 0.0, -1.0, 1.0]])
```

---

[3]For details, see http://docs.scipy.org/doc/numpy/reference/arrays.interface.html.

```
In [98]: b = np.array([1.0, 2.0, 3.0])
In [99]: A_ = cvxopt.matrix(A)
In [100]: b_ = cvxopt.matrix(b)
In [101]: c_ = cvxopt.matrix(c)
```

The cvxopt compatible matrices and vectors c_, A_, and b_ can now be passed to the linear programming solver cvxopt.solvers.lp:

```
In [102]: sol = cvxopt.solvers.lp(c_, A_, b_)
         Optimal solution found.
In [103]: sol
Out[103]: {'dual infeasibility': 1.4835979218054372e-16,
           'dual objective': -10.0,
           'dual slack': 0.0,
           'gap': 0.0,
           'iterations': 0,
           'primal infeasibility': 0.0,
           'primal objective': -10.0,
           'primal slack': -0.0,
           'relative gap': 0.0,
           'residual as dual infeasibility certificate': None,
           'residual as primal infeasibility certificate': None,
           's': <3x1 matrix, tc='d'>,
           'status': 'optimal',
           'x': <3x1 matrix, tc='d'>,
           'y': <0x1 matrix, tc='d'>,
           'z': <3x1 matrix, tc='d'>}
In [104]: x = np.array(sol['x'])
In [105]: x
Out[105]: array([[ 0.25],
                 [ 0.75],
                 [ 3.75]])
In [106]: sol['primal objective']
Out[106]: -10.0
```

The solution to the optimization problem is given in terms of the vector x, which in this particular example is $x = (0.25, 0.75, 3.75)$, which corresponds to the $f(x)$ value -10. With this method and the `cvxopt.solvers.lp` solver, linear programming problems with hundreds or thousands of variables can readily be solved. All that is needed is to write the optimization problem on the standard form and create the c, A, and b arrays.

Summary

Optimization – to select the best option from a set of alternatives – is fundamental in many applications in science and engineering. Mathematical optimization provides a rigorous framework for systematically treating optimization problems, if they can be formulated as a mathematical problem. Computational methods for optimization are the tools with which such optimization problems are solved in practice. In a scientific computing environment, optimization therefore plays a very important role. For scientific computing with Python, the SciPy library provides efficient routines for solving many standard optimization problems, which can be used to solve a vast variety of computational optimization problems. However, optimization is a large field in mathematics, requiring a different array of methods for solving different types of problems, and there are several optimization libraries for Python that provide specialized solvers for specific types of optimization problems. In general, the SciPy `optimize` module provides good and flexible general-purpose solvers for a wide variety of optimization problems, but for specific types of optimization problems, there are also many specialized libraries that provide better performance or more features. An example of such a library is `cvxopt`, which complements the general-purpose optimization routines in SciPy with efficient solvers for linear and quadratic problems.

Further Reading

For an accessible introduction to optimization, with more detailed discussions of the numerical properties of several of the methods introduced in this chapter, see, for example, Heath (2002). For a more rigorous and in-depth introduction to optimization, see, for example, E.K.P. Chong (2013). A thorough treatment of convex optimization is given by the creators of the `cvxopt` library in the excellent book (S. Boyd, 2004), which is also available online at `http://stanford.edu/~boyd/cvxbook`.

References

E.K.P. Chong, S. Z. (2013). *An Introduction to Optimization* (4th ed.). New York: Wiley.

Heath, M. (2002). *Scientific Computing: An introductory Survey* (2nd ed.). Boston: McGraw-Hill.

S. Boyd, L. V. (2004). *Convex Optimization.* Cambridge: Cambridge University Press.

CHAPTER 7

Interpolation

Interpolation is a mathematical method for constructing a function from a discrete set of data points. The interpolation function, or interpolant, should exactly coincide with the given data points, and it can also be evaluated for other intermediate input values within the sampled range. There are many applications of interpolation: A typical use-case that provides an intuitive picture is the plotting of a smooth curve through a given set of data points. Another use-case is to approximate complicated functions, which, for example, could be computationally demanding to evaluate. In that case, it can be beneficial to evaluate the original function only at a limited number of points and use interpolation to approximate the function when evaluating it for intermediary points.

Interpolation may at a first glance look a lot like least square fitting, which we saw already in both Chapter 5 (linear least square) and Chapter 6 (nonlinear least square). Indeed, there are many similarities between interpolation and curve fitting with least square methods, but there are also important conceptual differences that distinguish these two methods: In least square fitting, we are interested in approximately fitting a function to data points in manner that minimizes the sum of square errors, using many data points and an overdetermined system of equations. In interpolation, on the other hand, we require a function that exactly coincides with the given data points and only uses the number of data points that equals the number of free parameters in the interpolation function. Least square fitting is therefore more suitable for fitting a large number of data points to a model function, and interpolation is a mathematical tool for creating a functional representation for a given minimum number of data points. In fact, interpolation is an important component in many mathematical methods, including some of the methods for equation solving and optimization that we used in Chapters 5 and 6.

Extrapolation is a concept that is related to interpolation. It refers to evaluating the estimated function outside of the sampled range, while interpolation only refers to evaluating the function within the range that is spanned by the given data points. Extrapolation can often be riskier than interpolation, because it involves estimating a

© Robert Johansson 2019
R. Johansson, *Numerical Python*, https://doi.org/10.1007/978-1-4842-4246-9_7

function in a region where it has not been sampled. Here we are only concerned with interpolation. To perform interpolation in Python, we use the `polynomial` module from NumPy and the `interpolate` module from SciPy.

Importing Modules

Here we will continue with the convention of importing submodules from the SciPy library explicitly. In this chapter we need the `interpolate` module from SciPy, and also the `polynomial` module from NumPy, which provides functions and classes for polynomials. We import these modules as follows:

```
In [1]: from scipy import interpolate
In [2]: from numpy import polynomial as P
```

In addition, we also need the rest of the NumPy library, the linear algebra module `linalg` from SciPy, and the Matplotlib library for plotting:

```
In [3]: import numpy as np
In [4]: from scipy import linalg
In [5]: import matplotlib.pyplot as plt
```

Interpolation

Before we dive into the details of how to perform interpolation with NumPy and SciPy, we first state the interpolation problem in mathematical form. For notational brevity, here we only consider one-dimensional interpolation, which can be formulated as follows: for a given set of n data point $\{(x_i, y_i)\}_{i=1}^{n}$, find a function $f(x)$ such that $f(x_i) = y_i$, for $i \in [1, n]$. The function $f(x)$ is known as the interpolant, and it is not unique. In fact, there are an infinite number of functions that satisfy the interpolation criteria. Typically we can write the interpolant as a linear combination of some basis functions $\phi_j(x)$, such that $f(x) = \sum_{j=1}^{n} c_j \phi_j(x)$, where c_j are unknown coefficients. Substituting the given data points into this linear combination results in a linear equation system for the unknown coefficients: $\sum_{j=1}^{n} c_j \phi_j(x_i) = y_i$. This equation system can be written on explicit matrix form as

$$
\begin{bmatrix}
\phi_1(x_1) & \phi_2(x_1) & \cdots & \phi_n(x_1) \\
\phi_1(x_2) & \phi_2(x_2) & \cdots & \phi_n(x_2) \\
\vdots & \vdots & \ddots & \vdots \\
\phi_1(x_n) & \phi_2(x_n) & \cdots & \phi_n(x_n)
\end{bmatrix}
\begin{bmatrix}
c_1 \\ c_2 \\ \vdots \\ c_n
\end{bmatrix}
=
\begin{bmatrix}
y_1 \\ y_2 \\ \vdots \\ y_n
\end{bmatrix},
$$

or in a more compact implicit matrix form as $\Phi(x)c = y$, where the elements of the matrix $\Phi(x)$ are $\{\Phi(x)\}_{ij} = \phi_j(x_i)$. Note that here the number of basis functions is the same as the number of data points, and $\Phi(x)$ is therefore a square matrix. Assuming that this matrix has full rank, we can solve for the unique c-vector using the standard methods discussed in Chapter 5. If the number of data points is larger than the number of basis functions, then the system is overdetermined, and in general there is no solution that satisfies the interpolation criteria. Instead, in this situation it is more suitable to consider a least square fit than an exact interpolation; see Chapter 5.

The choice of basis functions affects the properties of the resulting equation system, and a suitable choice of basis depends on the properties of the data that is fitted. Common choices of basis functions for interpolation are various types of polynomials, for example, the power basis $\phi_i(x) = x^{i-1}$, or orthogonal polynomials such as Legendre polynomials $\phi_i(x) = P_{i-1}(x)$, Chebyshev polynomials $\phi_i(x) = T_{i-1}(x)$, or piecewise polynomials. Note that in general $f(x)$ is not unique, but for n data points, there is a unique interpolating polynomial of order $n - 1$, regardless of which polynomial basis we use. For power basis $\phi_i(x) = x^{i-1}$, the matrix $\Phi(x)$ is the Vandermonde matrix, which we already have seen applications of in least square fitting in Chapter 5. For other polynomial bases, $\Phi(x)$ are generalized Vandermonde matrices, for which each basis defines the matrix of the linear equation system that has to be solved in the interpolation problem. The structure of the $\Phi(x)$ matrix is different for different polynomial bases, and its condition number and the computational cost of solving the interpolation problem vary correspondingly. Polynomials thus play an important role in interpolation, and before we can start to solve interpolation problems, we need a convenient way of working with polynomials in Python. This is the topic of the following section.

Polynomials

The NumPy library contains the submodule `polynomial` (here imported as P), which provides functions and classes for working with polynomials. In particular, it provides implementations of many standard orthogonal polynomials. These functions and classes are useful when working with interpolation, and we therefore review how to use this module before looking at polynomial interpolation.

Note   There are two modules for polynomials in NumPy: `numpy.poly1d` and `numpy.polynomial`. There is a large overlap in functionality in these two modules, but they are not compatible with each other (specifically, the coordinate arrays have reversed order in the two representations). The `numpy.poly1d` module is older and has been superseded by `numpy.polynomial`, which is now recommended for new code. Here we only focus on `numpy.polynomial`, but it is worth being aware of `numpy.poly1d` as well.

The `np.polynomial` module contains a number of classes for representing polynomials in different polynomial bases. Standard polynomials, written in the usual power basis $\{x^i\}$, are represented with the `Polynomial` class. To create an instance of this class, we can pass a coefficient array to its constructor. In the coefficient array, the ith element is the coefficient of x^i. For example, we can create a representation of the polynomial $1+2x+3x^2$ by passing the list $[1, 2, 3]$ to the `Polynomial` class:

```
In [6]: p1 = P.Polynomial([1, 2, 3])
In [7]: p1
Out[7]: Polynomial([ 1.,   2.,   3.], domain=[-1,   1], window=[-1,   1])
```

Alternatively, we can also initialize a polynomial by specifying its roots using the class method `P.Polynomial.fromroots`. The polynomial with roots at $x = -1$ and $x = 1$, for example, can be created using:

```
In [8]: p2 = P.Polynomial.fromroots([-1, 1])
In [9]: p2
Out[9]: Polynomial([-1.,   0.,   1.], domain=[-1.,   1.], window=[-1.,   1.])
```

Here, the result is the polynomial with the coefficient array $[-1, 0, 1]$, which corresponds to $-1+x^2$. The roots of a polynomial can be computed using the `roots` method. For example, the roots of the two previously created polynomials are

```
In [10]: p1.roots()
Out[10]: array([-0.33333333-0.47140452j, -0.33333333+0.47140452j])
In [11]: p2.roots()
Out[11]: array([-1.,   1.])
```

As expected, the roots of the polynomial p2 are $x = -1$ and $x = 1$, as was requested when it was created using the `fromroots` class method.

In the preceding examples, the representation of a polynomial is on the form `Polynomial([-1., 0., 1.], domain=[-1., 1.], window=[-1., 1.])`. The first of the lists in this representation is the coefficient array. The second and third lists are the `domain` and `window` attributes, which can be used to map the input domain of a polynomial to another interval. Specifically, the input domain interval [`domain[0]`, `domain[1]`] is mapped to the interval [`window[0]`, `window[1]`] through a linear transformation (scaling and translation). The default values are `domain=[-1, 1]` and `window=[-1, 1]`, which correspond to an identity transformation (no change). The `domain` and `window` arguments are particularly useful when working with polynomials that are orthogonal with respect to a scalar product that is defined on a specific interval. It is then desirable to map the domain of the input data onto this interval. This is important when interpolating with orthogonal polynomials, such as the Chebyshev or Hermite polynomials, because performing this transformation can vastly improve the conditioning number of the Vandermonde matrix for the interpolation problem.

The properties of a `Polynomial` instance can be directly accessed using the `coeff`, `domain`, and `window` attributes. For example, for the p1 polynomial defined in the preceding example, we have

```
In [12]: p1.coef
Out[12]: array([ 1.,    2.,    3.])
In [13]: p1.domain
Out[13]: array([-1,    1])
In [14]: p1.window
Out[14]: array([-1,    1])
```

A polynomial that is represented as a `Polynomial` instance can easily be evaluated at arbitrary values of x by calling the class instance as a function. The x variable can be specified as a scalar, a list, or an arbitrary NumPy array. For example, to evaluate the polynomial p1 at the points $x = \{1.5, 2.5, 3.5\}$, we simply call the p1 class instance with an array of x values as the argument:

```
In [15]: p1(np.array([1.5, 2.5, 3.5]))
Out[15]: array([ 10.75,   24.75,   44.75])
```

Instances of Polynomial can be operated on using the standard arithmetic operators +, -, *, /, and so on. The // operator is used for polynomial division. To see how this works, consider the division of the polynomial $p_1(x) = (x - 3)(x - 2)(x - 1)$ with the polynomial $p_2(x) = (x - 2)$. The answer, which is obvious when written in the factorized form, is $(x - 3)(x - 1)$. We can compute and verify this using NumPy in the following manner: first create Polynomial instances for the p1 and p2, and then use the // operator to compute the polynomial division.

```
In [16]: p1 = P.Polynomial.fromroots([1, 2, 3])
In [17]: p1
Out[17]: Polynomial([ -6.,   11.,   -6.,    1.], domain=[-1.,   1.],
            window=[-1.,   1.])
In [18]: p2 = P.Polynomial.fromroots([2])
In [19]: p2
Out[19]: Polynomial([-2.,   1.], domain=[-1.,   1.], window=[-1.,   1.])
In [20]: p3 = p1 // p2
In [21]: p3
Out[21]: Polynomial([ 3., -4.,   1.], domain=[-1.,   1.], window=[-1.,   1.])
```

The result is a new polynomial with coefficient array [3, -4, 1], and if we compute its roots, we find that they are 1 and 3, so this polynomial is indeed $(x - 3)(x - 1)$:

```
In [22]: p3.roots()
Out[22]: array([ 1.,   3.])
```

In addition to the Polynomial class for polynomials in the standard power basis, the polynomial module also has classes for representing polynomials in Chebyshev, Legendre, Laguerre, and Hermite bases, with the names Chebyshev, Legendre, Laguerre, Hermite (Physicists'), and HermiteE (Probabilists'), respectively. For example, the Chebyshev polynomial with coefficient list [1, 2, 3], that is, the polynomial $1T_0(x)+2T_1(x)+3T_2(x)$, where $T_i(x)$ is the Chebyshev polynomial of order i, can be created using:

```
In [23]: c1 = P.Chebyshev([1, 2, 3])
In [24]: c1
Out[24]: Chebyshev([ 1.,   2.,   3.], domain=[-1,   1], window=[-1,   1])
```

and its roots can be computed using the roots attribute:

```
In [25]: c1.roots()
Out[25]: array([-0.76759188,  0.43425855])
```

All the polynomial classes have the same methods, attributes, and operators as the Polynomial class discussed above, and they can all be used in the same manner. For example, to create the Chebyshev and Legendre representations of the polynomial with roots $x = -1$ and $x = 1$, we can use the fromroots attribute, in the same way as we did previously with the Polynomial class:

```
In [26]: c1 = P.Chebyshev.fromroots([-1, 1])
In [27]: c1
Out[27]: Chebyshev([-0.5,  0. ,  0.5], domain=[-1.,  1.],
         window=[-1.,  1.])
In [28]: l1 = P.Legendre.fromroots([-1, 1])
In [29]: l1
Out[29]: Legendre([-0.66666667,  0.  ,  0.66666667], domain=[-1.,  1.],
         window=[-1.,  1.])
```

Note that the same polynomials, here with the roots at $x = -1$ and $x = 1$ (which is a unique polynomial), have different coefficient arrays when represented in different bases, but when evaluated at specific values of x, the two gives the same results (as expected):

```
In [30]: c1(np.array([0.5, 1.5, 2.5]))
Out[30]: array([-0.75,  1.25,  5.25])
In [31]: l1(np.array([0.5, 1.5, 2.5]))
Out[31]: array([-0.75,  1.25,  5.25])
```

Polynomial Interpolation

The polynomial classes discussed in the previous section all provide useful functions for polynomial interpolation. For instance, recall the linear equation for the polynomial interpolation problem: $\Phi(x)c = y$, where x and y are vectors containing the x_i and y_i data points and c is the unknown coefficient vector. To solve the interpolation problem, we need to first evaluate the matrix $\Phi(x)$ for a given basis and then solve the resulting linear equation system. Each of the polynomial classes in `polynomial` conveniently provides a function for computing the (generalized) Vandermonde matrix for the corresponding basis. For example, for polynomials in the power basis, we can use `np.polynomial.polynomial.polyvander`; for polynomials in the Chebyshev basis, we can use the corresponding `np.polynomial.chebyshev.chebvander` function; and so on. See the docstrings for `np.polynomial` and its submodules for the complete list of generalized Vandermonde matrix functions for the various polynomial bases.

Using the abovementioned functions for generating the Vandermonde matrices, we can easily perform a polynomial interpolation in different bases. For example, consider the data points $(1, 1)$, $(2, 3)$, $(3, 5)$, and $(4, 4)$. We begin with creating NumPy array for the x and y coordinates for the data points.

```
In [32]: x = np.array([1, 2, 3, 4])
In [33]: y = np.array([1, 3, 5, 4])
```

To interpolate a polynomial through these points, we need to use a polynomial of third degree (number of data points minus one). For interpolation in the power basis, we seek the coefficient c_i such that $f(x) = \sum_{i=1}^{4} c_i x^{i-1} = c_1 x^0 + c_2 x^1 + c_3 x^2 + c_4 x^3$, and to find this coefficient, we evaluate the Vandermonde matrix and solve the interpolation equation system:

```
In [34]: deg = len(x) - 1
In [35]: A = P.polynomial.polyvander(x, deg)
In [36]: c = linalg.solve(A, y)
In [37]: c
Out[37]: array([ 2. , -3.5,  3. , -0.5])
```

The sought coefficient vector is [2, -3.5, 3, -0.5], and the interpolation polynomial is thus $f(x) = 2 - 3.5x + 3x^2 - 0.5x^3$. Given the coefficient array c, we can now create a polynomial representation that can be used for interpolation:

```
In [38]: f1 = P.Polynomial(c)
In [39]: f1(2.5)
Out[39]: 4.1875
```

To perform this polynomial interpolation in another polynomial basis, all that we need to change is the name of the function that was used to generate the Vandermonde matrix A in the previous example. For example, to interpolate using the Chebyshev basis polynomials, we can do:

```
In [40]: A = P.chebyshev.chebvander(x, deg)
In [41]: c = linalg.solve(A, y)
In [42]: c
Out[42]: array([ 3.5  , -3.875,  1.5  , -0.125])
```

As expected, the coefficient array has different values in this basis, and the interpolation polynomial in the Chebyshev basis is $f(x) = 3.5T_0(x) - 3.875T_1(x) + 1.5T_2(x) - 0.125T_3(x)$. However, regardless of the polynomial basis, the interpolation polynomial is unique, and evaluating the interpolant will always result in the same values:

```
In [43]: f2 = P.Chebyshev(c)
In [44]: f2(2.5)
Out[44]: 4.1875
```

We can demonstrate that the interpolation with the two bases indeed results in the same interpolation function by plotting the f1 and f2 together with the data points (see Figure 7-1):

```
In [45]: xx = np.linspace(x.min(), x.max(), 100)  # supersampled [x[0],
         x[-1]] interval
In [45]: fig, ax = plt.subplots(1, 1, figsize=(12, 4))
    ...: ax.plot(xx, f1(xx), 'b', lw=2, label='Power basis interp.')
    ...: ax.plot(xx, f2(xx), 'r--', lw=2, label='Chebyshev basis interp.')
    ...: ax.scatter(x, y, label='data points')
    ...: ax.legend(loc=4)
    ...: ax.set_xticks(x)
    ...: ax.set_ylabel(r"$y$", fontsize=18)
    ...: ax.set_xlabel(r"$x$", fontsize=18)
```

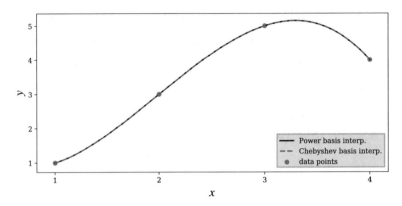

Figure 7-1. *Polynomial interpolation of four data points, using power basis and the Chebyshev basis*

While interpolation with different polynomial bases is convenient due to the functions for the generalized Vandermonde matrices, there is an even simpler and better method available. Each polynomial class provides a class method `fit` that can be used to compute an interpolation polynomial.[1] The two interpolation functions that were computed manually in the previous example could therefore instead be computed in the following manner: using the power basis and its `Polynomial` class, we obtain:

```
In [46]: f1b = P.Polynomial.fit(x, y, deg)
In [47]: f1b
Out[47]: Polynomial([ 4.1875,  3.1875, -1.6875, -1.6875],
         domain=[ 1.,  4.], window=[-1.,  1.])
```

and by using the class method `fit` from the `Chebyshev` class instead, we obtain:

```
In [48]: f2b = P.Chebyshev.fit(x, y, deg)
In [49]: f2b
Out[49]: Chebyshev([ 3.34375 ,  1.921875, -0.84375 , -0.421875],
         domain=[ 1.,  4.], window=[-1.,  1.])
```

Note that with this method, the `domain` attribute of the resulting instances are automatically set to the appropriate x values of the data points (in this example, the input range is $[1, 4]$), and the coefficients are adjusted accordingly. As mentioned previously, mapping the interpolation data into the range that is most suitable for a specific basis

---

[1] If the requested polynomial degree of the interpolant is smaller than the number of data points minus one, then a least square fit is computed rather than an exact interpolation.

can significantly improve the numerical stability of the interpolation. For example, using the Chebyshev basis with x values that are scaled such that $x \in [-1, 1]$, rather than the original x values in the previous example, reduces the condition number from almost 4660 to about 1.85:

```
In [50]: np.linalg.cond(P.chebyshev.chebvander(x, deg))
Out[50]: 4659.7384241399586
In [51]: np.linalg.cond(P.chebyshev.chebvander((2*x-5)/3.0, deg))
Out[51]: 1.8542033440472896
```

Polynomial interpolation of a few data points is a powerful and useful mathematical tool, which is an important part of many mathematical methods. When the number of data points increases, we need to use increasingly high-order polynomials for exact interpolation, and this is problematic in several ways. To begin with, it becomes increasingly demanding to both determine and evaluate the interpolant for increasing polynomial order. However, a more serious issue is that high-order polynomial interpolation can have undesirable behavior between the interpolation points. Although the interpolation is exact at the given data points, a high-order polynomial can vary wildly between the specified points. This is famously illustrated by polynomial interpolation of Runge's function $f(x) = 1/(1+25x^2)$ using evenly spaced sample points in the interval $[-1, 1]$. The result is an interpolant that nearly diverges between the data points near the end of the interval.

To illustrate this behavior, we create a Python function runge that implements Runge's function, and a function runge_interpolate that interpolates an nth order polynomial, in the power basis, to the Runge's function at evenly spaced sample points:

```
In [52]: def runge(x):
   ...:     return 1/(1 + 25 * x**2)
In [53]: def runge_interpolate(n):
   ...:     x = np.linspace(-1, 1, n + 1)
   ...:     p = P.Polynomial.fit(x, runge(x), deg=n)
   ...:     return x, p
```

Next we plot Runge's function together with the 13th and 14th order polynomial interpolations, at supersampled x values in the $[-1, 1]$ interval. The resulting plot is shown in Figure 7-2.

```
In [54]: xx = np.linspace(-1, 1, 250)
In [55]: fig, ax = plt.subplots(1, 1, figsize=(8, 4))
    ...: ax.plot(xx, runge(xx), 'k', lw=2, label="Runge's function")
    ...: # 13th order interpolation of the Runge function
    ...: n = 13
    ...: x, p = runge_interpolate(n)
    ...: ax.plot(x, runge(x), 'ro')
    ...: ax.plot(xx, p(xx), 'r', label='interp. order %d' % n)
    ...: # 14th order interpolation of the Runge function
    ...: n = 14
    ...: x, p = runge_interpolate(n)
    ...: ax.plot(x, runge(x), 'go')
    ...: ax.plot(xx, p(xx), 'g', label='interp. order %d' % n)
    ...:
    ...: ax.legend(loc=8)
    ...: ax.set_xlim(-1.1, 1.1)
    ...: ax.set_ylim(-1, 2)
    ...: ax.set_xticks([-1, -0.5, 0, 0.5, 1])
    ...: ax.set_ylabel(r"$y$", fontsize=18)
    ...: ax.set_xlabel(r"$x$", fontsize=18)
```

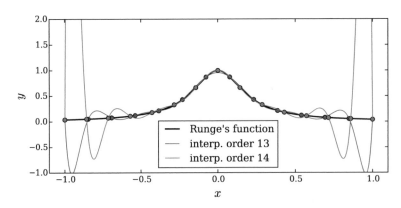

Figure 7-2. *The Runge's function together with two high-order polynomial interpolations*

We note that in Figure 7-2, the interpolants exactly agree with Runge's function at the sample points, but between these points they oscillate wildly near the ends of the interval. This is an undesirable property of an interpolant, and it defeats the purpose of the interpolation. A solution to this problem is to use piecewise low-order polynomials when interpolating with a large number of data points. In other words, instead of fitting all the data points to a single high-order polynomial, a different low-order polynomial is used to describe each subinterval bracketed by two consecutive data points. This is the topic of the following section.

Spline Interpolation

For a set of n data points $\{x_i, y_i\}$, there are $n-1$ subintervals $[x_i, x_{i+1}]$ in the full range of the data $[x_0, x_{n-1}]$. An interior data point that connects two such subintervals is known as a *knot* in the terminology of piecewise polynomial interpolation. To interpolate the n data points using piecewise polynomials of degree k on each of the subintervals, we must determine $(k+1)(n-1)$ unknown parameters. The values at the knots give $2(n-1)$ equations. These equations, by themselves, are only sufficient to determine a piecewise polynomial of order one, i.e., a piecewise linear function. However, additional equations can be obtained by requiring that also derivatives and higher-order derivatives are continuous at the knots. This condition ensures that the resulting piecewise polynomial has a smooth appearance.

A spline is a special type of piecewise polynomial interpolant: a piecewise polynomial of degree k is a spline if it is continuously differentiable $k-1$ times. The most popular choice is the third-order spline, $k=3$, which requires $4(n-1)$ parameters. For this case, the continuity of two derivatives at the $n-2$ knots gives $2(n-2)$ additional equations, bringing the total number of equations to $2(n-1)+2(n-2) = 4(n-1)-2$. There are therefore two remaining undetermined parameters, which must be determined by other means. A common approach is to additionally require that the second-order derivatives at the endpoints are zero (resulting in the *natural* spline). This gives two more equations, which closes the equation system.

The SciPy `interpolate` module provides several functions and classes for performing spline interpolation. For example, we can use the `interpolate.interp1d` function, which takes x and y arrays for the data points as first and second arguments.

The optional keyword argument kind can be used to specify the type and order of the interpolation. In particular, we can set kind=3 (or, equivalently, kind='cubic') to compute the cubic spline. This function returns a class instance that can be called like a function and which can be evaluated for different values of x using function calls. An alternative spline function is interpolate.InterpolatedUnivariateSpline, which also takes x and y arrays as the first and second arguments, but which uses the keyword argument k (instead of kind) to specify the order of the spline interpolation.

To see how the interpolate.interp1d function can be used, consider again Runge's function, and we now want to interpolate this function with a third-order spline polynomial. To this end, we first create NumPy arrays for the x and y coordinates of the sample points. Next we call the interpolate.interp1d function with kind=3 to obtain the third-order spline for the given data:

```
In [56]: x = np.linspace(-1, 1, 11)
In [57]: y = runge(x)
In [58]: f_i = interpolate.interp1d(x, y, kind=3)
```

To evaluate how good this spline interpolation is (here represented by the class instance f_i), we plot the interpolant together with the original Runge's function and the sample points. The result is shown in Figure 7-3.

```
In [59]: xx = np.linspace(-1, 1, 100)
In [60]: fig, ax = plt.subplots(figsize=(8, 4))
    ...: ax.plot(xx, runge(xx), 'k', lw=1, label="Runge's function")
    ...: ax.plot(x, y, 'ro', label='sample points')
    ...: ax.plot(xx, f_i(xx), 'r--', lw=2, label='spline order 3')
    ...: ax.legend()
    ...: ax.set_xticks([-1, -0.5, 0, 0.5, 1])
    ...: ax.set_ylabel(r"$y$", fontsize=18)
    ...: ax.set_xlabel(r"$x$", fontsize=18)
```

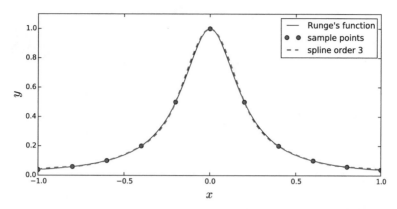

Figure 7-3. *Runge's function with a third-order Spline interpolation using 11 data points*

Here we used 11 data points and a spline of the third order. We note that the interpolant agrees very well with the original function in Figure 7-3. Typically spline interpolation of order three or less does not suffer from the same type of oscillations that we observed with high-order polynomial interpolation, and normally it is sufficient to use splines of order three if we have a sufficient number of data points.

To illustrate the effect of the order of a spline interpolation, consider the problem of interpolating the data $(0, 3)$, $(1, 4)$, $(2, 3.5)$, $(3, 2)$, $(4, 1)$, $(5, 1.5)$, $(6, 1.25)$, and $(7, 0.9)$ with splines of increasing order. We first define the x and y arrays, and then loop over the required spline orders, computing the interpolation and plotting it for each order:

```
In [61]: x = np.array([0, 1, 2, 3, 4, 5, 6, 7])
In [62]: y = np.array([3, 4, 3.5, 2, 1, 1.5, 1.25, 0.9])
In [63]: xx = np.linspace(x.min(), x.max(), 100)
In [64]: fig, ax = plt.subplots(figsize=(8, 4))
    ...: ax.scatter(x, y)
    ...:
    ...: for n in [1, 2, 3, 5]:
    ...:     f = interpolate.interp1d(x, y, kind=n)
    ...:     ax.plot(xx, f(xx), label='order %d' % n)
    ...:
    ...: ax.legend()
    ...: ax.set_ylabel(r"$y$", fontsize=18)
    ...: ax.set_xlabel(r"$x$", fontsize=18)
```

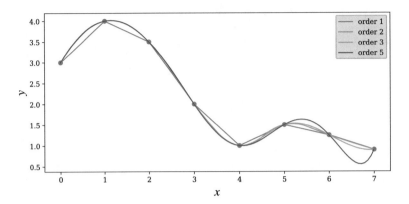

Figure 7-4. _Spline interpolations of different orders_

From the spline interpolation shown in Figure 7-4, it is clear that spline order two or three already provides a rather good interpolation, with relatively small errors between the original function and the interpolant function. For higher-order splines, the same problem as we saw for high-order polynomial interpolation resurfaces. In practice, it is therefore often suitable to use third-order spline interpolation.

Multivariate Interpolation

Polynomial and spline interpolations can be straightforwardly generalized to multivariate situations. In analogy with the univariate case, we seek a function whose values are given at a set of specified points and that can be evaluated for intermediary points within the sampled range. SciPy provides several functions and classes for multivariate interpolation, and in the following two examples, we explore two of the most useful functions for bivariate interpolation: the `interpolate.interp2d` and `interpolate.griddata` functions, respectively. See the docstring for the `interpolate` module and its reference manual for further information on other interpolation options.

We begin by looking at `interpolate.interp2d`, which is a straightforward generalization of the `interp1d` function that we previously used. This function takes the *x* and *y* coordinates of the available data points as separate one-dimensional arrays, followed by a two-dimensional array of values for each combination of *x* and *y* coordinates. This presumes that the data points are given on a regular and uniform grid of *x* and *y* coordinates.

To illustrate how the `interp2d` function can be used, we simulate noisy measurements by adding random noise to a known function, which in the following example is taken to be $f(x, y) = \exp\left(-(x+1/2)^2 - 2(y+1/2)^2\right) - \exp\left(-(x-1/2)^2 - 2(y-1/2)^2\right)$. To form an interpolation problem, we sample this function at 10 points in the interval $[-2, 2]$, along the x and y coordinates, and then add a small normal-distributed noise to the exact values. We first create NumPy arrays for the x and y coordinates of the sample points and define a Python function for $f(x, y)$:

```
In [65]: x = y = np.linspace(-2, 2, 10)
In [66]: def f(x, y):
    ...:     return np.exp(-(x + .5)**2 - 2*(y + .5)**2) -
             np.exp(-(x - .5)**2 - 2*(y - .5)**2)
```

Next we evaluate the function at the sample points and add the random noise to simulate uncertain measurements:

```
In [67]: X, Y = np.meshgrid(x, y)
In [68]: # simulate noisy data at fixed grid points X, Y
    ...: Z = f(X, Y) + 0.05 * np.random.randn(*X.shape)
```

At this point, we have a matrix of data points Z with noisy data, which is associated with exactly known and regularly spaced coordinates x and y. To obtain an interpolation function that can be evaluated for intermediary x and y values, within the sampled range, we can now use the interp2d function:

```
In [69]: f_i = interpolate.interp2d(x, y, Z, kind='cubic')
```

Note that here x and y are one-dimensional arrays (of length 10), and Z is a two-dimensional array of shape $(10, 10)$. The `interp2d` function returns a class instance, here f_i, that behaves as a function that we can evaluate at arbitrary x and y coordinates (within the sampled range). A supersampling of the original data, using the interpolation function, can therefore be obtained in the following way:

```
In [70]: xx = yy = np.linspace(x.min(), x.max(), 100)
In [71]: ZZi = f_i(xx, yy)
In [72]: XX, YY = np.meshgrid(xx, yy)
```

Here, XX and YY are coordinate matrices for the supersampled points, and the corresponding interpolated values are ZZi. These can, for example, be used to plot a smoothed function describing the sparse and noisy data. The following code plots contours of both the original function and the interpolated data. See Figure 7-5 for the resulting contour plot.

```
In [73]: fig, axes = plt.subplots(1, 2, figsize=(12, 5))
    ...: # for reference, first plot the contours of the exact function
    ...: c = axes[0].contourf(XX, YY, f(XX, YY), 15, cmap=plt.cm.RdBu)
    ...: axes[0].set_xlabel(r"$x$", fontsize=20)
    ...: axes[0].set_ylabel(r"$y$", fontsize=20)
    ...: axes[0].set_title("exact / high sampling")
    ...: cb = fig.colorbar(c, ax=axes[0])
    ...: cb.set_label(r"$z$", fontsize=20)
    ...: # next, plot the contours of the supersampled interpolation of the
           noisy data
    ...: c = axes[1].contourf(XX, YY, ZZi, 15, cmap=plt.cm.RdBu)
    ...: axes[1].set_ylim(-2.1, 2.1)
    ...: axes[1].set_xlim(-2.1, 2.1)
    ...: axes[1].set_xlabel(r"$x$", fontsize=20)
    ...: axes[1].set_ylabel(r"$y$", fontsize=20)
    ...: axes[1].scatter(X, Y, marker='x', color='k')
    ...: axes[1].set_title("interpolation of noisy data / low sampling")
    ...: cb = fig.colorbar(c, ax=axes[1])
    ...: cb.set_label(r"$z$", fontsize=20)
```

Figure 7-5. *Contours of the exact function (left) and a bivariate cubic spline interpolation (right) of noisy samples from the function on a regular grid (marked with crosses)*

With relatively sparsely spaced data points, we can thus construct an approximation of the underlying function by using the `interpolate.interp2d` to compute the bivariate cubic spline interpolation. This gives a smoothed approximation for the underplaying function, which is frequently useful when dealing with data obtained from measurements or computations that are costly, in time or other resources. For higher-dimensional problems, there is a function `interpolate.interpnd`, which is a generalization to N-dimensional problems.

Another common situation that requires multivariate interpolation occurs when sampled data is given on an irregular coordinate grid. This situation frequently arises (e.g., in experiments or other data collection processes) when the exact values at which the observations are collected cannot be directly controlled. To be able to easily plot and analyze such data with existing tools, it may be desirable to interpolate it onto a regular coordinate grid. In SciPy we can use the `interpolate.griddata` for exactly this task. This function takes as first argument. a tuple of one-dimensional coordinate vectors (`xdata`, `ydata`) for the data values `zdata`, which are passed to the function in matrix form as third argument. The fourth argument is a tuple (`X`, `Y`) of coordinate vectors or coordinate matrices for the new points at which the interpolant is to be evaluated. Optionally, we can also set the interpolation method using the `method` keyword argument (`'nearest'`, `'linear'`, or `'cubic'`):

```
In [74]: Zi = interpolate.griddata((xdata, ydata), zdata, (X, Y),
method='cubic')
```

To demonstrate how to use the `interpolate.griddata` function for interpolating data at unstructured coordinate points, we take the function $f(x, y) = \exp(-x^2 - y^2) \cos 4x \sin 6y$ and randomly select sampling points in the interval $[-1, 1]$ along the x and y coordinates. The resulting $\{x_i, y_i, z_i\}$ data is then interpolated and evaluated on a supersampled regular grid spanning the $x, y \in [-1, 1]$ region. To this end, we first define a Python function for $f(x, y)$ and then generate the randomly sampled data:

```
In [75]: def f(x, y):
   ...:     return np.exp(-x**2 - y**2) * np.cos(4*x) * np.sin(6*y)
In [76]: N = 500
In [77]: xdata = np.random.uniform(-1, 1, N)
In [78]: ydata = np.random.uniform(-1, 1, N)
In [79]: zdata = f(xdata, ydata)
```

To visualize the function and the density of the sampling points, we plot a scatter plot for the sampling locations overlaid on a contour graph of $f(x, y)$. The result is shown in Figure 7-6.

```
In [80]: x = y = np.linspace(-1, 1, 100)
In [81]: X, Y = np.meshgrid(x, y)
In [82]: Z = f(X, Y)
In [83]: fig, ax = plt.subplots(figsize=(8, 6))
    ...: c = ax.contourf(X, Y, Z, 15, cmap=plt.cm.RdBu);
    ...: ax.scatter(xdata, ydata, marker='.')
    ...: ax.set_ylim(-1,1)
    ...: ax.set_xlim(-1,1)
    ...: ax.set_xlabel(r"$x$", fontsize=20)
    ...: ax.set_ylabel(r"$y$", fontsize=20)
    ...: cb = fig.colorbar(c, ax=ax)
    ...: cb.set_label(r"$z$", fontsize=20)
```

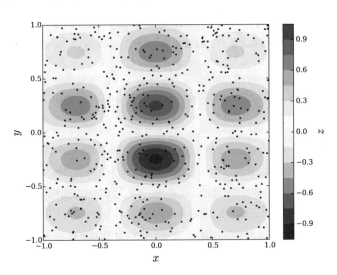

Figure 7-6. *Exact contour plot of a randomly sampled function. The 500 sample points are marked with black dots.*

From the contour graph and scatter plots in Figure 7-6, it appears that the randomly chosen sample points cover the coordinate region of interest fairly well, and it is plausible that we should be able to reconstruct the function $f(x, y)$ relatively accurately by interpolating the data. Here we would like to interpolate the data on the finely spaced (supersampled) regular grid described by the X and Y coordinate arrays. To compare different interpolation methods, and the effect of increasing number of sample points, we define the function z_interpolate that interpolates the given data points with the nearest data point, a linear interpolation, and a cubic spline interpolation:

```
In [84]: def z_interpolate(xdata, ydata, zdata):
    ...:     Zi_0 = interpolate.griddata((xdata, ydata), zdata, (X, Y),
                 method='nearest')
    ...:     Zi_1 = interpolate.griddata((xdata, ydata), zdata, (X, Y),
                 method='linear')
    ...:     Zi_3 = interpolate.griddata((xdata, ydata), zdata, (X, Y),
                 method='cubic')
    ...:     return Zi_0, Zi_1, Zi_3
```

Finally we plot a contour graph of the interpolated data for the three different interpolation methods applied to three subsets of the total number of sample points that use 50, 150, and all 500 points, respectively. The result is shown in Figure 7-7.

```
In [85]: fig, axes = plt.subplots(3, 3, figsize=(12, 12), sharex=True,
             sharey=True)
    ...:
    ...: n_vec = [50, 150, 500]
    ...: for idx, n in enumerate(n_vec):
    ...:     Zi_0, Zi_1, Zi_3 = z_interpolate(xdata[:n], ydata[:n],
             zdata[:n])
    ...:     axes[idx, 0].contourf(X, Y, Zi_0, 15, cmap=plt.cm.RdBu)
    ...:     axes[idx, 0].set_ylabel("%d data points\ny" % n, fontsize=16)
    ...:     axes[idx, 0].set_title("nearest", fontsize=16)
    ...:     axes[idx, 1].contourf(X, Y, Zi_1, 15, cmap=plt.cm.RdBu)
    ...:     axes[idx, 1].set_title("linear", fontsize=16)
    ...:     axes[idx, 2].contourf(X, Y, Zi_3, 15, cmap=plt.cm.RdBu)
    ...:     axes[idx, 2].set_title("cubic", fontsize=16)
    ...:
    ...: for m in range(len(n_vec)):
    ...:     axes[idx, m].set_xlabel("x", fontsize=16)
```

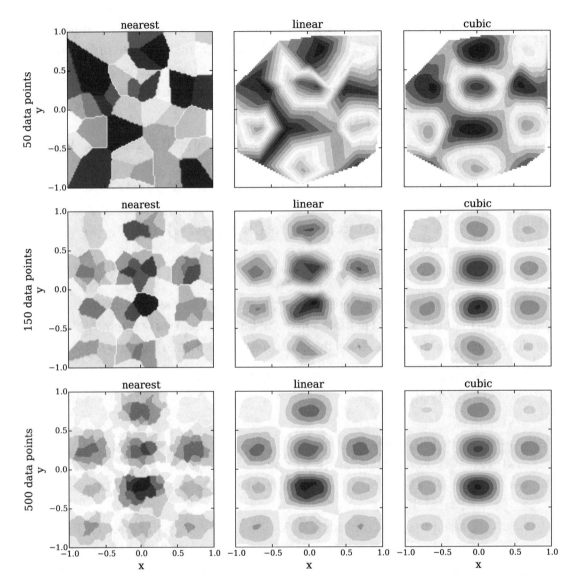

Figure 7-7. *Bivariate interpolation of randomly sampled values, for increasing interpolation order (left to right) and increasing number of sample points (top to bottom)*

Figure 7-7 shows that it is possible to reconstruct a function fairly well from interpolation of unstructured samples, as long as the region of interest is well covered. In this example, and quite generally for other situations as well, it is clear that the cubic spline interpolation is vastly superior to nearest-point and linear interpolation, and although it is more computationally demanding to compute the spline interpolation, it is often worthwhile.

Summary

Interpolation is a fundamental mathematical tool that has significant applications throughout scientific and technical computing. In particular, interpolation is a crucial part of many mathematical methods and algorithms. It is also a practical tool in itself, which is useful when plotting or analyzing data that are obtained from experiments, observations, or resource-demanding computations. The combination of the NumPy and SciPy libraries provides good coverage of numerical interpolation methods, in one or more independent variables. For most practical interpolation problems that involve a large number of data points, cubic spline interpolation is the most useful technique, although polynomial interpolation of low degree is most commonly used as a tool in other numerical methods (such as root finding, optimization, numerical integration). In this chapter we have explored how to use NumPy's `polynomial` and SciPy's `interpolate` modules to perform interpolation on given datasets with one and two independent variables. Mastering these techniques is an important skill of a computational scientist, and I encourage further exploring the content in `scipy.interpolate` that was not covered here by studying the docstrings for this module and its many functions and classes.

Further Reading

Interpolation is covered in most texts on numerical methods, and for a more thorough theoretical introduction to the subject, see, for example, J. Stoer (1992) or Hamming (1987).

References

Hamming, R. (1987). *Numerical Methods for Scientists and Engineers*. New York: Dover Publications.

J. Stoer, R. B. (1992). *Introduction to Numerical Analysis*. New York: Springer.

CHAPTER 8

Integration

In this chapter we cover different aspects of integration, with the main focus on numerical integration. For historical reasons, numerical integration is also known as *quadrature*. Integration is significantly more difficult than its inverse operation – differentiation – and while there are many examples of integrals that can be calculated analytically, in general we have to resort to numerical methods. Depending on the properties of the integrand (the function being integrated) and the integration limits, it can be easy or difficult to numerically compute an integral. Integrals of continuous functions and with finite integration limits can in most cases be computed efficiently in one dimension, but integrable functions with singularities or integrals with infinite integration limits are examples of cases that can be difficult to handle numerically, even in a single dimension. Two-dimensional integrals (double integrals) and higher-order integrals can be numerically computed with repeated single-dimension integration or using methods that are multidimensional generalizations of the techniques used to solve single-dimensional integrals. However, the computational complexity grows quickly with the number of dimensions to integrate over, and in practice such methods are only feasible for low-dimensional integrals, such as double integrals or triple integrals. Integrals of higher dimension than that often require completely different techniques, such as Monte Carlo sampling algorithms.

In addition to numerical evaluation of integrals with definite integration limits, which gives a single number as a result, integration also has other important applications. For example, equations where the integrand of an integral is the unknown quantity are called integral equations, and such equations frequently appear in science and engineering applications. Integral equations are usually difficult to solve, but they can often be recast into linear equation systems by discretizing the integral. However, we do not cover this topic here, but we will see examples of this type of problem in Chapter 10. Another important application of integration is integral transforms, which

267

© Robert Johansson 2019
R. Johansson, *Numerical Python*, https://doi.org/10.1007/978-1-4842-4246-9_8

are techniques for transforming functions and equations between different domains. At the end of this chapter, we briefly discuss how SymPy can be used to compute some integral transforms, such as Laplace transforms and Fourier transforms.

To carry out symbolic integration, we can use SymPy, as briefly discussed in Chapter 3, and to compute numerical integration, we mainly use the `integrate` module in SciPy. However, SymPy (through the multiple-precision library `mpmath`) also have routines for numerical integration, which complement those in SciPy, for example, by offering arbitrary-precision integration. In this chapter we look into both these options and discuss their pros and cons. We also briefly look at Monte Carlo integrations using the `scikit-monaco` library.

Scikit-monaco   `Scikit-monaco` is a small and recent library that makes Monte Carlo integration convenient and easily accessible. At the time of writing, the most recent version of scikit-monaco is 0.2.1. See `http://scikit-monaco.readthedocs.org` for more information.

Importing Modules

In this chapter we require, as usual, the NumPy and the Matplotlib libraries for basic numerical and plotting support, and on top of that, we use the `integrate` module from SciPy, the SymPy library, and the arbitrary-precision math library mpmath. Here we assume that these modules are imported as follows:

```
In [1]: import numpy as np
In [2]: import matplotlib.pyplot as plt
   ...: import maplotlib as mpl
In [3]: from scipy import integrate
In [4]: import sympy
In [5]: import mpmath
```

In addition, for nicely formatted output from SymPy, we also need to set up its printing system:

```
In [6]: sympy.init_printing()
```

Numerical Integration Methods

Here we are concerned with evaluating definite integrals on the form $I(f) = \int_a^b f(x) \, dx$, with given integration limits a and b. The interval $[a, b]$ can be finite, semi-infinite (where either $a = -\infty$ or $b = \infty$), or infinite (where both $a = -\infty$ and $b = \infty$). The integral $I(f)$ can be interpreted as the area between the curve of the integrand $f(x)$ and the x axis, as illustrated in Figure 8-1.

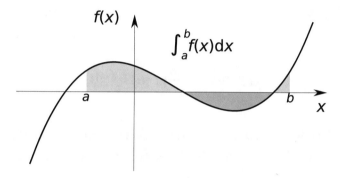

Figure 8-1. *Interpretation of an integral as the area between the curve of the integrand and the x axis, where the area is counted as positive where f(x) > 0 (green/light) and negative otherwise (red/dark)*

A general strategy for numerically evaluating an integral $I(f)$, on the form given above, is to write the integral as a discrete sum that approximates the value of the integral:

$$I(f) = \sum_{i=1}^{n} w_i f(x_i) + r_n.$$

Here w_i are the weights of n evaluations of $f(x)$ at the points $x_i \in [a, b]$, and r_n is the residual due to the approximation. In practice we assume that r_n is small and can be neglected, but it is important to have an estimate of r_n to known how accurately the integral is approximated. This summation formula for $I(f)$ is known as an n-point *quadrature rule*, and the choice of the number of points n, their locations in $[a, b]$, and the weight factors w_i influence the accuracy and the computational complexity of its evaluation. Quadrature rules can be derived from interpolations of $f(x)$ on the interval $[a, b]$. If the points x_i are evenly spaced in the interval $[a, b]$, and a polynomial interpolation is used, then the resulting quadrature rule is known as a *Newton-Cotes quadrature rule.* For instance, approximating *f*(*x*) with a zeroth-order polynomial (constant value) using the midpoint value $x_0 = (a+b)/2$, we obtain

$$\int_a^b f(x)\,dx \approx f\left(\frac{a+b}{2}\right)\int_a^b dx = (b-a)f\left(\frac{a+b}{2}\right).$$

This is known as the *midpoint rule,* and it integrates polynomials of up to order one (linear functions) exactly, and it is therefore said to be of polynomial degree one. Approximating *f*(*x*) by a polynomial of degree one, evaluated at the endpoints of the interval, results in

$$\int_a^b f(x)\,dx \approx \frac{b-a}{2}\big(f(a)+f(b)\big).$$

This is known as the *trapezoid rule,* and it is also of polynomial degree one. Using an interpolation polynomial of second order results in *Simpson's rule,*

$$\int_a^b f(x)\,dx \approx \frac{b-a}{6}\left(f(a)+4f\left(\frac{a+b}{2}\right)+f(b)\right),$$

which uses function evaluations at the endpoints and the midpoint. This method is of polynomial degree three, meaning that it integrates exactly polynomials up to order three. The method of arriving at this formula can easily be demonstrated using SymPy: First we define symbols for the variables *a, b,* and *x,* as well as the function *f.*

```
In [7]: a, b, X = sympy.symbols("a, b, x")
In [8]: f = sympy.Function("f")
```

Next we define a tuple x that contains the sample points (the endpoints and the middle point of the interval [*a, b*]) and a list w of weight factors to be used in the quadrature rule, corresponding to each sample point:

```
In [9]: x = a, (a+b)/2, b  # for Simpson's rule
In [10]: w = [sympy.symbols("w_%d" % i) for i in range(len(x))]
```

Given x and w, we can now construct a symbolic expression for the quadrature rule:

```
In [11]: q_rule = sum([w[i] * f(x[i]) for i in range(len(x))])
In [12]: q_rule
```
Out[12]: $w_0 f(a) + w_1 f\left(\dfrac{a}{2} + \dfrac{b}{2}\right) + w_2 f(b)$

To compute the appropriate values of the weight factors w_i, we choose the polynomial basis functions $\{\phi_n(x) = x^n\}_{n=0}^{2}$ for the interpolation of $f(x)$, and here we use the sympy.Lambda function to create symbolic representations for each of these basis functions:

```
In [13]: phi = [sympy.Lambda(X, X**n) for n in range(len(x))]
In [14]: phi
```
Out[14]: $\left[(x \mapsto 1),(x \mapsto x),(x \mapsto x^2)\right]$

The key to finding the weight factors in the quadrature expression (Out[12]) is that the integral $\int_a^b \phi_n(x)\,dx$ can be computed analytically for each of the basis functions $\phi_n(x)$. By substituting the function $f(x)$ with each of the basis functions $\phi_n(x)$ in the quadrature rule, we obtain an equation system for the unknown weight factors:

$$\sum_{i=0}^{2} w_i \phi_n(x_i) = \int_a^b \phi_n(x)\,dx,$$

These equations are equivalent to requiring that the quadrature rule exactly integrates all the basis functions and therefore also (at least) all functions that are spanned by the basis. This equation system can be constructed with SymPy using

```
In [15]: eqs = [q_rule.subs(f, phi[n]) - sympy.integrate(phi[n](X), (X, a, b))
   ...:         for n in range(len(phi))]
In [16]: eqs
```
Out[16]: $\left[a - b + w_0 + w_1 + w_2,\; \dfrac{a^2}{2} + aw_0 - \dfrac{b^2}{2} + bw_2 + w_1\left(\dfrac{a}{2} + \dfrac{b}{2}\right),\; \dfrac{a^3}{3} + a^2 w_0 - \dfrac{b^3}{3} + \right.$

$\left. b^2 w_2 + w_1\left(\dfrac{a}{2} + \dfrac{b}{2}\right)^2 \right]$

Solving this linear equation system gives analytical expressions for the weight factors,

```
In [17]: w_sol = sympy.solve(eqs, w)
In [18]: w_sol
```

Out[18]: $\left\{ w_0 : -\dfrac{a}{6}+\dfrac{b}{6},\ w_1 : -\dfrac{2a}{3}+\dfrac{2b}{3},\ w_2 : -\dfrac{a}{6}+\dfrac{b}{6} \right\}$

and by substituting the solution into the symbolic expression for the quadrature rule, we obtain

```
In [19]: q_rule.subs(w_sol).simplify()
```

Out[19]: $-\dfrac{1}{6}(a-b)\left(f(a)+f(b)+4f\left(\dfrac{a}{2}+\dfrac{b}{2}\right) \right)$

We recognize this result as Simpson's quadrature rule given in the preceding section. Choosing different sample points (the x tuple in this code) results in different quadrature rules.

Higher-order quadrature rules can similarly be derived using higher-order polynomial interpolation (more sample points in the $[a, b]$ interval). However, high-order polynomial interpolation can have undesirable behavior between the sample points, as discussed in Chapter 7. Rather than using higher-order quadrature rules, it is therefore often better to divide the integration interval $[a, b]$ into subintervals $[a = x_0, x_1], [x_1, x_2],...,[x_{N-1}, x_N = b]$ and use a low-order quadrature rule in each of these subintervals. Such methods are known as *composite quadrature rules*. Figure 8-2 shows the three lowest-order Newton-Cotes quadrature rules for the function $f(x) = 3+x+x^2+x^3+x^4$ on the interval $[-1, 1]$ and the corresponding composite quadrature rules with four subdivisions of the original interval.

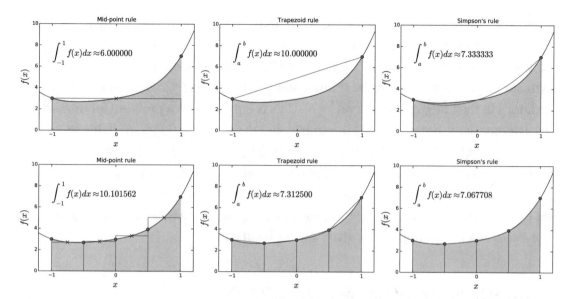

Figure 8-2. *Visualization of quadrature rules (top panel) and composite quadrature rules (bottom panel) of orders zero (the midpoint rule), one (the trapezoid rule), and two (Simpson's rule)*

An important parameter that characterizes composite quadrature rules is the subinterval length $h = (b - a)/N$. Estimates for the errors in an approximate quadrature rule, and the scaling of the error with respect to h, can be obtained from Taylor series expansions of the integrand and the analytical integration of the term in the resulting series. An alternative technique is to simultaneously consider quadrature rules of different orders or of different subinterval lengths h. The difference between two such results can often be shown to give estimates of the error, and this is the basis for how many quadrature routines produce an estimate of the error in addition to the estimate of the integral, as we will see in the examples in the following section.

We have seen that the Newton-Cotes quadrature rules use evenly spaced sample points of the integrand $f(x)$. This is often convenient, especially if the integrand is obtained from measurements or observations at prescribed points, and cannot be evaluated at arbitrary points in the interval $[a, b]$. However, this is not necessarily the most efficient choice of quadrature nodes, and if the integrand is given as a function that easily can be evaluated at arbitrary values of $x \in [a, b]$, then it can be advantageous to use quadrature rules that do not use evenly spaced sample points. An example of such a method is *Gaussian quadrature*, which also uses polynomial interpolation to determine the values of the weight factors in the quadrature rule but where the quadrature nodes x_i are chosen to maximize order of polynomials that can be integrated exactly (the polynomial degree) given a fixed number of quadrature points. It turns out that choices x_i that satisfy these criteria are the roots of different orthogonal polynomials, and the sample points x_i are typically located at irrational locations in the integration interval $[a, b]$. This is typically not a problem for numerical implementations, but practically it requires that the function $f(x)$ is available to be evaluated at arbitrary points that are decided by the integration routine, rather than given as tabulated or precomputed data at regularly spaced x values. Gaussian quadrature rules are typically superior if $f(x)$ can be evaluated at arbitrary values, but for the reason just mentioned, the Newton-Cotes quadrature rules also have important use-cases when the integrand is given as tabulated data.

Numerical Integration with SciPy

The numerical quadrature routines in the SciPy integrate module can be categorized into two types: routines that take the integrand as a Python function and routines that take arrays with samples of the integrand at given points. The functions of the first type use Gaussian quadrature (quad, quadrature, fixed_quad), while functions of the second type use Newton-Cotes methods (trapz, simps, and romb).

The quadrature function is an adaptive Gaussian quadrature routine that is implemented in Python. The quadrature repeatedly calls the fixed_quad function, for Gaussian quadrature of fixed order, with increasing order until the required accuracy is reached. The quad function is a wrapper for routines from the *FORTRAN* library QUADPACK, which has superior performance in terms of speed and has more features (such as support for infinite integration limits). It is therefore usually preferable to use quad, and in the following, we use this quadrature function. However, all these functions take similar arguments and can often be replaced with each other. They take as a first argument the function that implements the integrand, and the second and third arguments are the lower and upper integration limits. As a concrete example, consider the numerical evaluation of the integral $\int_{-1}^{1} e^{-x^2} dx$. To evaluate this integral using SciPy's quad function, we first define a function for the integrand and then call the quad function:

```
In [20]: def f(x):
    ...:     return np.exp(-x**2)
In [21]: val, err = integrate.quad(f, -1, 1)
```

```
In [22]: val
Out[22]: 1.493648265624854
In [23]: err
Out[23]: 1.6582826951881447e-14
```

The quad function returns a tuple that contains the numerical estimate of the integral, val, and an estimate of the absolute error, err. The tolerances for the absolute and the relative errors can be set using the optional epsabs and epsrel keyword arguments, respectively. If the function f takes more than one variable, the quad routine integrates the function over its first argument. We can optionally specify the values of additional arguments by passing those values to the integrand function via the keyword argument args to the quad function. For example, if we wish to evaluate $\int_{-1}^{1} ae^{-(x-b)^2/c^2}dx$

for the specific values of the parameters $a = 1$, $b = 2$, and $c = 3$, we can define a function for the integrand that takes all these additional arguments and then specify the values of a, b, and c by passing args=(1, 2, 3) to the quad function:

```
In [24]: def f(x, a, b, c):
   ...:     return a * np.exp(-((x - b)/c)**2)
In [25]: val, err = integrate.quad(f, -1, 1, args=(1, 2, 3))
In [26]: val
Out[26]: 1.2763068351022229
In [27]: err
Out[27]: 1.4169852348169507e-14
```

When working with functions where the variable we want to integrate over is not the first argument, we can reshuffle the arguments by using a lambda function. For example, if we wish to compute the integral $\int_0^5 J_0(x)dx$, where the integrand $J_0(x)$ is the zeroth-order Bessel function of the first kind, it would be convenient to use the function jv from the scipy.special module as integrand. The function jv takes the arguments v and x and is the Bessel function of the first kind for the real-valued order v and evaluated at x. To be able to use the jv function as integrand for quad, we therefore need to reshuffle the arguments of jv. With a lambda function, we can do this in the following manner:

```
In [28]: from scipy.special import jv
In [29]: f = lambda x: jv(0, x)
In [30]: val, err = integrate.quad(f, 0, 5)
```

```
In [31]: val
Out[31]: 0.7153119177847678
In [32]: err
Out[32]: 2.47260738289741e−14
```

With this technique we can arbitrarily reshuffle arguments of any function and always obtain a function where the integration variable is the first argument, so that the function can be used as integrand for quad.

The quad routine supports infinite integration limits. To represent integration limits that are infinite, we use the floating-point representation of infinity, float('inf'), which is conveniently available in NumPy as np.inf. For example, consider the integral $\int_{-\infty}^{\infty} e^{-x^2} dx$. To evaluate it using quad, we can do

```
In [33]: f = lambda x: np.exp(-x**2)
In [34]: val, err = integrate.quad(f, -np.inf, np.inf)
In [35]: val
Out[35]: 1.7724538509055159
In [36]: err
Out[36]: 1.4202636780944923e−08
```

However, note that the quadrature and fixed_quad functions only support finite integration limits.

With a bit of extra guidance, the quad function is also able to handle many integrals with integrable singularities. For example, consider the integral $\int_{-1}^{1} \frac{1}{\sqrt{|x|}} dx$. The integrand diverges at $x = 0$, but the value of the integral does not diverge, and its value is 4. Naively trying to compute this integral using quad may fail because of the diverging integrand:

```
In [37]: f = lambda x: 1/np.sqrt(abs(x))
In [38]: a, b = -1, 1
In [39]: integrate.quad(f, a, b)
Out[39]: (inf, inf)
```

In situations like these, it can be useful to graph the integrand to get insights into how it behaves, as shown in Figure 8-3.

```
In [40]: fig, ax = plt.subplots(figsize=(8, 3))
    ...: x = np.linspace(a, b, 10000)
    ...: ax.plot(x, f(x), lw=2)
```

```
...: ax.fill_between(x, f(x), color='green', alpha=0.5)
...: ax.set_xlabel("$x$", fontsize=18)
...: ax.set_ylabel("$f(x)$", fontsize=18)
...: ax.set_ylim(0, 25)
...: ax.set_xlim(-1, 1)
```

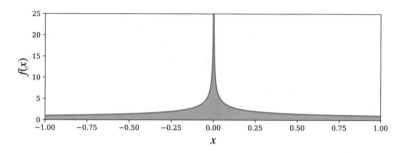

Figure 8-3. *Example of a diverging integrand with finite integral (green/shaded area) that can be computed using the* quad *function*

In this case the evaluation of the integral fails because the integrand diverges exactly at one of the sample points in the Gaussian quadrature rule (the midpoint). We can guide the quad routine by specifying a list of points that should be avoided using the points keyword arguments, and using points=[0] in the current example allows quad to correctly evaluate the integral:

```
In [41]: integrate.quad(f, a, b, points=[0])
Out[41]: (4.0,5.684341886080802e-14)
```

Tabulated Integrand

We have seen that the quad routine is suitable for evaluating integrals when the integrand is specified using a Python function that the routine can evaluate at arbitrary points (which is determined by the specific quadrature rule). However, in many situations we may have an integrand that is only specified at predetermined points, such as evenly spaced points in the integration interval $[a, b]$. This type of situation can occur, for example, when the integrand is obtained from experiments or observations that cannot be controlled by the particular integration routine. In this case we can use a Newton-Cotes quadrature, such as the midpoint rule, trapezoid rule, or Simpson's rule that were described earlier in this chapter.

In the SciPy `integrate` module, the composite trapezoid rule and Simpson's rule are implemented in the `trapz` and `simps` functions. These functions take as first argument an array y with values of the integrand at a set of points in the integration interval, and they optionally take as second argument an array x that specifies the x values of the sample points, or alternatively the spacing dx between each sample (if uniform). Note that the sample points do not necessarily need to be evenly spaced, but they must be known in advance.

To see how to evaluate an integral of a function that is given by sampled values, let's evaluate the integral $\int_0^2 \sqrt{x}\,\mathrm{d}x$ by taking 25 samples of the integrand in the integration interval $[0, 2]$, as shown in Figure 8-4:

```
In [42]: f = lambda x: np.sqrt(x)
In [43]: a, b = 0, 2
In [44]: x = np.linspace(a, b, 25)
In [45]: y = f(x)
In [46]: fig, ax = plt.subplots(figsize=(8, 3))
    ...: ax.plot(x, y, 'bo')
    ...: xx = np.linspace(a, b, 500)
    ...: ax.plot(xx, f(xx), 'b-')
    ...: ax.fill_between(xx, f(xx), color='green', alpha=0.5)
    ...: ax.set_xlabel(r"$x$", fontsize=18)
    ...: ax.set_ylabel(r"$f(x)$", fontsize=18)
```

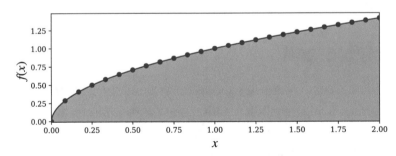

Figure 8-4. *Integrand given as tabulated values marked with dots. The integral corresponds to the shaded area.*

To evaluate the integral, we can pass the x and y arrays to the trapz or simps methods. Note that the y array must be passed as the first argument:

```
In [47]: val_trapz = integrate.trapz(y, x)
In [48]: val_trapz
Out[48]: 1.88082171605
In [49]: val_simps = integrate.simps(y, x)
In [50]: val_simps
Out[50]: 1.88366510245
```

The trapz and simps functions do not provide any error estimates, but for this particular example, we can compute the integral analytically and compare to the numerical values computed with the two methods:

```
In [51]: val_exact = 2.0/3.0 * (b-a)**(3.0/2.0)
In [52]: val_exact
Out[52]: 1.8856180831641267
In [53]: val_exact - val_trapz
Out[53]: 0.00479636711328
In [54]: val_exact - val_simps
Out[54]: 0.00195298071541
```

Since all the information we have about the integrand is the given sample points, we also cannot ask either of trapz or simps to compute more accurate solutions. The only options for increasing the accuracy are to increase the number of sample points (which might be difficult if the underlying function is not known) or possibly to use a higher-order method.

The integrate module also provides an implementation of the Romberg method with the romb function. The Romberg method is a Newton-Cotes method but one that uses Richardson extrapolation to accelerate the convergence of the trapezoid method; however this method does require that the sample points are evenly spaced and also that there are 2^n+1 sample points, where n is an integer. Like the trapz and simps methods, romb takes an array with integrand samples as first argument, but the second argument must (if given) be the sample-point spacing dx:

```
In [55]: x = np.linspace(a, b, 1 + 2**6)
In [56]: len(x)
Out[56]: 65
In [57]: y = f(x)
In [58]: dx = x[1] - x[0]
In [59]: val_exact - integrate.romb(y, dx=dx)
Out[59]: 0.000378798422913
```

Among the SciPy integration functions that we have discussed here, `simps` is perhaps the most useful one overall, since it provides a good balance between ease of use (no constraints on the sample points) and relatively good accuracy.

Multiple Integration

Multiple integrals, such as double integrals $\int_a^b\int_c^d f(x,y)dxdy$ and triple integrals $\int_a^b\int_c^d\int_e^f f(x,y,z)dxdydz$, can be evaluated using the `dblquad` and `tplquad` functions from the SciPy integrate module. Also, integration over n variables $\int\ldots\int_D f(x)dx$, over some domain D, can be evaluated using the `nquad` function. These functions are wrappers around the single-variable quadrature function `quad`, which is called repeatedly along each dimension of the integral.

Specifically, the double integral routine `dblquad` can evaluate integrals on the form

$$\int_a^b\int_{g(x)}^{h(x)} f(x,y)dxdy,$$

and it has the function signature `dblquad(f, a, b, g, h)`, where `f` is a Python function for the integrand, `a` and `b` are constant integration limits along the x dimension, and `g` and `f` are Python functions (taking `x` as argument) that specify the integration limits along the y dimension. For example, consider the integral $\int_0^1\int_0^1 e^{-x^2-y^2}dxdy$. To evaluate this we first define the function `f` for the integrand and graph the function and the integration region, as shown in Figure 8-5:

```
In [60]: def f(x, y):
    ...:     return np.exp(-x**2 - y**2)
In [61]: fig, ax = plt.subplots(figsize=(6, 5))
```

```
...: x = y = np.linspace(-1.25, 1.25, 75)
...: X, Y = np.meshgrid(x, y)
...: c = ax.contour(X, Y, f(X, Y), 15, cmap=mpl.cm.RdBu, vmin=-1,
     vmax=1)
...: bound_rect = plt.Rectangle((0, 0), 1, 1, facecolor="grey")
...: ax.add_patch(bound_rect)
...: ax.axis('tight')
...: ax.set_xlabel('$x$', fontsize=18)
...: ax.set_ylabel('$y$', fontsize=18)
```

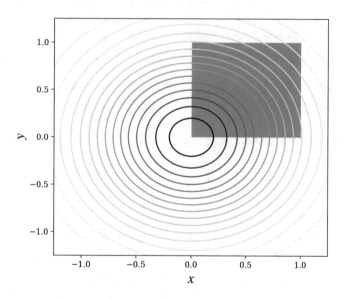

Figure 8-5. *Two-dimensional integrand as contour plot with integration region shown as a shaded area*

In this example the integration limits for both the x and y variables are constants, but since dblquad expects functions for the integration limits for the y variable, we must also define the functions h and g, even though in this case they only evaluate to constants regardless of the value of x.

```
In [62]: a, b = 0, 1
In [63]: g = lambda x: 0
In [64]: h = lambda x: 1
```

Now, with all the arguments prepared, we can call dblquad to evaluate the integral:

```
In [65]: integrate.dblquad(f, a, b, g, h)
Out[65]: (0.5577462853510337, 6.1922276789587025e-15)
```

Note that we could also have done the same thing a bit more concisely, although slightly less readable, by using inline lambda function definitions:

```
In [66]: integrate.dblquad(lambda x, y: np.exp(-x**2-y**2), 0, 1, lambda
         x: 0, lambda x: 1)
Out[66]: (0.5577462853510337, 6.1922276789587025e-15)
```

Because g and h are functions, we can compute integrals with x-dependent integration limits along the y dimension. For example, with $g(x) = x - 1$ and $h(x) = 1 - x$, we obtain:

```
In [67]: integrate.dblquad(f, 0, 1, lambda x: -1 + x, lambda x: 1 - x)
Out[67]: (0.7320931000008094, 8.127866157901059e-15)
```

The tplquad function can compute integrals on the form

$$\int_a^b \int_{g(x)}^{h(x)} \int_{q(x,y)}^{r(x,y)} f(x,y,z)\mathrm{d}x\mathrm{d}y\mathrm{d}z,$$

which is a generalization of the double integral expression computed with dblquad. It additionally takes two Python functions as arguments, which specifies the integration limits along the z dimension. These functions take two arguments, x and y, but note that g and h still only take one argument (x). To see how tplquad can be used, consider the generalization of the previous integral to three variables: $\int_0^1 \int_0^1 \int_0^1 e^{-x^2-y^2-z^2}\mathrm{d}x\mathrm{d}y\mathrm{d}z$.

We compute this integral using a similar method compared to the dblquad example. That is, we first define functions for the integrand and the integration limits and the call to the tplquad function:

```
In [68]: def f(x, y, z):
   ...:     return np.exp(-x**2-y**2-z**2)
In [69]: a, b = 0, 1
In [70]: g, h = lambda x: 0, lambda x: 1
In [71]: q, r = lambda x, y: 0, lambda x, y: 1
In [72]: integrate.tplquad(f, 0, 1, g, h, q, r)
Out[72]: (0.4165383858866382, 4.624505066515441e-15)
```

For an arbitrary number of integrations, we can use the nquad function. It also takes the integrand as a Python function as the first argument. The integrand function should have the function signature f(x1, x2, ..., xn). In contrast to dplquad and tplquad, the nquad function expects a list of integration limit specifications, as the second argument. The list should contain a tuple with integration limits for each integration variable or a callable function that returns such a limit. For example, to compute the integral that we previously computed with tplquad, we could use

```
In [73]: integrate.nquad(f, [(0, 1), (0, 1), (0, 1)])
Out[73]: (0.4165383858866382, 8.291335287314424e-15)
```

For increasing number of integration variables, the computational complexity of a multiple integral grows quickly, for example, when using nquad. To see this scaling trend, consider the following generalized version of the integrand studied with dplquad and tplquad.

```
In [74]: def f(*args):
    ...:     """
    ...:     f(x1, x2, ... , xn) = exp(-x1^2 - x2^2 - ... - xn^2)
    ...:     """
    ...:     return np.exp(-np.sum(np.array(args)**2))
```

Next, we evaluate the integral for varying number of dimensions (ranging from 1 up to 5). In the following examples, the length of the list of integration limits determines the number of the integrals. To see a rough estimate of the computation time, we use the IPython command %time:

```
In [75]: %time integrate.nquad(f, [(0,1)] * 1)
CPU times: user 398 µs, sys: 63 µs, total: 461 µs
Wall time: 466 µs
Out[75]: (0.7468241328124271,8.291413475940725e-15)

In [76]: %time integrate.nquad(f, [(0,1)] * 2)
CPU times: user 6.31 ms, sys: 298 µs, total: 6.61 ms
Wall time: 6.57 ms
Out[76]: (0.5577462853510337,8.291374381535408e-15)
```

```
In [77]: %time integrate.nquad(f, [(0,1)] * 3)
CPU times: user 123 ms, sys: 2.46 ms, total: 126 ms
Wall time: 125 ms
Out[77]: (0.4165383858866382,8.291335287314424e-15)

In [78]: %time integrate.nquad(f, [(0,1)] * 4)
CPU times: user 2.41 s, sys: 11.1 ms, total: 2.42 s
Wall time: 2.42 s
Out[78]: (0.31108091882287664,8.291296193277774e-15)

In [79]: %time integrate.nquad(f, [(0,1)] * 5)
CPU times: user 49.5 s, sys: 169 ms, total: 49.7 s
Wall time: 49.7 s
Out[79]: (0.23232273743438786,8.29125709942545e-15)
```

Here we see that increasing the number of integrations from one to five increases the computation time from hundreds of microseconds to nearly a minute. For an even larger number of integrals, it may become impractical to use direct quadrature routines, and other methods, such as Monte Carlo sampling techniques, can often be superior, especially if the required precision is not that high. Monte Carlo integration is a simple but powerful technique that is based on sampling the integrand at randomly selected points in the domain of the integral and gradually forming an estimate of the integral. Due to the stochastic nature of the algorithm, the conversion rate is typically relatively slow, and it is difficult to achieve very high accuracy. However, Monte Carlo integration scales very well with dimensionality, and it is often a competitive method for high-dimensional integrals.

To compute an integral using Monte Carlo sampling, we can use the mcquad function from the skmonaco library (known as scikit-monaco). As first argument it takes a Python function for the integrand, as second argument it takes a list of lower integration limits, and as third argument it takes a list of upper integration limits. Note that the way the integration limits are specified is not exactly the same as for the quad function in SciPy's integrate module. We begin by importing the skmonaco (scikit-monaco) module:

```
In [80]: import skmonaco
```

Once the module is imported, we can use the skmonaco.mcquad function for performing a Monte Carlo integration. In the following example, we compute the same integral as in the previous example using nquad:

```
In [81]: %time val, err = skmonaco.mcquad(f, xl=np.zeros(5), xu=np.ones(5),
         npoints=100000)
CPU times: user 1.43 s, sys: 100 ms, total: 1.53 s
Wall time: 1.5 s
In [82]: val, err
Out[82]: (0.231322502809, 0.000475071311272)
```

While the error is not comparable to the result given by nquad, the computation time is much shorter. By increasing the number of sample points, which we can specify using the npoints argument, we can increase the accuracy of the result. However, the convergence of Monte Carlo integration is very slow, and it is most suitable when high accuracy is not required. However, the beauty of Monte Carlo integration is that its computational complexity is independent of the number of integrals. This is illustrated in the following example, which computes a ten-variable integration at the same time and with comparable error level as the previous example with a five-variable integration:

```
In [83]: %time val, err = skmonaco.mcquad(f, xl=np.zeros(10), xu=np.
ones(10), npoints=100000)
CPU times: user 1.41 s, sys: 64.9 ms, total: 1.47 s
Wall time: 1.46 s
In [84]: val, err
Out[84]: (0.0540635928549, 0.000171155166006)
```

Symbolic and Arbitrary-Precision Integration

In Chapter 3, we already saw examples of how SymPy can be used to compute definite and indefinite integrals of symbolic functions, using the sympy.integrate function. For example, to compute the integral $\int_{-1}^{1} 2\sqrt{1-x^2}\,dx$, we first create a symbol for x and define expressions for the integrand and the integration limits $a = -1$ and $b = 1$,

```
In [85]: x = sympy.symbols("x")
In [86]: f = 2 * sympy.sqrt(1-x**2)
In [87]: a, b = -1, 1
```

285

after which we can compute the closed-form expression for the integral using

```
In [88]: val_sym = sympy.integrate(f, (x, a, b))
In [89]: val_sym
Out[89]: π
```

For this example, SymPy is able to find the analytic expression for the integral: π. As pointed out earlier, this situation is the exception, and in general we will not be able to find an analytical closed-form expression. We then need to resort to numerical quadrature, for example, using SciPy's `integrate.quad`, as discussed earlier in this chapter. However, the mpmath library,[1] which is closely integrated with SymPy, provides an alternative implementation of numerical quadrature, using arbitrary-precision computations. With this library, we can evaluate an integral to arbitrary precision, without being restricted to the limitations of floating-point numbers. However, the downside is, of course, that arbitrary-precision computations are significantly slower than floating-point computations. But when we require precision beyond what the SciPy quadrature functions can provide, this multiple-precision quadrature provides a solution.

For example, to evaluate the integral $\int_{-1}^{1} 2\sqrt{1-x^2}\,dx$ to a given precision,[2] we can use the mpmath.quad function, which takes a Python function for the integrand as first argument, and the integration limits as a tuple (a, b) as second argument. To specify the precision, we set the variable mpmath.mp.dps to the required number of accurate decimal places. For example, if we require 75 accurate decimal places, we set

```
In [90]: mpmath.mp.dps = 75
```

The integrand must be given as a Python function that uses math functions from the mpmath library to compute the integrand. From a SymPy expression, we can create such a function using `sympy.lambdify` with `'mpmath'` as third argument, which indicates that we want an mpmath compatible function. Alternatively, we can directly implement a Python function using the math functions from the mpmath module in SymPy, which in this case would be `f_mpmath = lambda x: 2 * mpmath.sqrt(1 - x**2)`. However, here we use `sympy.lambdify` to automate this step:

```
In [91]: f_mpmath = sympy.lambdify(x, f, 'mpmath')
```

---

[1]For more information about the multiprecision (arbitrary precision) math library mpmath, see the project's web page at http://mpmath.org.

[2]Here we deliberately choose to work with an integral that has a known analytical value, so that we can compare the multiprecision quadrature result with the known exact value.

Next we can compute the integral using mpmath.quad and display the resulting value:

```
In [92]: val = mpmath.quad(f_mpmath, (a, b))
In [93]: sympy.sympify(val)
Out[93]: 3.14159265358979323846264338327950288419716939937510582097494459230
         781640629
```

To verify that the numerically computed value is accurate to the required number of decimal places (75), we compare the result with the known analytical value (π). The error is indeed very small:

```
In [94]: sympy.N(val_sym, mpmath.mp.dps+1) - val
Out[94]: 6.908934844075555700309081490240319656892800291549025108
         01896277613487344253e-77
```

This level of precision cannot be achieved with the quad function in SciPy's integrate module, since it is limited by the precision of floating-point numbers.

The mpmath library's quad function can also be used to evaluate double and triple integrals. To do so, we only need to pass to it an integrand function that takes multiple variables as arguments, and pass tuples with integration limits for each integration variable. For example, to compute the double integral

$$\int_0^1\int_0^1 \cos(x)\cos(y)e^{-x^2-y^2}\mathrm{d}x\mathrm{d}y$$

and the triple integral

$$\int_0^1\int_0^1\int_0^1 \cos(x)\cos(y)\cos(z)e^{-x^2-y^2-z^2}\mathrm{d}x\mathrm{d}y\mathrm{d}z$$

to 30 significant decimals (this example cannot be solved symbolically with SymPy), we could first create SymPy expressions for the integrands, and then use sympy.lambdify to create the corresponding mpmath expressions:

```
In [95]: x, y, z = sympy.symbols("x, y, z")
In [96]: f2 = sympy.cos(x) * sympy.cos(y) * sympy.exp(-x**2 - y**2)
In [97]: f3 = sympy.cos(x) * sympy.cos(y) * sympy.cos(z) * sympy.
         exp(-x**2 - y**2 - z**2)
In [98]: f2_mpmath = sympy.lambdify((x, y), f2, 'mpmath')
In [99]: f3_mpmath = sympy.lambdify((x, y, z), f3, 'mpmath')
```

The integrals can then be evaluated to the desired accuracy by setting `mpmath.mp.dps` and calling `mpmath.quad`:

```
In [100]: mpmath.mp.dps = 30
In [101]: mpmath.quad(f2_mpmath, (0, 1), (0, 1))
Out[101]: mpf('0.430564794306099099242308990195783')
In [102]: res = mpmath.quad(f3_mpmath, (0, 1), (0, 1), (0, 1))
In [103]: sympy.sympify(res)
Out[103]: 0.282525579518426896867622772405
```

Again, this gives access to levels of accuracy that is beyond what `scipy.integrate.quad` can achieve, but this additional accuracy comes with a hefty increase in computational cost. Note that the type of the object returned by `mpmath.quad` is a multiprecision float (`mpf`). It can be cast into a SymPy type using `sympy.sympify`.

Line Integrals

SymPy can also be used to compute line integrals on the form $\int_C f(x, y)ds$, where C is a curve in the x–y plane, using the `line_integral` function. This function takes the integrand, as a SymPy expression, as first argument, a `sympy.Curve` instance as second argument, and a list of integration variables as third argument. The path of the line integral is specified by the `Curve` instance, which describes a parameterized curve for which the x and y coordinates are given as a function of an independent parameter, say t. To create a `Curve` instance that describes a path along the unit circle, we can use:

```
In [104]: t, x, y = sympy.symbols("t, x, y")
In [105]: C = sympy.Curve([sympy.cos(t), sympy.sin(t)], (t, 0, 2 * sympy.pi))
```

Once the integration path is specified, we can easily compute the corresponding line integral for a given integrand using `line_integral`. For example, with the integrand $f(x, y) = 1$, the result is the circumference of the unit circle:

```
In [106]: sympy.line_integrate(1, C, [x, y])
Out[106]: 2π
```

The result is less obvious for a nontrivial integrand, such as in the following example where we compute the line integral with the integrand $f(x, y) = x^2y^2$:

```
In [107]: sympy.line_integrate(x**2 * y**2, C, [x, y])
Out[107]: π/4
```

Integral Transforms

The last application of integrals that we discuss in this chapter is integral transforms. An integral transform is a procedure that takes a function as input and outputs another function. Integral transforms are the most useful when they can be computed symbolically, and here we explore two examples of integral transforms that can be performed using SymPy: the Laplace transform and the Fourier transform. There are numerous applications of these two transformations, but the fundamental motivation is to transform problems into a form that is more easily handled. It can, for example, be a transformation of a differential equation into an algebraic equation, using Laplace transforms, or a transformation of a problem from the time domain to the frequency domain, using Fourier transforms.

In general, an integral transform of a function $f(t)$ can be written as

$$T_f(u) = \int_{t_1}^{t_2} K(t,u) f(t) \, \mathrm{d}t,$$

where $T_f(u)$ is the transformed function. The choice of the kernel $K(t, u)$ and the integration limits determine the type of integral transform. The inverse of the integral transform is given by

$$f(u) = \int_{u_1}^{u_2} K^{-1}(u,t) T_f(u) \, \mathrm{d}u,$$

where $K^{-1}(u, t)$ is the kernel of the inverse transform. SymPy provides functions for several types of integral transform, but here we focus on the Laplace transform

$$L_f(s) = \int_0^\infty e^{-st} f(t) \, \mathrm{d}t,$$

with the inverse transform

$$f(t) = \frac{1}{2\pi i} \int_{c-i\infty}^{c+i\infty} e^{st} L_f(s) \, \mathrm{d}s,$$

and the Fourier transform

$$F_f(\omega) = \frac{1}{\sqrt{2\pi}} \int_{-\infty}^{\infty} e^{-i\omega t} f(t) dt,$$

with the inverse transform

$$f(t) = \frac{1}{\sqrt{2\pi}} \int_{-\infty}^{\infty} e^{i\omega t} F_f(\omega) d\omega.$$

With SymPy, we can perform these transforms with the `sympy.laplace_transform` and `sympy.fourier_transform`, respectively, and the corresponding inverse transforms can be computed with the `sympy.inverse_laplace_transform` and `sympy.inverse_fourier_transform`. These functions take a SymPy expression for the function to transform as first argument, and the symbol for independent variable of the expression to transform as second argument (e.g., t), and as third argument they take the symbol for the transformation variable (e.g., s). For example, to compute the Laplace transform of the function $f(t) = sin(at)$, we begin by defining SymPy symbols for the variables a, t, and s and a SymPy expression for the function $f(t)$:

```
In [108]: s = sympy.symbols("s")
In [109]: a, t = sympy.symbols("a, t", positive=True)
In [110]: f = sympy.sin(a*t)
```

Once we have SymPy objects for the variables and the function, we can call the `laplace_transform` function to compute the Laplace transform:

```
In [111]: sympy.laplace_transform(f, t, s)
```
Out[111]: $\left(\dfrac{a}{a^2 + s^2}, -\infty, 0 < \Re s \right)$

By default, the `laplace_transform` function returns a tuple containing the resulting transform; the value A from convergence condition of the transform, which takes the form $A < \Re s$; and lastly additional conditions that are required for the transform to be well defined. These conditions typically depend on the constraints that are specified when symbols are created. For example, here we used positive=True when creating the symbols a and t, to indicate that they represent real and positive numbers. Often we are only interested in the transform itself, and we can then use the noconds=True keyword argument to suppress the conditions in the return result:

```
In [112]: F = sympy.laplace_transform(f, t, s, noconds=True)
In [113]: F
```

Out[113]: $\dfrac{a}{a^2+s^2}$

The inverse transformation can be used in a similar manner, except that we need to reverse the roles of the symbols s and t. The Laplace transform is a unique one-to-one mapping, so if we compute the inverse Laplace transform of the previously computed Laplace transform, we expect to recover the original function:

```
In [114]: sympy.inverse_laplace_transform(F, s, t, noconds=True)
```

Out[114]: $\sin(at)$

SymPy can compute the transforms for many elementary mathematical functions and for a wide variety of combinations of such functions. When solving problems using Laplace transformations by hand, one typically searches for matching functions in reference tables with known Laplace transformations. Using SymPy, this process can conveniently be automated in many, but not all, cases. The following examples show a few additional examples of well-known functions that one finds in Laplace transformation tables. Polynomials have simple Laplace transformation:

```
In [115]: [sympy.laplace_transform(f, t, s, noconds=True) for f in
          [t, t**2, t**3, t**4]]
```

Out[115]: $[\dfrac{1}{s^2}, \dfrac{2}{s^3}, \dfrac{6}{s^4}, \dfrac{24}{s^5}]$

and we can also compute the general result with an arbitrary integer exponent:

```
In [116]: n = sympy.symbols("n", integer=True, positive=True)
In [117]: sympy.laplace_transform(t**n, t, s, noconds=True)
```

Out[117]: $\dfrac{\Gamma(n+1)}{s^{n+1}}$

The Laplace transform of composite expressions can also be computed, as in the following example which computes the transform of the function $f(t) = (1 - at)e^{-at}$:

```
In [118]: sympy.laplace_transform((1 - a*t) * sympy.exp(-a*t), t, s,
          noconds=True)
```

Out[118]: $\dfrac{s}{(a+s)^2}$

The main application of Laplace transforms is to solve differential equations, where the transformation can be used to bring the differential equation into a purely algebraic form, which can then be solved and transformed back to the original domain by applying the inverse Laplace transform. In Chapter 9 we will see concrete examples of this method. Fourier transforms can also be used for the same purpose.

The Fourier transform function, `fourier_tranform`, and its inverse, `inverse_fourier_transform`, are used in much the same way as the Laplace transformation functions. For example, to compute the Fourier transform of $f(t) = e^{-at^2}$, we would first define SymPy symbols for the variables a, t, and ω, and the function $f(t)$, and then compute the Fourier transform by calling the sympy.fourier_transform function:

```
In [119]: a, t, w = sympy.symbols("a, t, omega")
In [120]: f = sympy.exp(-a*t**2)
In [121]: F = sympy.fourier_transform(f, t, w)
In [122]: F
Out[122]: √π/ae^{-π²ω²/a}
```

As expected, computing the inverse transformation for F recovers the original function:

```
In [123]: sympy.inverse_fourier_transform(F, w, t)
Out[123]: e^{-at²}
```

SymPy can be used to compute a wide range of Fourier transforms symbolically, but unfortunately it does not handle well transformations that involve Dirac delta functions, in either the original function or the resulting transformation. This currently limits its usability, but nonetheless, for problems that do not involve Dirac delta functions, it is a valuable tool.

Summary

Integration is one of the fundamental tools in mathematical analysis. Numerical quadrature, or numerical evaluation of integrals, has important applications in many fields of science, because integrals that occur in practice often cannot be computed analytically and expressed as a closed-form expression. Their computation then requires numerical techniques. In this chapter we have reviewed basic techniques and methods for numerical quadrature and introduced the corresponding functions in the SciPy `integrate` module that can be used for the evaluation of integrals in practice. When the integrand is given as a function that can be evaluated at arbitrary points, we typically prefer Gaussian quadrature rules. On the other hand, when the integrand is defined as a tabulated data, the simpler Newton-Cotes quadrature rules can be used. We also studied symbolic integration and arbitrary-precision quadrature, which can complement floating-point quadrature for specific integrals that can be computed symbolically or when additional precision is required. As usual, a good starting point is to begin to analyze a problem symbolically, and if a particular integral can be solved symbolically by finding its antiderivative, that is generally the most desirable situation. When symbolic integration fails, we need to resort to numerical quadrature, which should first be explored with floating-point-based implementations, like the ones provided by the SciPy `integrate` module. If additional accuracy is required, we can fall back on arbitrary-precision quadrature. Another application of symbolic integration is integral transformation, which can be used to transform problems, such as differential equations, between different domains. Here we briefly looked at how to perform Laplace and Fourier transforms symbolically using SymPy, and in the following chapter, we continue to explore this for solving certain types of differential equations.

Further Reading

Numerical quadrature is discussed in many introductory textbooks on numerical computing, such as Heath (2002) and J. Stoer (1992). Detailed discussions on many quadrature methods, together with example implementations, are available in W. H. Press (2002). The theory of integral transforms, such as the Fourier transform and the Laplace transform, is introduced; see, for example, Folland (1992).

References

Folland, G.B. *Fourier Analysis and Its Applications.* American Mathematical Society, 1992.
Heath, M.T. *Scientific Computing An introductory survey.* 2nd. New York: McGrawHill, 2002.
    J. Stoer, R. Bulirsch. *Introduction to Numerical Analysis.* New York: Springer, 1992.
W. H. Press, S. A. Teukolsky, W. T. Vetterling, B. P. Flannery. *Numerical Recipes in C.* Cambridge: Cambridge University Press, 2002.

CHAPTER 9

Ordinary Differential Equations

Equations wherein the unknown quantity is a function, rather than a variable, and that involve derivatives of the unknown function, are known as differential equations. An *ordinary* differential equation is a special case where the unknown function has only one independent variable with respect to which derivatives occur in the equation. If, on the other hand, derivatives of more than one variable occur in the equation, then it is known as a *partial* differential equation, and that is the topic of Chapter 11. Here we focus on ordinary differential equations (in the following abbreviated as ODEs), and we explore both symbolic and numerical methods for solving this type of equations in this chapter. Analytical closed-form solutions to ODEs often do not exist, but for many special types of ODEs, there are analytical solutions, and in those cases, there is a chance that we can find solutions using symbolic methods. If that fails, we must as usual resort to numerical techniques.

Ordinary differential equations are ubiquitous in science and engineering, as well as in many other fields, and they arise, for example, in studies of dynamical systems. A typical example of an ODE is an equation that describes the time evolution of a process where the rate of change (the derivative) can be related to other properties of the process. To learn how the process evolves in time, given some initial state, we must solve, or integrate, the ODE that describes the process. Specific examples of applications of ODEs are the laws of mechanical motion in physics, molecular reactions in chemistry and biology, and population modeling in ecology, just to mention a few.

In this chapter we will explore both symbolic and numerical approaches to solving ODE problems. For symbolic methods we use the SymPy module, and for numerical integration of ODEs, we use functions from the `integrate` module in SciPy.

295

© Robert Johansson 2019
R. Johansson, *Numerical Python*, https://doi.org/10.1007/978-1-4842-4246-9_9

Importing Modules

Here we require the NumPy and Matplotlib libraries for basic numerical and plotting purposes, and for solving ODEs, we need the SymPy library and SciPy's `integrate` module. As usual, we assume that these modules are imported in the following manner:

```
In [1]: import numpy as np
In [2]: import matplotlib.pyplot as plt
In [3]: from scipy import integrate
In [4]: import sympy
```

For nicely displayed output from SymPy, we need to initialize its printing system:

```
In [5]: sympy.init_printing()
```

Ordinary Differential Equations

The simplest form of an ordinary differential equation is $\dfrac{dy(x)}{dx} = f(x, y(x))$, where $y(x)$ is the unknown function and $f(x,y(x))$ is known. It is a differential equation because the derivative of the unknown function $y(x)$ occurs in the equation. Only the first derivative occurs in the equation, and it is therefore an example of a first-order ODE. More generally, we can write an ODE of nth order in *explicit* form as

$$\frac{d^n y}{dx^n} = f\left(x, y, \frac{dy}{dx}, \ldots, \frac{d^{n-1} y}{dx^{n-1}}\right), \text{ or in } \textit{implicit} \text{ form as } F\left(x, y, \frac{dy}{dx}, \ldots, \frac{d^n y}{dx^n}\right) = 0,$$

where f and F are known functions.

An example of a first-order ODE is Newton's law of cooling $\dfrac{dT(t)}{dt} = -k\left(T(t) - T_a\right)$, which describes the temperature $T(t)$ of a body in a surrounding with temperature T_a. The solution to this ODE is $T(t) = T_0 + (T_0 - T_a)e^{-kt}$, where T_0 is the initial temperature of the body. An example of a second-order ODE is Newton's second law of motion $F = ma$, or more explicitly $F\left(x(t)\right) = m\dfrac{d^2 x(t)}{dt^2}$. This equation describes the position $x(t)$ of an object with mass m, when subjected to a position-dependent force $F(x(t))$. To completely specify a solution to this ODE, we would, in addition to finding its general solution, also have to give the initial position and velocity of the object. Similarly, the general solution of an nth order ODE has n free parameters that we need to specify, for example, as initial conditions for the unknown function and $n-1$ of its derivatives.

An ODE can always be rewritten as a system of first-order ODEs. Specifically, the nth order ODE on the explicit form $\dfrac{d^n y}{dx^n} = g\left(x, y, \dfrac{dy}{dx}, ..., \dfrac{d^{n-1} y}{dx^{n-1}}\right)$ can be written in the

standard form by introducing n new functions $y_1 = y$, $y_2 = \dfrac{dy}{dx}$, ..., $y_n = \dfrac{d^{n-1} y}{dx^{n-1}}$. This gives the following system of first-order ODEs

$$\frac{d}{dx}\begin{bmatrix} y_1 \\ y_2 \\ \vdots \\ y_{n-1} \\ y_n \end{bmatrix} = \begin{bmatrix} y_2 \\ y_3 \\ \vdots \\ y_n \\ g(x, y_1, ..., y_n) \end{bmatrix},$$

which also can be written in a more compact vector form: $\dfrac{d}{dx} \boldsymbol{y}(x) = f(x, \boldsymbol{y}(x))$.

This canonical form is particularly useful for numerical solutions of ODEs, and it is common that numerical methods for solving ODEs take the function $f = (f_1, f_2, ..., f_n)$, which in the current case is $f = (y_2, y_3, ..., g)$, as the input that specifies the ODE.

For example, the second-order ODE for Newton's second law of motion, $F(x) = m\dfrac{d^2 x}{dt^2}$, can be written on the standard form using $\boldsymbol{y} = \left[y_1 = x, y_2 = \dfrac{dx}{dt} \right]^T$, giving

$$\frac{d}{dt}\begin{bmatrix} y_1 \\ y_2 \end{bmatrix} = \begin{bmatrix} y_2 \\ F(y_1)/m \end{bmatrix}.$$

If the functions $f_1, f_2, ..., f_n$ are all linear, then the corresponding system of ODEs can be written on the simple form $\dfrac{d\boldsymbol{y}(x)}{dx} = A(x)\boldsymbol{y}(x) + \boldsymbol{r}(x)$, where $A(x)$ is an $n \times n$ matrix and $\boldsymbol{r}(x)$ is an n-vector that only depends on x. In this form, the $\boldsymbol{r}(x)$ is known as the *source term*, and the linear system is known as *homogeneous* if $\boldsymbol{r}(x) = 0$ and *nonhomogeneous* otherwise. Linear ODEs are an important special case that can be solved, for example, using eigenvalue decomposition of $A(x)$. Likewise, for certain properties and forms of the function $f(x, \boldsymbol{y}(x))$, there may be known solutions and special methods for solving the corresponding ODE problem, but there is no general method for an arbitrary $f(x, \boldsymbol{y}(x))$, other than approximate numerical methods.

In addition to the properties of the function $f(x, \boldsymbol{y}(x))$, the boundary conditions for an ODE also influence the solvability of the ODE problem, as well as which numerical approaches are available. Boundary conditions are needed to determine the values of the integration constants that appear in a solution. There are two main types of boundary conditions for ODE problems: *initial value conditions* and *boundary value conditions*. For initial value problems, the value of the function and its derivatives are given at a starting point, and the problem is to evolve the function forward in the independent variable (e.g., representing time or position) from this starting point. For boundary value problems, the value of the unknown function, or its derivatives, is given at fixed points. These fixed points are frequently the endpoints of the domain of interest. In this chapter we mostly focus on initial value problem, and methods that are applicable to boundary value problems are discussed in Chapter 10 on partial differential equations.

Symbolic Solution to ODEs

SymPy provides a generic ODE solver `sympy.dsolve`, which is able to find analytical solutions to many elementary ODEs. The `sympy.dsolve` function attempts to automatically classify a given ODE, and it may attempt a variety of techniques to find its solution. It is also possible to give hints to the `dsolve` function, which can guide it to the most appropriate solution method. While `dsolve` can be used to solve many simple ODEs symbolically, as we will see in the following, it is worth keeping in mind that most ODEs cannot be solved analytically. Typical examples of ODEs where one can hope to find a symbolic solution are ODEs of first or second order or linear systems of first-order ODEs with only a few unknown functions. It also helps greatly if the ODE has special symmetries or properties, such as being separable, having constant coefficients, or is in a special form for which there exist known analytical solutions. While these types of ODEs are exceptions and special cases, there are many important applications of such ODEs, and for these cases SymPy's `dsolve` can be a very useful complement to traditional analytical methods. In this section we will explore how to use SymPy and its `dsolve` function to solve simple but commonly occurring ODEs.

To illustrate the method for solving ODEs with SymPy, we begin with a simple problem and gradually look at more complicated situations. The first example is the simple first-order ODE for Newton's cooling law, $\dfrac{dT(t)}{dt} = -k\big(T(t) - T_a\big)$, with the initial value $T(0) = T_0$. To approach this problem using SymPy, we first need to define symbols for the variables t, k, T_0, and T_a, and to represent the unknown function $T(t)$, we can use a `sympy.Function` object:

```
In [6]: t, k, T0, Ta = sympy.symbols("t, k, T_0, T_a")
In [7]: T = sympy.Function("T")
```

Next, we can define the ODE very naturally by simply creating a SymPy expression for the left-hand side of the ODE when written on the form $\dfrac{dT(t)}{dt} + k\big(T(t) - T_a\big) = 0$. Here, to represent the function $T(t)$, we can now use the Sympy function object T. Applying the symbol t to it, using the function call syntax T(t), results in an applied function object that we can take derivatives of using either sympy.diff or the diff method on the T(t) expression:

```
In [8]: ode = T(t).diff(t) + k*(T(t) - Tₐ)
In [9]: sympy.Eq(ode)
```
Out[9]: $k\big(-T_a + T(t)\big) + \dfrac{dT(t)}{dt} = 0$

Here we used sympy.Eq to display the equation including the equality sign and a right-hand side that is zero. Given this representation of the ODE, we can directly pass it to sympy.dsolve, which will attempt to automatically find the general solution of the ODE.

```
In [10]: ode_sol = sympy.dsolve(ode)
In [11]: ode_sol
```
Out[11]: $T(t) = C_1 e^{-kt} + T_a$

For this ODE problem, the sympy.dsolve function indeed finds the general solution, which here includes an unknown integration constant C_1 that we have to determine from the initial conditions of the problem. The return value from the sympy.dsolve is an instance of sympy.Eq, which is a symbolic representation of an equality. It has the attributes lhs and rhs for accessing the left-hand side and the right-hand side of the equality object:

```
In [12]: ode_sol.lhs
```
Out[12]: $T(t)$
```
In [13]: ode_sol.rhs
```
Out[13]: $C_1 e^{-kt} + T_a$

Once the general solution has been found, we need to use the initial conditions to find the values of the yet-to-be-determined integration constants. Here the initial condition is $T(0) = T_0$. To this end, we first create a dictionary that describes the initial condition, ics = {T(0): T0}, that we can use with SymPy's subs method to apply the initial condition to the solution of the ODE. This results in an equation for the unknown integration constant C_1:

```
In [14]: ics = {T(0): T0}
In [15]: ics
Out[15]: {T(0): T₀}
In [16]: C_eq = ode_sol.subs(t, 0).subs(ics)
In [17]: C_eq
Out[17]: T₀ = C₁ + Tₐ
```

In the present example, the equation for C_1 is trivial to solve, but for the sake of generality, here we solve it using sympy.solve. The result is a list of solutions (in this case a list of only one solution). We can substitute the solution for C_1 into the general solution of the ODE problem to obtain the particular solution that corresponds to the given initial conditions:

```
In [18]: C_sol = sympy.solve(C_eq)
In [19]: C_sol
Out[19]: [{C₁:T₀ - Tₐ}]
In [20]: ode_sol.subs(C_sol[0])
Out[20]: T(t) = Tₐ + (T₀ - Tₐ)e⁻ᵏᵗ
```

By carrying out these steps, we have completely solved the ODE problem symbolically, and we obtained the solution $T(t) = T_a + (T_0 - T_a)e^{-kt}$. The steps involved in this process are straightforward, but applying the initial conditions and solving for the undetermined integration constants can be slightly tedious, and it is worthwhile to collect these steps in a reusable function. The following function apply_ics is a basic implementation that generalizes these steps to a differential equation of arbitrary order.

```
In [21]: def apply_ics(sol, ics, x, known_params):
    ....:      """
    ....:      Apply the initial conditions (ics), given as a dictionary on
    ....:      the form ics = {y(0): y0, y(x).diff(x).subs(x, 0): yp0, ...},
    ....:      to the solution of the ODE with independent variable x.
    ....:      The undetermined integration constants C1, C2, ... are extracted
```

```
....:         from the free symbols of the ODE solution, excluding symbols in
....:         the known_params list.
....:         """
....:         free_params = sol.free_symbols - set(known_params)
....:         eqs = [(sol.lhs.diff(x, n) - sol.rhs.diff(x, n))
               .subs(x, 0).subs(ics) for n in range(len(ics))]
....:         sol_params = sympy.solve(eqs, free_params)
....:         return sol.subs(sol_params)
```

With this function, we can more conveniently single out a particular solution to an ODE that satisfies a set of initial conditions, given the general solution to the same ODE. For our previous example, we get

```
In [22]: ode_sol
Out[22]: T(t) = C₁e⁻ᵏᵗ + Tₐ
In [23]: apply_ics(ode_sol, ics, t, [k, Ta])
Out[23]: T(t) = Tₐ + (T₀ - Tₐ)e⁻ᵏᵗ
```

The example we looked at so far is almost trivial, but the same method can be used to approach any ODE problem, although there is of course no guarantee that a solution will be found. As an example of a slightly more complicated problem, consider the ODE for a damped harmonic oscillator, which is a second-order ODE on the form

$$\frac{d^2 x(t)}{dt^2} + 2\gamma\omega_0 \frac{dx(t)}{dt} + \omega_0^2 x(t) = 0, \text{ where } x(t) \text{ is the position of the oscillator at time } t,$$

ω_0 is the frequency for the undamped case, and γ is the damping ratio. We first define the required symbols and construct the ODE and then ask SymPy to find the general solution by calling sympy.dsolve:

```
In [24]: t, omega0, gamma= sympy.symbols("t, omega_0, gamma",
         positive=True)
In [25]: x = sympy.Function("x")
In [26]: ode = x(t).diff(t, 2) + 2 * gamma * omega0 * x(t).diff(t) +
         omega0**2 * x(t)
In [27]: sympy.Eq(ode)
```

$$Out[27]: \quad \frac{d^2 x(t)}{dt^2} + 2\gamma\omega_0 \frac{dx(t)}{dt} + \omega_0^2 x(t) = 0$$

```
In [28]: ode_sol = sympy.dsolve(ode)
```

301

```
In [29]: ode_sol
```

$$\text{Out[29]: } x(t) = C_1 e^{\omega_0 t\left(-\gamma - \sqrt{\gamma^2 - 1}\right)} + C_2 e^{\omega_0 t\left(-\gamma + \sqrt{\gamma^2 - 1}\right)}$$

Since this is a second-order ODE, there are two undetermined integration constants in the general solution. We need to specify initial conditions for both the position $x(0)$ and the velocity $\left.\dfrac{dx(t)}{dt}\right|_{t=0}$ to single out a particular solution to the ODE. To do this we create a dictionary with these initial conditions and apply it to the general ODE solution using apply_ics:

```
In [30]: ics = {x(0): 1, x(t).diff(t).subs(t, 0): 0}
In [31]: ics
```

$$\text{Out[31]: } \left\{ x(0):1, \; \left.\frac{dx(t)}{dt}\right|_{t=0} : 0 \right\}$$

```
In [32]: x_t_sol = apply_ics(ode_sol, ics, t, [omega0, gamma])
In [33]: x_t_sol
```

$$\text{Out[33]: } x(t) = \left(-\frac{\gamma}{2\sqrt{\gamma^2 - 1}} + \frac{1}{2} \right) e^{\omega_0 t\left(-\gamma - \sqrt{\gamma^2 - 1}\right)} + \left(\frac{\gamma}{2\sqrt{\gamma^2 - 1}} + \frac{1}{2} \right) e^{\omega_0 t\left(-\gamma + \sqrt{\gamma^2 - 1}\right)}$$

This is the solution for the dynamics of the oscillator for arbitrary values of t, ω_0, and γ, where we used the initial condition $x(0) = 1$ and $\left.\dfrac{dx(t)}{dt}\right|_{t=0} = 0$. However, substituting $\gamma = 1$, which corresponds to critical damping, directly into this expression results in a division by zero error, and for this particular choice of γ, we need to be careful and compute the limit where $\gamma \to 1$.

```
In [34]: x_t_critical = sympy.limit(x_t_sol.rhs, gamma, 1)
In [35]: x_t_critical
```

$$\text{Out[35]: } \frac{\omega_0 t + 1}{e^{\omega_0 t}}$$

Finally, we plot the solutions for $\omega_0 = 2\pi$ and a sequence of different values of the damping ratio γ:

```
In [36]: fig, ax = plt.subplots(figsize=(8, 4))
    ...: tt = np.linspace(0, 3, 250)
    ...: w0 = 2 * sympy.pi
    ...: for g in [0.1, 0.5, 1, 2.0, 5.0]:
    ...:     if g == 1:
```

```
...:            x_t = sympy.lambdify(t, x_t_critical.subs({omega0: w0}), 'numpy')
...:        else:
...:            x_t = sympy.lambdify(t, x_t_sol.rhs.subs({omega0: w0,
             gamma: g}), 'numpy')
...:        ax.plot(tt, x_t(tt).real, label=r"$\gamma = %.1f$" % g)
...: ax.set_xlabel(r"$t$", fontsize=18)
...: ax.set_ylabel(r"$x(t)$", fontsize=18)
...: ax.legend()
```

The solution to the ODE for the damped harmonic oscillator is graphed in Figure 9-1. For $\gamma < 1$, the oscillator is underdamped, and we see oscillatory solutions. For $\gamma > 1$ the oscillator is overdamped and decays monotonically. The crossover between these two behaviors occurs at the critical damping ratio $\gamma = 1$.

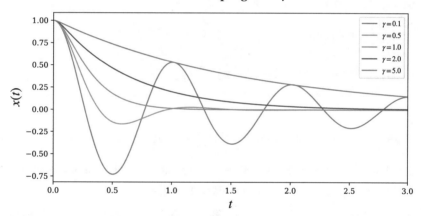

Figure 9-1. *Solutions to the ODE for a damped harmonic oscillator, for a sequence of damping ratios*

The two examples of ODEs we have looked at so far could both be solved exactly by analytical means, but this is far from always the case. Even many first-order ODEs cannot be solved exactly in terms of elementary functions. For example, consider $\dfrac{dy(x)}{dx} = x + y(x)^2$, which is an example of an ODE that does not have any closed-form solution. If we try to solve this equation using sympy.dsolve, we obtain an approximate solution, in the form of a power series:

```
In [37]: x = sympy.symbols("x")
In [38]: y = sympy.Function("y")
In [39]: f = y(x)**2 + x
```

```
In [40]: sympy.Eq(y(x).diff(x), f)
```

Out[40]: $\dfrac{dy(x)}{dx} = x + y(x)^2$

```
In [41]: sympy.dsolve(y(x).diff(x) - f)
```

Out[41]: $y(x) = C_1 + C_1 x + \dfrac{1}{2}(2C_1 + 1)x^2 + \dfrac{7C_1}{6}x^3 + \dfrac{C_1}{12}(C_1 + 5)x^4$

$$+ \dfrac{1}{60}\left(C_1^2(C_1 + 45) + 20C_1 + 3\right)x^5 + \mathcal{O}(x^6)$$

For many other types of equations, SymPy outright fails to produce any solution at all. For example, if we attempt to solve the second-order ODE $\dfrac{d^2 y(x)}{dx^2} = x + y(x)^2$, we obtain the following error message:

```
In [42]: sympy.Eq(y(x).diff(x, x), f)
```

Out[42]: $\dfrac{d^2 y(x)}{dx^2} = x + y(x)^2$

```
In [43]: sympy.dsolve(y(x).diff(x, x) - f)
---------------------------------------------------------------------------
...
NotImplementedError: solve: Cannot solve -x - y(x)**2 + Derivative(y(x), x, x)
```

This type of result can mean that there actually is no analytical solution to the ODE or, just as likely, simply that SymPy is unable to handle it.

The dsolve function accepts many optional arguments, and it can frequently make a difference if the solver is guided by giving hints about which methods should be used to solve the ODE problem at hand. See the docstring for sympy.dsolve for more information about the available options.

Direction Fields

A *direction field graph* is a simple but useful technique to visualize possible solutions to arbitrary first-order ODEs. It is made up of short lines that show the slope of the unknown function on a grid in the x–y plane. This graph can be easily produced because the slope of $y(x)$ at arbitrary points of the x–y plane is given by the definition of the ODE: $\dfrac{dy(x)}{dx} = f(x, y(x))$. That is, we only need to iterate over the x and y values on the coordinate grid of interest and evaluate $f(x, y(x))$ to know the slope of $y(x)$ at that point. The reason why the direction field graph is useful is that smooth and continuous curves that tangent the slope lines (at every point) in the direction field graph are possible solutions to the ODE.

The following function plot_direction_field produces a direction field graph for a first-order ODE, given the independent variable x, the unknown function $y(x)$, and the right-hand side function $f(x, y(x))$. It also takes optional ranges for the x and y axes (x_lim and y_lim, respectively) and an optional Matplotlib axis instance to draw the graph on.

```
In [44]: def plot_direction_field(x, y_x, f_xy, x_lim=(-5, 5),
          y_lim=(-5, 5), ax=None):
   ...:      f_np = sympy.lambdify((x, y_x), f_xy, 'numpy')
   ...:      x_vec = np.linspace(x_lim[0], x_lim[1], 20)
   ...:      y_vec = np.linspace(y_lim[0], y_lim[1], 20)
   ...:
   ...:      if ax is None:
   ...:          _, ax = plt.subplots(figsize=(4, 4))
   ...:
   ...:      dx = x_vec[1] - x_vec[0]
   ...:      dy = y_vec[1] - y_vec[0]
   ...:
   ...:      for m, xx in enumerate(x_vec):
   ...:          for n, yy in enumerate(y_vec):
   ...:              Dy = f_np(xx, yy) * dx
   ...:              Dx = 0.8 * dx**2 / np.sqrt(dx**2 + Dy**2)
   ...:              Dy = 0.8 * Dy*dy / np.sqrt(dx**2 + Dy**2)
   ...:              ax.plot([xx - Dx/2, xx + Dx/2],
   ...:                      [yy - Dy/2, yy + Dy/2], 'b', lw=0.5)
   ...:      ax.axis('tight')
   ...:      ax.set_title(r"$%s$" %
   ...:                  (sympy.latex(sympy.Eq(y(x).diff(x), f_xy))),
   ...:                  fontsize=18)
   ...:      return ax
```

With this function we can produce the direction field graphs for the ODEs on the form $\dfrac{dy(x)}{dx} = f(x, y(x))$. For example, the following code generates the direction field graphs for $f(x, y(x)) = y(x)^2 + x$, $f(x, y(x)) = -x/y(x)$, and $f(x, y(x)) = y(x)^2/x$. The result is shown in Figure 9-2.

```
In [45]: x = sympy.symbols("x")
In [46]: y = sympy.Function("y")
In [47]: fig, axes = plt.subplots(1, 3, figsize=(12, 4))
    ...: plot_direction_field(x, y(x), y(x)**2 + x, ax=axes[0])
    ...: plot_direction_field(x, y(x), -x / y(x), ax=axes[1])
    ...: plot_direction_field(x, y(x), y(x)**2 / x, ax=axes[2])
```

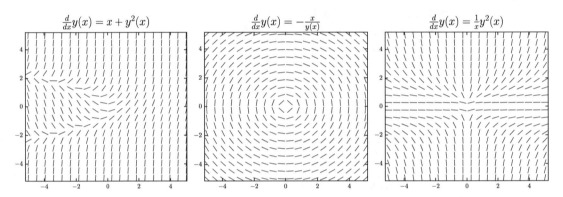

Figure 9-2. *Direction fields for three first-order differential equations*

The direction lines in the graphs in Figure 9-2 suggest how the curves that are solutions to the corresponding ODE behave, and direction field graphs are therefore a useful tool for visualizing solutions to ODEs that cannot be solved analytically. To illustrate this point, consider again the ODE $\dfrac{dy(x)}{dx} = x + y(x)^2$ with the initial condition $y(0) = 0$, which we previously saw can be solved inexactly as an approximate power series. Like before, we solve this problem again by defining the symbol x and the function y(x), which we in turn use to construct and display the ODE:

```
In [48]: x = sympy.symbols("x")
In [49]: y = sympy.Function("y")
In [50]: f = y(x)**2 + x
In [51]: sympy.Eq(y(x).diff(x), f)
```
$$\text{Out[51]: } \frac{dy(x)}{dx} = x + y(x)^2$$

Now we want to find the specific power-series solution that satisfies the initial condition, and for this problem, we can specify the initial condition directly using the ics keyword argument to the dsolve function[1]:

```
In [52]: ics = {y(0): 0}
In [53]: ode_sol = sympy.dsolve(y(x).diff(x) - f, ics=ics)
In [54]: ode_sol
```

Out[54]: $y(x) = \dfrac{x^2}{2} + \dfrac{x^5}{20} + \mathcal{O}(x^6)$

Plotting the solution together with the direction field for the ODE is a quick and simple way to get an idea of the validity range of the power-series approximation. The following code plots the approximate solution and the direction field (Figure 9-3, left panel). A solution with extended validity range is also obtained by repeatedly solving the ODE with initial conditions at increasing values of x, taken from a previous power-series solution (Figure 9-3, right panel).

```
In [55]: fig, axes = plt.subplots(1, 2, figsize=(8, 4))
    ...: # left panel
    ...: plot_direction_field(x, y(x), f, ax=axes[0])
    ...: x_vec = np.linspace(-3, 3, 100)
    ...: axes[0].plot(x_vec, sympy.lambdify(x, ode_sol.rhs.removeO())
         (x_vec), 'b', lw=2)
    ...: axes[0].set_ylim(-5, 5)
    ...:
    ...: # right panel
    ...: plot_direction_field(x, y(x), f, ax=axes[1])
    ...: x_vec = np.linspace(-1, 1, 100)
    ...: axes[1].plot(x_vec, sympy.lambdify(x, ode_sol.rhs.removeO())
         (x_vec), 'b', lw=2)
    ...: # iteratively resolve the ODE with updated initial conditions
    ...: ode_sol_m = ode_sol_p = ode_sol
    ...: dx = 0.125
    ...: # positive x
```

---

[1]In the current version of SymPy, the ics keyword argument is only recognized by the power-series solver in dsolve. Solvers for other types of ODEs ignore the ics argument and hence the need for the apply_ics function we defined and used earlier in this chapter.

```
...: for x0 in np.arange(1, 2., dx):
...:     x_vec = np.linspace(x0, x0 + dx, 100)
...:     ics = {y(x0): ode_sol_p.rhs.removeO().subs(x, x0)}
...:     ode_sol_p = sympy.dsolve(y(x).diff(x) - f, ics=ics, n=6)
...:     axes[1].plot(x_vec, sympy.lambdify(x, ode_sol_p.rhs.removeO())
        (x_vec), 'r', lw=2)
...: # negative x
...: for x0 in np.arange(-1, -5, -dx):
...:     x_vec = np.linspace(x0, x0 - dx, 100)
...:     ics = {y(x0): ode_sol_m.rhs.removeO().subs(x, x0)}
...:     ode_sol_m = sympy.dsolve(y(x).diff(x) - f, ics=ics, n=6)
...:     axes[1].plot(x_vec, sympy.lambdify(x, ode_sol_m.rhs.removeO())
        (x_vec), 'r', lw=2)
```

In the left panel of Figure 9-3, we see that the approximate solution curve aligns well with the direction field lines near $x = 0$ but starts to deviate for $|x| \gtrsim 1$, suggesting that the approximate solution is no longer valid. The solution curve shown in the right panel aligns better with the direction field throughout the plotted range. The blue (dark gray) curve segment is the original approximate solution, and the red (light gray) curves are continuations obtained from resolving the ODE with an initial condition sequence that starts where the blue (dark gray) curves end.

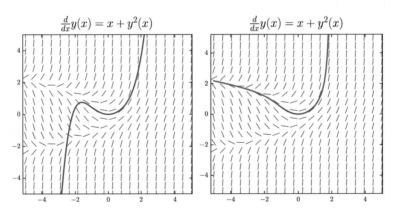

Figure 9-3. *Direction field graph of the ODE $\dfrac{dy(x)}{dx} = y(x)^2 + x$, with the fifth-order power-series solutions around $x = 0$ (left), and consecutive power-series expansions around x between –5 and 2, with a 0.125 spacing (right)*

Solving ODEs Using Laplace Transformations

An alternative to solving ODEs symbolically with SymPy's "black-box" solver[2] dsolve is to use the symbolic capabilities of SymPy to assist in a more manual approach to solving ODEs. A technique that can be used to solve certain ODE problems is to Laplace transform the ODE, which for many problems results in an algebraic equation that is easier to solve. The solution to the algebraic equation can then be transformed back to the original domain with an inverse Laplace transform, to obtain the solution to the original problem. The key to this method is that the Laplace transform of the derivative of a function is an algebraic expression in the Laplace transform of the function itself: $\mathcal{L}\big[y'(t)\big]=s\mathcal{L}\big[y(t)\big]-y(0)$. However, while SymPy is good at Laplace transforming many types of elementary functions, it does not recognize how to transform derivatives of an unknown function. But defining a function that performs this task easily amends this shortcoming.

For example, consider the following differential equation for a driven harmonic oscillator:

$$\frac{d^2}{dt^2}y(t)+2\frac{d}{dt}y(t)+10y(t)=2\sin 3t.$$

To work with this ODE, we first create SymPy symbols for the independent variable t and the function $y(t)$ and then use them to construct the symbolic expression for the ODE:

```
In [56]: t = sympy.symbols("t", positive=True)
In [57]: y = sympy.Function("y")
In [58]: ode = y(t).diff(t, 2) + 2 * y(t).diff(t) + 10 * y(t) - 2 * sympy.
         sin(3*t)
In [59]: sympy.Eq(ode)
```
Out[59]: $10y(t)-2\sin(3t)+2\dfrac{d}{dt}y(t)+\dfrac{d^2}{dt^2}y(t)=0$

---

[2] Or "white-box" solver, since SymPy is open source and the inner workings of dsolve are readily available for inspection.

Laplace transforming this ODE should yield an algebraic equation. To pursue this approach using SymPy and its function sympy.laplace_transform, we first need to create a symbol *s*, to be used in the Laplace transformation. At this point we also create a symbol *Y* for later use.

```
In [60]: s, Y = sympy.symbols("s, Y", real=True)
```

Next we proceed to Laplace transform the unknown function y(t), as well as the entire ODE equation:

```
In [61]: L_y = sympy.laplace_transform(y(t), t, s)
In [62]: L_y
```
Out[62]: $\mathcal{L}_t\left[y(t)\right](s)$
```
In [63]: L_ode = sympy.laplace_transform(ode, t, s, noconds=True)
In [64]: sympy.Eq(L_ode)
```
Out[64]: $10\mathcal{L}_t\left[y(t)\right](s)+2\mathcal{L}_t\left[\dfrac{d}{dt}y(t)\right](s)+\mathcal{L}_t\left[\dfrac{d^2}{dt^2}y(t)\right](s)-\dfrac{6}{s^2+9}=0$

When Laplace transforming the unknown function y(t), we get the undetermined result $\mathcal{L}_t\left[y(t)\right](s)$, which is to be expected. However, applying sympy.laplace_transform on a derivative of y(t), such as $\dfrac{d}{dt}y(t)$, results in the unevaluated expression, $\mathcal{L}_t\left[\dfrac{d}{dt}y(t)\right](s)$. This is not the desired result, and we need to work around this issue to obtain the sought-after algebraic equation. The Laplace transformation of the derivative of an unknown function has a well-known form that involves the Laplace transform of the function itself, rather than its derivatives: for the *n*th derivative of a function y(t), the formula is

$$\mathcal{L}_t\left[\frac{d^n}{dt^n}y(t)\right](s)=s^n\mathcal{L}_t\left[y(t)\right](s)-\sum_{m=0}^{n-1}s^{n-m-1}\frac{d^m}{dt^m}y(t)\bigg|_{t=0}.$$

By iterating through the SymPy expression tree for L_ode, and replacing the occurrences of $\mathcal{L}_t\left[\dfrac{d^n}{dt^n}y(t)\right](s)$ with expressions of the form given by this formula, we can obtain the algebraic form of the ODE that we seek. The following function takes a Laplace-transformed ODE and performs the substitution of the unevaluated Laplace transforms of the derivatives of $y(t)$:

```
In [65]: def laplace_transform_derivatives(e):
    ...:         """
    ...:         Evaluate laplace transforms of derivatives of functions
    ...:         """
    ...:         if isinstance(e, sympy.LaplaceTransform):
    ...:             if isinstance(e.args[0], sympy.Derivative):
    ...:                 d, t, s = e.args
    ...:                 n = len(d.args) - 1
    ...:                 return ((s**n) * sympy.LaplaceTransform(d.args[0], t, s) -
    ...:                         sum([s**(n-i) * sympy.diff(d.args[0], t, i-1).
    ...:                         subs(t, 0) for i in range(1, n+1)]))
    ...:
    ...:         if isinstance(e, (sympy.Add, sympy.Mul)):
    ...:             t = type(e)
    ...:             return t(*[laplace_transform_derivatives(arg) for arg in
    ...:             e.args])
    ...:
    ...:         return e
```

Applying this function on the Laplace-transformed ODE equation, L_ode, yields:

```
In [66]: L_ode_2 = laplace_transform_derivatives(L_ode)
In [67]: sympy.Eq(L_ode_2)
```

$$\text{Out[67]: } s^2 \mathcal{L}_t\big[y(t)\big](s) + 2s\mathcal{L}_t\big[y(t)\big](s) - sy(0)$$
$$+ 10\mathcal{L}_t\big[y(t)\big](s) - 2y(0) - \frac{d}{dt}y(t)\Big|_{t=0} - \frac{6}{s^2+9} = 0$$

To simplify the notation, we now substitute the expression $\mathcal{L}_t\left[y(t)\right](s)$ for the symbol Y:

```
In [68]: L_ode_3 = L_ode_2.subs(L_y, Y)
In [69]: sympy.Eq(L_ode_3)
```

Out[69]: $s^2 Y + 2sY - sy(0) + 10Y - 2y(0) - \dfrac{d}{dt}y(t)\Big|_{t=0} - \dfrac{6}{s^2+9} = 0$

At this point we need to specify the boundary conditions for the ODE problem. Here we use $y(0) = 1$ and $y'(t) = 0$, and after creating a dictionary that contains these boundary conditions, we use it to substitute the values into the Laplace-transformed ODE equation:

```
In [70]: ics = {y(0): 1, y(t).diff(t).subs(t, 0): 0}
In [71]: ics
```

Out[71]: $\left\{ y(0):1, \dfrac{d}{dt}y(t)\Big|_{t=0} :0 \right\}$

```
In [72]: L_ode_4 = L_ode_3.subs(ics)
In [73]: sympy.Eq(L_ode_4)
```

Out[74]: $Ys^2 + 2Ys + 10Y - s - 2 - \dfrac{6}{s^2+9} = 0$

This is an algebraic equation that can be solved for Y:

```
In [75]: Y_sol = sympy.solve(L_ode_4, Y)
In [76]: Y_sol
```

Out[76]: $\left[\dfrac{s^3 + 2s^2 + 9s + 24}{s^4 + 2s^3 + 19s^2 + 18s + 90} \right]$

The result is a list of solutions, which in this case contains only one element. Performing the inverse Laplace transformation on this expression gives the solution to the original problem in the time domain:

```
In [77]: y_sol = sympy.inverse_laplace_transform(Y_sol[0], s, t)
In [78]: sympy.simplify(y_sol)
```

Out[78]: $\dfrac{1}{111e^t}\left(6(\sin 3t - 6\cos 3t)e^t + 43\sin 3t + 147\cos 3t\right)$

This technique of Laplace transforming an ODE, solving the corresponding algebraic equation, and inverse Laplace transforming the result to obtain the solution to the original problem can be applied to solve many important ODE problems that arise in, for example, electrical engineering and process control applications. Although these problems can be solved by hand with the help of Laplace transformation tables, using SymPy has the potential of significantly simplifying the process.

Numerical Methods for Solving ODEs

While some ODE problems can be solved with analytical methods, as we have seen in the examples in the previous sections, it is much more common with ODE problems that cannot be solved analytically. In practice, ODE problems are therefore mainly solved with numerical methods. There are many approaches to solving ODEs numerically, and most of them are designed for problems that are formulated as a system of first-order ODEs on the standard form[3] $\dfrac{dy(x)}{dx} = f(x,y(x))$, where $y(x)$ is a vector of unknown functions of x. SciPy provides functions for solving this kind of problems, but before we explore how to use those functions, we briefly review the fundamental concepts and introduce the terminology used for the numerical integration of ODE problems.

The basic idea of many numerical methods for ODEs is captured in Euler method. This method can, for example, be derived from a Taylor series expansion of $y(x)$ around the point x

$$y(x+h) = y(x) + \frac{dy(x)}{dx}h + \frac{1}{2}\frac{d^2y(x)}{dx^2}h^2 + \ldots,$$

where for notational simplicity, we consider the case when $y(x)$ is a scalar function. By dropping terms of second order or higher, we get the approximate equation $y(x+h) \approx y(x) + f(x, y(x))h$, which is accurate to first order in the stepsize h. This equation can be turned into an iteration formula by discretizing the x variables, $x_0, x_1, \ldots, x_k$, choosing the stepsize $h_k = x_{k+1} - x_k$, and denoting $y_k = y(x_k)$. The resulting iteration formula $y_{k+1} \approx y_k + f(x_k, y_k)h_k$ is known as the *forward Euler method*, and it is said to be an *explicit* form because given the value of the y_k, we can directly compute y_{k+1} using the

---

[3]Recall that any ODE problem can be written as a system of first-order ODEs on this standard form.

formula. The goal of the numerical solution of an initial value problem is to compute $y(x)$ at some points x_n, given the initial condition $y(x_0) = y_0$. An iteration formula like the forward Euler method can therefore be used to compute successive values of y_k, starting from y_0. There are two types of errors involved in this approach: First, the truncation of the Taylor series gives an error that limits the *accuracy* of the method. Second, using the approximation of y_k given by the previous iteration when computing y_{k+1} gives an additional error that may accumulate over successive iterations and that can affect the *stability* of the method.

An alternative form, which can be derived in a similar manner, is the *backward Euler method*, given by the iteration formula $y_{k+1} \approx y_k + f(x_{k+1}, y_{k+1})h_k$. This is an example of a *backward differentiation formula (BDF)*, which is *implicit*, because y_{k+1} occurs on both sides of the equation. To compute y_{k+1}, we therefore need to solve an algebraic equation (e.g., using Newton's method, see Chapter 5). Implicit methods are more complicated to implement than explicit methods, and each iteration requires more computational work. However, the advantage is that implicit methods generally have larger stability region and better accuracy, which means that larger stepsize h_k can be used while still obtaining an accurate and stable solution. Whether explicit or implicit methods are more efficient depends on the particular problem that is being solved. Implicit methods are often particularly useful for *stiff* problems, which loosely speaking are ODE problems that describe dynamics with multiple disparate timescales (e.g., dynamics that includes both fast and slow oscillations).

There are several methods to improve upon the first-order Euler forward and backward methods. One strategy is to keep higher-order terms in the Taylor series expansion of $y(x+h)$, which gives higher-order iteration formulas that can have better accuracy, such as the second-order method $y_{k+1} \approx y(x_k) + f(x_{k+1}, y_{k+1})h_k + \frac{1}{2}y_k''(x)h_k^2$. However, such methods require evaluating higher-order derivatives of $y(x)$, which may be a problem if $f(x, y(x))$ is not known in advance (and not given in symbolic form). Ways around this problem include approximating the higher-order derivatives using finite-difference approximations of the derivatives or sampling the function $f(x, y(x))$ at intermediary points in the interval $[x_k, x_{k+1}]$. An example of this type of method is the well-known Runge-Kutta method, which is a single-step method that uses additional evaluations of $f(x, y(x))$. The most well-known Runge-Kutta method is the fourth-order scheme

$$y_{k+1} = y_k + \frac{1}{6}(k_1 + 2k_2 + 2k_3 + k_4),$$

where

$$k_1 = f\left(t_k, y_k\right)h_k,$$

$$k_2 = f\left(t_k + \frac{h_k}{2}, y_k + \frac{k_1}{2}\right)h_k,$$

$$k_3 = f\left(t_k + \frac{h_k}{2}, y_k + \frac{k_2}{2}\right)h_k,$$

$$k_4 = f\left(t_k + h_k, y_k + k_3\right)h_k.$$

Here, k_1 to k_4 are four different evaluations of the ODE function $f(x, y(x))$ that are used in the explicit formula for y_{k+1} given in the preceding text. The resulting estimate of y_{k+1} is accurate to fourth order, with an error of fifth order. Higher-order schemes that use more function evaluations can also be constructed. By combining two methods of different orders, it can be possible to also estimate the error in the approximation. A popular combination is the Runge-Kutta fourth- and fifth-order schemes, which results in a fourth-order accurate method with error estimates. It is known as RK45 or the Runge-Kutta-Fehlberg method. The Dormand-Prince method is another example of a higher-order method, which additionally uses adaptive stepsize control. For example, the 8-5-3 method combines third- and fifth-order schemes to produce an eighth-order method. An implementation of this method is available in SciPy, which we will see in the next section.

An alternative method is to use more than one previous value of y_k to compute y_{k+1}. Such methods are known as multistep methods and can in general be written in the form

$$y_{k+s} = \sum_{n=0}^{s-1} a_n y_{k+n} + h\sum_{n=0}^{s} b_n f\left(x_{k+n}, y_{k+n}\right).$$

With this formula, to compute y_{k+s}, the previous s values of y_k and $f(x_k, y_k)$ are used (known as an s-step method). The choices of the coefficients a_n and b_n give rise to different multistep methods. Note that if $b_s = 0$, then the method is explicit, and if $b_s \neq 0$, it is implicit.

For example, $b_0 = b_1 = \ldots = b_{s-1} = 0$ gives the general formula for an s-step BDF formula, where a_n and b_n are chosen to maximize the order of the accuracy by requiring that the method is exact for polynomials up to as high order as possible. This gives an equation system that can be solved for the unknown coefficients a_n and b_n. For example, the one-step BDF method with $b_1 = a_0 = 1$ reduces to the backward Euler method, $y_{k+1} = y_k + hf(x_{k+1}, y_{k+1})$, and the two-step BDF method, $y_{k+2} = a_0 y_k + a_1 y_{k+1} + hb_2 f(x_{k+2}, y_{k+2})$, when solved for the coefficients (a_0, a_1, and b_2), becomes $y_{k+2} = -\frac{1}{3} y_k + \frac{4}{3} y_{k+1} + \frac{2}{3} hf\left(x_{k+2}, y_{k+2}\right)$. Higher-order BDF methods can also be constructed. SciPy provides a BDF solver that is recommended for stiff problems, because of its good stability properties.

Another family of multistep methods is the Adams methods, which result from the choice $a_0 = a_1 = ... = a_{s-2} = 0$ and $a_{s-1} = 1$, where again the remaining unknown coefficients are chosen to maximize the order of the method. Specifically, the explicit methods with $b_s = 0$ are known as Adams-Bashforth methods, and the implicit methods with $b_s \neq 0$ are known as Adams-Moulton methods. For example, the one-step Adams-Bashforth and Adams-Moulton methods reduce to the forward and backward Euler methods, respectively, and the two-step methods are

$$y_{k+2} = y_{k+1} + h\left(-\frac{1}{2} f\left(x_k, y_k\right) + \frac{3}{2} f\left(x_{k+1}, y_{k+1}\right)\right) \text{ and } y_{k+1} = y_k + \frac{1}{2} h\left(f\left(x_k, y_k\right) + f\left(x_{k+1}, y_{k+1}\right)\right),$$

respectively. Higher-order explicit and implicit methods can also be constructed in this way. Solvers using these Adams methods are also available in SciPy.

In general explicit methods are more convenient to implement and less computational demanding to iterate than implicit methods, which in principle requires solving (a potentially nonlinear) equation in each iteration with an initial guess for the unknown y_{k+1}. However, as mentioned earlier, implicit methods often are more accurate and have superior stability properties. A compromise that retains some of the advantages of both methods is to combine explicit and implicit methods in the following way: first compute y_{k+1} using an explicit method, and then use this y_{k+1} as an initial guess for solving the equation for y_{k+1} given by an implicit method. This equation does not need to be solved exactly, and since the initial guess from the explicit method should be quite good, it could be sufficient with a small number of iterations, using, for example, Newton's method. Methods like these, where the result from an explicit method is used to predict y_{k+1} and an implicit method is used to *correct* the prediction, are called *predictor-corrector* methods.

Finally, an important technique that is employed by many advanced ODE solvers is *adaptive stepsize* or *stepsize control*: the accuracy and stability of an ODE are strongly related to the stepsize h_k used in the iteration formula for an ODE method, and so is the computational cost of the solution. If the error in y_{k+1} can be estimated together with the computation of y_{k+1} itself, then it is possible to automatically adjust the stepsize h_k so that the solver uses large economical stepsizes when possible and smaller stepsizes when required. A related technique, which is possible with some methods, is to automatically adjust the order of the method, so that a lower-order method is used when possible and a higher-order method is used when necessary. The Adams methods are examples of methods where the order can be changed easily.

There exist a vast variety of high-quality implementations of ODE solvers, and rarely should it be necessary to reimplement any of the methods discussed here. In fact, doing so would probably be a mistake, unless it is for educational purposes or if one's primary interest is research on methods for numerical ODE solving. For practical purposes, it is advisable to use one of the many highly tuned and thoroughly tested ODE suites that already exist, most of which are available for free and as open source and packaged into libraries such as SciPy. However, there are a large number of solvers to choose from, and to be able to make an informed decision on which one to use for a particular problem, and to understand many of their options, it is important to be familiar with the basic ideas and methods and the terminology that is used to discuss them.

Numerical Integration of ODEs Using SciPy

After the review of numerical methods for solving ODEs given in the previous section, we are now ready to explore the ODE solvers that are available in SciPy and how to use them. The `integrate` module of SciPy provides two ODE solver interfaces: `integrate.odeint` and `integrate.ode`. The `odeint` function is an interface to the LSODA solver from ODEPACK,[4] which automatically switches between an Adams predictor-corrector method for nonstiff problems and a BDF method for stiff problems. In contrast, the `integrate.ode` class provides an object-oriented interface to a number of different solvers: the VODE and ZVODE solvers[5] (ZVODE is a variant of VODE for complex-valued functions), the LSODA solver, and dopri5 and dop853 which are fourth- and eighth-order Dormand-Prince methods (i.e., types of Runge-Kutta methods) with adaptive stepsize. While the object-oriented interface provided by `integrate.ode` is more flexible, the `odeint` function is in many cases simpler and more convenient to use. In the following we look at both these interfaces, starting with the `odeint` function.

---

[4]More information about ODEPACK is available at `http://computation.llnl.gov/casc/odepack`.

[5]The VODE and ZVODE solvers are available at netlib: `http://www.netlib.org/ode`.

The odeint function takes three mandatory arguments: a function for evaluating the right-hand side of the ODE on the standard form, an array (or scalar) that specifies the initial condition for the unknown functions, and an array with values of the independent variable where an unknown function is to be computed. The function for the right-hand side of the ODE takes two mandatory arguments and an arbitrary number of optional arguments. The required arguments are the array (or scalar) for the vector $y(x)$ as first argument and the value of x as second argument. For example, consider again the scalar ODE $y'(x) = f(x, y(x)) = x+y(x)^2$. To be able to plot the direction field for this ODE again, this time together with a specific solution obtained by numerical integration using odeint, we first define the SymPy symbols required to construct a symbolic expression for $f(x, y(x))$:

```
In [79]: x = sympy.symbols("x")
In [80]: y = sympy.Function("y")
In [81]: f = y(x)**2 + x
```

To be able to solve this ODE with SciPy's odeint, we first and foremost need to define a Python function for $f(x, y(x))$ that takes Python scalars or NumPy arrays as input. From the SymPy expression f, we can generate such a function using sympy.lambdify with the 'numpy' argument[6]:

```
In [82]: f_np = sympy.lambdify((y(x), x), f)
```

Next we need to define the initial value y0 and a NumPy array with the discrete values of x for which to compute the function $y(x)$. Here we solve the ODE starting at $x = 0$ in both the positive and negative directions, using the NumPy arrays xp and xm, respectively. Note that to solve the ODE in the negative direction, we only need to create a NumPy array with negative increments. Now that we have set up the ODE function f_np, initial value y0, and array of x coordination, for example, xp, we can integrate the ODE problem by calling integrate.odeint(f_np, y0, xp):

```
In [83]: y0 = 0
In [84]: xp = np.linspace(0, 1.9, 100)
In [85]: yp = integrate.odeint(f_np, y0, xp)
```

---

[6]In this particular case, with a scalar ODE, we could also use the 'math' argument, which produces a scalar function using functions from the standard math library, but more frequently we will need array-aware functions, which we obtain by using the 'numpy' argument to sympy. lambdify.

```
In [86]: xm = np.linspace(0, -5, 100)
In [87]: ym = integrate.odeint(f_np, y0, xm)
```

The results are two one-dimensional NumPy arrays ym and yp, of the same length as the corresponding coordinate arrays xm and xp (i.e., 100), which contain the solution to the ODE problem at the specified points. To visualize the solution, we next plot the ym and yp arrays together with the direction field for the ODE. The result is shown in Figure 9-4. It is apparent that the solution aligns with (tangents) the lines in the direction field at every point in the graph, as expected.

```
In [88]: fig, ax = plt.subplots(1, 1, figsize=(4, 4))
    ...: plot_direction_field(x, y(x), f, ax=ax)
    ...: ax.plot(xm, ym, 'b', lw=2)
    ...: ax.plot(xp, yp, 'r', lw=2)
```

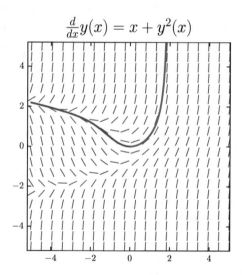

Figure 9-4. *The direction field of the ODE y'(x) = x+y(x)² and the specific solution that satisfies y(0) = 0*

In the previous example, we solved a scalar ODE problem. More often we are interested in vector-valued ODE problems (systems of ODEs). To see how we can solve that kind of problems using odeint, consider the Lotka-Volterra equations for the dynamics of a population of predator and prey animals (a classic example of coupled ODEs). The equations are $x'(t) = ax - bxy$ and $y'(t) = cxy - dy$, where $x(t)$ is the number of prey animals and $y(t)$ is the number of predator animals, and the coefficients a, b, c, and d describe the rates of the processes in the model. For example, a is the rate at which prey animals are born, and d is the rate at which predators die. The b and c coefficients are the rates at which predators consume prey and the rate at which the predator population grows at the expense of the prey population, respectively. Note that this is a nonlinear system of ODEs, because of the xy terms.

To solve this problem with odeint, we first need to write a function for the right-hand side of the ODE in vector form. For this case we have $f(t, [x, y]^T) = [ax - bxy, cxy - dy]^T$, which we can implement as a Python function in the following way:

```
In [89]: a, b, c, d = 0.4, 0.002, 0.001, 0.7
In [90]: def f(xy, t):
    ...:     x, y = xy
    ...:     return [a * x - b * x * y, c * x * y - d * y]
```

Here we have also defined variables and values for the coefficients a, b, c, and d. Note that here the first argument of the ODE function f is an array containing the current values of $x(t)$ and $y(t)$. For convenience, we first unpack these variables into separate variables x and y, which makes the rest of the function easier to read. The return value of the function should be an array, or list, that contains the values of the derivatives of $x(t)$ and $y(t)$. The function f must also take the argument t, with the current value of the independent coordinate. However, t is not used in this example. Once the f function is defined, we also need to define an array xy0 with the initial values $x(0)$ and $y(0)$ and an array t for the points at which we wish to compute the solution to the ODE. Here we use the initial conditions $x(0) = 600$ and $y(0) = 400$, which corresponds to 600 prey animals and 400 predators at the beginning of the simulation.

```
In [91]: xy0 = [600, 400]
In [92]: t = np.linspace(0, 50, 250)
In [93]: xy_t = integrate.odeint(f, xy0, t)
In [94]: xy_t.shape
Out[94]: (250,2)
```

Calling integrate.odeint(f, xy0, t) integrates the ODE problem and returns an array of shape (250, 2), which contains $x(t)$ and $y(t)$ for each of the 250 values in t. The following code plots the solution as a function of time and in phase space. The result is shown in Figure 9-5.

```
In [95]: fig, axes = plt.subplots(1, 2, figsize=(8, 4))
   ...: axes[0].plot(t, xy_t[:,0], 'r', label="Prey")
   ...: axes[0].plot(t, xy_t[:,1], 'b', label="Predator")
   ...: axes[0].set_xlabel("Time")
   ...: axes[0].set_ylabel("Number of animals")
   ...: axes[0].legend()
   ...: axes[1].plot(xy_t[:,0], xy_t[:,1], 'k')
   ...: axes[1].set_xlabel("Number of prey")
   ...: axes[1].set_ylabel("Number of predators")
```

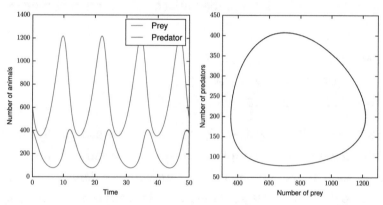

***Figure** 9-5.* *A solution to the Lotka-Volterra ODE for predator-prey populations, as a function of time (left) and in phase space (right)*

In the previous two examples, the function for the right-hand side of the ODE was implemented without additional arguments. In the example with the Lotka-Volterra equation, the function f used globally defined coefficient variables. Rather than using global variables, it is often convenient and elegant to implement the f function in such a way that it takes arguments for all its coefficient or parameters. To illustrate this point, let's consider another famous ODE problem: the Lorenz equations, which are the following system of three coupled nonlinear ODEs, $x'(t) = \sigma(y - x)$, $y'(t) = x(\rho - z) - y$, and $z'(t) = xy - \beta z$. These equations are known for their chaotic solutions, which sensitively depend on the values of the parameters σ, ρ, and β. If we wish to solve these equations for different values of these parameters, it is useful to write the ODE function so that it additionally takes the values of these variables as arguments. In the following implementation of f, the three arguments sigma, rho, and beta, for the correspondingly named parameters, have been added after the mandatory $y(t)$ and t arguments:

```
In [96]: def f(xyz, t, sigma, rho, beta):
    ...:     x, y, z = xyz
    ...:     return [sigma * (y - x),
    ...:             x * (rho - z) - y,
    ...:             x * y - beta * z]
```

Next, we define variables with specific values of the parameters, the array with t values to compute the solution for, and the initial conditions for the functions $x(t)$, $y(t)$, and $z(t)$.

```
In [97]: sigma, rho, beta = 8, 28, 8/3.0
In [98]: t = np.linspace(0, 25, 10000)
In [99]: xyz0 = [1.0, 1.0, 1.0]
```

This time when we call integrate.odeint, we need to also specify the args argument, which needs to be a list, tuple, or array with the same number of elements as the number of additional arguments in the f function we defined in the preceding section. In this case there are three parameters, and we pass a tuple with the values of these parameters via the args argument when calling integrate.odeint. In the following we solve the ODE for three different sets of parameters (but the same initial conditions).

```
In [100]: xyz1 = integrate.odeint(f, xyz0, t, args=(sigma, rho, beta))
In [101]: xyz2 = integrate.odeint(f, xyz0, t, args=(sigma, rho, 0.6*beta))
In [102]: xyz3 = integrate.odeint(f, xyz0, t, args=(2*sigma, rho, 0.6*beta))
```

The solutions are stored in the NumPy arrays xyz1, xyz2, and xyz3. In this case these arrays have the shape (10000, 3), because the t array have 10000 elements and there are three unknown functions in the ODE problem. The three solutions are plotted in 3D graphs in the following code, and the result is shown in Figure 9-6. With small changes in the system parameters, the resulting solutions can vary greatly.

```
In [103]: from mpl_toolkits.mplot3d.axes3d import Axes3D
In [104]: fig, (ax1, ax2, ax3) = plt.subplots(1, 3, figsize=(12, 4),
     ...:                                     subplot_kw={'projection':'3d'})
```

```
...: for ax, xyz, c in [(ax1, xyz1, 'r'), (ax2, xyz2, 'b'),
    (ax3, xyz3, 'g')]:
...:     ax.plot(xyz[:,0], xyz[:,1], xyz[:,2], c, alpha=0.5)
...:     ax.set_xlabel('$x$', fontsize=16)
...:     ax.set_ylabel('$y$', fontsize=16)
...:     ax.set_zlabel('$z$', fontsize=16)
...:     ax.set_xticks([-15, 0, 15])
...:     ax.set_yticks([-20, 0, 20])
...:     ax.set_zticks([0, 20, 40])
```

The three examples we have looked at so far all use the odeint solver. This function takes a large number of optional arguments that can be used to fine-tune the solver, including options for the maximum number of allowed steps (hmax) and the maximum order for the Adams (mxordn) and BDF (mxords) methods, just to mention a few. See the docstring of odeint for further information.

Figure 9-6. *The dynamics for the Lorenz ODE, for three different sets of parameters*

The alternative to odeint in SciPy is the object-oriented interface provided by the integrate.ode class. Like with the odeint function, to use the integrate.ode class, we first need to define the right-hand side function for the ODE and define the initial state array and an array for the values of the independent variable at which we want to compute the solution. However, one small but important difference is that while the function for $f(x, y(x))$ to be used with odeint had to have the function signature f(y, x, ...), the corresponding function to be used with integrate.ode must have the function signature f(x, y, ...) (i.e., the order of x and y is reversed).

The `integrate.ode` class can work with a collection of different solvers, and it has specific options for each solver. The docstring of `integrate.ode` describes the available solvers and their options. To illustrate how to use the `integrate.ode` interface, we first look at the following sets of coupled second-order ODEs:

$$m_1 x_1''(t) + \gamma_1\, x_1'(t) + k_1 x_1 - k_2\left(x_2 - x_1\right) = 0,$$

$$m_2 x_2''(t) + \gamma_2 x_2'(t) + k_2\left(x_2 - x_1\right) = 0.$$

These equations describe the dynamics of two coupled springs, where $x_1(t)$ and $x_2(t)$ are the displacements of two objects, with masses m_1 and m_2, from their equilibrium positions. The object at x_1 is connected to a fixed wall via a spring with spring constant k_1 and connected to the object at x_2 via a spring with spring constant k_2. Both objects are subject to damping forces characterized by γ_1 and γ_2, respectively. To solve this kind of problem with SciPy, we first have to write it in standard form, which we can do by introducing $y_0(t) = x_1(t)$, $y_1(t) = x_1'(t)$, $y_2(t) = x_2(t)$, and $y_3(t) = x_2'(t)$, which results in four coupled first-order equations:

$$\frac{d}{dt}\begin{bmatrix} y_0(t) \\ y_1(t) \\ y_2(t) \\ y_3(t) \end{bmatrix} = f(t, \mathbf{y}(t)) = \begin{bmatrix} y_1(t) \\ \left(-\gamma_1 y_1(t) - k_1 y_0(t) - k_2 y_0(t) + k_2 y_2(t)\right)/m_1 \\ y_3(t) \\ \left(-\gamma_2 y_3(t) - k_2 y_2(t) + k_2 y_0(t)\right)/m_2 \end{bmatrix}$$

The first task is to write a Python function that implements the function $f(t, \mathbf{y}(t))$, which also takes the problem parameters as additional arguments. In the following implementation, we bunch all the parameter into a tuple that is passed to the function as a single argument and unpack on the first line of the function body:

```
In [105]: def f(t, y, args):
     ...:     m1, k1, g1, m2, k2, g2 = args
     ...:     return [y[1], - k1/m1 * y[0] + k2/m1 * (y[2] - y[0]) -
             g1/m1 * y[1],
     ...:             y[3], - k2/m2 * (y[2] - y[0]) - g2/m2 * y[3]]
```

The return value of the function f is a list of length 4, whose elements are the derivatives of the ODE functions $y_0(t)$ to $y_3(t)$. Next we create variables with specific values for the parameters and pack them into a tuple `args` that can be passed to the function f. Like before, we also need to create arrays for the initial condition y0 and for the *t* values that we want to compute the solution to the ODE, t.

```
In [106]: m1, k1, g1 = 1.0, 10.0, 0.5
In [107]: m2, k2, g2 = 2.0, 40.0, 0.25
In [108]: args = (m1, k1, g1, m2, k2, g2)
In [109]: y0 = [1.0, 0, 0.5, 0]
In [110]: t = np.linspace(0, 20, 1000)
```

The main difference between using `integrate.odeint` and `integrate.ode` starts at this point. Instead of calling the `odeint` function, we now need to create an instance of the class `integrate.ode`, passing the ODE function f as an argument:

```
In [111]: r = integrate.ode(f)
```

Here we store the resulting solver instance in the variable r. Before we can start using it, we need to configure some of its properties. At a minimum, we need to set the initial state using the `set_initial_value` method, and if the function f takes additional arguments, we need to configure those using the `set_f_params` method. We can also select solver using `set_integrator` method, which accepts the following solver names as the first argument: `vode`, `zvode`, `lsoda`, `dopri5`, and `dop853`. Each solver takes additional optional arguments. See the docstring for `integrate.ode` for details. Here we use the LSODA solver and set the initial state and the parameters to the function f:

```
In [112]: r.set_integrator('lsoda');
In [113]: r.set_initial_value(y0, t[0]);
In [114]: r.set_f_params(args);
```

Once the solver is created and configured, we can start solving the ODE step by step by calling r.integrate method, and the status of the integration can be queried using the method r.successful (which returns True as long as the integration is proceeding fine). We need to keep track of which point to integrate to, and we need to store results by ourselves:

```
In [115]: dt = t[1] - t[0]
     ...: y = np.zeros((len(t), len(y0)))
     ...: idx = 0
     ...: while r.successful() and r.t < t[-1]:
     ...:     y[idx, :] = r.y
     ...:     r.integrate(r.t + dt)
     ...:     idx += 1
```

This is arguably not as convenient as simply calling the odeint, but it offers extra flexibility that sometimes is exactly what is needed. In this example we stored the solution in the array y for each corresponding element in t, which is similar to what odeint would have returned. The following code plots the solution, and the result is shown in Figure 9-7.

```
In [116]: fig = plt.figure(figsize=(10, 4))
     ...: ax1 = plt.subplot2grid((2, 5), (0, 0), colspan=3)
     ...: ax2 = plt.subplot2grid((2, 5), (1, 0), colspan=3)
     ...: ax3 = plt.subplot2grid((2, 5), (0, 3), colspan=2, rowspan=2)
     ...: # x_1 vs time plot
     ...: ax1.plot(t, y[:, 0], 'r')
     ...: ax1.set_ylabel('$x_1$', fontsize=18)
     ...: ax1.set_yticks([-1, -.5, 0, .5, 1])
     ...: # x2 vs time plot
     ...: ax2.plot(t, y[:, 2], 'b')
     ...: ax2.set_xlabel('$t$', fontsize=18)
     ...: ax2.set_ylabel('$x_2$', fontsize=18)
     ...: ax2.set_yticks([-1, -.5, 0, .5, 1])
     ...: # x1 and x2 phase space plot
     ...: ax3.plot(y[:, 0], y[:, 2], 'k')
     ...: ax3.set_xlabel('$x_1$', fontsize=18)
     ...: ax3.set_ylabel('$x_2$', fontsize=18)
     ...: ax3.set_xticks([-1, -.5, 0, .5, 1])
     ...: ax3.set_yticks([-1, -.5, 0, .5, 1])
     ...: fig.tight_layout()
```

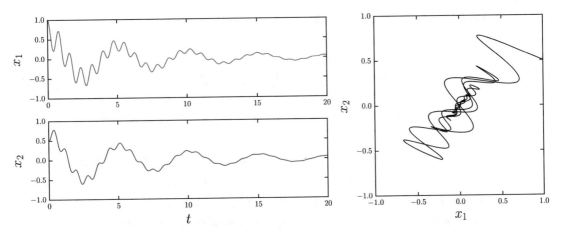

Figure 9-7. *The solution of the ODE for two coupled damped oscillators*

In addition to providing a Python function for the ODE function $f(t, y(t))$, we can also provide a Python function that computes the Jacobian matrix for a given t and $y(t)$. The solver can, for example, use the Jacobian to solve more efficiently the system of equations that arise in implicit methods. To use a Jacobian function jac, like the one defined below for the current problem, we need to pass it to the integrate.ode class when it is created, together with the f function. If the Jacobian function jac takes additional arguments, those also have to be configured using the set_jac_params method in the resulting integrate.ode instance:

```
In [117]: def jac(t, y, args):
     ...:     m1, k1, g1, m2, k2, g2 = args
     ...:     return [[0, 1, 0, 0],
     ...:             [- k1/m2 - k2/m1, - g1/m1 * y[1], k2/m1, 0],
     ...:             [0, 0, 1, 0],
     ...:             [k2/m2, 0, - k2/m2, - g2/m2]]
In [118]: r = integrate.ode(f, jac)
In [119]: r.set_jac_params(args);
```

Python functions for both $f(t, y(t))$ and its Jacobian can conveniently be generated using SymPy's lambdify, provided that the ODE problem first can be defined as a SymPy expression. This symbolic-numeric hybrid approach is a powerful method for solving ODE problems. To illustrate this approach, consider the rather complicated system of two coupled second-order and nonlinear ODEs for a double pendulum.

The equations of motion for the angular deflection, $\theta_1(t)$ and $\theta_2(t)$, for the first and the second pendulum, respectively, are[7]

$$(m_1+m_2)l_1\theta_1''(t)+m_2l_2\theta_2''(t)\cos(\theta_1-\theta_2)+m_2l_2(\theta_2'(t))^2\sin(\theta_1-\theta_2)+g(m_1+m_2)\sin\theta_1=0,$$

$$m_2l_2\theta_2''(t)+m_2l_1\theta_1''\cos(\theta_1-\theta_2)-m_2l_1(\theta_1'(t))^2\sin(\theta_1-\theta_2)+m_2g\sin\theta_2=0.$$

The first pendulum is attached to a fixed support, and the second pendulum is attached to the first pendulum. Here m_1 and m_2 are the masses and l_1 and l_2 the lengths of the first and second pendulums, respectively. We begin by defining SymPy symbols for the variables and the functions in the problem and then constructing the ode expressions:

```
In [120]: t, g, m1, l1, m2, l2 = sympy.symbols("t, g, m_1, l_1, m_2, l_2")
In [121]: theta1, theta2 = sympy.symbols("theta_1, theta_2", cls=sympy.
          Function)
In [122]: ode1 = sympy.Eq((m1+m2)*l1 * theta1(t).diff(t,t) +
   ...:                   m2*l2 * theta2(t).diff(t,t) * sympy.
                          cos(theta1(t)-theta2(t)) +
   ...:                   m2*l2 * theta2(t).diff(t)**2 * sympy.
                          sin(theta1(t)-theta2(t)) +
   ...:                   g*(m1+m2) * sympy.sin(theta1(t)))
   ...: ode1
```

Out[122]: $g(m_1+m_2)\sin\theta_1(t)+l_1(m_1+m_2)\dfrac{d^2}{dt^2}\theta_1(t)+l_2m_2\sin(\theta_1(t)-\theta_2(t))\left(\dfrac{d}{dt}\theta_2(t)\right)^2$

$$+l_2m_2\dfrac{d^2}{dt^2}\theta_2(t)\cos(\theta_1(t)-\theta_2(t))=0$$

```
In [123]: ode2 = sympy.Eq(m2*l2 * theta2(t).diff(t,t) +
   ...:                   m2*l1 * theta1(t).diff(t,t) * sympy.
                          cos(theta1(t)-theta2(t)) -
   ...:                   m2*l1 * theta1(t).diff(t)**2 * sympy.
                          sin(theta1(t) - theta2(t)) +
   ...:                   m2*g * sympy.sin(theta2(t)))
   ...: ode2
```

---

[7]See http://scienceworld.wolfram.com/physics/DoublePendulum.html for details.

Out[123]: $gm_2\sin\theta_2(t) - l_1 m_2 \sin(\theta_1(t) - \theta_2(t))\left(\dfrac{d}{dt}\theta_1(t)\right)^2$

$$+ l_1 m_2 \cos(\theta_1(t) - \theta_2(t))\dfrac{d^2}{dt^2}\theta_1(t) + l_2 m_2 \dfrac{d^2}{dt^2}\theta_2(t) = 0$$

Now ode1 and ode2 are SymPy expressions for the two second-order ODE equations. Trying to solve these equations with sympy.dsolve is fruitless, and to proceed we need to use a numerical method. However, the equations as they stand here are not in a form that is suitable for numerical solution with the ODE solvers that are available in SciPy. We first have to write the system of two second-order ODEs as a system of four first-order ODEs on the standard form. Rewriting the equations on the standard form is not difficult but can be tedious to do by hand. Fortunately we can leverage the symbolic capabilities of SymPy to automate this task. To this end we need to introduce new functions $y_1(t) = \theta_1(t)$ and $y_2(t) = \theta_1'(t)$ and $y_3(t) = \theta_2(t)$ and $y_4(t) = \theta_2'(t)$ and rewrite the ODEs in terms of these functions. By creating a dictionary for the variable change, and using the SymPy function subs to perform the substitution using this dictionary, we can easily obtain the equations for $y_2'(t)$ and $y_4'(t)$:

```
In [124]: y1, y2, y3, y4 = sympy.symbols("y_1, y_2, y_3, y_4", cls=sympy.
          Function)
In [125]: varchange = {theta1(t).diff(t, t): y2(t).diff(t),
     ...:              theta1(t): y1(t), ,
     ...:              theta2(t).diff(t, t): y4(t).diff(t),
     ...:              theta2(t): y3(t)}
In [126]: ode1_vc = ode1.subs(varchange)
In [127]: ode2_vc = ode2.subs(varchange)
```

We also need to introduce two more ODEs for $y_1'(t)$ and $y_3'(t)$:

```
In [128]: ode3 = y1(t).diff(t) - y2(t)
In [129]: ode4 = y3(t).diff(t) - y4(t)
```

At this point we have four coupled first-order ODEs for the functions y_1 to y_4. It only remains to solve for the derivatives of these functions to obtain the ODEs in standard form. We can do this using sympy.solve:

```
In [130]: y = sympy.Matrix([y1(t), y2(t), y3(t), y4(t)])
In [131]: vcsol = sympy.solve((ode1_vc, ode2_vc, ode3, ode4), y.diff(t),
          dict=True)
In [132]: f = y.diff(t).subs(vcsol[0])
```

Now f is SymPy expression for the ODE function $f(t, y(t))$. We can display the ODEs using sympy.Eq(y.diff(t), f), but the result is rather lengthy, and in the interest of space, we do not show the output here. The main purpose of constructing f here is to convert it to a NumPy-aware function that can be used with integrate.odeint or integrate.ode. The ODEs are now on a form that we can create such a function using sympy.lambdify. Also, since we have a symbolic representation of the problem so far, it is easy to also compute the Jacobian and create a NumPy-aware function for it too. When using sympy.lambdify to create functions for odeint and ode, we have to be careful to put t and y in the correct order in the tuple that is passed to sympy.lambdify. Here we will use integrate.ode, so we need a function with the signature f(t, y, *args), and thus we pass the tuple (t, y) as first argument to sympy.lambdify, and we wrap the resulting function with a lambda function in order to be able to receive the additional argument args, which is not used in the SymPy expression.

```
In [133]: params = {m1: 5.0, l1: 2.0, m2: 1.0, l2: 1.0, g: 10.0}
In [134]: _f_np = sympy.lambdify((t, y), f.subs(params), 'numpy')
In [135]: f_np = lambda _t, _y, *args: _f_np(_t, _y)
In [136]: jac = sympy.Matrix([[fj.diff(yi) for yi in y] for fj in f])
In [137]: _jac_np = sympy.lambdify((t, y), jac.subs(params), 'numpy')
In [138]: jac_np = lambda _t, _y, *args: _jac_np(_t, _y)
```

Here we have also substituted specific values of the system parameters before calling sympy.lambdify. The first pendulum is made twice as long and five times as heavy as the second pendulum. With the functions f_np and jac_np, we are now ready to solve the ODE using integrate.ode in the same manner as in the previous examples. Here we take the initial state to be $\theta_1(0) = 2$ and $\theta_2(0) = 0$, and with the derivatives zero to zero, and we solve for the time interval $[0, 20]$ with 1000 steps:

```
In [139]: y0 = [2.0, 0, 0, 0]
In [140]: tt = np.linspace(0, 20, 1000)
In [141]: r = integrate.ode(f_np, jac_np).set_initial_value(y0, tt[0])
In [142]: dt = tt[1] - tt[0]
```

```
...: yy = np.zeros((len(tt), len(y0)))
...: idx = 0
...: while r.successful() and r.t < tt[-1]:
...:     yy[idx, :] = r.y
...:     r.integrate(r.t + dt)
...:     idx += 1
```

The solution to the ODEs is now stored in the array yy, which have the shape (1000, 4). When visualizing this solution, it is more intuitive to plot the positions of the pendulums in the x – y plane rather than their angular deflections. The transformations between the angular variables θ_1 and θ_2 and x and y coordinates are $x_1 = l_1 \sin \theta_1$, $y_1 = l_1 \cos \theta_1$, $x_2 = x_1 + l_2 \sin \theta_2$, and $y_2 = y_1 + l_2 \cos \theta_2$:

```
In [143]: theta1_np, theta2_np = yy[:, 0], yy[:, 2]
In [144]: x1 = params[l1] * np.sin(theta1_np)
     ...: y1 = -params[l1] * np.cos(theta1_np)
     ...: x2 = x1 + params[l2] * np.sin(theta2_np)
     ...: y2 = y1 - params[l2] * np.cos(theta2_np)
```

Finally we plot the dynamics of the double pendulum as a function of time and in the x–y plane. The result is shown in Figure 9-8. As expected, pendulum 1 is confined to move on a circle (because of its fixed anchor point), while pendulum 2 have a much more complicated trajectory.

```
In [145]: fig = plt.figure(figsize=(10, 4))
     ...: ax1 = plt.subplot2grid((2, 5), (0, 0), colspan=3)
     ...: ax2 = plt.subplot2grid((2, 5), (1, 0), colspan=3)
     ...: ax3 = plt.subplot2grid((2, 5), (0, 3), colspan=2, rowspan=2)
     ...:
     ...: ax1.plot(tt, x1, 'r')
     ...: ax1.plot(tt, y1, 'b')
     ...: ax1.set_ylabel('$x_1, y_1$', fontsize=18)
     ...: ax1.set_yticks([-3, 0, 3])
     ...:
     ...: ax2.plot(tt, x2, 'r')
     ...: ax2.plot(tt, y2, 'b')
     ...: ax2.set_xlabel('$t$', fontsize=18)
```

```
   ...: ax2.set_ylabel('$x_2, y_2$', fontsize=18)
   ...: ax2.set_yticks([-3, 0, 3])
   ...:
   ...: ax3.plot(x1, y1, 'r')
   ...: ax3.plot(x2, y2, 'b', lw=0.5)
   ...: ax3.set_xlabel('$x$', fontsize=18)
   ...: ax3.set_ylabel('$y$', fontsize=18)
   ...: ax3.set_xticks([-3, 0, 3])
   ...: ax3.set_yticks([-3, 0, 3])
```

Figure 9-8. *The dynamics of the double pendulum*

Summary

In this chapter we have explored various methods and tools for solving ordinary differential equations (ODEs) using the scientific computing packages for Python. ODEs show up in many areas of science and engineering – in particular in modeling and the description of dynamical systems – and mastering the techniques to solve ODE problems is therefore a crucial part of the skillset of a computational scientist. In this chapter, we first looked at solving ODEs symbolically using SymPy, either with the sympy.dsolve function or using a Laplace transformation method. The symbolic approach is often a good starting point, and with the symbolic capabilities of SymPy, many fundamental ODE problems can be solved analytically. However, for most practical problems, there is no analytical solution, and the symbolic methods are then doomed to fail. Our remaining option is then to fall back on numerical techniques. Numerical integration of ODEs is a vast field in mathematics, and there exist numerous reliable methods for solving ODE problems. In this chapter we briefly reviewed methods for integrating ODEs, with the intent to introduce the concepts and ideas behind the Adams and BDF multistep methods that are used in the solvers provided by SciPy. Finally, we looked at how the `odeint` and `ode` solvers, available through the SciPy `integrate` module, can be used by solving a few example problems. Although most ODE problems eventually require numerical integration, there can be great advantages in using a hybrid symbolic-numerical approach, which uses features from both SymPy and SciPy. The last example of this chapter is devoted to demonstrating this approach.

Further Reading

An accessible introduction to many methods for numerically solving ODE problems is given in Heath (2002). For a review of ordinary differential equations with code examples, see Chapter 11 in *Numerical Recipes* (W.H. Press 2007). For a more detailed survey of numerical methods for ODEs, see, for example, (Kendall Atkinson (2009). The main implementations of ODE solvers that are used in SciPy are the VODE and LSODA solvers. The original source code for these methods is available from netlib at `www.netlib.org/ode/vode.f` and `www.netlib.org/odepack`, respectively. In addition to these solvers, there is also a well-known suite of solvers called sundials, which is provided by the Lawrence Livermore National Laboratory and available at `http://computation.llnl.gov/casc/sundials/main.html`. This suite also includes solvers of differential-algebraic equations (DAEs). A Python interface for the sundials solvers is provided by the `scikit.odes` library, which can be obtained from `http://github.com/bmcage/odes`. The `odespy` library also provides a unified interface to many different ODE solvers. For more information about `odespy`, see the project's web site at `http://hplgit.github.io/odespy/doc/web/index.html`.

References

Heath, M. T. *Scientific Computing*. 2nd. New York: McGraw-Hill, 2002.

Kendall Atkinson, Weimin Han, David Stewart. *Numerical solution of ordinary differential equations*. New Jersey: Wiley, 2009.

W.H. Press, S.A. Teukolsky, W.T. Vetterling, B.P. Flannery. *Numerical Recipes*. 3rd. New York: Cambridge University Press, 2007.

CHAPTER 10

Sparse Matrices and Graphs

We have already seen numerous examples of arrays and matrices being the essential entities in many aspects of numerical computing. So far we have represented arrays with the NumPy `ndarray` data structure, which is a heterogeneous representation that stores all the elements of the array that it represents. In many cases, this is the most efficient way to represent an object such as a vector, matrix, or a higher-dimensional array. However, notable exceptions are matrices where most of the elements are zeros. Such matrices are known as *sparse matrices*, and they occur in many applications, for example, in connection networks (such as circuits) and in large algebraic equation systems that arise, for example, when solving partial differential equations (see Chapter 11 for examples).

For matrices that are dominated by elements that are zero, it is inefficient to store all the zeros in the computer's memory, and it is more suitable to store only the nonzero values with additional information about their locations. For nonsparse matrices, known as *dense matrices*, such a representation is less efficient than storing all values consecutively in the memory, but for large sparse matrices, it can be vastly superior.

There are several options for working with sparse matrices in Python. Here we mainly focus on the sparse matrix module in SciPy, `scipy.sparse`, which provides a feature-rich and easy-to-use interface for representing sparse matrices and carrying out linear algebra operations on such objects. Another option is PySparse[1], which provides similar functionality. For very large-scale problems, the PyTrilinos[2] and PETSc[3] packages

---

[1] http://pysparse.sourceforge.net

[2] http://trilinos.org/packages/pytrilinos

[3] See http://www.mcs.anl.gov/petsc and https://bitbucket.org/petsc/petsc4py for its Python bindings.

R. Johansson, *Numerical Python*, https://doi.org/10.1007/978-1-4842-4246-9_10

have powerful parallel implementations of many sparse matrix operations. However, using these packages requires more programming, and they have a steeper learning curve and are more difficult to install and set up. For most basic use-cases, SciPy's `sparse` module is the most suitable option or at least a suitable starting point.

Toward the end of the chapter, we also briefly explore representing and processing graphs, using the SciPy `sparse.csgraph` module and the NetworkX library. Graphs can be represented as adjacency matrices, which in many applications are very sparse. Graphs and sparse matrices are therefore closely connected topics.

Importing Modules

The main module that we work with in this chapter is the `sparse` module in SciPy library. We assume that this module is included under the name sp, and in addition we need to explicitly import its submodule `linalg`, to make this module accessible through `sp.linalg`.

```
In [1]: import scipy.sparse as sp
In [2]: import scipy.sparse.linalg
```

We also need the NumPy library, which we, as usual, import under the name np, and the Matplotlib library for plotting:

```
In [3]: import numpy as np
In [4]: import matplotlib.pyplot as plt
```

And in the last part of this chapter, we use the `networkx` module, which we import under the name nx:

```
In [5]: import networkx as nx
```

Sparse Matrices in SciPy

The basic idea of sparse matrix representation is to avoid storing the excessive amount of zeros in a sparse matrix. In dense matrix representation, where all elements of an array are stored consecutively, it is sufficient to store the values themselves, since the row and column indices for each element are known implicitly from the position in the array. However, if we store only the nonzero elements, we clearly also need to store the row and column indices for each element. There are numerous approaches to organizing the storage of the nonzero elements and their corresponding row and column indices. These approaches have different advantages and disadvantages, for example, in terms how easy it is to create the matrices and, perhaps more importantly, how efficiently they can be used in implementations of mathematical operations on the sparse matrices. A summary and comparison of sparse matrix formats that are available in the SciPy `sparse` module is given in Table 10-1.

Table 10-1. *Summary and Comparison of Methods to Represent Sparse Matrices*

Type	Description	Pros	Cons
Coordinate list (COO, `sp.coo_matrix`)	Nonzero values are stored in a list together with their row and column.	Simple to construct and efficient to add new elements.	Inefficient element access. Not suitable for mathematical operations, such as matrix multiplication.
List of lists (LIL, `sp.lil_matrix`)	Stores a list of column indices for nonzero elements for each row and a list of the corresponding values.	Supports slicing operations.	Not ideal for mathematical operations.
Dictionary of keys (DOK, `sp.dok_matrix`)	Nonzero values are stored in a dictionary with a tuple of (row, column) as key.	Simple to construct and fast to add, remove and access elements.	Not ideal for mathematical operations.
Diagonal matrix (DIA, `sp.dia_matrix`)	Stores lists of diagonals of the matrix.	Efficient for diagonal matrices.	Not suitable for nondiagonal matrices.
Compressed sparse column (CSC, `sp.csc_matrix`) and compressed sparse row (CSR, `sp.csr_matrix`)	Stores the values together with arrays with column or row indices.	Relatively complicated to construct.	Efficient matrix-vector multiplication.
Block-sparse matrix (BSR, `cp.bsr_matrix`)	Similar to CSR, but for sparse matrices with dense submatrices.	Efficient for their specific intended purpose.	Not suitable for general-purpose use.

A simple and intuitive approach for storing sparse matrices is to simply store lists with column indices and row indices together with the list of nonzero values. This format is called *coordinate list format*, and it is abbreviated as COO in SciPy. The class `sp.coo_matrix` is used to represent sparse matrices in this format. This format is particularly easy to initialize. For instance, with the matrix

$$A = \begin{bmatrix} 0 & 1 & 0 & 0 \\ 0 & 0 & 0 & 2 \\ 0 & 0 & 3 & 0 \\ 4 & 0 & 0 & 0 \end{bmatrix},$$

we can easily identify the nonzero values $[A_{01} = 1, A_{13} = 2, A_{22} = 3, A_{30} = 4]$ and their corresponding rows $[0, 1, 2, 3]$ and columns $[1, 3, 2, 0]$ (note that here we have used Python's zero-based indexing). To create a `sp.coo_matrix` object, we can create lists (or arrays) for the values, row indices, and column indices and pass them to `sp.coo_matrix`. Optionally, we can also specify the shape of the array using the `shape` argument, which is useful when the nonzero elements do not span the entire matrix (i.e., if there are columns or rows containing only zeros, so that the shape cannot be correctly inferred from the row and column arrays):

```
In [6]: values = [1, 2, 3, 4]
In [7]: rows = [0, 1, 2, 3]
In [8]: cols = [1, 3, 2, 0]
In [9]: A = sp.coo_matrix((values, (rows, cols)), shape=[4, 4])
In [10]: A
Out[10]: <4x4 sparse matrix of type '<type 'numpy.int64'>'
            with 4 stored elements in Coordinate format>
```

The result is a data structure that represents the sparse matrix. All sparse matrix representations in SciPy's `sparse` module share several common attributes, many of which are derived from NumPy's `ndarray` object. Examples of such attributes are `size`, `shape`, `dtype`, and `ndim`, and common to all sparse matrix representations are the `nnz` (number of nonzero elements) and `data` (the nonzero values) attributes:

```
In [11]: A.shape, A.size, A.dtype, A.ndim
Out[11]: ((4, 4), 4, dtype('int64'), 2)
In [12]: A.nnz, A.data
Out[12]: (4, array([1, 2, 3, 4]))
```

In addition to the shared attributes, each type of sparse matrix representation also has attributes that are specific to its way of storing the positions for each nonzero value. For the case of sp.coo_matrix objects, there are row and col attributes for accessing the underlying row and column arrays:

```
In [13]: A.row
Out[13]: array([0, 1, 2, 3], dtype=int32)
In [14]: A.col
Out[14]: array([1, 3, 2, 0], dtype=int32)
```

There are also a large number of methods available for operating on sparse matrix objects. Many of these methods are for applying mathematical functions on the matrix, for example, elementwise math methods like sin, cos, arcsin, etc.; aggregation methods like min, max, sum, etc.; mathematical array operators such as conjugate (conj) and transpose (transpose), etc.; and dot for computing the dot product between sparse matrices or a sparse matrix and a dense vector (the * operator also denotes matrix multiplication for sparse matrices). For further details, see the docstring for the sparse matrix classes (summarized in Table 10-1). Another important family of methods is used to convert sparse matrices between different formats: for example, tocoo, tocsr, tolil, etc. There are also methods for converting a sparse matrix to NumPy ndarray and NumPy matrix objects (i.e., dense matrix representations): toarray and todense, respectively.

For example, to convert the sparse matrix A from COO format to CSR format, and to a NumPy array, respectively, we can use the following:

```
In [15]: A.tocsr()
Out[15]: <4x4 sparse matrix of type '<type 'numpy.int64'>'
         with 4 stored elements in Compressed Sparse Row format>
In [16]: A.toarray()
Out[16]: array([[0, 1, 0, 0],
                [0, 0, 0, 2],
                [0, 0, 3, 0],
                [4, 0, 0, 0]])
```

The obvious way to access elements in a matrix, which we have used in numerous different contexts so far, is using the indexing syntax, for example, A[1,2], as well as the slicing syntax, for example, A[1:3, 2], and so on. We can often use this syntax with sparse matrices too, but not all representations support indexing and slicing, and if it is supported, it may not be an efficient operation. In particular, assigning values to zero-valued elements can be a costly operation, as it may require to rearrange the underlying data structures, depending on which format is used. To incrementally add new elements to a sparse matrix, the LIL (sp.lil_matrix) format is a suitable choice, but this format is on the other hand not suitable for arithmetic operations.

When working with sparse matrices, it is common to face the situation that different tasks – such as construction, updating, and arithmetic operations – are most efficiently handled in different formats. Converting between different sparse formats is relatively efficient, so it is useful to switch between different formats in different parts of an application. Efficient use of sparse matrices therefore requires an understanding of how different formats are implemented and what they are suitable for. Table 10-1 briefly summarizes the pros and cons of the sparse matrix formats available in SciPy's sparse module, and using the conversion methods, it is easy to switch between different formats. For a more in-depth discussion of the merits of the various formats, see the "Sparse Matrices"[4] section in the SciPy reference manual.

For computations, the most important sparse matrix representations in SciPy's sparse module are the CSR (Compressed Sparse Row) and CSC (Compressed Sparse Column) formats, because they are well suited for efficient matrix arithmetic and linear algebra applications. Other formats, like COO, LIL, and DOK, are mainly used for constructing and updated sparse matrices, and once a sparse matrix is ready to be used in computations, it is best to convert it to either CSR or CSC format, using the tocsr or tocsc methods, respectively.

In the CSR format, the nonzero values (data) are stored along with an array that contains the column indices of each value (indices) and another array that stores the offsets of the column index array of each row (indptr). For instance, consider the matrix

$$A = \begin{bmatrix} 1 & 2 & 0 & 0 \\ 0 & 3 & 4 & 0 \\ 0 & 0 & 5 & 6 \\ 7 & 0 & 8 & 9 \end{bmatrix},$$

Here the nonzero values are [1, 2, 3, 4, 5, 6, 7, 8, 9] (data), and the *column indices* corresponding to the nonzero values in the first row are [0, 1], the second row [1, 2], the third row [2, 3], and the fourth row [0, 2, 3]. Concatenating all of these column index lists gives the indices array [0, 1, 1, 2, 2, 3, 0, 2, 3]. To keep track of which row entries in this

---

[4]http://docs.scipy.org/doc/scipy/reference/sparse.html

column index array correspond to, we can store the starting position in for each row in a second array. The column indices of the first row are elements 0 to 1, the second row elements 2 to 3, the third row elements 4 to 5, and finally the fourth row elements 6 to 9. Collecting the starting indices in an array gives [0, 2, 4, 6]. For convenience in the implementation, we also add at the end of this array the total number of nonzero elements, which results in the indptr array [0, 2, 4, 6, 9]. In the following code, we create a dense NumPy array corresponding to the matrix A, convert it to a CSR matrix using sp. csr_matrix, and then display the data, indices, and indptr attributes:

```
In [17]: A = np.array([[1, 2, 0, 0], [0, 3, 4, 0], [0, 0, 5, 6], [7, 0, 8, 9]])
    ...: A
Out[17]: array([[1, 2, 0, 0],
                [0, 3, 4, 0],
                [0, 0, 5, 6],
                [7, 0, 8, 9]])
In [18]: A = sp.csr_matrix(A)
In [19]: A.data
Out[19]: array([1, 2, 3, 4, 5, 6, 7, 8, 9], dtype=int64)
In [20]: A.indices
Out[20]: array([0, 1, 1, 2, 2, 3, 0, 2, 3], dtype=int32)
In [21]: A.indptr
Out[21]: array([0, 2, 4, 6, 9], dtype=int32)
```

With this storage scheme, the nonzero elements in the row with index i are stored in the data array between index indptr[i] and indptr[i+1]-1, and the column indices for these elements are stored at the same indices in the indices array. For example, the elements in the third row, with index i=2, starts at indptr[2]=4 and ends at indptr[3]-1=5, which gives the element values data[4]=5 and data[5]=6 and column indices indices[4]=2 and indices[5]=3. Thus, $A[2, 2] = 5$ and $A[2, 3] = 6$ (in zero-index-based notation):

```
In [22]: i = 2
In [23]: A.indptr[i], A.indptr[i+1]-1
Out[23]: (4, 5)
In [24]: A.indices[A.indptr[i]:A.indptr[i+1]]
Out[24]: array([2, 3], dtype=int32)
In [25]: A.data[A.indptr[i]:A.indptr[i+1]]
```

```
Out[25]: array([5, 6], dtype=int64)
In [26]: A[2, 2], A[2,3]  # check
Out[26]: (5, 6)
```

While the CSR storage method is not as intuitive as COO, LIL, or DOK, it turns out that it is well suited for use in the implementation of matrix arithmetic and for linear algebra operations. Together with the CSC format, it is therefore the main format for use in sparse matrix computations. The CSC format is essentially identical to CSR, except that instead of column indices and row pointers, row indices and column pointers are used (i.e., the role of columns and rows is reversed).

Functions for Creating Sparse Matrices

As we have seen in the examples earlier in this chapter, one way of constructing sparse matrices is to prepare the data structures for a specific sparse matrix format and pass these to the constructor of the corresponding sparse matrix class. While this method is sometimes suitable, it is often more convenient to compose sparse matrices from predefined template matrices. The sp.sparse module provides a variety of functions for generating such matrices, for example, sp.eye for creating diagonal sparse matrices with ones on the diagonal (optionally offset from the main diagonal), sp.diags for creating diagonal matrices with a specified pattern along the diagonal, sp.kron for calculating the Kronecker (tensor) product of two sparse matrices, and bmat, vstack, and hstack, for building sparse matrices from sparse block matrices, and by stacking sparse matrices vertically and horizontally, respectively.

For example, in many applications sparse matrices have a diagonal form. To create a sparse matrix of size 10×10 with a main diagonal and an upper and lower diagonal, we can use three calls to sp.eye, using the k argument to specify the offset from the main diagonal:

```
In [27]: N = 10
In [28]: A = sp.eye(N, k=1) - 2 * sp.eye(N) + sp.eye(N, k=-1)
In [29]: A
Out[29]: <10x10 sparse matrix of type '<class 'numpy.float64'>'
                  with 28 stored elements in Compressed Sparse Row format>
```

By default the resulting object is sparse matrix in the CSR format, but using the `format` argument, we can specify any other sparse matrix format. The value of the format argument should be a string such as `'csr'`, `'csc'`, `'lil'`, etc. All functions for creating sparse matrices in `sp.sparse` accept this argument. For example, in the previous example, we could have produced the same matrix using `sp.diags`, by specifying the pattern [1, -2, 1] (the coefficients to the `sp.eye` functions in the previous expression), and the corresponding offsets from the main diagonal [1, 0, -1]. If we additionally want the resulting sparse matrix in CSC format, we can set `format='csc'`:

```
In [30]: A = sp.diags([1, -2, 1], [1, 0, -1], shape=[N, N], format='csc')
In [31]: A
Out[31]: <10x10 sparse matrix of type '<class 'numpy.float64'>'
                with 28 stored elements in Compressed Sparse Column format>
```

The advantages of using sparse matrix formats rather than dense matrices only manifest themselves when working with large matrices. Sparse matrices are by their nature therefore large, and hence it can be difficult to visualize a matrix by, for example, printing its elements in the terminal. Matplotlib provides the function `spy`, which is a useful tool for visualizing the structure of a sparse matrix. It is available as a function in `pyplot` module or as a method for Axes instances. When using it on the previously defined A matrix, we obtain the results shown in Figure 10-1.

```
In [32]: fig, ax = plt.subplots()
    ...: ax.spy(A)
```

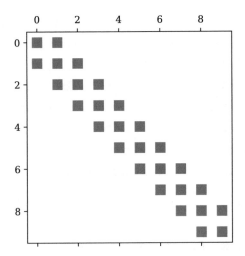

Figure 10-1. *Structure of the sparse matrix with nonzero elements only on the two diagonals closest to the main diagonal and the main diagonal itself*

Sparse matrices are also often associated with tensor product spaces. For such cases we can use the sp.kron function to compose a sparse matrix from its smaller components. For example, to create a sparse matrix for the tensor product between A and the matrix

$$B = \begin{bmatrix} 0 & 1 & 0 \\ 1 & 0 & 1 \\ 0 & 1 & 0 \end{bmatrix},$$

we can use sp.kron(A, B):

```
In [33]: B = sp.diags([1, 1], [-1, 1], shape=[3,3])
In [34]: C = sp.kron(A, B)
In [35]: fig, (ax_A, ax_B, ax_C) = plt.subplots(1, 3, figsize=(12, 4))
    ...: ax_A.spy(A)
    ...: ax_B.spy(B)
    ...: ax_C.spy(C)
```

For comparison, we also plotted the sparse matrix structure of *A*, *B*, and *C*, and the result is shown in Figure 10-2. For more detailed information on ways to build sparse matrices with the sp.sparse module, see its docstring and the "Sparse Matrices" section in the SciPy reference manual.

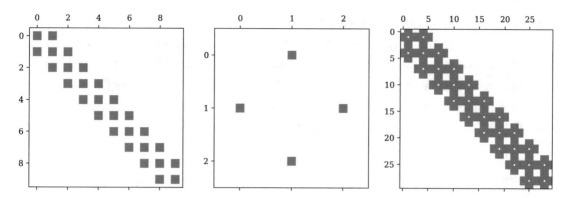

Figure 10-2. *The sparse matrix structures of two matrices A (left) and B (middle) and their tensor product (right)*

Sparse Linear Algebra Functions

The main application of sparse matrices is to perform linear algebra operations on large matrices that are intractable or inefficient to treat using dense matrix representations. The SciPy `sparse` module contains a module `linalg` that implements many linear algebra routines. Not all linear algebra operations are suitable for sparse matrices, and in some cases the behavior of the sparse matrix version of operations needs to be modified compared to the dense counterparts. Consequently, there are a number of differences between the sparse linear algebra module `scipy.sparse.linalg` and the dense linear algebra module `scipy.linalg`. For example, the eigenvalue solvers for dense problems typically compute and return all eigenvalues and eigenvectors. For sparse matrices this is not manageable, because storing all eigenvectors of a sparse matrix A of size $N \times N$ usually amounts to storing a dense matrix of size $N \times N$. Instead, sparse eigenvalue solvers typically give *a few* eigenvalues and eigenvectors, for example, those with the smallest or largest eigenvalues. In general, for sparse matrix methods to be efficient, they must retain the sparsity of matrices involved in the computation. An example of an operation where the sparsity usually is *not* retained is the matrix inverse, and it should therefore be avoided when possible.

Linear Equation Systems

The most important application of sparse matrices is arguably to solve linear equation system on the form $Ax = b$, where A is a sparse matrix and x and b are dense vectors. The SciPy `sparse.linalg` module has both direct and iterative solver for this type of problem (`sp.linalg.spsolve`) and methods to factor a matrix A, using, for example, LU factorization (`sp.linalg.splu`) and incomplete LU factorization (`sp.linalg.spilu`). For example, consider the problem $Ax = b$ where A is the tridiagonal matrix considered in the preceding text and b is a dense vector filled with negative ones (see Chapter 11 for a physical interpretation of this equation). To solve this problem for the system size 10×10, we first create the sparse matrix A and the dense vector b:

```
In [36]: N = 10
In [37]: A = sp.diags([1, -2, 1], [1, 0, -1], shape=[N, N], format='csc')
In [38]: b = -np.ones(N)
```

Now, to solve the equation system using the direct solver provided by SciPy, we can use:

```
In [39]: x = sp.linalg.spsolve(A, b)
In [40]: x
Out[40]: array([  5.,    9.,   12.,   14.,   15.,   15.,   14.,   12.,    9.,    5.])
```

The solution vector is a dense NumPy array. For comparison, we can also solve this problem using dense direct solver in NumPy `np.linalg.solve` (or, similarly, using `scipy.linalg.solve`). To be able to use the dense solver, we need to convert the sparse matrix A to a dense array using `A.todense()`:

```
In [41]: np.linalg.solve(A.todense(), b)
Out[41]: array([  5.,    9.,   12.,   14.,   15.,   15.,   14.,   12.,    9.,    5.])
```

As expected, the result agrees with what we obtained from the sparse solver. For small problems like this one, there is not much to gain using sparse matrices, but for increasing system size, the merits of using sparse matrices and sparse solvers become apparent. For this particular problem, the threshold system size beyond which using sparse methods outperforms dense methods is approximately $N = 100$, as shown in Figure 10-3. While the exact threshold varies from problem to problem, as well as hardware and software versions, this behavior is typical for problems where the matrix A is sufficiently sparse[5].

---

[5]For a discussion of techniques and methods to optimize Python code, see Chapter 21.

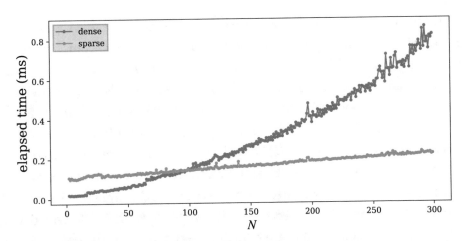

Figure 10-3. *Performance comparison between sparse and dense methods to solve the one-dimensional Poisson problem as a function of problem size*

An alternative to the spsolve interface is to explicitly compute the LU factorization using sp.sparse.splu or sp.sparse.spilu (incomplete LU factorization). These functions return an object that contains the *L* and *U* factors and that has a method that solves *LUx* = *b* for a given vector *b*. This is of course particularly useful when the *Ax* = *b* has to be solved for multiple vectors *b*. For example, the LU factorization of the matrix *A* used previously is computed using:

```
In [42]: lu = sp.linalg.splu(A)
In [43]: lu.L
Out[43]: <10x10 sparse matrix of type '<class 'numpy.float64'>'
                with 20 stored elements in Compressed Sparse Column format>
In [44]: lu.U
Out[44]: <10x10 sparse matrix of type '<class 'numpy.float64'>'
                with 20 stored elements in Compressed Sparse Column format>
```

Once the LU factorization is available, we can efficiently solve the equation *LUx* = *b* using the solve method for the lu object:

```
In [45]: x = lu.solve(b)
In [46]: x
Out[46]: array([  5.,    9.,   12.,   14.,   15.,   15.,   14.,   12.,    9.,    5.])
```

An important consideration that arises with sparse matrices is that the LU factorization of A may introduce new nonzero elements in L and U compared to the matrix A and therefore make L and U less sparse. Elements that exist in L or U, but not in A, are called fill-ins. If the amount of fill-ins is large, the advantage of using sparse matrices may be lost. While there is no complete solution to eliminate fill-ins, it is often possible to reduce fill-in by permuting the rows and columns in A, so that the LU factorization takes the form $P_r A P_c = LU$, where P_r and P_c are row and column permutation matrices, respectively. Several such methods for permutations methods are available. The spsolve, splu, and spilu functions all take the argument permc_spec, which can take the values NATURAL, MMD_ATA, MMD_AT_PLUS_A, or COLAMD, which indicates different permutation methods that are built into these methods. The object returned by splu and spilu accounts for such permutations, and the permutation vectors are available via the perm_c and perm_r attributes. Because of these permutations, the product of lu.L and lu.U is not directly equal to A, and to reconstruct A from lu.L and lu.U, we also need to undo the row and column permutations:

```
In [47]: def sp_permute(A, perm_r, perm_c):
    ...:     """ permute rows and columns of A """
    ...:     M, N = A.shape
    ...:     # row permutation matrix
    ...:     Pr = sp.coo_matrix((np.ones(M), (perm_r, np.arange(N)))).tocsr()
    ...:     # column permutation matrix
    ...:     Pc = sp.coo_matrix((np.ones(M), (np.arange(M), perm_c))).tocsr()
    ...:     return Pr.T * A * Pc.T
In [48]: lu.L * lu.U - # != 0
Out[48]: <10x10 sparse matrix of type '<class 'numpy.float64'>'
              with 8 stored elements in Compressed Sparse Column format>
In [49]: sp_permute(lu.L * lu.U, lu.perm_r, lu.perm_c) - A  # == 0
Out[49]: <10x10 sparse matrix of type '<class 'numpy.float64'>'
              with 0 stored elements in Compressed Sparse Column format>
```

By default, the direct sparse linear solver in SciPy uses the SuperLU[6] package. An alternative sparse matrix solver that also can be used in SciPy is the UMFPACK[7] package, although this package is *not* bundled with SciPy and requires that the scikit-umfpack

---

[6]http://crd-legacy.lbl.gov/~xiaoye/SuperLU/
[7]http://faculty.cse.tamu.edu/davis/suitesparse.html

Python library is installed. If `scikit-umfpack` is available, and if the `use_umfpack` argument to the `sp.linalg.spsolve` function is `True`, then the UMFPACK is used instead of SuperLU. Whether SuperLU or UMFPACK gives better performance varies from problem to problem, so it is worth having both installed and testing both for any given problem.

The `sp.spsolve` function is an interface to direct solvers, which internally performs matrix factorization. An alternative approach is to use iterative methods that originate in optimization. The SciPy `sparse.linalg` module contains several functions for the iterative solution of sparse linear problems: for example, `bicg` (biconjugate gradient method), `bicgstab` (biconjugate gradient stabilized method), `cg` (conjugate gradient), `gmres` (generalized minimum residual), and `lgmres` (loose generalized minimum residual method). All of these functions (and a few others) can be used to solve the problem $Ax = b$ by calling the function with A and b as arguments, and they all return a tuple `(x, info)` where x is the solution and `info` contains additional information about the solution process (`info=0` indicates success, and it is positive for convergence error and negative for input error). For example:

```
In [50]: x, info = sp.linalg.bicgstab(A, b)
In [51]: x
Out[51]: array([  5.,   9.,  12.,  14.,  15.,  15.,  14.,  12.,   9.,   5.])
In [52]: x, info = sp.linalg.lgmres(A, b)
In [53]: x
Out[53]: array([  5.,   9.,  12.,  14.,  15.,  15.,  14.,  12.,   9.,   5.])
```

In addition, each iterative solver takes its own solver-dependent arguments. See the docstring for each function for details. Iterative solver may have an advantage over direct solvers for very large problems, where direct solvers may require excessive memory usage due to undesirable fill-ins. In contrast, iterative solvers only require to evaluate sparse matrix-vector multiplications and therefore do not suffer from fill-in problems, but on the other hand, they might have slow convergence for many problems, especially if not properly preconditioned.

Eigenvalue Problems

Sparse eigenvalue and singular-value problems can be solved using the `sp.linalg.eigs` and `sp.linalg.svds` functions, respectively. For real symmetric or complex Hermitian matrices, the eigenvalues (which in this case are real) and eigenvectors can also be computed using `sp.linalg.eigsh`. These functions do not compute all eigenvalues or singular values, but rather compute a given number of eigenvalues and vectors (the default is six). Using the keyword argument k with these functions, we can define how many eigenvalues and vectors should be computed. Using the which keyword argument, we can specify which k values are to be computed. The options for eigs are the largest magnitude LM, smallest magnitude SM, largest real part LR, smallest real part SR, largest imaginary part LI, and smallest imaginary part SI. For svds only LM and SM are available.

For example, to compute the lowest four eigenvalues for the sparse matrix of the one-dimensional Poisson problem (of system size 10x10), we can use sp.linalg. eigs(A, k=4, which='LM'):

```
In [54]: N = 10
In [55]: A = sp.diags([1, -2, 1], [1, 0, -1], shape=[N, N], format='csc')
In [56]: evals, evecs = sp.linalg.eigs(A, k=4, which='LM')
In [57]: evals
Out[57]: array([-3.91898595+0.j, -3.68250707+0.j, -3.30972147+0.j,
                -2.83083003+0.j])
```

The return value of sp.linalg.eigs (and sp.linalg.eigsh) is a tuple (evals, evecs) whose first element is an array of eigenvalues (evals), and the second element is an array (evecs) of shape $N \times k$, whose columns are the eigenvectors corresponding to the k eigenvalues in evals. Thus, we expect that the dot product between A and a column in evecs is equal to the same column in evecs scaled by the corresponding eigenvalue in evals. We can directly confirm that this is indeed the case:

```
In [58]: np.allclose(A.dot(evecs[:,0]), evals[0] * evecs[:,0])
Out[58]: True
```

For this particular example, the sparse matrix A is symmetric, so instead of sp. linalg.eigs, we could use sp.linalg.eigsh, and in doing so we obtain an eigenvalue array with real-valued elements:

```
In [59]: evals, evecs = sp.linalg.eigsh(A, k=4, which='LM')
In [60]: evals
Out[60]: array([-3.91898595, -3.68250707, -3.30972147, -2.83083003])
```

By changing the argument which='LM' (for largest magnitude) to which='SM' (smallest magnitude), we obtain a different set of eigenvalues and vector (those with the smallest magnitude).

```
In [61]: evals, evecs = sp.linalg.eigs(A, k=4, which='SM')
In [62]: evals
Out[62]: array([-0.08101405+0.j, -0.31749293+0.j, -0.69027853+0.j,
          -1.16916997+0.j])
In [63]: np.real(evals).argsort()
Out[63]: array([3, 2, 1, 0])
```

Note that although we requested and obtained the four eigenvalues with the smallest magnitude in the previous example, those eigenvalues and vectors are not necessarily sorted within each other (although they are in this particular case). Obtaining sorted eigenvalues is often desirable, and this is easily achieved with a small but convenient wrapper function that sorts the eigenvalues using NumPy's argsort method. Here we give such a function, sp_eigs_sorted, which returns the eigenvalues and eigenvectors sorted by the real part of the eigenvalue.

```
In [64]: def sp_eigs_sorted(A, k=6, which='SR'):
    ...:     """ compute and return eigenvalues sorted by the real part """
    ...:     evals, evecs = sp.linalg.eigs(A, k=k, which=which)
    ...:     idx = np.real(evals).argsort()
    ...:     return evals[idx], evecs[idx]
In [65]: evals, evecs = sp_eigs_sorted(A, k=4, which='SM')
In [66]: evals
Out[66]: array([-1.16916997+0.j, -0.69027853+0.j, -0.31749293+0.j,
          -0.08101405+0.j])
```

As a less trivial example using sp.linalg.eigs and the wrapper function sp_eigs_sorted, consider the spectrum of lowest eigenvalues of the linear combination $(1 - x)M_1 + xM_2$ of random sparse matrices M_1 and M_2. We can use the sp.rand function to generate two random sparse matrices, and by repeatedly using sp_eigs_sorted to find the smallest 25 eigenvalues of the $(1 - x)M_1 + xM_2$ matrix for different values of x, we can build a matrix (evals_mat) that contains the eigenvalues as a function of x. Below we use 50 values of x in the interval $[0, 1]$:

```
In [67]: N = 100
In [68]: x_vec = np.linspace(0, 1, 50)
In [69]: M1 = sp.rand(N, N, density=0.2)
```

```
In [70]: M2 = sp.rand(N, N, density=0.2)
In [71]: evals_mat = np.array([sp_eigs_sorted((1-x)*M1 + x*M2, k=25)[0] for
         x in x_vec])
```

Once the matrix `evals_mat` of eigenvalues as a function of x is computed, we can plot the eigenvalue spectrum. The result is shown in Figure 10-4, which is a complicated eigenvalue spectrum due to the randomness of the matrices M_1 and M_2.

```
In [72]: fig, ax = plt.subplots(figsize=(8, 4))
    ...: for idx in range(evals_mat.shape[1]):
    ...:     ax.plot(x_vec, np.real(evals_mat[:,idx]), lw=0.5)
    ...: ax.set_xlabel(r"$x$", fontsize=16)
    ...: ax.set_ylabel(r"eig.vals. of $(1-x)M_1+xM_2$", fontsize=16)
```

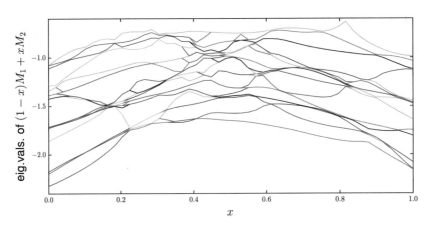

Figure 10-4. *The spectrum of the lowest 25 eigenvalues of the sparse matrix*
$(1 - x)M_1+xM_2$, as a function of x, where M_1 and M_2 are two random matrices

Graphs and Networks

Representing graphs as adjacency matrices is another important application of sparse matrices. In an adjacency matrix, an element describes which nodes in a graph are connected to each other. Consequently, if each node is only connected to a small set of other nodes, the adjacency matrix is sparse. The `csgraph` module in the SciPy `sparse` module provides functions for processing such graphs, including methods for traversing a graph using different methods (e.g., breadth-first and depth-first traversals) and for computing shortest paths between nodes in a graph, and so on. For more information about this module, refer to its docstring: `help(sp.csgraph)`.

For a more comprehensive framework for working with graphs, there is the NetworkX Python library. It provides utilities for creating and manipulating undirected and directed graphs and also implements many graph algorithms, such as finding minimum paths between nodes in a graph. Here we assume that the Networkx library is imported under the name nx. Using this library, we can, for example, create an undirected graph by initiating an object of the class nx.Graph. Any hashable Python object can be stored as nodes in a graph object, which makes it a very flexible data structure. However, in the following examples, we only use graph objects with integers and strings as node labels. See Table 10-2 for a summary of functions for creating graphs and for adding nodes and edges to graph objects.

Table 10-2. *Summary of Objects and Methods for Basic Graph Construction Using NetworkX*

Object/Method	Description
nx.Graph	Class for representing undirected graphs.
nx.DiGraph	Class for representing directed graphs.
nx.MultiGraph	Class for representing undirected graphs with support for multiple edges.
nx.MultiDiGraph	Class for representing directed graphs with support for multiple edges.
add_node	Add a node to the graph. Expects a node label (e.g., a string, or in general a hashable object) as argument.
add_nodes_from	Adds multiple nodes. Expects a list (or iterable) of node labels as argument.
add_edge	Add an edge. Expects two node arguments as arguments and creates an edge between those nodes.
add_edges_from	Adds multiple edges. Expects a list (or iterable) of tuples of node labels.
add_weighted_edges_from	Adds multiple edges with weight factors. Expects a list (or iterable) of tuples each containing two node labels and the weight factor.

For example, we can create a simple graph with node data that are integers using nx.Graph(), and the add_node method, or add_nodes_from to add multiple nodes in one go. The nodes method returns an iterator object for the nodes, called a NodeView:

```
In [73]: g = nx.Graph()
In [74]: g.add_node(1)
In [75]: g.nodes()
Out[75]: NodeView((1,))
In [76]: g.add_nodes_from([3, 4, 5])
In [77]: g.nodes()
Out[77]: NodeView((1, 3, 4, 5))
```

To connect nodes we can add edges, using add_edge. We pass the labels of the two nodes we want to connect as arguments. To add multiple edges, we can use add_edges_from and pass to it a list of tuples of nodes to connect. The edges method returns an iterator object for the edges, called EdgeView:

```
In [78]: g.add_edge(1, 2)
In [79]: g.edges()
Out[79]: EdgeView([(1, 2)])
In [80]: g.add_edges_from([(3, 4), (5, 6)])
In [81]: g.edges()
Out[81]: EdgeView([(1, 2), (3, 4), (5, 6)])
```

To represent edges between nodes that have weights associated with them (e.g., a distance), we can use add_weighted_edges_from, to which we pass a list of tuples that also contains the weight factor for each edge, in addition to the two nodes. When calling the edges method, we can additionally give argument data=True to indicate that also the edge data should be included in the resulting view.

```
In [82]: g.add_weighted_edges_from([(1, 3, 1.5), (3, 5, 2.5)])
In [83]: g.edges(data=True)
Out[83]: EdgeDataView([(1, 2, {}),
                       (1, 3, {'weight': 1.5}),
                       (3, 4, {}),
                       (3, 5, {'weight': 2.5}),
                       (5, 6, {})])
```

Note that if we add edges between nodes that do not yet exist in the graph, they are seamlessly added. For example, in the following code, we add a weighted edge between nodes 6 and 7. Node 7 does not previously exist in the graph, but when adding an edge to it, it is automatically created and added to the graph:

```
In [84]: g.add_weighted_edges_from([(6, 7, 1.5)])
In [85]: g.nodes()
Out[85]: NodeView((1, 3, 4, 5, 2, 6, 7))
In [86]: g.edges()
Out[86]: EdgeView([(1, 2), (1, 3), (3, 4), (3, 5), (5, 6), (6, 7)])
```

With these basic fundamentals in place, we are already prepared to look at a more complicated example of a graph. In the following we will build a graph from a dataset stored in a JSON file called tokyo-metro.json (available together with the code listings), which we load using the Python standard library module json[8]:

```
In [87]: import json
In [88]: with open("tokyo-metro.json") as f:
   ...:         data = json.load(f)
```

The result of loading the JSON file is a dictionary data that contains metro line descriptions. For each line, there is a list of travel times between stations (travel_times), a list of possible transfer points to other lines (transfer), as well as the line color:

```
In [89]: data.keys()
Out[89]: dict_keys(['C', 'T', 'N', 'F', 'Z', 'M', 'G', 'Y', 'H'])
In [90]: data["C"]
Out[90]: {'color': '#149848',
          'transfers': [['C3', 'F15'], ['C4', 'Z2'], ...],
          'travel_times': [['C1', 'C2', 2], ['C2', 'C3', 2], ...]}
```

Here the format of the travel_times list is [['C1', 'C2', 2], ['C2', 'C3', 2], ...], indicating that it takes 2 minutes to travel between the stations C1 and C2, and 2 minutes to travel between C2 and C3, etc. The format of the transfers list is [['C3', 'F15'], ...], indicating that it is possible to transfer from the C line to the F line at station C3 to station F15. The travel_times and transfers are directly suitable for feeding to

---

[8]For more information about the JSON format and the json module, see Chapter 18.

add_weighed_edges_from and add_edges_from, and we can therefore easily create a graph for representing the metro network by iterating over each metro line dictionary and call these methods:

```
In [91]: g = nx.Graph()
    ...: for line in data.values():
    ...:     g.add_weighted_edges_from(line["travel_times"])
    ...:     g.add_edges_from(line["transfers"])
```

The line transfer edges do not have edge weights, so let's first mark all transfer edges by adding a new Boolean attribute transfer to each edge:

```
In [92]: for n1, n2 in g.edges():
    ...:     g[n1][n2]["transfer"] = "weight" not in g[n1][n2]
```

Next, for plotting purposes, we create two lists of edges containing transfer edges and on-train edges, and we also create a list with colors corresponding to each node in the network:

```
In [93]: on_foot = [e for e in g.edges() if g.get_edge_data(*e)["transfer"]]
In [94]: on_train = [e for e in g.edges () if not g.get_edge_data(*e)
            ["transfer"]]
In [95]: colors = [data[n[0].upper()]["color"] for n in g.nodes()]
```

To visualize the graph, we can use the Matplotlib-based drawing routines in the Networkx library: we use nx.draw to draw each node, nx.draw_networkx_labels to draw the labels to the nodes, and nx.draw_network_edges to draw the edges. We call nx.draw_network_edges twice, with the edge lists for transfers (on_foot) and on-train (on_train) connections, and color the links as blue and black, respectively, using the edge_color argument. The layout of the graph is determined by the pos argument to the drawing functions. Here we used the graphviz_layout function from networkx. drawing.nx_agraph to lay out the nodes. All drawing functions also accept a Matplotlib axes instance via the ax argument. The resulting graph is shown in Figure 10-5.

```
In [96]: fig, ax = plt.subplots(1, 1, figsize=(14, 10))
    ...: pos = nx.drawing.nx_agraph.graphviz_layout(g, prog="neato")
    ...: nx.draw(g, pos, ax=ax, node_size=200, node_color=colors)
    ...: nx.draw_networkx_labels(g, pos=pos, ax=ax, font_size=6)
```

```
...: nx.draw_networkx_edges(g, pos=pos, ax=ax, edgelist=on_train, width=2)
...: nx.draw_networkx_edges(g, pos=pos, ax=ax, edgelist=on_foot, edge_
    color="blue")
```

Figure 10-5. *Network graph for the Tokyo Metro stations*

Once the network has been constructed, we can use the many graph algorithms provided by the NetworkX library to analyze the network. For example, to compute the degree (i.e., the number of connections to a node) of each node, we can use the degree method (here the output is truncated at ... to save space):

```
In [97]: g.degree()
Out[97]: DegreeView({'Y8': 3,  'N18': 2,  'M24': 2,  'G15': 3,  'C18': 3,
        'N13': 2, ... })
```

For this graph, the degree of a node can be interpreted as the number of connections to a station: the more metro lines that connect at a station, the higher the degree of the corresponding node. We can easily search for the most highly connected station in the network by using the degree method and the max function to find the highest degree in the network. Next we iterate over the result of the degree method and select out the nodes with the maximal degree (which is 6 in this network):

```
In [98]: d_max = max(d for (n, d) in g.degree())
In [99]: [(n, d) for (n, d) in g.degree() if d == d_max]
Out[99]: [('N7', 6), ('G5', 6), ('Y16', 6), ('M13', 6), ('Z4', 6)]
```

The result tells us that the most highly connected stations are station numbers 7 on the N line, 5 on the G line, and so on. All these lines intercept at the same station (the Nagatachou station). We can also compute the closest path between two points in the network using nx.shortest_path. For example, the optimal traveling route (assuming no waiting time and instantaneous transfer) for traveling between Y24 and C19 is

```
In [100]: p = nx.shortest_path(g, "Y24", "C19")
In [101]: p
Out[101]: ['Y24', 'Y23', 'Y22', 'Y21', 'Y20', 'Y19', 'Y18', 'C9', 'C10', 'C11',
           'C12', 'C13', 'C14', 'C15', 'C16', 'C17', 'C18', 'C19']
```

Given a path on this form, we can also directly evaluate the travel time by summing up the weight attributes of neighboring nodes in the path:

```
In [102]: np.sum([g[p[n]][p[n+1]]["weight"]
     ...:         for n in range(len(p)-1) if "weight" in g[p[n]][p[n+1]]])
Out[102]: 35
```

The result suggests that it takes 35 minutes to travel from Y24 to C19. Since the transfer nodes do not have a weight associated with them, the train transfers are effectively assumed to be instantaneous. It may be reasonable to assume that a train transfer takes about 5 minutes, and to take this into account in the shortest path and travel time computation, we can update the transfer nodes and add a weight of 5 to each of them. To do this we create a copy of the graph using the copy method and iterate through the edges and update those with transfer attribute set to True:

```
In [103]: h = g.copy()
In [104]: for n1, n2 in h.edges():
```

```
  ...:         if h[n1][n2]["transfer"]:
  ...:             h[n1][n2]["weight"] = 5
```

Recomputing the path and the traveling time with the new graph gives a more realistic estimate of the traveling time:

```
In [105]: p = nx.shortest_path(h, "Y24", "C19")
In [106]: p
Out[106]: ['Y24', 'Y23', 'Y22', 'Y21', 'Y20', 'Y19', 'Y18', 'C9', 'C10',
          'C11', 'C12', 'C13', 'C14', 'C15', 'C16', 'C17', 'C18', 'C19']
In [107]: np.sum([h[p[n]][p[n+1]]["weight"] for n in range(len(p)-1)])
Out[107]: 40
```

With this method, we can of course compute the optimal path and travel time between arbitrary nodes in the network. As another example, we also compute the shortest path and traveling time between Z1 and H16 (32 minutes):

```
In [108]: p = nx.shortest_path(h, "Z1", "H16")
In [109]: np.sum([h[p[n]][p[n+1]]["weight"] for n in range(len(p)-1)])
Out[109]: 32
```

The NetworkX representation of a graph can be converted to an adjacency matrix in the form of a SciPy sparse matrix using the nx.to_scipy_sparse_matrix, after which we can also analyze the graph with the routines in the sp.csgraph module. As an example of this, we convert the Tokyo Metro graph to an adjacency matrix and compute its reverse Cuthill-McKee ordering (using sp.csgraph.reverse_cuthill_mckee, which is a reordering that reduces the maximum distance of the matrix elements from the diagonal) and permute the matrix with this ordering. We plot the result of both matrices using Matplotlib's spy function, and the result is shown in Figure 10-6.

```
In [110]: A = nx.to_scipy_sparse_matrix(g)
In [111]: A
Out[111]: <184x184 sparse matrix of type '<class 'numpy.int64'>'
                    with 486 stored elements in Compressed Sparse
                    Row format>
In [112]: perm = sp.csgraph.reverse_cuthill_mckee(A)
In [113]: fig, (ax1, ax2) = plt.subplots(1, 2, figsize=(8, 4))
     ...: ax1.spy(A, markersize=2)
     ...: ax2.spy(sp_permute(A, perm, perm), markersize=2)
```

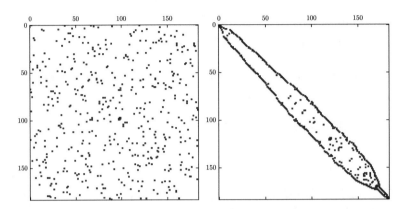

Figure 10-6. *The adjacency matrix of the Tokyo metro graph (left) and the same after RCM ordering (right)*

Summary

In this chapter we have briefly introduced common methods of storing sparse matrices and reviewed how these can be represented using the sparse matrix classes in the SciPy `sparse` module. We also reviewed the sparse matrix construction functions available in the SciPy `sparse` module and the sparse linear algebra routines available in `sparse.linalg`. To complement the linear algebra routines built into SciPy, we also discussed briefly the `scikit.umfpack` extension package, which makes the UMFPACK solver available to SciPy. The sparse matrix library in SciPy is versatile and very convenient to work with, and because it uses efficient low-level libraries for linear algebra routines (SuperLU or UMFPACK), it also offers good performance. For large-scale problems that require parallelization to distribute the workload to multiple cores or even multiple computers, the PETSc and Trilinos frameworks, which both have Python interfaces, provide routes for using sparse matrices and sparse linear algebra with Python in high-performance applications. We also briefly introduced graph representations and processing using the SciPy `sparse.csgraph` and NetworkX libraries.

Further Reading

A good and accessible introduction to sparse matrices and direct solvers for sparse linear equation systems is given in Davis (2006). A fairly detailed discussion of sparse matrices and methods is also given in W. H. Press (2007). For a thorough introduction to network and graph theory, see Newman (2010).

References

Davis, T. (2006). *Direct Methods for Sparse Linear Systems.* Philadelphia: SIAM.

Newman, M. (2010). *Networks: An introduction.* New York: Oxford.

W. H. Press, S. A. (2007). *Numerical Recipes in C: The Art of Scientific Computing.* Cambridge: Cambridge University Press.

Partial Differential Equations

Partial differential equations (PDEs) are multivariate differential equations where derivatives of more than one dependent variable occur. That is, the derivatives in the equation are *partial* derivatives. As such they are generalizations of ordinary differential equations, which were covered in Chapter 9. Conceptually, the difference between ordinary and partial differential equations is not that big, but the computational techniques required to deal with ODEs and PDEs are very different, and solving PDEs is typically much more computationally demanding. Most techniques for solving PDEs numerically are based on the idea of discretizing the problem in each independent variable that occurs in the PDE, thereby recasting the problem into an algebraic form. This usually results in very large-scale linear algebra problems. Two common techniques for recasting PDEs into algebraic form are the finite-difference methods (FDMs), where the derivatives in the problem are approximated with their finite-difference formula, and the finite-element methods (FEMs), where the unknown function is written as linear combination of simple basis functions that can be differentiated and integrated easily. The unknown function is described by a set of coefficients for the basis functions in this representation, and by a suitable rewriting of the PDEs, we can obtain algebraic equations for these coefficients.

With both FDMs and FEMs, the resulting algebraic equation system is usually very large, and in the matrix form, such equation systems are usually very sparse. Both FDM and FEM therefore heavily rely on sparse matrix representation for the algebraic linear equations, as discussed in Chapter 10. Most general-purpose frameworks for PDEs are based on FEM, or some variant thereof, as this method allows for solving very general problems on complicated problem domains.

© Robert Johansson 2019

R. Johansson, *Numerical Python*, https://doi.org/10.1007/978-1-4842-4246-9_11

Solving PDE problems can be far more resource demanding compared to other types of computational problems that we have covered so far (e.g., compared to ODEs). It can be resource demanding partly because of the number of points required to discretize a region of space scale exponentially with the number of dimensions. If a one-dimensional problem requires 100 points to describe, a two-dimensional problem with similar resolution requires $100^2 = 10^4$ points, and a three-dimensional problem requires $100^3 = 10^6$ points. Since each point in the discretized space corresponds to an unknown variable, it is easy to imagine that PDE problems can result in very large equation systems. Defining PDE problems programmatically can also be complicated. One reason for this is that the possible forms of a PDE vastly outnumber the more limited possible forms of ODEs. Another reason is geometry: while an interval in one-dimensional space is uniquely defined by two points, an area in two-dimensional problems and a volume in three-dimensional problems can have arbitrarily complicated geometries enclosed by curves and surfaces. To define the problem domain of a PDE and its discretization in a mesh of coordinate points can therefore require advanced tools, and there is a large amount of freedom in how boundary conditions can be defined as well. In contrast to ODE problems, there is no standard form on which any PDE problem can be defined.

For these reasons, the PDE solvers for Python are only available through libraries and frameworks that are specifically dedicated to PDE problems. For Python, there are at least three significant libraries for solving PDE problems using the FEM method: the FiPy library, the SfePy library, and the FEniCS library. All of these libraries are extensive and feature-rich, and going into the details of using either of these libraries is beyond the scope of this book. Here we can only give a brief introduction to PDE problems and survey prominent examples of PDE libraries that can be used from Python and go through a few examples that illustrate some of the features of one of these libraries (FEniCS). The hope is that this can give the reader who is interested in solving PDE problems with Python a bird's-eye overview of the available options and some useful pointers on where to look for further information.

Importing Modules

For basic numerical and plotting usage, in this chapter too, we require the NumPy and Matplotlib libraries. For 3D plotting we need to explicitly import the `mplot3d` module from the Matplotlib toolkit library `mpl_toolkits`. As usual, we assume that these libraries are imported in the following manner:

```
In [1]: import numpy as np
In [2]: import matplotlib.pyplot as plt
In [3]: import matplotlib as mpl
In [4]: import mpl_toolkits.mplot3d
```

We also use the `linalg` and the `sparse` modules from SciPy, and to use the `linalg` submodule of the `sparse` module, we also need to import it explicitly:

```
In [5]: import scipy.sparse as sp
In [6]: import scipy.sparse.linalg
In [7]: import scipy.linalg as la
```

With these imports, we can access the dense linear algebra module as `la`, while the sparse linear algebra module is accessed as `sp.linalg`. Furthermore, later in this chapter, we will also use the FEniCS FEM framework, and we require that its `dolfin` and `mshr` libraries be imported in the following manner:

```
In [8]: import dolfin
In [9]: import mshr
```

Partial Differential Equations

The unknown quantity in a PDE is a multivariate function, here denoted as u. In an N-dimensional problem, the function u depends on n-independent variables: $u(x_1, x_2, ..., x_n)$. A general PDE can formally be written as

$$F\left(x_1, x_2, ..., x_n, u, \left\{\frac{\partial u}{\partial x_{i_1}}\right\}_{1 \le i_1 \le n}, \left\{\frac{\partial^2 u}{\partial x_{i_1} x_{i_2}}\right\}_{1 \le i_1, i_2 \le n}, ...\right) = 0, \boldsymbol{x} \in \Omega,$$

where $\left\{\dfrac{\partial u}{\partial x_{i_1}}\right\}_{1 \le i_1 \le n}$ denotes all first-order derivatives with respect to the independent variables $x_1, ..., x_n$, $\left\{\dfrac{\partial^2 u}{\partial x_{n_1} x_{n_2}}\right\}_{1 \le i_1, i_2 \le n}$ denotes all second-order derivatives, and so on. Here F is a known function that describes the form of the PDE, and Ω is the domain of the PDE problem. Many PDEs that occur in practice only contain up to second-order derivatives, and we typically deal with problems in two or three spatial dimensions and possibly time. When working with PDEs, it is common to simplify the notation by denoting the partial derivatives with respect to an independent variable x using the subscript notation: $u_x = \dfrac{\partial u}{\partial x}$. Higher-order derivatives are denoted with multiple indices: $u_{xx} = \dfrac{\partial^2 u}{\partial x^2}$, $u_{xy} = \dfrac{\partial^2 u}{\partial x \partial y}$, and so on. An example of a typical PDE is the heat equation, which in a two-dimensional Cartesian coordinate system takes the form $u_t = \alpha(u_{xx} + u_{yy})$. Here the function $u = u(t, x, y)$ describes the temperature at the spatial point (x, y) at time t, and α is the thermal diffusivity coefficient.

To fully specify a particular solution to a PDE, we need to define its boundary conditions, which are known values of the function or a combination of its derivatives along the boundary of the problem domain Ω, as well as the initial values if the problem is time-dependent. The boundary is often denoted as Γ or $\partial\Omega$, and in general different boundary conditions can be given for different parts of the boundary. Two important types of boundary conditions are Dirichlet boundary conditions, which specify the value of the function at the boundary, $u(\boldsymbol{x}) = h(\boldsymbol{x})$ for $\boldsymbol{x} \in \Gamma_D$, and Neumann boundary conditions, which specify the normal derivative on the boundary, $\dfrac{\partial u(\boldsymbol{x})}{\partial \boldsymbol{n}} = g(\boldsymbol{x})$ for

$\boldsymbol{x} \in \Gamma_N$, where $\boldsymbol{n}$ is the outward normal from the boundary. Here $h(\boldsymbol{x})$ and $g(\boldsymbol{x})$ are arbitrary functions.

Finite-Difference Methods

The basic idea of the finite-difference method is to approximate the derivatives that occur in a PDE with their finite-difference formulas on a discretized space. For example, the finite-difference formula for the ordinary derivative $\dfrac{du(x)}{dx}$ on a discretization of the continuous variable x into discrete points $\{x_n\}$ can be approximated with the forward difference formula $\dfrac{du(x_n)}{dx} \approx \dfrac{u(x_{n+1}) - u(x_n)}{x_{n+1} - x_n}$, the backward difference formula $\dfrac{du(x_n)}{dx} \approx \dfrac{u(x_n) - u(x_{n-1})}{x_n - x_{n-1}}$, or the centered difference formula $\dfrac{du(x_n)}{dx} \approx \dfrac{u(x_{n+1}) - u(x_{n-1})}{x_{n+1} - x_{n-1}}$. Similarly, we can also construct finite-difference formulas for higher-order derivatives, such as the second-order derivative $\dfrac{d^2 u(x_n)}{dx^2} \approx \dfrac{u(x_{n+1}) - 2u(x_n) + u(x_{n-1})}{(x_n - x_{n-1})^2}$. Assuming that the discretization of the continuous variable x into discrete points is fine enough, these finite-difference formulas can give good approximations of the derivatives. Replacing derivatives in an ODE or PDE with their finite-difference formulas recasts the equations from differential equations to algebraic equations. If the original ODE or PDE is linear, the algebraic equations are also linear and can be solved with standard linear algebra methods.

To make this method more concrete, consider the ODE problem $u_{xx} = -5$ in the interval $x \in [0, 1]$ and with boundary conditions $u(x = 0) = 1$ and $u(x = 1) = 2$, which, for example, arises from the steady-state heat equation in one dimension. In contrast to the ODE initial value problem considered in Chapter 9, this is a boundary value problem because the value of u is specified at both $x = 0$ and $x = 1$. The methods for initial value problems are therefore not applicable here. Instead we can treat this problem by dividing the interval $[0, 1]$ into discrete points x_n, and the problem is then to find the function $u(x_n) = u_n$ at these points. Writing the ODE problem in finite-difference form gives an equation $(u_{n-1} - 2u_n + u_{n+1})/\Delta x^2 = -5$ for every interior point n and the boundary conditions $u_0 = 1$ and $u_{N+1} = 2$. Here the interval $[0, 1]$ is discretized into $N+2$ evenly spaced points, including the boundary points, with separation $\Delta x = 1/(N+1)$. Since the function is known at the two boundary points, there are N unknown variables u_n corresponding to the function values at the interior points. The set of equations for the interior points can be written in a matrix form as $Au = b$, where $u = [u_1, ..., u_N]^T$,

$$b = \left[-5 - \frac{u_0}{\Delta x^2}, -5, ..., -5, -5 - \frac{u_{N+1}}{\Delta x^2} \right]^T, \text{ and}$$

$$A = \frac{1}{\Delta x^2} \begin{bmatrix} -2 & 1 & 0 & 0 & \cdots \\ 1 & -2 & 1 & 0 & \cdots \\ 0 & 1 & -2 & 1 & 0 \\ 0 & 0 & 1 & -2 & \ddots \\ \vdots & \vdots & 0 & \ddots & \ddots \end{bmatrix}.$$

Here the matrix A describes the coupling of the equations for u_n to values at neighboring points due to the finite-difference formula that was used to approximate the second-order derivative in the ODE. The boundary values are included in the b vector, which also contains the constant right-hand side of the original ODE (the source term). At this point, we can straightforwardly solve the linear equation system $Au = b$ for the unknown vector of u and thereby obtain the approximate values of the function $u(x)$ at the discrete points $\{x_n\}$.

In Python code, we can set up and solve this problem in the following way: First, we define variables for the number of interior points N, the values of the function at the boundaries u0 and u1, as well as the spacing between neighboring points dx.

```
In [10]: N = 5
In [11]: u0, u1 = 1, 2
In [12]: dx = 1.0 / (N + 1)
```

Next we construct the matrix A as described in the preceding section. For this we can use the eye function from NumPy, which creates a two-dimensional array with ones on the diagonal or on the upper or lower diagonal that is shifted from the main diagonal by the number given by the argument k.

```
In [13]: A = (np.eye(N, k=-1) - 2 * np.eye(N) + np.eye(N, k=1)) / dx**2
In [14]: A
Out[14]: array([[-72.,  36.,   0.,   0.,   0.],
                [ 36., -72.,  36.,   0.,   0.],
                [  0.,  36., -72.,  36.,   0.],
                [  0.,   0.,  36., -72.,  36.],
                [  0.,   0.,   0.,  36., -72.]])
```

Next we need to define an array for the vector b, which corresponds to the source term -5 in the differential equation, as well as the boundary condition. The boundary conditions enter into the equations via the finite-difference expressions for the derivatives of the first and the last equation (for u_1 and u_N), but these terms are missing from the expression represented by the matrix A and must therefore be added to the vector b.

```
In [15]: b = -5 * np.ones(N)
    ...: b[0] -= u0 / dx**2
    ...: b[N-1] -= u1 / dx**2
```

Once the matrix A and the vector b are defined, we can proceed to solve the equation system using the linear equation solver from SciPy (we could also use the one provided by NumPy, np.linalg.solve).

```
In [16]: u = la.solve(A, b)
```

This completes the solution of this ODE problem. To visualize the solution, here we first create an array x that contains the discrete coordinate points for which we have solved the problem, including the boundary points, and we also create an array U that combines the boundary values and the interior points in one array. The result is then plotted and shown in Figure 11-1.

```
In [17]: x = np.linspace(0, 1, N+2)
In [18]: U = np.hstack([[u0], u, [u1]])
In [19]: fig, ax = plt.subplots(figsize=(8, 4))
    ...: ax.plot(x, U)
    ...: ax.plot(x[1:-1], u, 'ks')
    ...: ax.set_xlim(0, 1)
    ...: ax.set_xlabel(r"$x$", fontsize=18)
    ...: ax.set_ylabel(r"$u(x)$", fontsize=18)
```

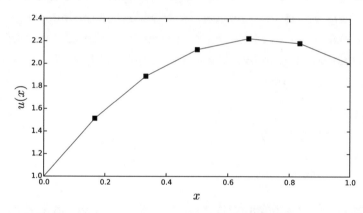

Figure 11-1. *Solution to the second-order ODE boundary value problem introduced in the text*

The finite-difference method can easily be extended to higher dimensions by using the finite-difference formula along each discretized coordinate. For a two-dimensional problem, we have a two-dimensional array u for the unknown interior function values, and when using the finite differential formula, we obtain a system of coupled equations for the elements in u. To write these equations on the standard matrix-vector form, we can rearrange the u array into a vector and assemble the corresponding matrix A from the finite-difference equations.

As an example, consider the following two-dimensional generalization of the previous problem: $u_{xx} + u_{yy} = 0$, with the boundary conditions $u(x = 0) = 3$, $u(x = 1) = -1$, $u(y = 0) = -5$, and $u(y = 1) = 5$. Here there is no source term, but the boundary conditions in a two-dimensional problem are more complicated than in the one-dimensional problem we solved earlier. In finite-difference form, we can write the PDE as $(u_{m-1,n} - 2u_{m,n} + u_{m+1,n})/\Delta x^2 + (u_{m,n-1} - 2u_{m,n} + u_{m,n+1})/\Delta y^2 = 0$. If we divide the x and y intervals into N interior points ($N+2$ points including the boundary points), then $\Delta x = \Delta y = \dfrac{1}{N+1}$, and u is an NxN matrix. To write the equation on the standard form $Av = b$, we can rearrange the matrix u by stacking its rows or columns into a vector of size $N^2 \times 1$. The matrix A is then of size $N^2 \times N^2$, which can be very big if we need to use a fine discretization of the x and y coordinates. For example, using 100 points along both x and y gives an equation system that has 10^4 unknown values u_{mn}, and the matrix A has $100^4 = 10^8$ elements. Fortunately, since the finite-difference formula only couples neighboring points, the matrix A turns out to be very sparse, and here we can benefit greatly from working with sparse matrices, as we will see in the following.

To solve this PDE problem with Python and the finite-element method, we start by defining variables for the number of interior points and the values along the four boundaries of the unit square:

```
In [20]: N = 100
In [21]: u0_t, u0_b = 5, -5
In [22]: u0_l, u0_r = 3, -1
In [23]: dx = 1. / (N+1)
```

We also computed the separation dx between the uniformly spaced coordinate points in the discretization of x and y (assumed equal). Because the finite-difference formula couples both neighboring rows and columns, it is slightly more involved to construct the matrix A for this example. However, a relatively direct approach is to first define the matrix A_1d that corresponds to the one-dimensional formula along one of the coordinates (say x or the index m in $u_{m,n}$). To distribute this formula along each row, we can take the tensor product of the identity matrix of size $N \times N$ with the A_1d matrix. The result describes all derivatives along the m-index for all indices n. To cover the terms that couple the equation for $u_{m,n}$ to $u_{m,n+1}$ and $u_{m,n-1}$, that is, the derivatives along the index n, we need to add diagonals that are separated from the main diagonal by N positions. In the following we perform these steps to construct A using the eye and kron functions from the scipy.sparse module. The result is a sparse matrix A that describes the finite-difference equation system for the two-dimensional PDE we are considering here:

```
In [24]: A_1d = (sp.eye(N, k=-1) + sp.eye(N, k=1) - 4 * sp.eye(N))/dx**2
In [25]: A = sp.kron(sp.eye(N), A_1d) + (sp.eye(N**2, k=-N) + sp.eye(N**2,
            k=N))/dx**2
In [26]: A
Out[26]: <10000x10000 sparse matrix of type '<type 'numpy.float64'>'
            with 49600 stored elements in Compressed Sparse Row format>
```

The printout of A shows that it is a sparse matrix with 10^8 elements with 49600 nonzero elements, so that only 1 out of about 2000 elements is nonzero, and A is indeed very sparse. To construct the vector b from the boundary conditions, it is convenient to create a $N \times N$ array of zeros and assign the boundary condition to edge elements of this array (which are the corresponding elements in u that are coupled to the boundaries, i.e., the interior points that are neighbors to the boundary). Once this $N \times N$ array is created and assigned, we can use the reshape method to rearrange it into a $N^2 \times 1$ vector that can be used in the $Av = b$ equation:

```
In [27]: b = np.zeros((N, N))
    ...: b[0, :] += u0_b    # bottom
    ...: b[-1, :] += u0_t  # top
    ...: b[:, 0] += u0_l    # left
    ...: b[:, -1] += u0_r  # right
    ...: b = - b.reshape(N**2) / dx**2
```

When the A and b arrays are created, we can proceed to solve the equation system for the vector v and use the reshape method to arrange it back into the $N \times N$ matrix u:

```
In [28]: v = sp.linalg.spsolve(A, b)
In [29]: u = v.reshape(N, N)
```

For plotting purposes, we also create a matrix U that combines the u matrix with the boundary conditions. Together with the coordinate matrices X and Y, we then plot a colormap graph and a 3D surface view of the solution. The result is shown in Figure 11-2.

```
In [30]: U = np.vstack([np.ones((1, N+2)) * u0_b,
    ...:                np.hstack([np.ones((N, 1)) * u0_l, u, np.ones
                        ((N, 1)) * u0_r]),
    ...:                np.ones((1, N+2)) * u0_t])
In [31]: x = np.linspace(0, 1, N+2)
```

```
In [32]: X, Y = np.meshgrid(x, x)
In [33]: fig = plt.figure(figsize=(12, 5.5))
    ...: cmap = mpl.cm.get_cmap('RdBu_r')
    ...:
    ...: ax = fig.add_subplot(1, 2, 1)
    ...: c = ax.pcolor(X, Y, U, vmin=-5, vmax=5, cmap=cmap)
    ...: ax.set_xlabel(r"$x_1$", fontsize=18)
    ...: ax.set_ylabel(r"$x_2$", fontsize=18)
    ...:
    ...: ax = fig.add_subplot(1, 2, 2, projection='3d')
    ...: p = ax.plot_surface(X, Y, U, vmin=-5, vmax=5, rstride=3, cstride=3,
    ...:                     linewidth=0, cmap=cmap)
    ...: ax.set_xlabel(r"$x_1$", fontsize=18)
    ...: ax.set_ylabel(r"$x_2$", fontsize=18)
    ...: cb = plt.colorbar(p, ax=ax, shrink=0.75)
    ...: cb.set_label(r"$u(x_1, x_2)$", fontsize=18)
```

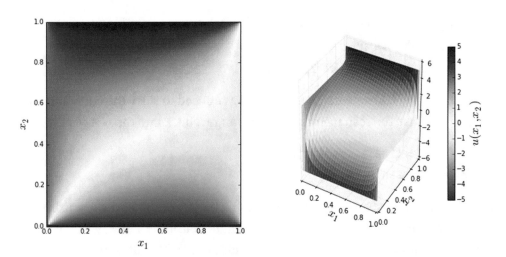

Figure 11-2. *The solution to the two-dimensional heat equation with Dirichlet boundary conditions defined in the text*

As mentioned in the preceding section, FDM methods result in matrices A that are very sparse, and using sparse matrix data structures, such as those provided by scipy. sparse, can give significant performance improvements compared to using dense

NumPy arrays. To illustrate in concrete terms the importance of using sparse matrices for this type of problems, we can compare the time required for solving of the $Av = b$ equation using the IPython command %timeit, for the two cases where A is a sparse and a dense matrix:

```
In [34]: A_dense = A.todense()
In [35]: %timeit la.solve(A_dense, b)
1 loops, best of 3: 10.8 s per loop
In [36]: %timeit sp.linalg.spsolve(A, b)
10 loops, best of 3: 31.9 ms per loop
```

From these results, we see that using sparse matrices for the present problem results in a speedup of several orders of magnitude (in this particular case, we have a speedup of a factor $10.8/0.0319 \approx 340$).

The finite-difference method that we used in the last two examples is a powerful and relatively simple method for solving ODE boundary value problems and PDE problems with simple geometries. However, it is not so easily adapted to problems on more complicated domains or problems on nonuniform coordinate grids. For such problems finite-element methods are typically more flexible and convenient to work with, and although FEMs are conceptually more complicated than FDMs, they can be computationally efficient and adapt well to complicated problem domains and more involved boundary conditions.

Finite-Element Methods

The finite-element method is a powerful and universal method for converting PDEs into algebraic equations. The basic idea of this method is to represent the domain on which the PDE is defined with a finite set of discrete regions, or *elements*, and to approximate the unknown function as a linear combination of basis functions with local support on each of these elements (or on a small group of neighboring elements). Mathematically, this approximation solution, u_h, represents a projection of the exact solution u in the function space V (e.g., continuous real-valued functions) onto a finite subspace $V_h \subset V$ that is related to the discretization of the problem domain. If V_h is a suitable subspace of V, then it can be expected that u_h can be a good approximation to u.

To be able to solve the approximate problem on the simplified function space V_h, we can first rewrite the PDE from its original formulation, which is known as the *strong form*, to its corresponding variational form, also known as the *weak form*. To obtain the weak form, we multiply the PDE with an arbitrary function v and integrate over the entire problem domain. The function v is called a *test function*, and it can in general be defined on function space $\widehat{V}$ that differs from V and V_h.

For example, consider the steady-state heat equation (also known as the Poisson equation) that we solved using the FDM earlier in this chapter: the strong form of this equation is $-\Delta u(\boldsymbol{x}) = f(\boldsymbol{x})$, where we have used the vector operator notation. By multiplying this equation with a test function v and integrating over the domain $\boldsymbol{x} \in \Omega$, we obtain the weak form:

$$-\int_\Omega \Delta u\, v\, \mathrm{d}x = \int_\Omega f\, v\, \mathrm{d}x.$$

Since the exact solution u satisfies the strong form, it also satisfies the weak form of the PDE for any reasonable choice of v. The reverse does not necessarily hold true, but if a function u_h, which is called a *trial function* in this context, satisfies the weak form for a large class of suitably chosen test functions v, then it is plausible that it is a good approximation to the exact solution u.

To treat this problem numerically, we first need to make the transition from the infinite-dimensional function spaces V and $\widehat{V}$ to approximate finite-dimensional function spaces V_h and $\widehat{V}_h$:

$$-\int_\Omega \Delta u_h v_h\, \mathrm{d}x = \int_\Omega f v_h\, \mathrm{d}x,$$

where $u_h \in V_h$ and $v_h \in \widehat{V}_h$. The key point here is that V_h and $\widehat{V}_h$ are finite dimensional, so we can use a finite set of basis functions $\{\phi_i\}$ and $\{\hat{\phi}_i\}$ that spans the function spaces V_h and $\widehat{V}_h$, respectively, to describe the functions u_h and v_h. In particular, we can express u_h as a linear combination of the basis functions that span its function space, $u_h = \sum U_i \phi_i$. Inserting this linear combination in the weak form of the PDE and carrying out the integrals and differential operators on the basis functions, instead of directly over terms in the PDE, yields a set of algebraic equations.

To obtain an equation system on the simple form $AU = b$, we also must write the weak form of the PDE on bilinear form with respect to the u_h and v_h functions $a(u_h, v_h) = L(v_h)$, for some functions a and L. This is not always possible, but for the current example of the Poisson equation, we can obtain this form by integrating by parts:

$$-\int_\Omega \Delta u_h v_h \, dx = \int_\Omega \nabla u_h \cdot \nabla v_h \, dx - \int_\Omega \nabla \cdot (\nabla u_h v_h) dx = \int_\Omega \nabla u_h \cdot \nabla v_h \, dx - \int_{\partial\Omega} (\nabla u_h \cdot \boldsymbol{n}) v_h \, d\Gamma,$$

where in the second equality we have also applied Gauss' theorem to convert the second term to an integral over the boundary $\partial\Omega$ of the domain Ω. Here $\boldsymbol{n}$ is the outward normal vector of the boundary $\partial\Omega$. There is no general method for rewriting a PDE on the strong form to weak form, and each problem will have to be approached on a case-by-case basis. However, the technique used here, to integrate by part and rewrite the resulting integrals using integral identities, can be used for many frequently occurring PDEs.

To reach the bilinear form that can be approached with standard linear algebra methods, we also have to deal with the boundary term in the preceding weak form equation. To this end, assume that the problem satisfies the Dirichlet boundary condition on a part of $\partial\Omega$ denoted Γ_D and Neumann boundary conditions on the remaining part of $\partial\Omega$, denoted Γ_N: $\{u = h, x \in \Gamma_D\}$ and $\{\nabla u \cdot \boldsymbol{n} = g, x \in \Gamma_N\}$. Not all boundary conditions are of Dirichlet or Neumann type, but together these cover many physically motivated situations.

Since we are free to choose the test functions v_h, we can let v_h vanish on the part of the boundary that satisfies Dirichlet boundary conditions. In this case we obtain the following weak form of the PDE problem:

$$\int_\Omega \nabla u_h \cdot \nabla v_h \, dx = \int_\Omega f v_h \, dx + \int_{\Gamma_N} g \, v_h \, d\Gamma.$$

If we substitute the function u_k for its expression as a linear combination of basis functions, and substitute the test function with one of its basis functions, we obtain an algebraic equation:

$$\sum U_j \int_\Omega \nabla \phi_j \cdot \nabla \hat{\phi}_i \, dx = \int_\Omega f \hat{\phi}_i \, dx + \int_{\Gamma_N} g \hat{\phi}_i \, d\Gamma.$$

If there are N basis functions in V_k, then there are N unknown coefficients U_i, and we need N-independent test functions $\hat{\phi}_i$ to obtain a closed equation system. This equation system is on the form $AU = b$ with $A_{ij} = \int_\Omega \nabla \phi_j \cdot \nabla \hat{\phi}_i \, dx$ and $b_i = \int_\Omega f \hat{\phi}_i \, dx + \int_{\Gamma_N} g \hat{\phi}_i \, d\Gamma$.

Following this procedure we have therefore converted the PDE problem into a system of linear equations that can be readily solved using techniques discussed in previous chapters.

In practice, a very large number of basis functions can be required to obtain a good approximation to the exact solution, and the linear equation system generated by FEMs is therefore often very large. However, the fact that each basis function has support only at one or a few nearby elements in the discretization of the problem domain ensures that the matrix A is sparse, which makes it tractable to solve rather large-scale FEM problems. We also note that an important property of the basis functions ϕ_i and $\hat{\phi}_i$ is that it should be easy to compute the derivatives and integrals of the expression that occurs in the final weak form of the problem, so that the matrix A and vector b can be assembled efficiently. Typical examples of basis functions are low-order polynomial functions that are nonzero only within a single element. See Figure 11-3 for a one-dimensional illustration of this type of basis function, where the interval [0, 6] is discretized using five interior points, and a continuous function (black solid curve) is approximated as a piecewise linear function (dashed red/light-gray line) by suitably weighted triangular basic functions (blue/dark-gray solid lines).

Figure 11-3. *An example of possible basis functions (blue/dark-gray lines), with local support, for the one-dimensional domain [0, 6]*

When using FEM software for solving PDE problems, it is typically required to convert the PDE to weak form by hand and if possible rewrite it on the bilinear form $a(u, v) = L(v)$. It is also necessary to provide a suitable discretization of the problem domain. This discretization is called a mesh, and it is usually made up of triangular partitioning (or their higher-order generalizations) of the total domain. Meshing an intricate problem domain can in itself be a complicated process, and it may require using sophisticated software especially dedicated for mesh generation. For simple geometries there are tools for programmatically generating meshes, and we will see examples of this in the following section.

Once a mesh is generated and the PDE problem is written on a suitable weak form, we can feed the problem into a FEM framework, which then automatically assembles the algebraic equation system and applies suitable sparse equation solvers to find the solution. In this processes, we often have a choice of what type of basis functions to use, as well as which type of solver to use. Once the algebraic equation is solved, we can construct the approximation solution to the PDE with the help of the basis functions, and we can, for example, visualize the solution or post-process it in some other fashion.

In summary, solving a PDE using FEM typically involves the following steps:

1. Generate a mesh for the problem domain.

2. Write the PDE on weak form.

3. Program the problem in the FEM framework.

4. Solve the resulting algebraic equations.

5. Post-process and/or visualize the solution.

In the following section, we will review available FEM frameworks that can be used with Python and then look at a number of examples that illustrate some of the key steps in the PDE solution process using FEM.

Survey of FEM Libraries

For Python there are at least three significant FEM packages: FiPy, SfePy, and FEniCS. These are all rather full-featured frameworks, which are capable of solving a wide range of PDE problems. Technically, the FiPy library is not a FEM software, but rather a finite-volume method (FVM) software, but the gist of this method is quite similar to FEM. The FiPy framework can be obtained from http://www.ctcms.nist.gov/fipy. The SfePy library is a FEM software that takes a slightly different approach to define PDE problems, in that it uses Python files as configuration files for its FEM solver, rather programmatically setting up a FEM problem (although this mode of operation is technically also supported in SfePy). The SfePy library is available from http://sfepy.org. The third major framework for FEM with Python is FEniCS, which is written for C++ and Python. The FEniCS framework is my personal favorite when it comes to FEM software for Python, as it provides an elegant Python interface to a powerful FEM engine. Like FDM problem, FEM problems typically result in very large-scale equation systems that require using sparse matrix techniques to solve efficiently. A crucial part of a FEM framework is therefore to efficiently solve large-scale linear and nonlinear systems, using sparse matrices representation and direct or iterative solvers that work on sparse systems, possibly using parallelization. Each of the frameworks mentioned in the preceding section supports multiple backends for such low-level computations. For example, many FEM frameworks can use the PETSc and Trilinos frameworks.

Unfortunately we are not able to explore in depth how to use either of these FEM frameworks here, but in the following section, we will look at solving example problems with FEniCS and thereby introduce some of its basic features and usage. The hope is that the examples can give a flavor of how it is to work with FEM problems in Python and provide a starting point for the readers interested in learning more about FEM with Python.

Solving PDEs Using FEniCS

In this section we solve a series of increasingly complicated PDEs using the FEniCS framework, and in the process, we introduce the workflow and a few of the main features of this FEM software. For a thorough introduction to the FEniCS framework, see the documentation at the project web sites and the official FEniCS book (Anders Logg, 2012).

FEniCS   FEniCS is a highly capable FEM framework that is made up of a collection of libraries and tools for solving PDE problem. Much of FEniCS is programmed in C++, but it also provides an official Python interface. Because of the complexity of the many dependencies of the FEniCS libraries to external low-level numerical libraries, FEniCS is usually packaged and installed as an independent environment, although it can also be installed using conda on some platforms. For more information about the FEniCS, see the project's web site at `http://fenicsproject.org`. At the time of writing, the most recent version is 2018.1.0.

The Python interface to FEniCS is provided by a library named `dolfin`. For mesh generation we will also use the `mshr` library. In the following code, we assume that these libraries are imported in their entirety, as shown in the beginning of this chapter. For a summary of the most important functions and classes from these libraries, see Table 11-1 and Table 11-2.

Table 11-1. *Summary of Selected Functions and Classes in the dolfin Library*

Function/Class	Description	Example
parameters	Dictionary holding configuration parameters for the FEniCS framework.	dolfin.parameters ["reorder_dofs_serial"]
RectangleMesh	Object for generating a rectangular 2D mesh.	mesh = dolfin. RectangularMesh(dolfin. Point(0, 0),dolfin.Point (1, 1), 10, 10)
MeshFunction	Function defined over a given mesh.	dolfin. MeshFunction("size_t", mesh, mesh.topology(). dim()-1)
FunctionSpace	Object for representing a function space.	V = dolfin. FunctionSpace(mesh, 'Lagrange', 1)
TrialFunction	Object for representing a trial function defined in a given function space.	u = dolfin. TrialFunction(V)
TestFunction	Object for representing a test function defined in a given function space.	v = dolfin.TestFunction(V)
Function	Object for representing unknown functions appearing in the weak form of a PDE.	u_sol = dolfin.Function(V)
Constant	Object for representing a fixed constant.	c = dolfin.Constant(1.0)
Expression	Representation of a mathematical expression in terms of the spatial coordinates.	dolfin. Expression("x[0]*x[0] + x[1]*x[1]")
DirichletBC	Object for representing Dirichlet-type boundary conditions.	dolfin.DirichletBC(V, u0, u0_boundary)
Equation	Object for representing an equation, for example, generated by using the == operator with other FEniCS objects.	a == L

(continued)

Table 11-1. (*continued*)

Function/Class	Description	Example
`inner`	Symbolic representation of the inner product.	`dolfin.inner(u, v)`
`nabla_grad`	Symbolic representation of the gradient operator.	`dolfin.nabla_grad(u)`
`dx`	Symbolic representation of the volume measure for integration.	`f*v*dx`
`ds`	Symbolic representation of a line measure for integration.	`g_v * v * dolfin.ds(0, domain=mesh, subdomain_data=boundary_parts)`
`assemble`	Assemble the algebraic equations by carrying out the integrations over the basis functions.	`A = dolfin.assemble(a)`
`solve`	Solve an algebraic equation.	`dolfin.solve(A, u_sol.vector(), b)`
`plot`	Plot a function or expression.	`dolfin.plot(u_sol)`
`File`	Write a function to a file that can be opened with visualization software such as ParaView.	`dolfin.File('u_sol.pvd') << u_sol`
`refine`	Refine a mesh by splitting a selection of the existing mesh elements into smaller pieces.	`mesh = dolfin.refine (mesh, cell_markers)`
`AutoSubDomain`	Representation of a subset of a domain, selected from all elements by the indicator function passed to it as argument.	`dolfin.AutoSubDomain (v_boundary_func)`

Table 11-2. *Summary of Selected Functions and Classes in the* mshr *and* dolfin *Library*

Function/Class	Description
dolfin.Point	Representation of a coordinate point.
mshr.Circle	Representation of a geometrical object with the shape of a circle, which can be used to compose 2D domain.
mshr.Ellipse	Representation of a geometrical object with the shape of an ellipse.
mshr.Rectangle	Representation of a domain defined by a rectangle in 2D.
mshr.Box	Representation of a domain defined by a box in 3D.
mshr.Sphere	Representation of a domain defined by a sphere in 3D.
mshr.generate_mesh	Generate a mesh from a domain composed of geometrical objects, such as those listed in the preceding section.

Before we proceed to use FEniCS and the dolfin Python library, we need to set two configuration parameters via the dolfin.parameters dictionary to obtain the behavior that we need in the following examples:

```
In [37]: dolfin.parameters["reorder_dofs_serial"] = False
In [38]: dolfin.parameters["allow_extrapolation"] = True
```

To get started with FEniCS, we begin by reconsidering the steady-state heat equation in two dimensions that we already solved earlier in this chapter using the FDM. Here we consider the problem $u_{xx}+u_{yy} = f$, where f is a source function. To begin with we will assume that the boundary conditions are $u(x = 0, y) = u(x = 1, y) = 0$ and $u(x, y = 0) = u(x, y = 1) = 0$. In later examples we will see how to define Dirichlet and Neumann boundary conditions.

The first step in the solution of a PDE with FEM is to define a mesh that describes the discretization of the problem domain. In the current example, the problem domain is the unit square $x, y \in [0, 1]$. For simple geometries like this, there are functions in the dolfin library for generating the mesh. Here we use the RectangleMesh function, which as first two arguments takes the coordinate points (x_0, y_0) and (x_1, y_1), represented as dolfin. Point instances, where (x_0, y_0) is the coordinates of the lower-left corner of the rectangle and (x_1, y_1) of the upper-right corner. The fifth and sixth arguments are the numbers of elements along the x and y directions, respectively. The resulting mesh object is viewed in a Jupyter Notebook via its rich display system (here we generate a less fine mesh for the purpose of displaying the mesh structure), as shown in Figure 11-4:

```
In [39]: N1 = N2 = 75
In [40]: mesh = dolfin.RectangleMesh(dolfin.Point(0, 0), dolfin.Point
             (1, 1), N1, N2)
In [41]: dolfin.RectangleMesh(dolfin.Point(0, 0), dolfin.Point(1, 1), 10, 10)
             # for display
```

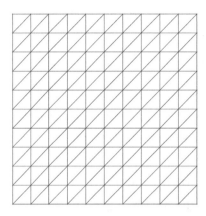

Figure 11-4. *A rectangular mesh generated using* `dolfin.RectangleMesh`

This mesh for the problem domain is the key to the discretization of the problem into a form that can be treated using numerical methods. The next step is to define a representation of the function space for the trial and the test functions, using the `dolfin.FunctionSpace` class. The constructor of this class takes at least three arguments: a mesh object, the name of the type of basis function, and the degree of the basis function. For our purposes we will use the Lagrange type of basis functions of degree one (linear basis functions):

```
In [42]: V = dolfin.FunctionSpace(mesh, 'Lagrange', 1)
```

Once the mesh and the function space objects are created, we need to create objects for the trial function u_h and the test function v_h, which we can use to define the weak form of the PDE of interest. In FEniCS, we use the `dolfin.TrialFunction` and `dolfin.TestFunction` classes for this purpose. They both require a function space object as first argument to their constructors:

```
In [43]: u = dolfin.TrialFunction(V)
In [44]: v = dolfin.TestFunction(V)
```

The purpose of defining representations of the function space V and the trial and test functions u and v is to be able to construct a representation of a generic PDE on the weak form. For the steady-state heat equation that we are studying here, the weak form was shown in the previous section to be (in the absence of Neumann boundary conditions)

$$\int_\Omega \nabla u \cdot \nabla v \, dx = \int_\Omega f v \, dx.$$

To arrive at this form usually requires rewriting and transforming by hand the direct integrals over the PDE, typically by performing integration by parts. In FEniCS, the PDE itself is defined using the integrands that appear in the weak form, including the integral measure (i.e., the dx). To this end, the dolfin library provides a number of functions acting on the trial and test function objects v and u that are used to represent operations on these function that commonly occur in the weak form of a PDE. For example, in the present case, the integrand of the left-hand side integral is $\nabla u \cdot \nabla v \, dx$. To represent this expression, we need a symbolic representation of the inner product, the gradients of u and v, and the integration measure dx. The names for these functions in the dolfin library are inner, nabla_grad, and dx, respectively, and using these functions we can create a representation of $a(u, v) = \nabla u \cdot \nabla v \, dx$ that the FEniCS framework understands and can work with:

```
In [45]: a = dolfin.inner(dolfin.nabla_grad(u), dolfin.nabla_grad(v)) *
         dolfin.dx
```

Likewise, for the right-hand side, we need a representation of $b(v) = f v \, dx$. At this point, we need to specify an explicit form of f (the source term in the original PDF) to be able to proceed with the solution of the problem. Here we look at two types of functions: $f = 1$ (a constant) and $f = x^2 + y^2$ (a function of x and y). To represent $f = 1$, we can use the dolfin.Constant object. It takes as its only argument the value of the constant that it represents:

```
In [46]: f1 = dolfin.Constant(1.0)
In [47]: L1 = f1 * v * dolfin.dx
```

If f is a function of x and y, we instead need to use the dolfin.Expression object to represent f. The constructor of this object takes a string as first argument that contains an expression that corresponds to the function. This expression must be defined in

C++ syntax, since the FEniCS framework automatically generates and compiles a C++ function for efficient evaluation of the expression. In the expression we have access to a variable x, which is an array of coordinates at a specific point, where x is accessed as x[0], y as x[1], and so on. For example, to write the expression for $f(x, y) = x^2 + y^2$, we can use "x[0]*x[0] + x[1]*x[1]". Note that because we need to use C++ syntax in this expression, we *cannot* use the Python syntax x[0]**2. The Expression class also takes the keyword argument degree that specifies the degree of the basis function or, alternatively, the keyword argument element that describes the finite elements, which, for example, can be obtained using the ufl_element method of the function space object V.

```
In [48]: f2 = dolfin.Expression("x[0]*x[0] + x[1]*x[1]", degree=1)
In [49]: L2 = f2 * v * dolfin.dx
```

At this point we have defined symbolic representations of the terms that occur in the weak form of the PDE. The next step is to define the boundary conditions. We begin with a simple uniform Dirichlet-type boundary condition. The dolfin library contains a class DirichletBC for representing this type of boundary conditions. We can use this class to represent arbitrary functions along the boundaries of the problem domain, but in this first example, consider the simple boundary condition $u = 0$ on the entire boundary. To represent the constant value on the boundary (zero in this case), we can again use the dolfin.Constant class.

```
In [50]: u0 = dolfin.Constant(0)
```

In addition to the boundary condition value, we also need to define a function (here called u0_boundary) that is used to select different parts of the boundary when creating an instance of the DirichletBC class. This function takes two arguments: a coordinate array x and a flag on_boundary that indicates if a point is on the physical boundary of the mesh, and it should return True if the point x belongs to the boundary and False otherwise. Since this function is evaluated for every vertex in the mesh, by customizing the function, one could pin down the function value at arbitrary parts of the problem domain to specific values or expressions. However, here we only need to select all the points that are on the physical boundary, so we can simply let the u0_boundary function return the on_boundary argument.

```
In [51]: def u0_boundary(x, on_boundary):
    ...:         return on_boundary
```

Once we have an expression for the value on the boundary, u0, and a function for selecting the boundary from the mesh vertices, u0_boundary, we can, with the function space object V, finally create the DirichletBC object:

```
In [52]: bc = dolfin.DirichletBC(V, u0, u0_boundary)
```

This completes the specification of the PDE problem, and our next step is to convert the problem into an algebraic form, by assembling the matrix and vector from the weak form representations of the PDE. We can do this explicitly using the dolfin.assemble function:

```
In [53]: A = dolfin.assemble(a)
In [54]: b = dolfin.assemble(L1)
In [55]: bc.apply(A, b)
```

which results in a matrix A and vector b that define the algebraic equation system for the unknown function. Here we have also used the apply method of the DirichletBC class instance bc, which modifies the A and b objects in such a way that the boundary condition is accounted for in the equations.

To finally solve the problem, we need to create a function object for storing the unknown solution and call the dolfin.solve function, providing the A matrix and the b vector, as well as the underlying data array of a Function object. We can obtain the data array for a Function instance by calling the vector method on the object.

```
In [56]: u_sol1 = dolfin.Function(V)
In [57]: dolfin.solve(A, u_sol1.vector(), b)
```

Here we named the Function object for the solution u_sol1, and the call to dolfin. solve function solves the equation system and fills in the values in the data array of the u_sol1 object. Here we solved the PDE problem by explicitly assembling the A and b matrices and passing the results to the dolfin.solve function. These steps can also be carried out automatically by the dolfin.solve function, by passing a dolfin.Equation object as first argument to the function, the Function object for the solution as second argument, and a boundary condition (or list of boundary conditions) as third argument. We can create the Equation object using, for example, a == L2:

```
In [58]: u_sol2 = dolfin.Function(V)
In [59]: dolfin.solve(a == L2, u_sol2, bc)
```

This is slightly more concise than that of the method we used to find u_sol1 using the equivalence of a == L1, but in some cases when a problem needs to be solved for multiple situations, it can be useful to use explicit assembling of the matrix A and/or the vector b, so it is worthwhile to be familiar with both methods.

With the solution available as a FEniCS Function object, there are a number ways we can proceed with post-processing and visualizing the solution. A straightforward way to plot the solution is to use the built-in dolfin.plot function, which can be used to plot mesh objects, function objects, as well as several other types of objects (see the docstring for dolfin.plot for more information). For example, to plot the solution u_sol2, we simply call dolfin.plot(u_sol2). The resulting graph window is shown in Figure 11-5.

```
In [60]: dolfin.plot(u_sol2)
```

Figure 11-5. *A graph of the mesh function u_sol2, produced by the plot function in the dolfin library*

Using dolfin.plot is a good way of quickly visualizing a solution or a grid, but for better control of the visualization, it is often necessary to export the data and plot it in dedicated visualization software, such as ParaView[1]. To save the solutions u_sol1 and u_sol2 in a format that can be opened with ParaView, we can use the dolfin.File object to generate PVD files (collections of VTK files) and append objects to the file using the << operator, in a C++ stream-like fashion:

```
In [61]: dolfin.File('u_sol1.pvd') << u_sol1
```

---

[1]http://www.paraview.org

We can also add multiple objects to a PVD file using this method:

```
In [62]: f = dolfin.File('u_sol_and_mesh.pvd')
    ...: f << mesh
    ...: f << u_sol1
    ...: f << u_sol2
```

Exporting data for FEniCS objects to files that can be loaded and visualized with external visualization software is a method that benefits from the many advantages of powerful visualization software, such as interactivity, parallel processing, and high level of control of the visualizations, just to mention a few. However, in many cases it might be preferable to work within, for example, the Jupyter Notebook also for visualization of the solutions and the mesh. For relatively simple problems in one, two, and, to some extent, three dimensions, we can use Matplotlib to visualize meshes and solution functions directly. To be able to use Matplotlib, we need to obtain a NumPy array with data corresponding to the FEniCS function object. There are several ways to construct such arrays. To begin with, the FEniCS function object can be called like a function, with an array (list) of coordinate values:

```
In [63]: u_sol1([0.21, 0.67])
Out[63]: 0.0466076997781351
```

This allows us to evaluate the solution at arbitrary points within the problem domain. We can also obtain the values of a function object like u_sol1 at the mesh vertices as a FEniCS vector using the `vector` method, which in turn can be converted to a NumPy array using the `np.array` function. The resulting NumPy arrays are flat (one-dimensional), and for the case of a two-dimensional rectangular mesh (like in the current example), it is sufficient to reshape the flat array to obtain a two-dimensional array that can be plotted with, for example, the `pcolor`, `contour`, or `plot_surface` functions from Matplotlib. In the following we follow these steps to convert the underlying data of the u_sol1 and u_sol2 function objects to NumPy arrays, which then is plotted using Matplotlib. The result is shown in Figure 11-6.

```
In [64]: u_mat1 = np.array(u_sol1.vector()).reshape(N1+1, N2+1)
In [65]: u_mat2 = np.array(u_sol2.vector()).reshape(N1+1, N2+1)
In [66]: X, Y = np.meshgrid(np.linspace(0, 1, N1+2), np.linspace(0, 1, N2+2))
In [67]: fig, (ax1, ax2) = plt.subplots(1, 2, figsize=(12, 5))
    ...:
```

```
...: c = ax1.pcolor(X, Y, u_mat1, cmap=mpl.cm.get_cmap('Reds'))
...: cb = plt.colorbar(c, ax=ax1)
...: ax1.set_xlabel(r"$x$", fontsize=18)
...: ax1.set_ylabel(r"$y$", fontsize=18)
...: cb.set_label(r"$u(x, y)$", fontsize=18)
...: cb.set_ticks([0.0, 0.02, 0.04, 0.06])
...:
...: c = ax2.pcolor(X, Y, u_mat2, cmap=mpl.cm.get_cmap('Reds'))
...: cb = plt.colorbar(c, ax=ax2)
...: ax1.set_xlabel(r"$x$", fontsize=18)
...: ax1.set_ylabel(r"$y$", fontsize=18)
...: cb.set_label(r"$u(x, x)$", fontsize=18)
...: cb.set_ticks([0.0, 0.02, 0.04])
```

Figure 11-6. *The solution of the steady-state heat equation on the unit square, with source terms f = 1 (left) and f = x²+y² (right), subject to the condition that the function u(x, y) is zero on the boundary*

The method used to produce Figure 11-6 is simple and convenient, but it only works for rectangular meshes. For more complicated meshes, the vertex coordinates are not organized in a structural manner, and a simple reshaping of the flat array data is not sufficient. However, the Mesh object that represents the mesh for the problem domain contains a list of the coordinates for each vertex. Together with values from a Function object, these can be combined into a form that can be plotted with Matplotlib triplot and tripcolor functions. To use these plot functions, we first need to create a Triangulation object from the vertex coordinates for the mesh:

```
In [68]: coordinates = mesh.coordinates()
    ...: triangles = mesh.cells()
    ...: triangulation = mpl.tri.Triangulation(coordinates[:, 0],
         coordinates[:, 1], triangles)
```

With the triangulation object defined, we can directly plot the array data for FEniCS functions using `triplot` and `tripcolor`, as shown in the following code. The resulting graph is shown in Figure 11-7.

```
In [69]: fig, (ax1, ax2) = plt.subplots(1, 2, figsize=(10, 4))
    ...: ax1.triplot(triangulation)
    ...: ax1.set_xlabel(r"$x$", fontsize=18)
    ...: ax1.set_ylabel(r"$y$", fontsize=18)
    ...: cmap = mpl.cm.get_cmap('Reds')
    ...: c = ax2.tripcolor(triangulation, np.array(u_sol2.vector()),
         cmap=cmap)
    ...: cb = plt.colorbar(c, ax=ax2)
    ...: ax2.set_xlabel(r"$x$", fontsize=18)
    ...: ax2.set_ylabel(r"$y$", fontsize=18)
    ...: cb.set_label(r"$u(x, y)$", fontsize=18)
    ...: cb.set_ticks([0.0, 0.02, 0.04])
```

Figure 11-7. *The same as Figure 11-6, except that this graph was produced with Matplotlib's triangulation functions. The mesh is plotted to the left and the solution of the PDE to the right*

To see how we can work with more complicated boundary conditions, consider again the heat equation, this time without a source term $u_{xx}+u_{yy} = 0$, but with the following boundary conditions: $u(x = 0) = 3$, $u(x = 1) = -1$, $u(y = 0) = -5$, and $u(y = 1) = 5$. This is the same problem as we solved with the FDM method earlier in this chapter. Here we solve this problem again using FEM. We begin, as in the previous example, by defining a mesh for the problem domain, the function space, and trial and test function objects:

```
In [70]: V = dolfin.FunctionSpace(mesh, 'Lagrange', 1)
In [71]: u = dolfin.TrialFunction(V)
In [72]: v = dolfin.TestFunction(V)
```

Next we define the weak form of the PDE. Here we set $f = 0$ using a dolfin.Constant object to represent f:

```
In [73]: a = dolfin.inner(dolfin.nabla_grad(u), dolfin.nabla_grad(v)) *
             dolfin.dx
In [74]: f = dolfin.Constant(0.0)
In [75]: L = f * v * dolfin.dx
```

Now it remains to define the boundary conditions according to the given specification. In this example we do not want a uniform boundary condition that applies to the entire boundary, so we need to use the first argument to the boundary selection function that is passed to the DirichletBC class, to single out different parts of the boundary. To this end, we define four functions that select the top, bottom, left, and right boundaries:

```
In [76]: def u0_top_boundary(x, on_boundary):
    ...:     # on boundary and y == 1 -> top boundary
    ...:     return on_boundary and abs(x[1]-1) < 1e-5
In [77]: def u0_bottom_boundary(x, on_boundary):
    ...:     # on boundary and y == 0 -> bottom boundary
    ...:     return on_boundary and abs(x[1]) < 1e-5
In [78]: def u0_left_boundary(x, on_boundary):
    ...:     # on boundary and x == 0 -> left boundary
    ...:     return on_boundary and abs(x[0]) < 1e-5
In [79]: def u0_right_boundary(x, on_boundary):
    ...:     # on boundary and x == 1 -> left boundary
    ...:     return on_boundary and abs(x[0]-1) < 1e-5
```

The values of the unknown function at each of the boundaries are simple constants that we can represent with instances of dolfin.Constant. Thus, we can create instances of DirichletBC for each boundary, and the resulting objects are collected in a list bcs:

```
In [80]: bc_t = dolfin.DirichletBC(V, dolfin.Constant(5), u0_top_boundary)
    ...: bc_b = dolfin.DirichletBC(V, dolfin.Constant(-5), u0_bottom_
         boundary)
    ...: bc_l = dolfin.DirichletBC(V, dolfin.Constant(3), u0_left_boundary)
    ...: bc_r = dolfin.DirichletBC(V, dolfin.Constant(-1), u0_right_
         boundary)
In [81]: bcs = [bc_t, bc_b, bc_r, bc_l]
```

With this specification of the boundary conditions, we can continue to solve the PDE problem by calling dolfin.solve. The resulting vector converted to a NumPy array is used for plotting the solution using Matplotlib's pcolor function. The result is shown in Figure 11-8. By comparing to the result from the corresponding FDM computation, shown in Figure 11-2, we can conclude that the two methods indeed give the same results.

```
In [82]: u_sol = dolfin.Function(V)
In [83]: dolfin.solve(a == L, u_sol, bcs)
In [84]: u_mat = np.array(u_sol.vector()).reshape(N1+1, N2+1)
In [85]: x = np.linspace(0, 1, N1+2)
    ...: y = np.linspace(0, 1, N1+2)
    ...: X, Y = np.meshgrid(x, y)
In [86]: fig, ax = plt.subplots(1, 1, figsize=(8, 6))
    ...: c = ax.pcolor(X, Y, u_mat, vmin=-5, vmax=5, cmap=mpl.cm.get_
         cmap('RdBu_r'))
    ...: cb = plt.colorbar(c, ax=ax)
    ...: ax.set_xlabel(r"$x_1$", fontsize=18)
    ...: ax.set_ylabel(r"$x_2$", fontsize=18)
    ...: cb.set_label(r"$u(x_1, x_2)$", fontsize=18)
```

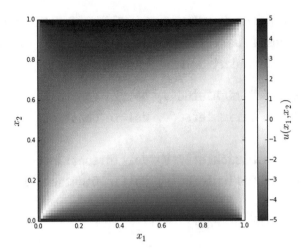

Figure 11-8. *The steady-state solution to the heat equation with different Dirichlet boundary condition on each of the sides of the unit square*

So far we have used FEM to solve the same kind of problems that we also solved with FDM, but the true strength of FEM becomes apparent first when PDE problem with more complicated problem geometries is considered. As an illustration of this, consider the heat equation on a unit circle perforated by five smaller circles, one centered at the origin and the other four smaller circles, as shown in the mesh figure below. To generate meshes for geometries like this one, we can use the mshr library that is distributed with FEniCS. It provides geometric primitives (Point, Circle, Rectangle, etc.) that can be used in algebraic (set) operations to compose mesh for the problem domain of interest. Here we first create a unit circle, centered at $(0, 0)$, using mshr.Circle, and subtract from it other Circle objects corresponding to the part of the mesh that should be removed. The resulting mesh is shown in Figure 11-9.

```
In [87]: r_outer = 1
    ...: r_inner = 0.25
    ...: r_middle = 0.1
    ...: x0, y0 = 0.4, 0.4
In [88]: domain = mshr.Circle(dolfin.Point(.0, .0), r_outer) \
    ...:        - mshr.Circle(dolfin.Point(.0, .0), r_inner) \
    ...:        - mshr.Circle(dolfin.Point( x0,  y0), r_middle) \
    ...:        - mshr.Circle(dolfin.Point( x0, -y0), r_middle) \
    ...:        - mshr.Circle(dolfin.Point(-x0,  y0), r_middle) \
    ...:        - mshr.Circle(dolfin.Point(-x0, -y0), r_middle)
In [89]: mesh = mshr.generate_mesh(domain, 10)
```

392

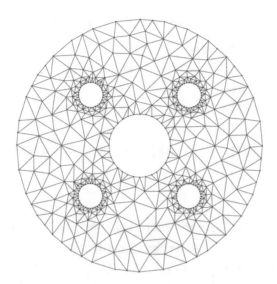

Figure 11-9. *A mesh object generated by the* mshr *library*

A physical interpretation of this mesh is that the geometry is a cross section of five pipes through a block of material, where, for example, the inner pipe carries a hot fluid and the outer pipes a cold fluid for cooling the material block (e.g., an engine cylinder surrounded by cooling pipes). With this interpretation in mind, we set the boundary condition of the inner pipe to a high value, $u_0(x,y)\big|_{x^2+y^2=r^2_{\text{outer}}} = 10$, and the smaller surrounding pipes to a lower value, $u_0(x,y)\big|_{(x-x_0)^2+(y-y_0)^2=r^2_{\text{inner}}} = 0$, where (x_0, y_0) is the center of each of the smaller pipes. We leave the outer boundary unspecified, which is equivalent to the special case of a Neumann boundary condition: $\dfrac{\partial u(\boldsymbol{x})}{\partial \boldsymbol{n}} = 0$. As before,

we define functions for singling out vertices on the boundary. Since we have different boundary conditions on different boundaries, here too we need to use the coordinate argument x to determine which vertices belong to which boundary.

```
In [90]: def u0_inner_boundary(x, on_boundary):
    ...:     x, y = x[0], x[1]
    ...:     return on_boundary and abs(np.sqrt(x**2 + y**2) - r_inner)
             < 5e-2
In [91]: def u0_middle_boundary(x, on_boundary):
    ...:     x, y = x[0], x[1]
    ...:     if on_boundary:
    ...:         for _x0 in [-x0, x0]:
    ...:             for _y0 in [-y0, y0]:
```

```
   ...:                  if abs(np.sqrt((x-_x0)**2 + (y-_y0)**2) - r_
                         middle) < 5e-2:
   ...:                          return True
   ...:      return False
In [92]: bc_inner = dolfin.DirichletBC(V, dolfin.Constant(10), u0_inner_
                  boundary)
   ...: bc_middle = dolfin.DirichletBC(V, dolfin.Constant(0), u0_middle_
                  boundary)
In [93]: bcs = [bc_inner, bc_middle]
```

Once the mesh and boundary conditions are specified, we can proceed as usual with defining the function space and the trial and test functions and constructing the weak form representation of the PDE problem:

```
In [94]: V = dolfin.FunctionSpace(mesh, 'Lagrange', 1)
In [95]: u = dolfin.TrialFunction(V)
In [96]: v = dolfin.TestFunction(V)
In [97]: a = dolfin.inner(dolfin.nabla_grad(u), dolfin.nabla_grad(v)) *
                  dolfin.dx
In [98]: f = dolfin.Constant(0.0)
In [99]: L = f * v * dolfin.dx
In [100]: u_sol = dolfin.Function(V)
```

Solving and visualizing the problem also follows the same pattern as before. The result of plotting the solution is shown in Figure 11-10.

```
In [101]: dolfin.solve(a == L, u_sol, bcs)
In [102]: coordinates = mesh.coordinates()
    ...: triangles = mesh.cells()
    ...: triangulation = mpl.tri.Triangulation(
    ...:     coordinates[:, 0], coordinates[:, 1], triangles)
In [103]: fig, (ax1, ax2) = plt.subplots(1, 2, figsize=(10, 4))
    ...: ax1.triplot(triangulation)
    ...: ax1.set_xlabel(r"$x$", fontsize=18)
    ...: ax1.set_ylabel(r"$y$", fontsize=18)
    ...: c = ax2.tripcolor(
    ...:     triangulation, np.array(u_sol.vector()), cmap=mpl.cm.get_
        cmap("Reds"))
```

```
...: cb = plt.colorbar(c, ax=ax2)
...: ax2.set_xlabel(r"$x$", fontsize=18)
...: ax2.set_ylabel(r"$y$", fontsize=18)
...: cb.set_label(r"$u(x, y)$", fontsize=18)
...: cb.set_ticks([0.0, 5, 10, 15])
```

Figure 11-10. *The solution to the heat equation on a perforated unit circle*

Problems with this kind of geometry are difficult to treat with FDM methods but can be handled with relative ease using FEM. Once we obtain a solution for a FEM problem, even for intricate problem boundaries, we can also with relative ease post-process the solution function in other ways than plotting it. For example, we might be interested in the value of the function along one of the boundaries. For instance, in the current problem, it is natural to look at the temperature along the outer radius of the problem domain, for example, to see how much the exterior temperature of the body decreases due to the four cooling pipes. In order to do this kind of analysis, we need a way of singling out the boundary values from the u_sol object. We can do this by defining an object that describes the boundary (here using dolfin.AutoSubDomain) and applying it to a new Function object that is used as a mask for selecting the desired elements from the u_sol and from mesh.coordinates(). In the following we call this mask function mask_outer.

```
In [104]: outer_boundary = dolfin.AutoSubDomain(
     ...:       lambda x, on_bnd: on_bnd and abs(np.sqrt(x[0]**2 + x[1]**2) -
           r_outer) < 5e-2)
In [105]: bc_outer = dolfin.DirichletBC(V, 1, outer_boundary)
In [106]: mask_outer = dolfin.Function(V)
```

```
In [107]: bc_outer.apply(mask_outer.vector())
In [108]: u_outer = u_sol.vector()[mask_outer.vector() == 1]
In [109]: x_outer = mesh.coordinates()[mask_outer.vector() == 1]
```

With these steps we have created the mask for the outer boundary condition and applied it to u_sol.vector() and mesh.coordinates() and thereby obtained the function values and the coordinates for the outer boundary points. Next we plot the boundary data as a function of the angle between the (x, y) point and the x axis. The result is shown in Figure 11-11.

```
In [110]: phi = np.angle(x_outer[:, 0] + 1j * x_outer[:, 1])
In [111]: order = np.argsort(phi)
In [112]: fig, ax = plt.subplots(1, 1, figsize=(8, 4))
     ...: ax.plot(phi[order], u_outer[order], 's-', lw=2)
     ...: ax.set_ylabel(r"$u(x,y)$ at $x^2+y^2=1$", fontsize=18)
     ...: ax.set_xlabel(r"$\phi$", fontsize=18)
     ...: ax.set_xlim(-np.pi, np.pi)
```

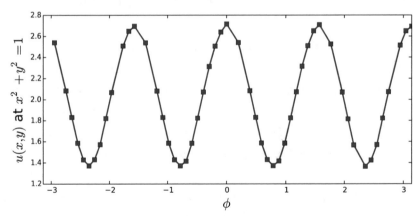

Figure 11-11. *Temperature distribution along the outer boundary of the perforated unit circle*

The accuracy of the solution to a PDE computed with FEM is intimately connected to the element sizes in the mesh that represent the problem domain: a finer mesh gives a more accurate solution. However, increasing the number of elements in the mesh also makes the problem more computationally demanding to solve. Thus, there is a trade-off between the accuracy of the mesh and the available computational resources that must be considered. An important tool for dealing with this trade-off is a mesh with nonuniformly distributed elements. With such a mesh, we can use smaller elements where the unknown function is expected to change in value quickly and fewer elements in less interesting regions. The dolfin library provides a simple way to refine a mesh, using the dolfin.refine function. It takes a mesh as first argument, and if no other arguments are given, it uniformly refines the mesh and returns a new mesh. However, the dolfin.refine function also accepts an optional a second argument that describes which parts of the mesh should be refined. This argument should be an instance of a Boolean-valued dolfin.MeshFunction, which acts as a mask that flags which elements (cells) should be divided. For example, consider a mesh for the unit circle less the part in the quadrant where $x > 0$ and $y < 0$. We can construct a mesh for this geometry using mshr.Circle and mshr.Rectangle:

```
In [113]: domain = mshr.Circle(dolfin.Point(.0, .0), 1.0) \
     ...:          - mshr.Rectangle(dolfin.Point(0.0, -1.0), dolfin.Point(1.0, 0.0))
In [114]: mesh = mshr.generate_mesh(domain, 10)
```

The resulting mesh is shown in the left part of Figure 11-12. It is often desirable to use meshes with finer structure near sharp corners in the geometry. For this example, it is reasonable to attempt to refine the mesh around the edge near the origin. To do this we need to create an instance of dolfin.MeshFunction; initialize all its elements to False, using the set_all method; iterate through the elements and mark those ones in the vicinity of the origin as True; and finally call the dolfin.refine function with the mesh and the MeshFunction instance as arguments. We can do this repeatedly until a sufficiently fine mesh is obtained. In the following we iteratively call dolfin.refine, with a decreasing number of cells marked for splitting:

```
In [115]: refined_mesh = mesh
     ...: for r in [0.5, 0.25]:
     ...:     cell_markers = dolfin.MeshFunction("bool", refined_mesh, dim=2)
     ...:     cell_markers.set_all(False)
     ...:     for cell in dolfin.cells(refined_mesh):
     ...:         if cell.distance(dolfin.Point(.0, .0)) < r:
     ...:             # mark cells within a radius r from the origin to be split
     ...:             cell_markers[cell] = True
     ...:     refined_mesh = dolfin.refine(refined_mesh, cell_markers)
```

The resulting mesh refined_mesh is a version of the original mesh that has finer element partitioning near the origin. The following code plots the two meshes for comparison, and the result is shown in Figure 11-12.

```
In [116]: def mesh_triangulation(mesh):
     ...:       coordinates = mesh.coordinates()
     ...:       triangles = mesh.cells()
     ...:       triangulation = mpl.tri.Triangulation(coordinates[:, 0],
     ...:       coordinates[:, 1],
     ...:       return triangulation
In [117]: fig, (ax1, ax2) = plt.subplots(1, 2, figsize=(8, 4))
     ...:
     ...: ax1.triplot(mesh_triangulation(mesh))
     ...: ax2.triplot(mesh_triangulation(refined_mesh))
     ...:
     ...: # hide axes and ticks
     ...: for ax in [ax1, ax2]:
     ...:       for side in ['bottom','right','top','left']:
     ...:             ax.spines[side].set_visible(False)
     ...:             ax.set_xticks([])
     ...:             ax.set_yticks([])
     ...:             ax.xaxis.set_ticks_position('none')
     ...:             ax.yaxis.set_ticks_position('none')
     ...:
     ...: ax.set_xlabel(r"$x$", fontsize=18)
     ...: ax.set_ylabel(r"$y$", fontsize=18)
```

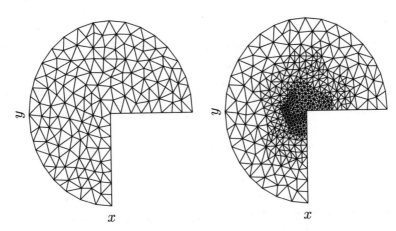

Figure 11-12. *The original and the refined meshes for three-quarters of the unit circle*

Refining a mesh using `dolfin.refine` is a practical technique for improving simple meshes that are constructed using expressions of geometrical primitives, like the one we have used in this chapter. As a final example of using FEniCS, we consider another example of the steady-state heat equation, using this refined mesh for the three-quarters of the unit circle, where we impose Neumann boundary conditions on the vertical and horizontal boundaries along the missing quarter of the unit circle: for the vertical edge, we assume an outflux of heat described by $\nabla u \cdot \boldsymbol{n} = -2$, $x = 0$, $y < 0$, and through the horizontal edge, we assume an influx of heat described by $\nabla u \cdot \boldsymbol{n} = 1$, $x > 0$, $y = 0$, while the outer radial boundary is assumed to be described by the Dirichlet boundary condition $u(x, y) = 0$, $x^2 + y^2 = 1$.

We begin, as usual, by creating objects for the function space, the test function, and the trial function:

```
In [118]: mesh = refined_mesh
In [119]: V = dolfin.FunctionSpace(mesh, 'Lagrange', 1)
In [120]: u = dolfin.TrialFunction(V)
In [121]: v = dolfin.TestFunction(V)
```

For problems with Neumann boundary conditions, we need to include the boundary condition in the weak form of the PDE. Recall that the weak form for the Poisson equation is $\int_{\Omega} \nabla u \cdot \nabla v \, dx = \int_{\Omega} f v \, dx + \int_{\Gamma_N} g v \, d\Gamma$, so compared to the earlier examples, we need to account for the additional term $\int_{\Gamma_N} g v \, d\Gamma$, which is an integral over the boundary with

Neumann boundary condition. To represent the integral measure for this integral in the weak form specification, we can use `dolfin.ds`, but to be able to distinguish different parts of the boundary, we first need to mark the boundary parts. One way to do this in FEniCS is to use a `dolfin.MeshFunction` object and assign to it a unique integer value for each distinct part of the boundary. To do this we first create a `dolfin.MeshFunction` instance:

```
In [122]: boundary_parts = dolfin.MeshFunction("size_t", mesh, mesh.
topology().dim()-1)
```

Next we define a function for selecting boundary points and a `dolfin.AutoSubDomain` instance that is initialized from the boundary selection function. The `AutoSubDomain` instance can then be used to mark the corresponding cells in the `MeshFunction` (here called `boundary_parts`) with an identifying integer value. The following lines of code perform these steps for the vertical edge of the mesh, where $x = 0$ and $y < 0$:

```
In [121]: def v_boundary_func(x, on_boundary):
     ...:      """ the vertical edge of the mesh, where x = 0 and y < 0 """
     ...:      x, y = x[0], x[1]
     ...:      return on_boundary and abs(x) < 1e-4 and y < 0.0
In [122]: v_boundary = dolfin.AutoSubDomain(v_boundary_func)
In [123]: v_boundary.mark(boundary_parts, 0)
```

We repeat the same procedure for the horizontal edge of the mesh, where $y = 0$ and $x > 0$:

```
In [124]: def h_boundary_func(x, on_boundary):
     ...:      """ the horizontal edge of the mesh, where y = 0 and x > 0 """
     ...:      x, y = x[0], x[1]
     ...:      return on_boundary and abs(y) < 1e-4 and x > 0.0
In [125]: h_boundary = dolfin.AutoSubDomain(h_boundary_func)
In [126]: h_boundary.mark(boundary_parts, 1)
```

We can also use the same method to define Dirichlet boundary conditions. Here we mark the part of the boundary that is described by the Dirichlet boundary condition and then use it in the creation of the `dolfin.DirichletBC` object:

```
In [127]: def outer_boundary_func(x, on_boundary):
     ...:      x, y = x[0], x[1]
     ...:      return on_boundary and abs(x**2 + y**2-1) < 1e-2
```

```
In [128]: outer_boundary = dolfin.AutoSubDomain(outer_boundary_func)
In [129]: outer_boundary.mark(boundary_parts, 2)
In [130]: bc = dolfin.DirichletBC(V, dolfin.Constant(0.0), boundary_parts, 2)
```

Once the boundaries are marked, we can proceed to create the weak form of the PDE. Since we use partitioned boundary here, we need to specify the domain and subdomain arguments to the integral measures dolfin.dx and dolfin.ds, using the mesh and boundary_parts objects.

```
In [131]: dx = dolfin.dx(domain=mesh, subdomain_data=boundary_parts)
In [132]: a = dolfin.inner(dolfin.nabla_grad(u), dolfin.nabla_grad(v)) * dx
In [133]: f = dolfin.Constant(0.0)
In [134]: g_v = dolfin.Constant(-2.0)
In [135]: g_h = dolfin.Constant(1.0)
In [136]: L = f * v * dolfin.dx(domain=mesh, subdomain_data=boundary_parts)
In [137]: L += g_v * v * dolfin.ds(0, domain=mesh, subdomain_data=boundary_parts)
In [138]: L += g_h * v * dolfin.ds(1, domain=mesh, subdomain_data=boundary_parts)
```

In the last two code cells, we have added new terms for the Neumann boundary conditions for the vertical and the horizontal edges in the mesh. These parts of the boundary are denoted by integers 0 and 1, respectively, as defined in the preceding section, and these integers are passed as an argument to the dolfin.ds to select integration over different parts of the boundaries.

```
In [139]: u_sol = dolfin.Function(V)
In [140]: dolfin.solve(a == L, u_sol, bc)
```

Once the representation of the weak form of the PDE is defined, we can go ahead and solve the problem using dolfin.solve, as we have done in earlier examples. Finally we plot the solution using Matplotlib's triangulation plot functions. The results are shown in Figure 11-13. From the graph we can see that, as expected, the solution has more structure near the edge at the origin. Using a mesh with smaller elements in this region is therefore a good way to obtain sufficient resolution in this region without inflicting excessive computational cost by using a uniformly fine-structured mesh.

```
In [141]: fig, (ax1, ax2) = plt.subplots(1, 2, figsize=(10, 4))
     ...: triangulation = mesh_triangulation(mesh)
     ...: ax1.triplot(triangulation)
     ...: ax1.set_xlabel(r"$x$", fontsize=18)
     ...: ax1.set_ylabel(r"$y$", fontsize=18)
     ...:
     ...: data = np.array(u_sol.vector())
     ...: norm = mpl.colors.Normalize(-abs(data).max(), abs(data).max())
     ...: c = ax2.tripcolor(triangulation, data, norm=norm, cmap=mpl.
          cm.get_cmap("RdBu_r"))
     ...: cb = plt.colorbar(c, ax=ax2)
     ...: ax2.set_xlabel(r"$x$", fontsize=18)
     ...: ax2.set_ylabel(r"$y$", fontsize=18)
     ...: cb.set_label(r"$u(x, y)$", fontsize=18)
     ...: cb.set_ticks([-.5, 0, .5])
```

Figure 11-13. *Solution to the heat equation on a quarter of the unit circle with Neumann and Dirichlet boundary conditions*

The examples we have explored in this section are merely a few simple demonstrations of the types of problems that the FEniCS framework can be used for. There are a vast number of features in FEniCS that we have not even been able to mention here. For the reader who is particularly interested in solving PDE problems, I recommend studying the FEniCS book (Anders Logg, 2012) and the many example applications that it contains. In particular, important aspects of solving PDEs with FEM that we have not been able to discuss here are nontrivial Neumann boundary conditions (which need to be included in the formulation of weak form of the PDE), PDEs for vector-valued functions, higher-dimensional PDE problems (e.g., the heat equation in three dimensions), and time-dependent PDE problems. All of these topics, and many others, are well supported in the FEniCS framework.

Summary

In this chapter we briefly surveyed methods for solving partial differential equations (PDEs) and how these methods can be employed within the scientific Python environment. Specifically, we introduced the finite-difference method (FDM) and the finite-element method (FEM) for solving PDE problems and used these methods to solve several example problems. The advantage of FDM is its simplicity, and for problems where it is easily applicable (simple problem domains, uniform discretization, etc.), it is a very practical method. For more complicated PDE problems, for example, where the problem domain is more complex, FEM is generally more suitable. However, the mathematical theory of the FEM is more involved, and the implementation is far more technical. While there are a number of advanced FEM frameworks that can be used from Python, in this chapter we focused on one prominent example: the FEniCS framework. FEniCS is a full-featured FEM software that can be used for a wide range of PDE problem. With the examples considered here, we have only scraped the surface of what can be achieved with the software. However, the hope is that the examples studied in this chapter give a general sense of the workflow when solving PDE problems with FEM and when using the FEniCS software in particular.

Further Reading

While we have discussed FDM and FEM in this chapter, there are also other successful and useful methods for numerically solving PDEs. For instance, the finite-volume method (FVM) is a variant of the FEM method that is often used for fluid dynamics calculations, as well as in other fields. The Python library FiPy provides a framework for solving PDE problems using this method, and a theoretical introduction to the method is given in, for example, Wesseling (2009). The theoretical background information about the FDM and FEM that is given in this chapter is very brief indeed, and it merely serves to introduce the terminology and notation used here. For serious work with the

FDM, and in particular the FEM method, it is important to thoroughly understand the fundamentals of these methods. Good introductions to FDM and FEM are given in, for example, Gockenbach (2011), Gockenbach (2006), Johnson (2009), and LeVeque (2007). The FEniCS book (Logg, 2012), which is available for free online from the FEniCS project's web site (`http://fenicsproject.org`), also contains a nice introduction to the FEM method, in addition to a detailed documentation of the FEniCS software itself.

References

Anders Logg, K.-A. M. (2012). Automated Solution of Differential Equations by the Finite Element Method. Springer.

Gockenbach, M. (2011). Partial Differential Equations. Philadelphia: SIAM.

Gockenbach, M. (2006). Understanding And Implementing the Finite Element Method. Philadelphia: SIAM.

Johnson, C. (2009). Numerical Solution of Partial Differential Equations by the Finite Element Method. Cambridge: Dover.

LeVeque, R. (2007). Finite Difference Methods for Ordinary and Partial Differential Equations: Steady-State and Time-Dependent Problems. Philadelphia: SIAM.

Wesseling, P. (2009). Principles of Computational Fluid Dynamics. Berlin: Springer.

CHAPTER 12

Data Processing and Analysis

In the last several chapters, we have covered the main topics of traditional scientific computing. These topics provide a foundation for most computational work. Starting with this chapter, we move on to explore data processing and analysis, statistics, and statistical modeling. As the first step in this direction, we look at the data analysis library pandas. This library provides convenient data structures for representing series and tables of data and makes it easy to transform, split, merge, and convert data. These are important steps in the process[1] of cleansing raw data into a tidy form that is suitable for analysis. The Pandas library builds on top of NumPy and complements it with features that are particularly useful when handling data, such as labeled indexing, hierarchical indices, alignment of data for comparison and merging of datasets, handling of missing data, and much more. As such, the pandas library has become a de facto standard library for high-level data processing in Python, especially for statistics applications. The pandas library itself contains only limited support for statistical modeling (namely, linear regression). For more involved statistical analysis and modeling, there are other packages available, such as statsmodels, patsy, and scikit-learn, which we cover in later chapters. However, also for statistical modeling with these packages, pandas can still be used for data representation and preparation. The pandas library is therefore a key component in the software stack for data analysis with Python.

---

[1]Also known as data munging or data wrangling

© Robert Johansson 2019
R. Johansson, *Numerical Python*, https://doi.org/10.1007/978-1-4842-4246-9_12

Pandas   The pandas library is a framework for data processing and analysis in Python. At the time of writing, the most recent version of Pandas is 0.23.4. For more information about the pandas library, and its official documentation, see the project's web site at `http://pandas.pydata.org`.

The main focus of this chapter is to introduce basic features and usage of the pandas library. Toward the end of the chapter, we also briefly explore the statistical visualization library Seaborn, which is built on top of Matplotlib. This library provides quick and convenient graphing of data represented as pandas data structure (or NumPy arrays). Visualization is a very important part of exploratory data analysis, and the Pandas library itself also provides functions for basic data visualization (which also builds on top of Matplotlib). The Seaborn library takes this further, by providing additional statistical graphing capabilities and improved styling: the Seaborn library is notable for generating good-looking graphics using default settings.

Seaborn   *The Seaborn library is a visualization library for statistical graphics. It builds on Matplotlib and provides easy-to-use functions for common statistical graphs. At the time of writing, the most recent version of Seaborn is 0.8.1. For more information about Seaborn, and its official documentation, see the project's web site at:* `http://stanford.edu/~mwaskom/software/seaborn`.

Importing Modules

In this chapter we mainly work with the pandas library, which we assume is imported under the name pd:

```
In [1]: import pandas as pd
```

We also require NumPy and Matplotlib, which we import as usual in the following way:

```
In [2]: import numpy as np
In [3]: import matplotlib.pyplot as plt
```

For more aesthetically pleasing appearance of Matplotlib figures produced by the pandas library, we select a style that is suitable for statistical graphs using the function `mpl.style.use`:

```
In [4]: import matplotlib as mpl
   ...: mpl.style.use('ggplot')
```

Later in this chapter, we will also require to import the seaborn module, which we will import under the name sns:

```
In [5]: import seaborn as sns
```

Introduction to Pandas

The main focus of this chapter is the pandas library for data analysis, and we begin here with an introduction to this library. The pandas library mainly provides data structures and methods for representing and manipulating data. The two main data structures in Pandas are the Series and DataFrame objects, which are used to represent data series and tabular data, respectively. Both of these objects have an index for accessing elements or rows in the data represented by the object. By default, the indices are integers starting from zero, like NumPy arrays, but it is also possible to use any sequence of identifiers as index.

Series

The merit of being able to index a data series with labels rather than integers is apparent even in the simplest of examples: Consider the following construction of a Series object. We give the constructor a list of integers, to create a Series object that represents the given data. Displaying the object in IPython reveals the data of the Series object together with the corresponding indices:

```
In [6]: s = pd.Series([909976, 8615246, 2872086, 2273305])
In [7]: s
Out[7]: 0    909976
        1    8615246
        2    2872086
        3    2273305
        dtype: int64
```

The resulting object is a `Series` instance with the data type (dtype) `int64`, and the elements are indexed by the integers 0, 1, 2, and 3. Using the `index` and `values` attributes, we can extract the underlying data for the index and the values stored in the series:

```
In [8]: list(s.index)
Out[8]: RangeIndex(start=0, stop=4, step=1)
In [9]: s.values
Out[9]: array([ 909976, 8615246, 2872086, 2273305], dtype=int64)
```

While using integer-indexed arrays or data series is a fully functional representation of the data, it is not descriptive. For example, if the data represents the population of four European capitals, it is convenient and descriptive to use the city names as indices rather than integers. With a `Series` object this is possible, and we can assign the `index` attribute of a `Series` object to a list with new indices to accomplish this. We can also set the `name` attribute of the Series object, to give it a descriptive name:

```
In [10]: s.index = ["Stockholm", "London", "Rome", "Paris"]
In [11]: s.name = "Population"
In [12]: s
Out[12]: Stockholm      909976
         London        8615246
         Rome          2872086
         Paris         2273305
         Name: Population, dtype: int64
```

It is now immediately obvious what the data represents. Alternatively, we can also set the `index` and `name` attributes through keyword arguments to the `Series` object when it is created:

```
In [13]: s = pd.Series([909976, 8615246, 2872086, 2273305], name="Population",
    ...:                 index=["Stockholm", "London", "Rome", "Paris"])
```

While it is perfectly possible to store the data for the populations of these cities directly in a NumPy array, even in this simple example, it is much clearer what the data represent when the data points are indexed with meaningful labels. The benefits of bringing the description of the data closer to the data are even greater when the complexity of the dataset increases.

We can access elements in a Series by indexing with the corresponding index (label) or directly through an attribute with the same name as the index (if the index label is a valid Python symbol name):

```
In [14]: s["London"]
Out[14]: 8615246
In [15]: s.Stockholm
Out[15]: 909976
```

Indexing a Series object with a list of indices gives a new Series object with a subset of the original data (corresponding to the provided list of indices):

```
In [16]: s[["Paris", "Rome"]]
Out[16]: Paris    2273305
         Rome     2872086
         Name: Population, dtype: int64
```

With a data series represented as a Series object, we can easily compute its descriptive statistics using the Series methods count (the number of data points), median (calculate the median), mean (calculate the mean value), std (calculate the standard deviation), min and max (minimum and maximum values), and the quantile (for calculating quantiles):

```
In [17]: s.median(), s.mean(), s.std()
Out[17]: (2572695.5, 3667653.25, 3399048.5005155364)
In [18]: s.min(), s.max()
Out[18]: (909976, 8615246)
In [19]: s.quantile(q=0.25), s.quantile(q=0.5), s.quantile(q=0.75)
Out[19]: (1932472.75, 2572695.5, 4307876.0)
```

All of the preceding data are combined in the output of the describe method, which provides a summary of the data represented by a Series object:

```
In [20]: s.describe()
Out[20]: count           4.000000
         mean      3667653.250000
         std       3399048.500516
         min        909976.000000
         25%       1932472.750000
```

```
50%        2572695.500000
75%        4307876.000000
max        8615246.000000
Name: Population, dtype: float64
```

Using the plot method, we can quickly and easily produce graphs that visualize the data in a Series object. The pandas library uses Matplotlib for plotting, and we can optionally pass a Matplotlib Axes instance to the plot method via the ax argument. The type of the graph is specified using the kind argument (valid options are line, hist, bar, barh, box, kde, density, area, and pie).

```
In [21]: fig, axes = plt.subplots(1, 4, figsize=(12, 3))
    ...: s.plot(ax=axes[0], kind='line', title='line')
    ...: s.plot(ax=axes[1], kind='bar', title='bar')
    ...: s.plot(ax=axes[2], kind='box', title='box')
    ...: s.plot(ax=axes[3], kind='pie', title='pie')
```

Figure 12-1. *Examples of plot styles that can be produced with Pandas using the* Series.plot *method*

DataFrame

As we have seen in the previous examples, a pandas Series object provides a convenient container for one-dimensional arrays, which can use descriptive labels for the elements and which provides quick access to descriptive statistics and visualization. For higher-dimensional arrays (mainly two-dimensional arrays, or tables), the corresponding data structure is the Pandas DataFrame object. It can be viewed as a collection of Series objects with a common index.

410

There are numerous ways to initialize a DataFrame. For simple examples, the easiest way is to pass a nested Python list or dictionary to the constructor of the DataFrame object. For example, consider an extension of the dataset we used in the previous section, where, in addition to the population of each city, we also include a column that specifies which state each city belongs to. We can create the corresponding DataFrame object in the following way:

```
In [22]: df = pd.DataFrame([[909976, "Sweden"],
   ...:                      [8615246, "United Kingdom"],
   ...:                      [2872086, "Italy"],
   ...:                      [2273305, "France"]])
In [23]: df
Out[23]:
```

	0	1
0	909976	Sweden
1	8615246	United Kingdom
2	2872086	Italy
3	2273305	France

The result is a tabular data structure with rows and columns. Like with a Series object, we can use labeled indexing for rows by assigning a sequence of labels to the index attribute, and, in addition, we can set the columns attribute to a sequence of labels for the columns:

```
In [24]: df.index = ["Stockholm", "London", "Rome", "Paris"]
In [25]: df.columns = ["Population", "State"]
In [26]: df
Out[26]:
```

	Population	State
Stockholm	909976	Sweden
London	8615246	United Kingdom
Rome	2872086	Italy
Paris	2273305	France

The index and columns attributes can also be set using the corresponding keyword arguments to the DataFrame object when it is created:

```
In [27]: df = pd.DataFrame([[909976, "Sweden"],
    ...:                     [8615246, "United Kingdom"],
    ...:                     [2872086, "Italy"],
    ...:                     [2273305, "France"]],
    ...:                     index=["Stockholm", "London", "Rome", "Paris"],
    ...:                     columns=["Population", "State"])
```

An alternative way to create the same data frame, which sometimes can be more convenient, is to pass a dictionary with column titles as keys and column data as values:

```
In [28]: df = pd.DataFrame({"Population": [909976, 8615246, 2872086, 2273305],
    ...:                     "State": ["Sweden", "United Kingdom", "Italy",
    ...:                     "France"]},
    ...:                     index=["Stockholm", "London", "Rome", "Paris"])
```

As before, the underlying data in a DataFrame can be obtained as a NumPy array using the values attribute and the index and column arrays through the index and columns attributes, respectively. Each column in a data frame can be accessed using the column name as attribute (or, alternatively, by indexing with the column label, e.g., df["Population"]):

```
In [29]: df.Population
Out[29]: Stockholm      909976
         London        8615246
         Rome          2872086
         Paris         2273305
         Name: Population, dtype: int64
```

The result of extracting a column from a DataFrame is a new Series object, which we can process and manipulate with the methods discussed in the previous section. Rows of a DataFrame instance can be accessed using the loc indexer attribute. Indexing this attribute also results in a Series object, which corresponds to a row of the original data frame:

```
In [30]: df.loc["Stockholm"]
Out[30]: Population     909976
         State          Sweden
         Name: Stockholm, dtype: object
```

Passing a list of row labels to the `loc` indexer results in a new `DataFrame` that is a subset of the original `DataFrame`, containing only the selected rows:

```
In [31]: df.loc[["Paris", "Rome"]]
Out[31]:
```

	Population	State
Paris	2273305	France
Rome	2872086	Italy

The loc indexer can also be used to select both rows and columns simultaneously, by first passing a row label (or a list thereof) and second a column label (or a list thereof). The result is a `DataFrame`, a `Series`, or an element value, depending on the number of columns and rows that are selected:

```
In [32]: df.loc[["Paris", "Rome"], "Population"]
Out[32]: Paris     2273305
         Rome      2872086
         Name: Population, dtype: int64
```

We can compute descriptive statistics using the same methods as we already used for Series objects. When invoking those methods (`mean`, `std`, `median`, `min`, `max`, etc.) for a DataFrame, the calculation is performed for each column with numerical data types:

```
In [33]: df.mean()
Out[33]: Population     3667653.25
         dtype: float64
```

In this case, only one of the two columns has a numerical data type (the one named Population). Using the DataFrame method `info` and the attribute `dtypes`, we can obtain a summary of the content in a DataFrame and the data types of each column:

```
In [34]: df.info()
<class 'pandas.core.frame.DataFrame'>
Index: 4 entries, Stockholm to Paris
Data columns (total 2 columns):
Population     4 non-null int64
State          4 non-null object
dtypes: int64(1), object(1)
```

```
memory usage: 96.0+ bytes
In [35]: df.dtypes
Out[35]: Population        int64
         State           object
         dtype: object
```

The real advantages of using pandas emerge when dealing with larger and more complex datasets than the examples we have used so far. Such data can rarely be defined as explicit lists or dictionaries, which can be passed to the DataFrame initializer. A more common situation is that the data must be read from a file or some other external sources. The pandas library supports numerous methods for reading data from files of different formats. Here we use the read_csv function to read in data and create a DataFrame object from a CSV file.[2] This function accepts a large number of optional arguments for tuning its behavior. See the docstring help(pd.read_csv) for details. Some of the most useful arguments are header (specifies which row, if any, contains a header with column names), skiprows (number of rows to skip before starting to read data, or a list of line numbers of lines to skip), delimiter (the character that is used as a delimiter between column values), encoding (the name of the encoding used in the file, e.g., utf-8), and nrows (number of rows to read). The first and only mandatory argument to the pd.read_csv function is a filename or a URL to the data source. For example, to read in a dataset stored in a file called european_cities.csv,[3] of which the first five lines are shown in the following code, we can simply call pd.read_csv("european_cities.csv"), since the default delimiter is "," and the header is by default taken from the first line. However, we could also write out all these options explicitly:

```
In [36]: !head -n 5 european_cities.csv
Rank,City,State,Population,Date of census
1,London, United Kingdom,"8,615,246",1 June 2014
2,Berlin, Germany,"3,437,916",31 May 2014
3,Madrid, Spain,"3,165,235",1 January 2014
4,Rome, Italy,"2,872,086",30 September 2014
In [37]: df_pop = pd.read_csv("european_cities.csv",
    ...:                          delimiter=",", encoding="utf-8", header=0)
```

---

[2]CSV, or comma-separated values, is a common text format where rows are stored in lines and columns are separated by a comma (or some other text delimiter). See Chapter 18 for more details about this and other file formats.

[3]This dataset was obtained from the Wiki page: http://en.wikipedia.org/wiki/Largest_cities_of_the_European_Union_by_population_within_city_limits

This dataset is similar to the example data we used earlier in this chapter, but here there are additional columns and many more rows for other cities. Once a dataset is read into a DataFrame object, it is useful to start by inspecting the summary given by the info method, to begin forming an idea of the properties of the dataset.

```
In [38]: df_pop.info()
<class 'pandas.core.frame.DataFrame'>
Int64Index: 105 entries, 0 to 104
Data columns (total 5 columns):
Rank             105 non-null int64
City             105 non-null object
State            105 non-null object
Population       105 non-null object
Date of census   105 non-null object
dtypes: int64(1), object(4) memory usage: 4.9+ KB
```

Here we see that there are 105 rows in this dataset and that it has five columns. Only the Rank column is of a numerical data type. In particular, the Population column is not yet of numeric data type because its values are of the format "8,615,246" and is therefore interpreted as string values by the read_csv function. It is also informative to display a tabular view of the data. However, this dataset is too large to display in full, and in situations like this, the head and tail methods are handy for creating a truncated dataset containing the first few and last few rows, respectively. Both of these functions take an optional argument that specifies how many rows to include in the truncated DataFrame. Note also that df.head(n) is equivalent to df[:n], where n is an integer.

```
In [39]: df_pop.head()
Out[39]:
```

	Rank	City	State	Population	Date of census
0	1	London	United Kingdom	8,615,246	1 June 2014
1	2	Berlin	Germany	3,437,916	31 May 2014
2	3	Madrid	Spain	3,165,235	1 January 2014
3	4	Rome	Italy	2,872,086	30 September 2014
4	5	Paris	France	2,273,305	1 January 2013

Displaying a truncated DataFrame gives a good idea of how the data look like and what remains to be done before the data is ready for analysis. It is common to have to transform columns in one way or another and to reorder the table by sorting by a specific column or by ordering the index. In the following we explore some methods for modifying DataFrame objects. First of all, we can create new columns and update columns in a DataFrame simply by assigning a Series object to the DataFrame indexed by the column name, and we can delete columns using the Python del keyword.

The apply method is a powerful tool to transform the content in a column. It creates and returns a new Series object for which a function passed to apply has been applied to each element in the original column. For example, we can use the apply method to transform the elements in the Population column from strings to integers, by passing a lambda function that removes the "," characters from the strings and casts the results to an integer. Here we assign the transformed column to a new column with the name NumericPopulation. Using the same method, we also tidy up the State values by removing extra white spaces in its elements using the string method strip.

```
In [40]: df_pop["NumericPopulation"] = df_pop.Population.apply(
    ...:        lambda x: int(x.replace(",", "")))
In [41]: df_pop["State"].values[:3]   # contains extra white spaces
Out[41]: array([' United Kingdom', ' Germany', ' Spain'], dtype=object)
In [42]: df_pop["State"] = df_pop["State"].apply(lambda x: x.strip())
In [43]: df_pop.head()
Out[43]:
```

	Rank	City	State	Population	Date of census	NumericPopulation
0	1	London	United Kingdom	8,615,246	1 June 2014	8615246
1	2	Berlin	Germany	3,437,916	31 May 2014	3437916
2	3	Madrid	Spain	3,165,235	1 January 2014	3165235
3	4	Rome	Italy	2,872,086	30 September 2014	2872086
4	5	Paris	France	2,273,305	1 January 2013	2273305

Inspecting the data types of the columns in the updated DataFrame confirms that the new column NumericPopulation is indeed of integer type (while the Population column is unchanged):

```
In [44]: df_pop.dtypes
Out[44]: Rank                 int64
         City                 object
         State                object
         Population           object
         Date of census       object
         NumericPopulation    int64
         dtype: object
```

We may also need to change the index to one of the columns of the DataFrame. In the current example, we may want to use the City column as index. We can accomplish this using the set_index method, which takes as argument the name of the column to use as index. The result is a new DataFrame object, and the original DataFrame is unchanged. Furthermore, using the sort_index method, we can sort the data frame with respect to the index:

```
In [45]: df_pop2 = df_pop.set_index("City")
In [46]: df_pop2 = df_pop2.sort_index()
In [47]: df_pop2.head()
Out[47]:
```

City	Rank	State	Population	Date of census	NumericPopulation
Aarhus	92	Denmark	326,676	1 October 2014	326676
Alicante	86	Spain	334,678	1 January 2012	334678
Amsterdam	23	Netherlands	813,562	31 May 2014	813562
Antwerp	59	Belgium	510,610	1 January 2014	510610
Athens	34	Greece	664,046	24 May 2011	664046

The sort_index method also accepts a list of column names, in which case a hierarchical index is created. A hierarchical index uses tuples of index labels to address rows in the data frame. We can use the sort_index method with the integer-valued argument level, to sort the rows in a DataFrame according to the nth level of the hierarchical index, where level=n. In the following example, we create a hierarchical index with State and City as indices, and we use the sort_index method to sort by the first index (State):

```
In [48]: df_pop3 = df_pop.set_index(["State", "City"]).sort_index(level=0)
In [49]: df_pop3.head(7)
Out[49]:
```

State	City	Rank	Population	Date of census
Austria	Vienna	7	1794770	1 January 2015
Belgium	Antwerp	59	510610	1 January 2014
	Brussels	16	1175831	1 January 2014
Bulgaria	Plovdiv	84	341041	31 December 2013
	Sofia	14	1291895	14 December 2014
	Varna	85	335819	31 December 2013
Croatia	Zagreb	24	790017	31 March 2011

A DataFrame with a hierarchical index can be partially indexed using only its zeroth-level index (df3.loc["Sweden"]) or completely indexed using a tuple of all hierarchical indices (df3.loc[("Sweden", "Gothenburg")]):

```
In [50]: df_pop3.loc["Sweden"]
Out[50]:
```

City	Rank	Population	Date of census	NumericPopulation
Gothenburg	53	528,014	31 March 2013	528014
Malmö	102	309,105	31 March 2013	309105
Stockholm	20	909,976	31 January 2014	909976

```
In [51]: df_pop3.loc[("Sweden", "Gothenburg")]
Out[51]: Rank                              53
         Population                   528,014
         Date of census         31 March 2013
         NumericPopulation             528014
         Name: (Sweden, Gothenburg), dtype: object
```

If we want to sort by a column rather than the index, we can use the sort_values method. It takes a column name, or a list of column names, with respect to which the DataFrame is to be sorted. It also accepts the keyword argument ascending, which is a Boolean or a list of Boolean values that specifies whether the corresponding column is to be sorted in ascending or descending order:

```
In [52]: df_pop.set_index("City").sort_values(["State", "NumericPopulation"],
    ...:                                       ascending=[False, True]).head()
Out[52]:
```

City	Rank	State	Population	Date of census	NumericPopulation
Nottingham	103	United Kingdom	308,735	30 June 2012	308735
Wirral	97	United Kingdom	320,229	30 June 2012	320229
Coventry	94	United Kingdom	323,132	30 June 2012	323132
Wakefield	91	United Kingdom	327,627	30 June 2012	327627
Leicester	87	United Kingdom	331,606	30 June 2012	331606

With categorical data such as the State column, it is frequently of interest to summarize how many values of each category a column contains. Such counts can be computed using the value_counts method (of the Series object). For example, to count the number of cities each country has on the list of the 105 largest cities in Europe, we can use:

```
In [53]: city_counts = df_pop.State.value_counts()
In [54]: city_counts.head()
Out[54]: Germany           19
         United Kingdom    16
         Spain             13
         Poland            10
         Italy             10
         dtype: int64
```

In this example, we see from the results that the state with the largest number of cities in the list is Germany, with 19 cities, followed by the United Kingdom with 16 cities, and so on. A related question is how large the total population of all cities within a state is. To answer this type of question, we can proceed in two ways: First, we can create a hierarchical index using State and City and use the sum method to reduce the DataFrame along one of the indices. In this case, we want to sum over all entries within the index level State, so we can use sum(level="State"), which eliminates the City index. For presentation we also sort the resulting DataFrame in descending order of the column NumericPopulation:

```
In [55]: df_pop3 = df_pop[["State", "City", "NumericPopulation"]].set_
         index(["State", "City"])
In [56]: df_pop4 = df_pop3.sum(level="State").sort_
         values("NumericPopulation", ascending=False)
In [57]: df_pop4.head()
Out[57]:
```

State	NumericPopulation
United Kingdom	16011877
Germany	15119548
Spain	10041639
Italy	8764067
Poland	6267409

Second, we can obtain the same results using the groupby method, which allows us to group rows of a DataFrame by the values of a given column, and apply a reduction function on the resulting object (e.g., sum, mean, min, max, etc.). The result is a new DataFrame with the grouped-by column as index. Using this method we can compute the total population of the 105 cities, grouped by state, in the following way:

```
In [58]: df_pop5 = (df_pop.drop("Rank", axis=1)
   ...:                     .groupby("State").sum()
   ...:                     .sort_values("NumericPopulation",
                              ascending=False))
```

Note that here we also used the drop method to remove the Rank column (hence the axis=1, use axis=0 to drop rows) from the DataFrame (since it is not meaningful to aggregate the rank by summation). Finally, we use the plot method of the Series object to plot bar graphs for the city count and the total population. The results are shown in Figure 12-2.

```
In [59]: fig, (ax1, ax2) = plt.subplots(1, 2, figsize=(12, 4))
   ...: city_counts.plot(kind='barh', ax=ax1)
   ...: ax1.set_xlabel("# cities in top 105")
   ...: df_pop5.NumericPopulation.plot(kind='barh', ax=ax2)
   ...: ax2.set_xlabel("Total pop. in top 105 cities")
```

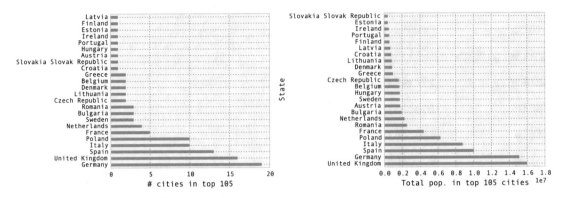

Figure 12-2. *The number of cities in the list of the top 105 most populated cities in Europe (left) and the total population in those cities (right), grouped by state*

Time Series

Time series are a common form of data in which a quantity is given, for example, at regularly or irregularly spaced timestamps or for fixed or variable time spans (periods). In pandas, there are dedicated data structures for representing these types of data. Series and DataFrame can have both columns and indices with data types describing timestamps and time spans. When dealing with temporal data, it is particularly useful to be able to index the data with time data types. Using pandas time-series indexers, DatetimeIndex and PeriodIndex, we can carry out many common date, time, period, and calendar operations, such as selecting time ranges and shifting and resampling of the data points in a time series.

To generate a sequence of dates that can be used as an index in a pandas Series or DataFrame objects, we can, for example, use the date_range function. It takes the starting point as a date and time string (or, alternatively, a datetime object from the Python standard library) as a first argument, and the number of elements in the range can be set using the periods keyword argument:

```
In [60]: pd.date_range("2015-1-1", periods=31)
Out[60]: <class 'pandas.tseries.index.DatetimeIndex'>
         [2015-01-01, ..., 2015-01-31]
         Length: 31, Freq: D, Timezone: None
```

To specify the frequency of the timestamps (which defaults to one day), we can use the freq keyword argument, and instead of using periods to specify the number of points, we can give both starting and ending points as date and time strings (or datetime objects) as the first and second arguments. For example, to generate hourly timestamps between 00:00 and 12:00 on 2015-01-01, we can use:

```
In [61]: pd.date_range("2015-1-1 00:00", "2015-1-1 12:00", freq="H")
Out[61]: <class 'pandas.tseries.index.DatetimeIndex'>
         [2015-01-01 00:00:00, ..., 2015-01-01 12:00:00]
         Length: 13, Freq: H, Timezone: None
```

The date_range function returns an instance of DatetimeIndex, which can be used, for example, as an index for a Series or DataFrame object:

```
In [62]: ts1 = pd.Series(np.arange(31), index=pd.date_range("2015-1-1",
             periods=31))
In [63]: ts1.head()
Out[63]: 2015-01-01    0
         2015-01-02    1
         2015-01-03    2
         2015-01-04    3
         2015-01-05    4
         Freq: D, dtype: int64
```

The elements of a DatetimeIndex object can, for example, be accessed using indexing with date and time strings. An element in a DatetimeIndex is of the type Timestamp, which is a pandas object that extends the standard Python datetime object (see the datetime module in the Python standard library).

```
In [64]: ts1["2015-1-3"]
Out[64]: 2
In [65]: ts1.index[2]
Out[65]: Timestamp('2015-01-03 00:00:00', offset='D')
```

In many aspects, a Timestamp and datetime object are interchangeable, and the Timestamp class has, like the datetime class, attributes for accessing time fields such as year, month, day, hour, minute, and so on. However, a notable difference between Timestamp and datetime is that Timestamp stores a timestamp with nanosecond resolution, while a datetime object only uses microsecond resolution.

```
In [66]: ts1.index[2].year, ts1.index[2].month, ts1.index[2].day
Out[66]: (2015, 1, 3)
In [67]: ts1.index[2].nanosecond
Out[67]: 0
```

We can convert a Timestamp object to a standard Python datetime object using the to_pydatetime method:

```
In [68]: ts1.index[2].to_pydatetime()
Out[68]: datetime.datetime(2015, 1, 3, 0, 0)
```

and we can use a list of datetime objects to create pandas time series:

```
In [69]: import datetime
In [70]: ts2 = pd.Series(np.random.rand(2),
    ...:                      index=[datetime.datetime(2015, 1, 1), datetime.
                             datetime(2015, 2, 1)])
In [71]: ts2
Out[71]: 2015-01-01    0.683801
         2015-02-01    0.916209
         dtype: float64
```

Data that is defined for sequences of time spans can be represented using Series and DataFrame objects that are indexed using the PeriodIndex class. We can construct an instance of the PeriodIndex class explicitly by passing a list of Period objects and then specify it as index when creating a Series or DataFrame object:

```
In [72]: periods = pd.PeriodIndex([pd.Period('2015-01'),
    ...:                            pd.Period('2015-02'),
    ...:                            pd.Period('2015-03')])
In [73]: ts3 = pd.Series(np.random.rand(3), index=periods)
In [74]: ts3
Out[74]: 2015-01    0.969817
         2015-02    0.086097
         2015-03    0.016567
         Freq: M, dtype: float64
In [75]: ts3.index
Out[75]: <class 'pandas.tseries.period.PeriodIndex'>
         [2015-01, ..., 2015-03]
         Length: 3, Freq: M
```

We can also convert a `Series` or `DataFrame` object indexed by a `DatetimeIndex` object to a `PeriodIndex` using the `to_period` method (which takes an argument that specifies the period frequency, here `'M'` for month):

```
In [76]: ts2.to_period('M')
Out[76]: 2015-01     0.683801
         2015-02     0.916209
         Freq: M, dtype: float64
```

In the remaining part of this section we explore select features of pandas time series through examples. We look at the manipulation of two time series that contain sequences of temperature measurements at given timestamps. We have one dataset for an indoor temperature sensor and one dataset for an outdoor temperature sensor, both with observations approximately every 10 minutes during most of 2014. The two data files, `temperature_indoor_2014.tsv` and `temperature_outdoor_2014.tsv`, are TSV (tab-separated values, a variant of the CSV format) files with two columns: the first column contains UNIX timestamps (seconds since Jan 1, 1970), and the second column is the measured temperature in degree Celsius. For example, the first five lines in the outdoor dataset are

```
In [77]: !head -n 5 temperature_outdoor_2014.tsv
1388530986     4.380000
1388531586     4.250000
1388532187     4.190000
1388532787     4.060000
1388533388     4.060000
```

We can read the data files using `read_csv` by specifying that the delimiter between columns is the TAB character: `delimiter="\t"`. When reading the two files, we also explicitly specify the column names using the `names` keyword argument, since the files in this example do not have header lines with the column names.

```
In [78]: df1 = pd.read_csv('temperature_outdoor_2014.tsv', delimiter="\t",
    ...:                    names=["time", "outdoor"])
In [79]: df2 = pd.read_csv('temperature_indoor_2014.tsv', delimiter="\t",
    ...:                    names=["time", "indoor"])
```

Once we have created `DataFrame` objects for the time-series data, it is informative to inspect the data by displaying the first few lines:

```
In [80]: df1.head()
Out[80]:
```

	time	outdoor
0	1388530986	4.38
1	1388531586	4.25
2	1388532187	4.19
3	1388532787	4.06
4	1388533388	4.06

The next step toward a meaningful representation of the time-series data is to convert the UNIX timestamps to date and time objects using to_datetime with the unit="s" argument. Furthermore, we localize the timestamps (assigning a time zone) using tz_localize and convert the time zone attribute to the Europe/Stockholm time zone using tz_convert. We also set the time column as index using set_index:

```
In [81]: df1.time = (pd.to_datetime(df1.time.values, unit="s")
    ...:                 .tz_localize('UTC').tz_convert('Europe/Stockholm'))
In [82]: df1 = df1.set_index("time")
In [83]: df2.time = (pd.to_datetime(df2.time.values, unit="s")
    ...:                 .tz_localize('UTC').tz_convert('Europe/Stockholm'))
In [84]: df2 = df2.set_index("time")
In [85]: df1.head()
Out[85]:
```

time	outdoor
2014-01-01 00:03:06+01:00	4.38
2014-01-01 00:13:06+01:00	4.25
2014-01-01 00:23:07+01:00	4.19
2014-01-01 00:33:07+01:00	4.06
2014-01-01 00:43:08+01:00	4.06

Displaying the first few rows of the data frame for the outdoor temperature dataset shows that the index now indeed is a date and time object. As we will see examples of such in the following, having the index of a time series represented as proper date and time objects (in contrast to using, e.g., integers representing the UNIX timestamps) allows us to easily perform many time-oriented operations. Before we proceed to explore the data in more detail, we first plot the two time series to obtain an idea of how the data looks like. For this we can use the DataFrame.plot method, and the results are shown in Figure 12-3. Note that there is data missing for a part of August. Imperfect data is a common problem, and handling missing data in a suitable manner is an important part of the mission statement of the pandas library.

```
In [86]: fig, ax = plt.subplots(1, 1, figsize=(12, 4))
    ...: df1.plot(ax=ax)
    ...: df2.plot(ax=ax)
```

Figure 12-3. *Plot of the time series for indoor and outdoor temperature*

It is also illuminating to display the result of the info method of the DataFrame object. Doing so tells us that there are nearly 50000 data points in this dataset and that it contains data points starting at 2014-01-01 00:03:06 and ending at 2014-12-30 23:56:35:

```
In [87]: df1.info()
<class 'pandas.core.frame.DataFrame'>
DatetimeIndex: 49548 entries, 2014-01-01 00:03:06+01:00 to 2014-12-30
23:56:35+01:00
Data columns (total 1 columns):
outdoor    49548 non-null float64
dtypes: float64(1) memory usage: 774.2 KB
```

A common operation on time series is to select and extract parts of the data. For example, from the full dataset that contains data for all of 2014, we may be interested in selecting out and analyzing only the data for the month of January. In pandas, we can accomplish this in a number of ways. For example, we can use Boolean indexing of a DataFrame to create a DataFrame for a subset of the data. To create the Boolean indexing mask that selects the data for January, we can take advantage of the pandas time-series features that allow us to compare the time-series index with string representations of a date and time. In the following code, the expressions like df1.index >= "2014-1-1", where df1.index is a time DateTimeIndex instance, result in a Boolean NumPy array that can be used as a mask to select the desired elements.

```
In [88]: mask_jan = (df1.index >= "2014-1-1") & (df1.index < "2014-2-1")
In [89]: df1_jan = df1[mask_jan]
In [90]: df1_jan.info()
<class 'pandas.core.frame.DataFrame'>
DatetimeIndex: 4452 entries, 2014-01-01 00:03:06+01:00 to 2014-01-31
23:56:58+01:00
Data columns (total 1 columns):
outdoor    4452 non-null float64
dtypes: float64(1) memory usage: 69.6 KB
```

Alternatively, we can use slice syntax directly with date and time strings:

```
In [91]: df2_jan = df2["2014-1-1":"2014-1-31"]
```

The results are two DataFrame objects, df1_jan and df2_jan, that contain data only for the month of January. Plotting this subset of the original data using the plot method results in the graph shown in Figure 12-4.

```
In [92]: fig, ax = plt.subplots(1, 1, figsize=(12, 4))
    ...: df1_jan.plot(ax=ax)
    ...: df2_jan.plot(ax=ax)
```

Figure 12-4. *Plot of the time series for indoor and outdoor temperature for a selected month (January)*

Like the datetime class in Python's standard library, the Timestamp class that is used in pandas to represent time values has attributes for accessing fields such as year, month, day, hour, minute, and so on. These fields are particularly useful when processing time series. For example, if we wish to calculate the average temperature for each month of the year, we can begin by creating a new column month, which we assign to the month field of the Timestamp values of the DatetimeIndex indexer. To extract the month field from each Timestamp value, we first call reset_index to convert the index to a column in the data frame (in which case the new DataFrame object falls back to using an integer index), after which we can use the apply function on the newly created time column.[4]

```
In [93]: df1_month = df1.reset_index()
In [94]: df1_month["month"] = df1_month.time.apply(lambda x: x.month)
In [95]: df1_month.head()
Out[95]:
```

	time	outdoor	month
0	2014-01-01 00:03:06+01:00	4.38	1
1	2014-01-01 00:13:06+01:00	4.25	1
2	2014-01-01 00:23:07+01:00	4.19	1
3	2014-01-01 00:33:07+01:00	4.06	1
4	2014-01-01 00:43:08+01:00	4.06	1

---

[4]We can also directly use the month method of the DatetimeIndex index object, but for the sake of demonstration, we use a more explicit approach here.

Next, we can group the DataFrame by the new month field and aggregate the grouped values using the mean function for computing the average within each group.

```
In [96]: df1_month = df1_month.groupby("month").aggregate(np.mean)
In [97]: df2_month = df2.reset_index()
In [98]: df2_month["month"] = df2_month.time.apply(lambda x: x.month)
In [99]: df2_month = df2_month.groupby("month").aggregate(np.mean)
```

After having repeated the same process for the second DataFrame (indoor temperatures), we can combine df1_month and df2_month into a single DataFrame using the join method:

```
In [100]: df_month = df1_month.join(df2_month)
In [101]: df_month.head(3)
Out[101]:
```

time	outdoor	indoor
1	-1.776646	19.862590
2	2.231613	20.231507
3	4.615437	19.597748

In only a few lines of code, we have here leveraged some of the many data processing capabilities of pandas to transform and compute with the data. It is often the case that there are many different ways to combine the tools provided by pandas to do the same, or a similar, analysis. For the current example, we can do the whole process in a single line of code, using the to_period and groupby methods and the concat function (which like join combines DataFrame into a single DataFrame):

```
In [102]: df_month = pd.concat([df.to_period("M").groupby(level=0).mean()
        for df in [df1, df2]],
    ...:                        axis=1)
In [103]: df_month.head(3)
```

Out[103]:

time	outdoor	indoor
2014-01	-1.776646	19.862590
2014-02	2.231613	20.231507
2014-03	4.615437	19.597748

To visualize the results, we plot the average monthly temperatures as a bar plot and a boxplot using the DataFrame method plot. The result is shown in Figure 12-5.

```
In [104]: fig, axes = plt.subplots(1, 2, figsize=(12, 4))
     ...: df_month.plot(kind='bar', ax=axes[0])
     ...: df_month.plot(kind='box', ax=axes[1])
```

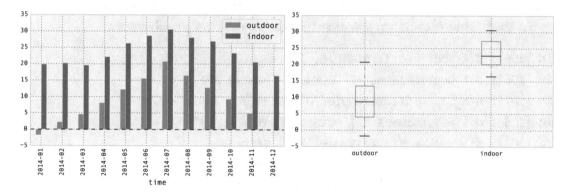

Figure 12-5. *Average indoor and outdoor temperatures per month (left) and a boxplot for monthly indoor and outdoor temperature (right)*

Finally, a very useful feature of the pandas time-series objects is the ability to up- and down-sample the time series using the resample method. Resampling means that the number of data points in a time series is changed. It can be either increased (up-sampling) or decreased (down-sampling). For up-sampling, we need to choose a method for filling in the missing values, and for down-sampling we need to choose a method for aggregating multiple sample points between each new sample point. The resample method expects as first argument a string that specifies the new period of data points in the resampled time series. For example, the string H represents a period of

431

one hour, the string D one day, the string M one month, and so on.[5] We can also combine these in simple expressions, such as 7D, which denotes a time period of seven days. The `resample` method returns a resampler object for which we can invoke aggregation methods such as `mean` and `sum`, in order to obtain the resampled data.

To illustrate the use of the `resample` method, consider the previous two time series with temperature data. The original sampling frequency is roughly 10 minutes, which amounts to a lot of data points over a year. For plotting purposes, or if we want to compare the two time series, which are sampled at slightly different timestamps, it is often necessary to down-sample the original data. This can give less busy graphs and regularly spaced time series that can be readily compared to each other. In the following code, we resample the outdoor temperature time series to four different sampling frequencies and plot the resulting time series. We also resample both the outdoor and indoor time series to daily averages that we subtract to obtain the daily average temperature difference between indoors and outdoors throughout the year. These types of manipulations are very handy when dealing with time series, and it is one of the many areas in which the pandas library really shines.

```
In [105]: df1_hour = df1.resample("H").mean()
In [106]: df1_hour.columns = ["outdoor (hourly avg.)"]
In [107]: df1_day = df1.resample("D").mean()
In [108]: df1_day.columns = ["outdoor (daily avg.)"]
In [109]: df1_week = df1.resample("7D").mean()
In [110]: df1_week.columns = ["outdoor (weekly avg.)"]
In [111]: df1_month = df1.resample("M").mean()
In [112]: df1_month.columns = ["outdoor (monthly avg.)"]
In [113]: df_diff = (df1.resample("D").mean().outdoor - df2.resample("D").
                mean().indoor)
In [114]: fig, (ax1, ax2) = plt.subplots(2, 1, figsize=(12, 6))
     ...: df1_hour.plot(ax=ax1, alpha=0.25)
     ...: df1_day.plot(ax=ax1)
     ...: df1_week.plot(ax=ax1)
     ...: df1_month.plot(ax=ax1)
     ...: df_diff.plot(ax=ax2)
     ...: ax2.set_title("temperature difference between outdoor and indoor")
     ...: fig.tight_layout()
```

---

[5]There are a large number of available time-unit codes. See the sections on "Offset aliases" and "Anchored offsets" in the Pandas reference manual for details.

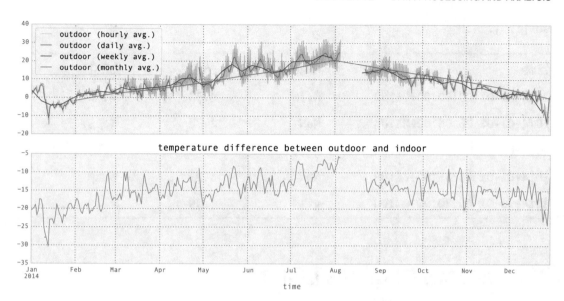

Figure 12-6. *Outdoor temperature, resampled to hourly, daily, weekly, and monthly averages (top). Daily temperature difference between outdoors and indoors (bottom)*

As an illustration of up-sampling, consider the following example where we resample the data frame df1 to a sampling frequency of 5 minutes, using three different aggregation methods (mean, ffill for forward-fill, and bfill for back-fill). The original sample frequency is approximately 10 minutes, so this resampling is indeed up-sampling. The result is three new data frames that we combine into a single DataFrame object using the concat function. The first five rows in the data frame are also shown in the following example. Note that every second data point is a new sample point, and depending on the value of the aggregation method, those values are filled (or not) according to the specified strategies. When no fill strategy is selected, the corresponding values are marked as missing using the NaN value.

```
In [115]: pd.concat(
    ...:         [df1.resample("5min").mean().rename(columns={"outdoor":
    ...:         'None'}),
    ...:         df1.resample("5min").ffill().rename(columns={"outdoor":
    ...:         'ffill'}),
    ...:         df1.resample("5min").bfill().rename(columns={"outdoor":
    ...:         'bfill'})],
    ...:         axis=1).head()
```

Out[115]:

time	None	ffill	bfill
2014-01-01 00:00:00+01:00	4.38	4.38	4.38
2014-01-01 00:05:00+01:00	NaN	4.38	4.25
2014-01-01 00:10:00+01:00	4.25	4.25	4.25
2014-01-01 00:15:00+01:00	NaN	4.25	4.19
2014-01-01 00:20:00+01:00	4.19	4.19	4.19

The Seaborn Graphics Library

The Seaborn graphics library is built on top of Matplotlib, and it provides functions for generating graphs that are useful when working with statistics and data analysis, including distribution plots, kernel-density plots, joint distribution plots, factor plot, heatmaps, facet plots, and several ways of visualizing regressions. It also provides methods for coloring data in graphs and numerous well-crafted color palettes. The Seaborn library is created with close attention to the aesthetics of the graphs it produces, and the graphs generated with the library tend to be both good looking and informative. The Seaborn library distinguishes itself from the underlying Matplotlib library in that it provides polished higher-level graph functions for a specific application domain, namely, statistical analysis and data visualization. The ease with which standard statistical graphs can be generated with the library makes it a valuable tool in exploratory data analysis.

To get started using the Seaborn library, we first set a style for the graphs it produces using the sns.set function. Here we choose to work with the style called darkgrid, which produces graphs with a gray background (also try the whitegrid style).

In [116]: sns.set(style="darkgrid")

Importing seaborn and setting a style for the library alters the default settings for how Matplotlib graphs appear, including graphs produced by the pandas library. For example, consider the following plot of the previously used indoor and outdoor temperature time series. The resulting graph is shown in Figure 12-7, and although the graph was produced using the pandas DataFrame method plot, using the sns.set function has changed the appearance of the graph (compare with Figure 12-3).

```
In [117]: df1 = pd.read_csv('temperature_outdoor_2014.tsv', delimiter="\t",
     ...:                     names=["time", "outdoor"])
     ...: df1.time = (pd.to_datetime(df1.time.values, unit="s")
     ...:                   .tz_localize('UTC').tz_convert('Europe/Stockholm'))
     ...: df1 = df1.set_index("time").resample("10min").mean()
In [118]: df2 = pd.read_csv('temperature_indoor_2014.tsv', delimiter="\t",
     ...:                     names=["time", "indoor"])
     ...: df2.time = (pd.to_datetime(df2.time.values, unit="s")
     ...:                   .tz_localize('UTC').tz_convert('Europe/Stockholm'))
     ...: df2 = df2.set_index("time").resample("10min").mean()
In [119]: df_temp = pd.concat([df1, df2], axis=1)
In [120]: fig, ax = plt.subplots(1, 1, figsize=(8, 4))
     ...: df_temp.resample("D").mean().plot(y=["outdoor", "indoor"], ax=ax)
```

Figure 12-7. *Time-series plot produced by Matplotlib using the Pandas library, with a plot style that is set up by the Seaborn library*

The main strength of the Seaborn library, apart from generating good-looking graphics, is its collection of easy-to-use statistical plots. Examples of these are the kdeplot and distplot, which plot a kernel-density estimate plot and a histogram plot with a kernel-density estimate overlaid on top of the histogram, respectively. For example, the following two lines of code produce the graph shown in Figure 12-8. The solid blue and green lines in this figure are the kernel-density estimate that can also be graphed separately using the function kdeplot (not shown here).

```
In [121]: sns.distplot(df_temp.to_period("M")["outdoor"]["2014-04"].
dropna().values, bins=50);
     ...: sns.distplot(df_temp.to_period("M")["indoor"]["2014-04"].
dropna().values, bins=50);
```

Figure 12-8. *The histogram (bars) and kernel-density plots (solid lines) for the subset of the indoor and outdoor datasets that corresponds to the month of April*

The kdeplot function can also operate on two-dimensional data, showing a contour graph of the joint kernel-density estimate. Relatedly, we can use the jointplot function to plot the joint distribution for two separate datasets. In the following example, we use the kdeplot and jointplot to show the correlation between the indoor and outdoor data series, which are resampled to hourly averages before visualized (we also drop missing values using dropna method, since the functions from the seaborn module do not accept arrays with missing data). The results are shown in Figure 12-9.

```
In [122]: sns.kdeplot(df_temp.resample("H").mean()["outdoor"].dropna().values,
     ...:             df_temp.resample("H").mean()["indoor"].dropna().
                 values, shade=False)
In [123]: with sns.axes_style("white"):
     ...:         sns.jointplot(df_temp.resample("H").mean()["outdoor"].values,
     ...:                 df_temp.resample("H").mean()["indoor"].values,
                 kind="hex")
```

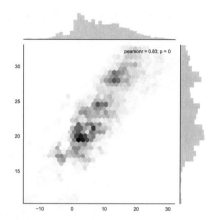

Figure 12-9. *Two-dimensional kernel-density estimate contours (left) and the joint distribution for the indoor and outdoor temperature datasets (right). The outdoor temperatures are shown on the x axis and the indoor temperatures on the y axis.*

The seaborn library also provides functions for working with categorical data. A simple example of a graph type that is often useful for datasets with categorical variables is the standard boxplot for visualizing the descriptive statistics (min, max, median, and quartiles) of a dataset. An interesting twist on the standard boxplot is violin plot, in which the kernel-density estimate is shown in the width of boxplot. The boxplot and violinplot functions can be used to produce such graphs, as shown in the following example, and the resulting graph is shown in Figure 12-10.

```
In [124]: fig, (ax1, ax2) = plt.subplots(1, 2, figsize=(8, 4))
     ...: sns.boxplot(df_temp.dropna(), ax=ax1, palette="pastel")
     ...: sns.violinplot(df_temp.dropna(), ax=ax2, palette="pastel")
```

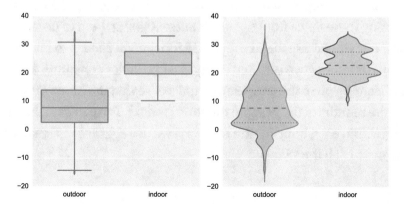

Figure 12-10. *A boxplot (left) and violin plot (right) for the indoor and outdoor temperature datasets*

437

As a further example of violin plots, consider the outdoor temperature dataset partitioned by the month, which can be produced by passing the month field of the index of the data frame as a second argument (used to group the data into categories). The resulting graph, which is shown in Figure 12-11, provides a compact and informative visualization of the distribution of temperatures for each month of the year.

```
In [125]: sns.violinplot(x=df_temp.dropna().index.month,
     ...:                      y=df_temp.dropna().outdoor, color="skyblue");
```

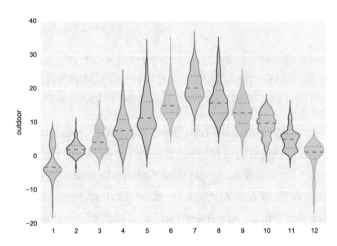

Figure 12-11. *Violin plot for the outdoor temperature grouped by month*

Heatmaps are another type of graph that is handy when dealing with categorical variables, especially for variables with a large number of categories. The Seaborn library provides the function heatmap for generating this type of graphs. For example, working with the outdoor temperature dataset, we can create two categorical columns month and hour by extracting those fields from the index and assigning them to new columns in the data field. Next we can use the pivot_table function in pandas to pivot the columns into a table (matrix) where two selected categorical variables constitute the new index and columns. Here we pivot the temperature dataset so that the hours of the day are the columns and the months of the year are the rows (index). To aggregate the multiple data points that fall within each hour-month category, we use aggfunc=np.mean argument to compute the mean of all the values:

```
In [126]: df_temp["month"] = df_temp.index.month
    ...: df_temp["hour"] = df_temp.index.hour
In [127]: table = pd.pivot_table(df_temp, values='outdoor',
              index=['month'], columns=['hour'],
    ...:                          aggfunc=np.mean)
```

Once we have created a pivot table, we can visualize it as a heatmap using the heatmap function in Seaborn. The result is shown in Figure 12-12.

```
In [128]: fig, ax = plt.subplots(1, 1, figsize=(8, 4))
    ...: sns.heatmap(table, ax=ax)
```

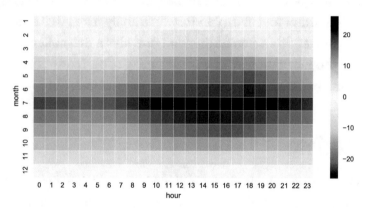

Figure 12-12. *A heatmap of the outdoor temperature data grouped by the hour of the day and month of the year*

The Seaborn library contains much more statistical visualization tools than what we have been able to survey here. However, I hope that looking at a few examples of what this library can do illustrates the essence of the Seaborn library – that it is a convenient tool for statistical analysis and exploration of data, which is able to produce many standard statistical graphs with a minimum of effort. In the following chapters, we will see further examples of applications of the Seaborn library.

Summary

In this chapter we have explored data representation and data processing using the Pandas library, and we briefly surveyed the statistical graphics tools provided by the Seaborn visualization library. The Pandas library provides the backend of much of data wrangling done with Python. It achieves this by adding a higher-level abstraction layer in the data representation on top of NumPy arrays, with additional methods for operating on the underlying data. The ease with which data can be loaded, transformed, and manipulated makes it an invaluable part of the data processing workflow in Python. The pandas library also contains basic functions for visualizing the data that is represented by its data structures. Being able to quickly visualize data represented as Pandas series and data frames is an important tool in exploratory data analytics as well as for presentation. The Seaborn library takes this a step further and provides a rich collection of statistical graphs that can be produced often with a single line of code. Many functions in the Seaborn library can operate directly on Pandas data structures.

Further Reading

A great introduction to the Pandas library is given by the original creator of the library in McKinney (2013), and it is also a rather detailed introduction to NumPy. The Pandas official documentation, available at `http://pandas.pydata.org/pandas-docs/stable`, also provides an accessible and very detailed description of the features of the library. Another good online resource for learning Pandas is `http://github.com/jvns/pandas-cookbook`. For data visualization, we have looked at the Seaborn library in this chapter, and it is well described in the documentation available on its web site. With respect to higher-level visualization tools, it is also worth exploring the `ggplot` library for Python, `http://ggplot.yhathq.com`, which is an implementation based on the renowned *The Grammar of Graphics* (L. Wilkinson, 2005). This library is also closely integrated with the Pandas library, and it provides statistical visualization tools that are convenient when analyzing data. For more information about visualization in Python, see, for example, Vaingast (2014).

References

L. Wilkinson, D. W. (2005). *The Grammar of Graphics.* Chicago: Springer.

McKinney, W. (2013). *Python for Data Analysis.* Sebastopol: O'Reilly.

Vaingast, S. (2014). *Beginning Python Visualization.* New York: Apress.

CHAPTER 13

Statistics

Statistics has long been a field of mathematics that is relevant to practically all applied disciplines of science and engineering, as well as business, medicine, and other fields where data is used for obtaining knowledge and making decisions. With the recent proliferation of data analytics, there has been a surge of renewed interest in statistical methods. Still, computer-aided statistics has a long history, and it is a field that traditionally has been dominated by domain-specific software packages and programming environments, such as the S language, and more recently its open source counterpart: the R language. The use of Python for statistical analysis has grown rapidly over the last several years, and by now there is a mature collection of statistical libraries for Python. With these libraries Python can match the performance and features of domain-specific languages in many areas of statistics, albeit not all, while also providing the unique advantages of the Python programming language and its environment. The Pandas library that we discussed in Chapter 12 is an example of a development within the Python community that was strongly influenced by statistical software, with the introduction of the data frame data structure to the Python environment. The NumPy and SciPy libraries provide computational tools for many fundamental statistical concepts, and higher-level statistical modeling and machine learning are covered by the statsmodels and scikit-learn libraries, which we will see more of in the following chapters.

In this chapter we focus on fundamental statistical applications using Python and in particular the stats module in SciPy. Here we discuss computing descriptive statistics, random numbers, random variables, distributions, and hypothesis testing. We defer more involved statistical modeling and machine-learning applications to the following chapters. Some fundamental statistical functions are also available through the NumPy library, such as its functions and methods for computing descriptive statistics and its module for generating random numbers. The SciPy stats module builds on top of NumPy and, for example, provides random number generators with more specialized distributions.

443

© Robert Johansson 2019
R. Johansson, *Numerical Python*, https://doi.org/10.1007/978-1-4842-4246-9_13

Importing Modules

In this chapter we mainly work with the `stats` module in SciPy, and following the convention to selectively import modules from SciPy, we here assume that this module, as well as the `optimize` module, is imported in the following way:

```
In [1]: from scipy import stats
   ...: from scipy import optimize
```

In addition, as usual we also require the NumPy and the Matplotlib libraries:

```
In [2]: import numpy as np
In [3]: import matplotlib.pyplot as plt
```

and for statistical graphs and styling, we use the Seaborn library:

```
In [4]: import seaborn as sns
In [5]: sns.set(style="whitegrid")
```

Review of Statistics and Probability

We begin with a brief review of statistics, in order to introduce some of the key concepts and the notation that we use in this and the following chapters. Statistics deals with the collection and analysis of data for the purpose of gaining insights, drawing conclusions, and supporting decision-making. Statistical methods are necessary when we have incomplete information about a phenomenon. Typically we have incomplete information because we are unable to collect data from all members of a *population* or if there is *uncertainty* in observations that we make (e.g., due to measurement noise). When we are unable to survey an entire population, a randomly chosen *sample* can be studied instead, and we can use statistical methods and compute descriptive statistics (parameters such as the mean and the variances) to make inferences about the properties of the entire population (also called *sample space*) in a systematic manner and with controlled risks of error.

Statistical methods are built on the foundation provided by probability theory, with which we can model uncertainty and incomplete information using probabilistic, random variables. For example, with *randomly* selected samples of a population, we can hope to obtain representative samples whose properties can be used to infer properties of the entire population. In probability theory, each possible outcome for an observation is given a probability, and the probability for all possible outcomes constitutes the probability distribution. Given the probability distribution, we can compute the properties of the population, such as its mean and variance, but for randomly selected samples, we only know the *expected*, or average, results.

In statistical analysis it is important to distinguish between population and sample statistics. Here we denote parameters of the population with Greek symbols and parameters of a sample with the corresponding population symbol with the added subscript x (or the symbol that is used to represent the sample). For example, the mean and the variance of a population are denoted with μ and σ^2, and the mean and the variance of a sample x are denoted as μ_x and σ_x^2. Furthermore, we denote variables representing a population (random variables) with capital letters, for example, X, and a set of sample elements is denoted with a lowercase letter, for example, x. A bar over a symbol denotes the average or mean, $\mu = \bar{X} = \dfrac{1}{N}\sum_{i=1}^{N} x_i$ and $\mu_x = \bar{x} = \dfrac{1}{n}\sum_{i=1}^{n} x_i$, where N is the number of elements in the population X and n is the number of elements in the sample x. The only difference between these two expressions is the number of elements in the sum ($N \geq n$). The situation is slightly more complex for the variance: the population variance is the mean of the squared distance from the mean, $\sigma^2 = \dfrac{1}{N}\sum_{i=1}^{N}(x_i - \mu)^2$, and the corresponding sample variance is $\sigma_x^2 = \dfrac{1}{n-1}\sum_{i=1}^{n}(x_i - \mu_x)^2$. In the latter expression, we have replaced the population mean μ with the sample mean μ_x and divided the sum with $n - 1$ rather than n. The reason for this is that one degree of freedom has been eliminated from the sample set when calculating the sample mean μ_x, so when computing the sample variance, only $n - 1$ degrees of freedom remains. Consequently, the way to compute the variance for a population and a sample is slightly different. This is reflected in functions we can use to compute these statistics in Python.

In Chapter 2 we have already seen that we can compute descriptive statistics for data using NumPy functions or the corresponding `ndarray` methods. For example, to compute the mean and the median of a dataset, we can use the NumPy functions `mean` and `median`:

```
In [6]: x = np.array([3.5, 1.1, 3.2, 2.8, 6.7, 4.4, 0.9, 2.2])
In [7]: np.mean(x)
Out[7]: 3.1
```

```
In [8]: np.median(x)
Out[8]: 3.0
```

Similarly, we can use min and max functions or ndarray methods to compute the minimum and maximum values in the array:

```
In [9]: x.min(), x.max()
Out[9]: (0.90, 6.70)
```

To compute the variance and the standard deviation for a dataset, we use the var and std methods. By default the formula for the population variance and standard deviation is used (i.e., it is assumed that the dataset is the entire population).

```
In [10]: x.var()
Out[10]: 3.07
In [11]: x.std()
Out[11]: 1.7521415467935233
```

However, to change this behavior, we can use the argument ddof (delta degrees of freedom). The denominator in the expression for the variance is the number of elements in the array minus ddof, so to calculate the unbiased estimate of the variance and standard deviation from a sample, we need to set ddof=1:

```
In [12]: x.var(ddof=1)
Out[12]: 3.5085714285714293
In [13]: x.std(ddof=1)
Out[13]: 1.8731181032095732
```

In the following sections, we look into more detail on how to use NumPy and SciPy's stats module to generate random numbers, represent random variables and distributions, and test hypotheses.

Random Numbers

The Python standard library contains the module random, which provides functions for generating single random numbers with a few basic distributions. The random module in the NumPy module provides similar functionality but offers functions that generate

NumPy arrays with random numbers, and it has support for a wider selection of probability distributions. Arrays with random numbers are often practical for computational purposes, so here we focus on the random module in NumPy, and later also the higher-level functions and classes in scipy.stats, which build on top of and extend NumPy.

Earlier in this book, we have already used np.random.rand, which generates uniformly distributed floating-point numbers in the half-open interval $[0, 1)$ (i.e., 0.0 is a possible outcome, but 1.0 is not). In addition to this function, the np.random module also contains a large collection of other functions for generating random numbers that cover different intervals, have different distributions, and take values of different types (e.g., floating-point numbers and integers). For example, the randn function produces random numbers that are distributed according to the *standard normal distribution* (the normal distribution with mean 0 and standard deviation 1), and the randint function generates uniformly distributed integers between a given low (inclusive) and high (exclusive) value. When the rand and randn functions are called without any arguments, they produce a single random number:

```
In [14]: np.random.rand()
Out[14]: 0.532833024789759
In [15]: np.random.randn()
Out[15]: 0.8768342101492541
```

However, passing the shape of the array as arguments to these functions produces arrays of random numbers. For example, here we generate a vector of length 5 using rand by passing a single argument 5 and a 2×4 array using randn by passing 2 and 4 as arguments (higher-dimensional arrays are generated by passing the length of each dimension as arguments):

```
In [16]: np.random.rand(5)
Out[16]: array([ 0.71356403,  0.25699895,  0.75269361,  0.88387918,  0.15489908])
In [17]: np.random.randn(2, 4)
Out[17]: array([[ 3.13325952,  1.15727052,  1.37591514,  0.94302846],
                [ 0.8478706 ,  0.52969142, -0.56940469,  0.83180456]])
```

To generate random integers using `randint` (see also `random_integers`), we need to either provide the upper limit for the random numbers (in which case the lower limit is implicitly zero) or provide both the lower and upper limits. The size of the generated array is specified using the `size` keyword arguments, and it can be an integer or a tuple that specifies the shape of a multidimensional array:

```
In [18]: np.random.randint(10, size=10)
Out[18]: array([0, 3, 8, 3, 9, 0, 6, 9, 2, 7])
In [19]: np.random.randint(low=10, high=20, size=(2, 10))
Out[19]: array([[12, 18, 18, 17, 14, 12, 14, 10, 16, 19],
                [15, 13, 15, 18, 11, 17, 17, 10, 13, 17]])
```

Note that the `randint` function generates random integers in the half-open interval [low, high). To demonstrate that the random numbers produced by `rand`, `randn`, and `randint`, indeed, are distributed differently, we can plot the histograms of say 10000 random numbers produced by each function. The result is shown in Figure 13-1. We note that the distributions for `rand` and `randint` appear uniform but have different ranges and types, while the distribution of the numbers produced by `randn` resembles a Gaussian curve centered at zero, as expected.

```
In [20]: fig, axes = plt.subplots(1, 3, figsize=(12, 3))
    ...: axes[0].hist(np.random.rand(10000))
    ...: axes[0].set_title("rand")
    ...: axes[1].hist(np.random.randn(10000))
    ...: axes[1].set_title("randn")
    ...: axes[2].hist(np.random.randint(low=1, high=10, size=10000),
        bins=9, align='left')
    ...: axes[2].set_title("randint(low=1, high=10)")
```

Figure 13-1. *Distributions for 10000 random numbers generated by the* rand, *randn, and* randint *functions in NumPy's* random *module*

In statistical analysis, it is often necessary to generate a unique list of integers. This corresponds to sampling (randomly selecting) items from a set (population) without replacement (so that we do not get the same item twice). From the NumPy random module, we can use the choice function to generate this type of random numbers. As the first argument, we can either provide a list (or array) with the values in the population or an integer that corresponds to the number of elements in the population. As the second argument, we give the number of values that are to be sampled. Whether the values are to be sampled with or without replacement can be specified using the replace keyword argument, which takes the Boolean values True or False. For example, to sample five unique (without replacement) items from the set of integers between 0 (inclusive) and 10 (exclusive), we can use

```
In [21]: np.random.choice(10, 5, replace=False)
Out[21]: array([9, 0, 5, 8, 1])
```

When working with random number generation, it can be useful to *seed* the random number generator. The seed is a number that initializes random number generator to a specific state, so that once it has been seeded with a specific number, it always generates the same sequence of random numbers. This can be useful when testing and for reproducing previous results and occasionally in applications that require reseeding the random number generator (e.g., after having forked a process). To seed the random number generator in NumPy, we can use the seed function, which takes an integer as argument:

```
In [22]: np.random.seed(123456789)
In [23]: np.random.rand()
Out[23]: 0.532833024789759
```

449

Note that after having seeded the random number generator with a specific number, here 123456789, the following calls to the random number generators always produce the same results:

```
In [24]: np.random.seed(123456789); np.random.rand()
Out[24]: 0.532833024789759
```

The seed of the random number generator is a global state of the np.random module. A finer level of control of the state of the random number generator can be achieved by using the RandomState class, which optionally takes a seed integer as argument to its initializer. The RandomState object keeps tracks of the state of the random number generator and allows maintaining several independent random number generators in the same program (which can be useful, e.g., when working with threaded applications). Once a RandomState object has been created, we can use methods of this object to generate random numbers. The RandomState class has methods that correspond to the functions that are available in np.random module. For example, we can use the method randn of the RandomState class to generate standard normal distributed random numbers:

```
In [25]: prng = np.random.RandomState(123456789)
In [26]: prng.randn(2, 4)
Out[26]: array([[ 2.212902,    2.1283978,   1.8417114,   0.08238248],
                [ 0.85896368, -0.82601643,  1.15727052,  1.37591514]])
```

Similarly, there are methods, rand, randint, rand_integers, and choice, which also correspond to the functions in the np.random module with the same name. It is considered good programming practice to use a RandomState instance rather than directly using the functions in the np.random module, because it avoids relying on a global state variable and improves the isolation of the code. This is an important consideration when developing library functions that use random numbers, but is perhaps less important in smaller applications and calculations.

In addition to the fundamental random number distributions we have looked at so far (discrete and continuous uniform distributions, randint and rand, and the standard normal distribution, randn), there are also functions, and RandomState methods, for a large number of probability distributions that occur in statistics. To mention just a few, there is the continuous χ^2 distribution (chisquare), the Student's t distribution (standard_t), and the F distribution (f):

```
In [27]: prng.chisquare(1, size=(2, 2))
Out[27]: array([[ 0.78631596,   0.19891367],
                [ 0.11741336,   2.8713997 ]])
In [28]: prng.standard_t(1, size=(2, 3))
Out[28]: array([[ 0.39697518, -0.19469463,   1.15544019],
                [-0.65730814, -0.55125015,   0.13578694]])
In [29]: prng.f(5, 2, size=(2, 4))
Out[29]: array([[  0.45471421,  17.64891848,   1.48620557,   2.55433261],
                [  1.21823269,   3.47619315,   0.50835525,   0.70599655]])
```

and the discrete binomial distribution (binomial) and the Poisson distribution (poisson):

```
In [30]: prng.binomial(10, 0.5, size=10)
Out[30]: array([4, 5, 6, 7, 3, 5, 7, 5, 4, 5])
In [31]: prng.poisson(5, size=10)
Out[31]: array([3, 5, 5, 5, 0, 6, 5, 4, 6, 3])
```

For a complete list of available distribution functions, see the docstrings for the np. random module, help(np.random), and the RandomState class. While it is possible to use the functions in np.random and methods in RandomState to draw random numbers from many different statistical distribution functions, when working with distributions there is a higher-level interface in the scipy.stats module that combines random number sampling with many other convenient functions for probability distributions. In the following section, we explore this in more detail.

Random Variables and Distributions

In probability theory, the set of possible outcomes of a random process is called the *sample space*. Each element in the sample space (i.e., an outcome of an experiment or an observation) can be assigned a probability, and the probabilities of all possible outcomes define the probability distribution. A *random variable* is a mapping from the sample space to the real numbers or to integers. For example, the possible outcomes of a coin toss are head and tail, so the sample space is {head, tail}, and a possible random variable takes the value 0 for head and 1 for tail. In general there are many ways to define random variables for the possible outcomes of a given random process. Random variables are a problem-independent representation of a random process. It is easier to work with random variables because they are described by numbers instead of outcomes from problem-specific sample spaces. A common step in statistical problem-solving is therefore to map outcomes to numerical values and figure out the probability distribution of those values.

Consequently, a random variable is characterized by its possible values and its probability distribution, which assigns a probability for each possible value. Each observation of the random variable results in a random number, and the distribution of the observed values are described by the probability distribution. There are two main types of distributions, discrete and continuous distributions, which are integer valued and real valued, respectively. When working with statistics, dealing with random variables is of central importance, and in practice this often means working with probability distributions. The SciPy `stats` module provides classes for representing random variables with a large number of probability distributions. There are two base classes for discrete and continuous random variables: `rv_discrete` and `rv_continuous`. These classes are not used directly, but rather used as base classes for random variables with specific distributions, and define a common interface for all random variable classes in SciPy `stats`. A summary of selected methods for discrete and continuous random variables is given in Table 13-1.

Table 13-1. *Selected Methods for Discrete and Continuous Random Variables in the SciPy* `stats` *Module*

Methods	Description
pdf/pmf	Probability distribution function (continuous) or probability mass function (discrete).
cdf	Cumulative distribution function.
sf	Survival function (1 – cdf).
ppf	Percent-point function (inverse of cdf).
moment	Noncentral moments of *n*th order.
stats	Statistics of the distribution (typically the mean and variance, sometimes additional statistics).

(continued)

Table 13-1. (*continued*)

Methods	Description
fit	Fit distribution to data using a numerical maximum likelihood optimization (for continuous distributions).
expect	Expectation value of a function with respect to the distribution.
interval	The endpoints of the interval that contains a given percentage of the distribution (confidence interval).
rvs	Random variable samples. Takes as argument the size of the resulting array of samples.
mean, median, std, var	Descriptive statistics: mean, median, standard deviation, and the variance of the distribution.

There are a large number of classes for the discrete and continuous random variable in the SciPy stats module. At the time of writing, there are classes for 13 discrete and 98 continuous distributions, and these include the most commonly encountered distributions (and many less common). For a complete reference, see the docstring for the stats module: help(stats). In the following we explore some of the more common distributions, but the usage of all the other distributions follows the same pattern.

The random variable classes in the SciPy stats module have several use-cases. They are both representations of the distribution, which can be used to compute descriptive statistics and for graphing, and they can be used to generate random numbers following the given distribution using the rvs (random variable sample) method. The latter use-case is similar to what we used the np.random module for earlier in this chapter.

As a demonstration of how to use the random variable classes in SciPy stats, consider the following example where we create a normal distributed random variable with mean 1.0 and standard deviation 0.5:

```
In [32]: X = stats.norm(1, 0.5)
```

Now X is an object that represents a random variable, and we can compute descriptive statistics of this random variable using, for example, the mean, median, std, and var methods:

```
In [33]: X.mean()
Out[33]: 1.0
```

```
In [34]: X.median()
Out[34]: 1.0
In [35]: X.std()
Out[35]: 0.5
In [36]: X.var()
Out[36]: 0.25
```

Noncentral moments of arbitrary order can be computed with the moment method:

```
In [37]: [X.moment(n) for n in range(5)]
Out[37]: [1.0, 1.0, 1.25, 1.75, 2.6875]
```

And we can obtain a distribution-dependent list of statistics using the stats method (here, for a normal distributed random variable, we get the mean and the variance):

```
In [38]: X.stats()
Out[38]: (array(1.0), array(0.25))
```

We can evaluate the probability distribution function, the cumulative distribution function, the survival function, etc., using the methods pdf, cdf, sf, etc. These all take a value, or an array of values, at which to evaluate the function:

```
In [39]: X.pdf([0, 1, 2])
Out[39]: array([ 0.10798193,  0.79788456,  0.10798193])
In [40]: X.cdf([0, 1, 2])
Out[40]: array([ 0.02275013,  0.5,         0.97724987])
```

The interval method can be used to compute the lower and upper values of x such that a given percentage of the probability distribution falls within the interval (lower, upper). This method is useful for computing confidence intervals and for selecting a range of x values for plotting:

```
In [41]: X.interval(0.95)
Out[41]: (0.020018007729972975, 1.979981992270027)
In [42]: X.interval(0.99)
Out[42]: (-0.28791465177445019, 2.2879146517744502)
```

To build intuition for the properties of a probability distribution, it is useful to graph it, together with the corresponding cumulative probability function and the percent-point function. To make it easier to repeat this for several distributions, we first create a function plot_rv_distribution that plots the result of pdf or pmf, the cdf and sf, and ppf methods of the SciPy stats random variable objects, over an interval that contains 99.9% of the probability distribution function. We also highlight the area that contains 95% of the probability distribution using the fill_between drawing method:

```
In [43]: def plot_rv_distribution(X, axes=None):
    ...:     """Plot the PDF or PMF, CDF, SF and PPF of a given random
             variable"""
    ...:     if axes is None:
    ...:         fig, axes = plt.subplots(1, 3, figsize=(12, 3))
    ...:
    ...:     x_min_999, x_max_999 = X.interval(0.999)
    ...:     x999 = np.linspace(x_min_999, x_max_999, 1000)
    ...:     x_min_95, x_max_95 = X.interval(0.95)
    ...:     x95 = np.linspace(x_min_95, x_max_95, 1000)
    ...:
    ...:     if hasattr(X.dist, "pdf"):
    ...:         axes[0].plot(x999, X.pdf(x999), label="PDF")
    ...:         axes[0].fill_between(x95, X.pdf(x95), alpha=0.25)
    ...:     else:
    ...:         # discrete random variables do not have a pdf method,
               instead we use pmf:
    ...:         x999_int = np.unique(x999.astype(int))
    ...:         axes[0].bar(x999_int, X.pmf(x999_int), label="PMF")
    ...:     axes[1].plot(x999, X.cdf(x999), label="CDF")
    ...:     axes[1].plot(x999, X.sf(x999), label="SF")
    ...:     axes[2].plot(x999, X.ppf(x999), label="PPF")
    ...:
    ...:     for ax in axes:
    ...:         ax.legend()
```

Next we use this function to graph a few examples of distributions: the normal distribution, the F distribution, and the discrete Poisson distribution. The result is shown in Figure 13-2.

```
In [44]: fig, axes = plt.subplots(3, 3, figsize=(12, 9))
    ...: X = stats.norm()
    ...: plot_rv_distribution(X, axes=axes[0, :])
    ...: axes[0, 0].set_ylabel("Normal dist.")
    ...: X = stats.f(2, 50)
    ...: plot_rv_distribution(X, axes=axes[1, :])
    ...: axes[1, 0].set_ylabel("F dist.")
    ...: X = stats.poisson(5)
    ...: plot_rv_distribution(X, axes=axes[2, :])
    ...: axes[2, 0].set_ylabel("Poisson dist.")
```

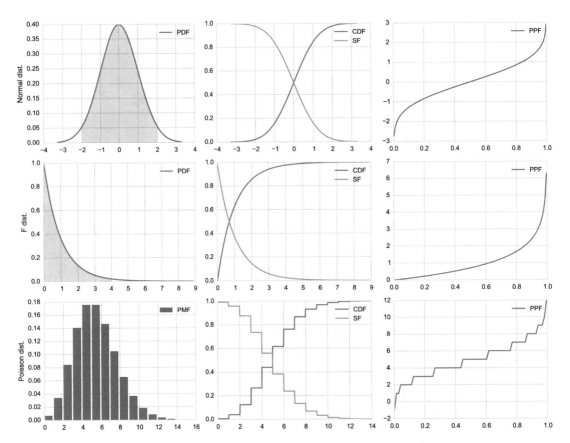

Figure 13-2. *Examples of probability distribution functions (PDF) or probability mass functions (PMFs), cumulative distribution functions (CDF), survival functions (SF), and percent-point functions (PPF) for a normal distribution (top), an F distribution (middle), and a Poisson distribution (bottom)*

In the examples so far, we have initiated an instance of a random variable class and computed statistics and other properties using method calls. An alternative way to use the random variable classes in SciPy's `stats` module is to use class methods, for example, `stats.norm.mean`, and pass the distribution parameters as arguments (often `loc` and `scale`, as in this example for normally distributed values):

```
In [45]: stats.norm.stats(loc=2, scale=0.5)
Out[45]: (array(2.0), array(0.25))
```

which gives the same result as first creating an instance and then calling the corresponding method:

```
In [46]: stats.norm(loc=1, scale=0.5).stats()
Out[46]: (array(1.0), array(0.25))
```

Most methods in the `rv_discrete` and `rv_continuous` classes can be used as class methods in this way.

So far we have only looked at properties of the distribution function of random variables. Note that although a distribution function describes a random variable, the distribution itself is fully deterministic. To draw random numbers that are distributed according to the given probability distribution, we can use the `rvs` (random variable sample) method. It takes as argument the shape of the required array (can be an integer for a vector or a tuple of dimension lengths for a higher-dimensional array). Here we use `rvs(10)` to generate a one-dimensional array with ten values:

```
In [47]: X = stats.norm(1, 0.5)
In [48]: X.rvs(10)
Out[48]: array([2.106451,    2.0641989,   1.9208557,  1.04119124,  1.42948184,
                0.58699179,  1.57863526,  1.68795757,  1.47151423,  1.4239353 ])
```

To see that the resulting random numbers indeed are distributed according to the corresponding probability distribution function, we can graph a histogram of a large number of samples of a random variable and compare it to the probability distribution function. Again, to be able to do this easily for samples of several random variables, we create a function `plot_dist_samples` for this purpose. This function uses the `interval` method to obtain a suitable plot range for a given random variable object.

```
In [49]: def plot_dist_samples(X, X_samples, title=None, ax=None):
    ...:        """ Plot the PDF and histogram of samples of a continuous
                random variable"""
    ...:        if ax is None:
    ...:            fig, ax = plt.subplots(1, 1, figsize=(8, 4))
    ...:
    ...:        x_lim = X.interval(.99)
    ...:        x = np.linspace(*x_lim, num=100)
    ...:
    ...:        ax.plot(x, X.pdf(x), label="PDF", lw=3)
    ...:        ax.hist(X_samples, label="samples", normed=1, bins=75)
    ...:        ax.set_xlim(*x_lim)
    ...:        ax.legend()
    ...:
    ...:        if title:
    ...:            ax.set_title(title)
    ...:        return ax
```

Note that in this function we have used the tuple unpacking syntax *x_lim, which distributes the elements in the tuple x_lim to different arguments for the function. In this case it is equivalent to np.linspace(x_lim[0], x_lim[1], num=100).

Next we use this function to visualize 2000 samples of three random variables with different distributions: here we use the Student's t distribution, the χ^2 distribution, and the exponential distribution, and the results are shown in Figure 13-3. Since 2000 is a fairly large sample, the histogram graphs of the samples coincide well with the probability distribution function. With an even larger number of samples, the agreement can be expected to be even better.

```
In [50]: fig, axes = plt.subplots(1, 3, figsize=(12, 3))
    ...: N = 2000
    ...: # Student's t distribution
    ...: X = stats.t(7.0)
    ...: plot_dist_samples(X, X.rvs(N), "Student's t dist.", ax=axes[0])
    ...: # The chisquared distribution
    ...: X = stats.chi2(5.0)
    ...: plot_dist_samples(X, X.rvs(N), r"$\chi^2$ dist.", ax=axes[1])
```

```
...: # The exponential distribution
...: X = stats.expon(0.5)
...: plot_dist_samples(X, X.rvs(N), "exponential dist.", ax=axes[2])
```

Figure 13-3. *Probability distribution function (PDF) together with histograms of 2000 random samples from the Student's t distribution (left), the χ^2 distribution (middle), and the exponential distribution (right)*

The opposite of drawing random samples from a known distribution function is to fit given probability distribution with unknown parameters to a set of data points. In such a fit, we typically wish to optimize the unknown parameters such that the likelihood of observing the given data is maximized. This is called a maximum likelihood fit. Many of the random variable classes in the SciPy stats module implements the method fit that performs such a fitting to given data. As a first example, consider drawing 500 random samples from the χ^2 distribution with five degrees of freedom (df=5) and then refitting the random variables to the χ^2 distribution using the fit method.

```
In [51]: X = stats.chi2(df=5)
In [52]: X_samples = X.rvs(500)
In [53]: df, loc, scale = stats.chi2.fit(X_samples)
In [54]: df, loc, scale
Out[54]: (5.2886783664198465, 0.0077028130326141243, 0.93310362175739658)
In [55]: Y = stats.chi2(df=df, loc=loc, scale=scale)
```

The fit method returns the maximum likelihood parameters of the distribution, for the given data. We can pass on those parameters to the initializer of the stats.chi2 to create a new random variable instance Y. The probability distribution of Y should resemble the probability distribution of the original random variable X. To verify this we can plot the probability distribution functions for both random variables. The resulting graph is shown in Figure 13-4.

```
In [56]:  fig, axes = plt.subplots(1, 2, figsize=(12, 4))
    ...: x_lim = X.interval(.99)
    ...: x = np.linspace(*x_lim, num=100)
    ...:
    ...: axes[0].plot(x, X.pdf(x), label="original")
    ...: axes[0].plot(x, Y.pdf(x), label="recreated")
    ...: axes[0].legend()
    ...:
    ...: axes[1].plot(x, X.pdf(x) - Y.pdf(x), label="error")
    ...: axes[1].legend()
```

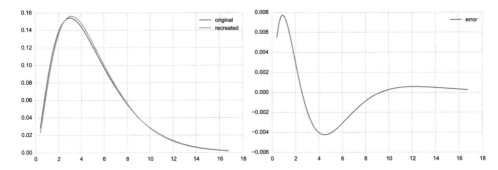

Figure 13-4. *Original and recreated probability distribution function (left) and the error (right), from a maximum likelihood fit of 500 random samples of the original distribution*

In this section we have explored how to use random variable objects from the SciPy stats model to describe random variables with various distributions and how they can be used to compute properties of the given distributions, as well as generating random variable samples and performing maximum likelihood fitting. In the following section, we see how to further use these random variable objects for hypothesis testing.

Hypothesis Testing

Hypothesis testing is a cornerstone of the scientific method, which requires that claims are investigated objectively and that a claim is rejected or accepted on the basis of factual observations. Statistical hypothesis testing has a more specific meaning. It is a systematic methodology for evaluating if a claim, or a hypothesis, is reasonable or not, on the basis of data. As such it is an important application of statistics. In this methodology, we formulate the hypothesis in terms of a null hypothesis, H_0, which represents the currently accepted state of knowledge, and an alternative hypothesis, H_A, which represents a new claim that challenges the current state of knowledge. The null hypothesis and the alternative hypothesis must be mutually exclusive and complementary, so that one and only one of the hypotheses is true.

Once H_0 and H_A are defined, the data that support the test must be collected, for example, through measurements, observations, or a survey. The next step is to find a test statistics that can be computed from the data and whose probability distribution function can be found under the null hypothesis. Next we can evaluate the data by computing the probability (the *p-value*) of obtaining the observed value of the test statistics (or a more extreme one) using the distribution function that is implied by the null hypothesis. If the *p*-value is smaller than a predetermined threshold, known as the significance level, and denoted by α (typically 5% or 1%), we can conclude that the observed data is unlikely to have been described by the distribution corresponding to the null hypothesis. In that case, we can therefore reject the null hypothesis in favor of the alternative hypothesis. The steps for carrying out a hypothesis test are summarized in the following list:

1. Formulate the null hypothesis and the alternative hypothesis.

2. Select a test statistics such that its sampling distribution under the null hypothesis is known (exactly or approximately).

3. Collect data.

4. Compute the test statistics from the data and calculate its *p*-value under the null hypothesis.

5. If the *p*-value is smaller than the predetermined significance level α, we reject the null hypothesis. If the *p*-value is larger, we fail to reject the null hypothesis.

Statistical hypothesis testing is a probabilistic method, which means that we cannot be certain in the decision to reject or not to reject the null hypothesis. There can be two types of error: we can mistakenly reject the null hypothesis when in reality it should not be rejected, and we can fail to reject the null hypothesis when it should be rejected. These are called type I and type II errors, respectively. By choosing the required significance level, we can balance the trade-off between these two types of error.

In general, the most challenging step in the method outlined in the preceding section is to know the sampling distribution of the test statistics. Fortunately many hypothesis tests fall in a few standard categories for which the probability distributions are known. A brief summary and overview of common hypothesis test cases and the corresponding distribution of their test statistics are given in Table 13-2. For motivations for why each of these tests is suitable for stated situations, and the full set of conditions for the validity of the tests, see statistics textbooks such as Wasserman (2004) or Rice (1995). The docstring for each listed functions in the SciPy `stats` module also contains further information about each test.

Table 13-2. *Summary of Common Hypothesis Test Cases with the Corresponding Distributions and SciPy Functions*

Null Hypothesis	Distributions	SciPy Functions for Test
Test if the mean of a population is a given value.	Normal distribution (`stats.norm`), or Student's *t* distribution (`stats.t`)	`stats.ttest_1samp`
Test if the means of two random variables are equal (independent or paired samples).	Student's *t* distribution (`stats.t`)	`stats.ttest_ind`, `stats.ttest_rel`
Test goodness of fit of a continuous distribution to data.	Kolmogorov-Smirnov distribution	`stats.kstest`
Test if categorical data occur with given frequency (sum of squared normally distributed variables).	χ^2 distribution (`stats.chi2`)	`stats.chisquare`
Test for the independence of categorical variables in a contingency table.	χ^2 distribution (`stats.chi2`)	`stats.chi2_contingency`
Test for equal variance in samples of two or more variables.	*F* distribution (`stats.f`)	`stats.barlett`, `stats.levene`
Test for noncorrelation between two variables.	Beta distribution (`stats.beta`, `stasts.mstats.betai`)	`stats.pearsonr`, `stats.spearmanr`
Test if two or more variables have the same population mean (ANOVA – analysis of variance) .	*F* distribution	`stats.f_oneway`, `stats.kruskal`

In the following we also look at examples of how the corresponding functions in SciPy `stats` module can be used to carry out steps 4 and 5 in the preceding procedure given: computing a test statistic and the corresponding *p*-value.

For example, a common null hypothesis is a claim that the mean μ of a population is a certain value μ_0. We can then sample the population and use the sample mean $\bar{x}$ to form a test statistic $z = \dfrac{\bar{x} - \mu_0}{\sigma / \sqrt{n}}$, where n is the sample size. If the population is large and the variance σ is known, then it is reasonable to use assume that the test statistic is normally distributed. If the variance is unknown, we can substitute σ^2 with the sample variance σ_x^2. The test statistic then follows the Student's *t* distribution, which in the limit of a large number of samples approaches the normal distribution. Regardless of which distribution we end up using, we can compute a *p*-value for the test statistics using the given distribution.

As an example of how this type of hypothesis test can be carried out using the functions provided by the SciPy `stats` module, consider a null hypothesis that claims that a random variable *X* has mean $\mu_0 = 1$. Given samples of *X*, we then wish to test if the sampled data is compatible with the null hypothesis. Here we simulate the samples by drawing 100 random samples from a distribution slightly different than that claimed by the null hypothesis (using $\mu = 0.8$):

```
In [57]: mu0, mu, sigma = 1.0, 0.8, 0.5
In [58]: X = stats.norm(mu, sigma)
In [59]: n = 100
In [60]: X_samples = X.rvs(n)
```

Given the sample data, `X_samples`, next we need to compute a test statistic. If the population standard deviation σ is known, as in this example, we can use $z = \dfrac{\bar{x} - \mu_0}{\sigma / \sqrt{n}}$, which is normally distributed.

```
In [61]: z = (X_samples.mean() - mu0)/(sigma/np.sqrt(n))
In [62]: z
Out[62]: -2.8338979550098298
```

If the population variance is not known, we can use the sample standard deviation instead: $t = \dfrac{\bar{x} - \mu}{\sigma_x / \sqrt{n}}$. However, in this case, the test statistics *t* follows the Student's *t* distribution instead rather than the normal distribution. To compute *t* in this case, we can use the NumPy method std with the ddof=1 argument to compute the sample standard deviation:

```
In [63]: t = (X_samples.mean() - muO)/(X_samples.std(ddof=1)/np.sqrt(n))
In [64]: t
Out[64]: -2.9680338545657845
```

In either case we get a test statistics that we can compare with the corresponding distribution to obtain a *p*-value. For example, for a normal distribution, we can use a stats.norm instance to represent a normal distributed random variable, and with its ppf method, we can look up the statistics value that corresponds to a certain significance level. For a two-sided hypothesis test of significance level 5% (2.5% on each side), the statistics threshold is

```
In [65]: stats.norm().ppf(0.025)
Out[65]: -1.9599639845400545
```

Since the observed statistics is about -2.83, which is smaller than the threshold value –1.96 for a two-sided test with significance level 5%, we have sufficient grounds to reject the null hypothesis in this case. We can explicitly compute the *p*-value for the observed test statistics using the cdf method (multiplied by two for a two-sided test). The resulting *p*-value is indeed rather small, which supports the rejection of the null hypothesis:

```
In [66]: 2 * stats.norm().cdf(-abs(z))
Out[66]: 0.0045984013290753566
```

If we would like to use the *t* distribution, we can use the stats.t class instead of stats.norm. After computing the sample mean, $\bar{x}$, only $n - 1$ degrees of freedom (df) remains in the sample data. The number of degrees of freedom is an important parameter for the *t* distribution, which we need to specify when we create the random variable instance:

```
In [67]: 2 * stats.t(df=(n-1)).cdf(-abs(t))
Out[67]: 0.0037586479674227209
```

The *p*-value is again very small, suggesting that we should reject the null hypothesis. Instead of explicitly carrying out these steps (computing the test statistics, then computing the *p*-value), there are built-in functions in SciPy's stats module for carrying out many common tests, as summarized in Table 13-2. For the test we have used here, we can directly compute the test statistics and the *p*-value using the stats.ttest_1samp function:

```
In [68]: t, p = stats.ttest_1samp(X_samples, mu)
In [69]: t
Out[69]: -2.9680338545657841
In [70]: p
Out[70]: 0.0037586479674227209
```

Again we see that the *p*-value is very small (the same value as in the preceding text) and that we should reject the null hypothesis. It is also illustrative to plot the distribution corresponding to the null hypothesis, together with the sampled data (see Figure 13-5):

```
In [71]: fig, ax = plt.subplots(figsize=(8, 3))
    ...: sns.distplot(X_samples, ax=ax)
    ...: x = np.linspace(*X.interval(0.999), num=100)
    ...: ax.plot(x, stats.norm(loc=mu, scale=sigma).pdf(x))
```

Figure 13-5. *Distribution function according to the null hypothesis (light green) and the sample estimated distribution function (dark blue)*

For another example, consider a two-variable problem, where the null hypothesis states that the population means of two random variables are equal (e.g., corresponding to independent subjects with and without treatment). We can simulate this type of test by creating two random variables with normal distribution, with a randomly chosen population means. Here we select 50 samples for each random variable.

```
In [72]: n, sigma = 50, 1.0
In [73]: mu1, mu2 = np.random.rand(2)
In [74]: X1 = stats.norm(mu1, sigma)
In [75]: X1_sample = X1.rvs(n)
In [76]: X2 = stats.norm(mu2, sigma)
In [77]: X2_sample = X2.rvs(n)
```

We are interested in evaluating if the observed samples provide sufficient evidence that the two population means are not equal (rejecting the null hypothesis). For this situation, we can use the *t* test for two independent samples, which is available in SciPy stats.ttext_ind, which returns the test statistics and the corresponding *p*-value:

```
In [78]: t, p = stats.ttest_ind(X1_sample, X2_sample)
In [79]: t
Out[79]: -1.4283175246005888
In [80]: p
Out[80]: 0.15637981059673237
```

Here the *p*-value is about 0.156, which is not small enough to support rejecting the null hypothesis that the two means are different. In this example the two population means are indeed different:

```
In [81]: mu1, mu2
Out[81]: (0.24764580637159606, 0.42145435527527897)
```

However, the particular samples drawn from these distributions did not statistically prove that these means are different (an error of type II). To increase the power of the statistical test, we would need to increase the number of samples from each random variable.

The SciPy stats module contains functions for common types of hypothesis testing (see the summary in Table 13-2), and their use closely followed what we have seen in the examples in this section. However, some tests require additional arguments for distribution parameters. See the docstrings for each individual test function for details.

Nonparametric Methods

So far we have described random variables with distributions that are completely determined by a few parameters, such as the mean and the variance for the normal distributions. Given the sampled data, we can fit a distribution function using maximum likelihood optimization with respect to the distribution parameters. Such distribution functions are called *parametric*, and statistical methods based on such distribution functions (e.g., a hypothesis test) are called *parametric methods*. When using those methods, we make a strong assumption that the sampled data is indeed described by the given distribution. An alternative approach to constructing a representation of an unknown distribution function is *kernel-density estimation* (KDE), which can be viewed as a smoothened version of the histogram of the sampled data (see, e.g., Figure 13-6). In this method, the probability distribution is estimated by a sum of the kernel function centered at each data point $\hat{f}(x) = \frac{1}{n \cdot \text{bw}} \sum_{i=0}^{n} K\left(\frac{x - x_i}{\text{bw}}\right)$, where bw is a free parameter known as the bandwidth and K is the kernel function (normalized so that it integrates to unity). The bandwidth is an important parameter that defines a scale for the influence of each term in the sum. A too broad bandwidth gives a featureless estimate of the probability distribution, and a too small bandwidth gives a noisy overly structured estimate (see the middle panel in Figure 13-6). Different choices of kernel functions are also possible. A Gaussian kernel is a popular choice, because of its smooth shape with local support, and it is relatively easy to perform computations with.

Figure 13-6. *Histogram (left), kernel-density estimation of the distribution function (middle), and both a histogram and the kernel-density estimate in the same graph (right)*

In SciPy's the KDE method using a Gaussian kernel is implemented in the function `stats.kde.gaussian_kde`. This function returns a callable object that behaves as, and can be used as, a probability distribution function. For example, consider a set of samples, `X_samples`, drawn from a random variable X with unknown distribution (here simulated using the χ^2 distribution with five degrees of freedom):

```
In [82]: X = stats.chi2(df=5)
In [83]: X_samples = X.rvs(100)
```

To compute the kernel-density estimate for the given data, we call the function `stats.kde.guassian_kde` with the array of sample points as argument:

```
In [84]: kde = stats.kde.gaussian_kde(X_samples)
```

By default, a standard method for computing a suitable bandwidth is used, which often gives acceptable results. However, if we wish, we could also specify a function for computing the bandwidth, or directly setting the bandwidth, using the bw_method argument. To set a smaller bandwidth, we can, for example, use

```
In [85]: kde_low_bw = stats.kde.gaussian_kde(X_samples, bw_method=0.25)
```

The gaussian_kde function returns an estimate of the distribution function, which we, for example, can graph or use for other applications. Here we plot a histogram of the data and the two kernel-density estimates (with default and explicitly set bandwidth). For reference, we also plot the true probability distribution function for the samples. The result is shown in Figure 13-6.

```
In [86]: x = np.linspace(0, 20, 100)
In [87]: fig, axes = plt.subplots(1, 3, figsize=(12, 3))
    ...: axes[0].hist(X_samples, normed=True, alpha=0.5, bins=25)
    ...: axes[1].plot(x, kde(x), label="KDE")
    ...: axes[1].plot(x, kde_low_bw(x), label="KDE (low bw)")
    ...: axes[1].plot(x, X.pdf(x), label="True PDF")
    ...: axes[1].legend()
    ...: sns.distplot(X_samples, bins=25, ax=axes[2])
```

The seaborn statistical graphics library provides a convenient function for plotting both a histogram and the kernel-density estimation for a set of data: distplot. A graph produced by this function is shown in the right panel of Figure 13-6.

Given kernel-density estimate, we can also use it to generate new random numbers using the resample method, which takes the number of data points as arguments:

```
In [88]: kde.resample(10)
Out[88]: array([[1.75376869,  0.5812183,  8.19080268,  1.38539326,  7.56980335,
                 1.16144715,  3.07747215,  5.69498716,  1.25685068,  9.55169736]])
```

The kernel-density estimate object does not directly contain methods for computing the cumulative distribution functions (CDF) and its inverse, the percent-point function (PPF). But there are several methods for integrating the kernel-density estimate of the probability distribution function. For example, for a one-dimensional KDE, we can use the integrate_box_1d to obtain the corresponding CDF:

```
In [89]: def _kde_cdf(x):
    ...:         return kde.integrate_box_1d(-np.inf, x)
In [90]: kde_cdf = np.vectorize(_kde_cdf)
```

and can use the SciPy `optimize.fsolve` function to find the inverse (the PPF):

```
In [91]: def _kde_ppf(q):
    ...:         return optimize.fsolve(lambda x, q: kde_cdf(x) - q, kde.
             dataset.mean(), args=(q,))[0]
    ...:
In [92]: kde_ppf = np.vectorize(_kde_ppf)
```

With the CDF and PPF for the kernel-density estimate, we can, for example, perform statistical hypothesis testing and compute confidence intervals. For example, using the kde_ppf function defined in the preceding section, we can compute an approximate 90% confidence interval for the mean of the population from which the sample was collected:

```
In [93]: kde_ppf([0.05, 0.95])
Out[93]: array([  0.39074674,  11.94993578])
```

As illustrated with this example, once we have a KDE that represents the probability distribution for a statistical problem, we can proceed with many of the same methods as we use in parametric statistics. The advantage of nonparametric methods is that we do not necessarily need to make assumptions about the shape of the distribution function. However, because nonparametric methods use less information (weaker assumptions) than parametric methods, their statistical power is lower. Therefore, if we can justify using a parametric method, then that is usually the best approach. Nonparametric methods offer a versatile generic approach that we can fall back on when parametric methods are not feasible.

Summary

In this chapter we have explored how NumPy and the SciPy `stats` module can be used in basic statistical applications, including random number generation, for representing random variables and probability distribution functions, maximum likelihood fitting of distributions to data, and using probability distributions and test statistics for hypothesis testing. We also briefly looked at kernel-density estimation of an unknown probability distribution, as an example of a nonparametric method. The concepts and methods discussed in this chapter are fundamental building blocks for working with statistics, and the computational tools introduced here also provide a foundation for many statistical applications. In the following chapters, we build on what has been discussed here and explore statistical modeling and machine learning in more depth.

Further Reading

Good introductions to the fundamentals of statistics and data analysis are given in Rice (1995) and Wasserman (2004). A computationally oriented introduction to statistics is given in Dalgaard (2008), which, although it uses the R language, is relevant for statistics in Python too. There are also free online resources about statistics, for example, OpenIntro Statistics, which is available from `www.openintro.org/stat/textbook.php`.

References

Dalgaard, P. (2008). *Introductory Statistics with R.* New York: Springer.

Rice, J. A. (1995). *Mathematical Statistics and Data Analysis.* Belmont: Duxbury Press.

Wasserman, L. (2004). *All of statistics.* New York: Springer.

CHAPTER 14

Statistical Modeling

In the previous chapter, we covered basic statistical concepts and methods. In this chapter we build on the foundation laid out in the previous chapter and explore statistical modeling, which deals with creating models that attempt to explain data. A model can have one or several parameters, and we can use a fitting procedure to find the values of the parameter that best explains the observed data. Once a model has been fitted to data, it can be used to predict the values of new observations, given the values of the independent variables of the model. We can also perform statistical analysis on the data and the fitted model and try to answer questions such as if the model accurately explains the data, which factors in the model are more relevant (predictive) than others, and if there are parameters that do not contribute significantly to the predictive power of the model.

In this chapter we mainly use the statsmodels library. It provides classes and functions for defining statistical models and fitting them to observed data, for calculating descriptive statistics, and for carrying out statistical tests. The statsmodels library has some overlap with the SciPy `stats` module that we covered in the previous chapter, but it is mostly an extension of what is available in SciPy[1]. In particular, the main focus of the statsmodels library is on fitting models to data rather than probability distributions and random variables, for which in many cases it relies on the SciPy `stats`.

---

statsmodels The statsmodels library provides a rich set of functionality related to statistical tests and statistical modeling, including linear regression, logistic regression, and time-series analysis. For more information about the project and its documentation, see the project's web page at `http://statsmodels. sourceforge.net`. At the time of writing, the latest version of statsmodels is 0.9.0.

---

[1]The statsmodels library originally started as a part of the SciPy `stats` module but was later moved to a project on its own. The SciPy `stats` library remains an important dependency for statsmodels.

© Robert Johansson 2019
R. Johansson, *Numerical Python*, https://doi.org/10.1007/978-1-4842-4246-9_14

The statsmodels library is closely integrated with the Patsy library, which allows us to write statistical models as simple formulas. The Patsy library is one of the dependencies of the statsmodels library but can also be used with other statistical libraries as well, such as scikit-learn that will be discussed in Chapter 15. However, here we will introduce the Patsy library in the context of using it together with the statsmodels library.

---

patsy   The patsy library provides features for defining statistical models with a simple formula language inspired by statistical software such as R. The patsy library is designed to be a companion library for statistical modeling packages, such as statsmodels. For more information about the project and its documentation, see the web page at `http://patsy.readthedocs.org`. At the time of writing, the most recent version of Patsy is 0.5.0.

Importing Modules

In this chapter we work extensively with the statsmodels library. This library encourages an import convention that is slightly different than other libraries we have used so far: It provides `api` modules that collect the publically accessible symbols that the library provides. Here we assume that the `statsmodels.api` is imported under the name `sm`, and `statsmodels.formula.api` is imported as the name `smf`. We also require the `statsmodels.graphics.api` module to be imported as the name `smg`:

```
In [1]: import statsmodels.api as sm
In [2]: import statsmodels.formula.api as smf
In [3]: import statsmodels.graphics.api as smg
```

Since the statsmodels library internally uses the Patsy library, it is normally not necessary to access this library's functions directly. However, here we directly use Patsy for demonstration purposes, and we therefore need to import the library explicitly:

```
In [4]: import patsy
```

As usual, we also require the Matplotlib, NumPy, and Pandas libraries to be imported as

```
In [5]: import matplotlib.pyplot as plt
In [6]: import numpy as np
In [7]: import pandas as pd
```

and the SciPy `stats` module as

In [8]: from scipy import stats

Introduction to Statistical Modeling

In this chapter we consider the following type of problem: For a set of response (dependent) variables Y and explanatory (independent) variables X, we wish to find a mathematical relationship (model) between Y and X. In general, we can write a mathematical model as a function $Y = f(X)$. Knowing the function $f(X)$ would allow us to compute the value of Y for any of values X. If we do not know the function $f(X)$, but we have access to data for observations $\{y_i, x_i\}$, we can parameterize the function $f(X)$ and fit the values of the parameters to the data. An example of a parameterization of $f(X)$ is the linear model $f(X) = \beta_0 + \beta_1 X$, where the coefficients β_0 and β_1 are the parameters of the model. Typically we have many more data points than the number of free parameters in the model. In such cases we can, for example, use a least square fit that minimizes the norm of the residual $r = Y - f(X)$, although other minimization objective functions can also be used[2], for example, depending on the statistical properties of the residual r. So far we have described a *mathematical* model. The essential component that makes a model *statistical* is that the data $\{y_i, x_i\}$ has an element of uncertainty, for example, due to measurement noise or other uncontrolled circumstances. The uncertainty in the data can be described in the model as random variables: For example, $Y = f(X) + \varepsilon$, where ε is a random variable. This is a statistical model because it includes random variables. Depending on how the random variables appear in the model and what distributions the random variables follow, we obtain different types of statistical models, which each may require different approaches to analyze and solve.

A typical situation where a statistical model can be used is to describe the observations y_i in an experiment, where x_i is a vector with control knobs that are recorded together with each observation. An element in x_i may or may not be relevant for predicting the observed outcome y_i, and an important aspect of statistical modeling is to determine which explanatory variables are relevant. It is of course also possible that there are relevant factors that are not included in the set of explanatory variables x_i, but which influence the outcome of the observation y_i. In this case, it might not be possible

---

[2]We will see examples of this later in Chapter 15, when we consider regularized regression.

to accurately explain the data with the model. Determining if a model accurately explains the data is another essential aspect of statistical modeling.

A widely used statistical model is $Y = \beta_0 + \beta_1 X + \varepsilon$, where β_0 and β_1 are model parameters and ε is normally distributed with zero mean and variance σ^2: $\varepsilon \sim N(0, \sigma^2)$. This model is known as *simple linear regression* if X is a scalar, *multiple linear regression* if X is a vector, and *multivariate linear regression* if Y is a vector. Because the residual ε is normally distributed, for all these cases, the model can be fitted to data using *ordinary least squares* (OLS). Relaxing the condition that the elements in Y, in the case of multivariate linear regression, must be independent and normally distributed with equal variance gives rise to variations of the model that can be solved with methods known as *generalized least squares* (GLS) and *weighted least squares* (WLS). All methods for solving statistical models typically have a set of assumptions that one has to be mindful of when applying the models. For standard linear regression, the most important assumption is that the residuals are independent and normally distributed.

The *generalized linear model* is an extension of the linear regression model that allows the errors in the response variable to have distributions other than the normal distribution. In particular, the response variable is assumed to be a function of a linear predictor and where the variance of the response variable can be a function of the variable's value. This provides a broad generalization of the linear model that is applicable in many situations. For example, this enables modeling important types of problems where the response variable takes discrete values, such as binary outcomes of count values. The errors in the response variables of such models may follow different statistical distributions (e.g., the binomial and/or the Poisson distribution). Examples of these types of models include *logistic regression* for binary outcomes and *Poisson regression* for positive integer outcomes.

In the following sections, we will explore how statistical models of these types can be defined and solved using the Patsy and statsmodels libraries.

Defining Statistical Models with Patsy

Common to all statistical modeling is that we need to make assumptions about the mathematical relation between the response variables Y and explanatory variables X. In the vast majority of cases, we are interested in linear models, such that Y can be written as a linear combination of the response variables X, or functions of the response variables, or models that have a linear component. For example, $Y = \alpha_1 X_1 + \ldots + \alpha_n X_n$, $Y = \alpha_1 X + \alpha_2 X^2 \ldots + \alpha_n X^n$, and $Y = \alpha_1 \sin X_1 + \alpha_2 \cos X_2$ are all examples of such linear models. Note that for the model to be linear, we only need the relation to be linear with respect to the unknown coefficients α and not necessarily in the known explanatory variables X. In contrast, an example of a nonlinear model is $Y = \exp(\beta_0 + \beta_1 X)$, since in this case Y is not a linear function with respect to β_0 and β_1. However, this model is *log-linear* in the sense that taking the logarithm of the relation yields a linear model: $\tilde{Y} = \beta_0 + \beta_1 X$ for $\tilde{Y} = \log Y$. Problems that can be transformed into a linear model in this manner are the type of problems that can be handled with the generalized linear model.

Once the mathematical form of the model has been established, the next step is often to construct the so-called design matrices y and X such that the regression problem can be written in matrix form as $y = X\beta + \varepsilon$, where y is the vector (or matrix) of observations, β is a vector of coefficients, and ε is the residual (error). The elements X_{ij} of the design matrix X are the values of the (functions of) explanatory variables corresponding to each coefficient β_j and observation y_i. Many solvers for statistical models in statsmodels and other statistical modeling libraries can take the design matrices X and y as input.

For example, if the observed values are $y = [1, 2, 3, 4, 5]$ with two independent variables with corresponding values $x_1 = [6, 7, 8, 9, 10]$ and $x_2 = [11, 12, 13, 14, 15]$, and if the linear model under consideration is $Y = \beta_0 + \beta_1 X_1 + \beta_2 X_2 + \beta_3 X_1 X_2$, then the design matrix for the right-hand side is $X = [\mathbf{1}, x_1, x_2, x_1 x_2]$. We can construct this design matrix using the NumPy vstack function:

```
In [9]: y = np.array([1, 2, 3, 4, 5])
In [10]: x1 = np.array([6, 7, 8, 9, 10])
In [11]: x2 = np.array([11, 12, 13, 14, 15])
In [12]: X = np.vstack([np.ones(5), x1, x2, x1*x2]).T
In [13]: X
Out[13]: array([[   1.,    6.,   11.,    66.],
                [   1.,    7.,   12.,    84.],
                [   1.,    8.,   13.,   104.],
                [   1.,    9.,   14.,   126.],
                [   1.,   10.,   15.,   150.]])
```

Given the design matrix X and observation vector y, we can solve for the unknown coefficient vector β, for example, using least square fit (see Chapters 5 and 6):

```
In [14]: beta, res, rank, sval = np.linalg.lstsq(X, y)
In [15]: beta
Out[15]: array([ -5.55555556e-01, 1.88888889e+00, -8.88888889e-01, -1.33226763e-15])
```

These steps are the essence of statistical modeling in its simplest form. However, variations and extensions to this basic method make statistical modeling a field in its own right and call for computational frameworks such as statsmodels for systematic analysis. For example, although constructing the design matrix X was straightforward in this simple example, it can be tedious for more involved models and if we wish to be able to easily change how the model is defined. This is where the Patsy library enters the picture. It offers a convenient (although not necessarily intuitive) formula language for defining a model and automatically constructing the relevant design matrices. To construct the design matrix for a Patsy formula, we can use the `patsy.dmatrices` function. It takes the formula as a string as the first argument and a dictionary-like object with data arrays for the response and explanatory variables as the second argument. The basic syntax for the Patsy formula is `"y ~ x1 + x2 + ..."`, which means that `y` is a linear combination of the explanatory variables `x1` and `x2` (explicitly including an intercept coefficient). For a summary of the Patsy formula syntax, see Table 14-1.

As an introductory example, consider again the linear model $Y = \beta_0 + \beta_1 X_1 + \beta_2 X_2 + \beta_3 X_1 X_2$ that we used earlier. To define this model with Patsy, we can use the formula `"y ~ 1 + x1 + x2 + x1*x2"`. Note that we leave out coefficients in the model formula, as it is implicitly assumed that each term in the formula has a model parameter as a coefficient. In addition to specifying the formula, we also need to create a dictionary `data` that maps the variable names to the corresponding data arrays:

```
In [16]: data = {"y": y, "x1": x1, "x2": x2}
In [17]: y, X = patsy.dmatrices("y ~ 1 + x1 + x2 + x1*x2", data)
```

The result is two arrays `y` and `X`, which are the design matrices for the given data arrays and the specified model formula:

```
In [18]: y
Out[18]: DesignMatrix with shape (5, 1)
```

```
          y
          1
          2
          3
          4
          5
        Terms:
          'y' (column 0)
In [19]: X
Out[19]: DesignMatrix with shape (5, 4)
          Intercept   x1   x2   x1:x2
                  1    6   11      66
                  1    7   12      84
                  1    8   13     106
                  1    9   14     126
                  1   10   15     150
        Terms:
          'Intercept' (column 0)
          'x1' (column 1)
          'x2' (column 2)
          'x1:x2' (column 3)
```

These arrays are of type DesignMatrix, which is a Patsy-supplied subclass of the standard NumPy array, which contains additional metadata and an altered printing representation.

```
In [20]: type(X)
Out[20]: patsy.design_info.DesignMatrix
```

Note that the numerical values of the DesignMatrix array are equal to those of the explicitly constructed array that we produced earlier using vstack.

As a subclass of the NumPy ndarray, the arrays of type DesignMatrix are fully compatible with code that expects NumPy arrays as input. However, we can also explicitly cast a DesignMatrix instance into an ndarray object using the np.array function, although this normally should not be necessary.

```
In [21]: np.array(X)
Out[21]: array([[    1.,     6.,    11.,     66.],
                [    1.,     7.,    12.,     84.],
                [    1.,     8.,    13.,    104.],
                [    1.,     9.,    14.,    126.],
                [    1.,    10.,    15.,    150.]])
```

Alternatively, we can set the return_type argument to "dataframe", in which case the patsy.dmatrices function returns design matrices in the form of Pandas DataFrame objects. Also note that since DataFrame objects behave as dictionary-like objects, so we can use data frames to specify the model data as the second argument to the patsy. dmatrices function.

```
In [22]: df_data = pd.DataFrame(data)
In [23]: y, X = patsy.dmatrices("y ~ 1 + x1 + x2 + x1:x2", df_data, return_
         type="dataframe")
In [24]: X
Out[24]:
```

	Intercept	x1	x2	x1:x2
0	1	6	11	66
1	1	7	12	84
2	1	8	13	104
3	1	9	14	126
4	1	10	15	150

With the help of Patsy, we have now automatically created the design matrices required for solving a statistical model, using, for example, the np.linalg.lstsq function (as we saw an example earlier), or using one of the many statistical model solvers provided by the statsmodels library. For example, to perform an ordinary linear regression (OLS), we can use the class OLS from the statsmodels library instead of using the lower-level method np.linalg.lstsq. Nearly all classes for statistical models in statsmodels take the design matrices y and X as the first and second arguments and return a class instance that represents the model. To actually fit the model to the data encoded in the design matrices, we need to invoke the fit method, which returns a result object that contains fitted parameters (among other attributes):

```
In [25]: model = sm.OLS(y, X)
In [26]: result = model.fit()
In [27]: result.params
Out[27]: Intercept    -5.555556e-01
         x1            1.888889e+00
         x2           -8.888889e-01
         x1:x2        -8.881784e-16
         dtype: float64
```

Note that the result is equivalent to the least square fitting that we computed earlier in this chapter. Using the statsmodels formula API (the module that we imported as smf), we can directly pass the Patsy formula for the model when we create a model instance, which completely eliminates the need for first creating the design matrices. Instead of passing y and X as arguments, we then pass the Patsy formula and the dictionary-like object (e.g., a Pandas data frame) that contains the model data.

```
In [28]: model = smf.ols("y ~ 1 + x1 + x2 + x1:x2", df_data)
In [29]: result = model.fit()
In [30]: result.params
Out[30]: Intercept    -5.555556e-01
         x1            1.888889e+00
         x2           -8.888889e-01
         x1:x2        -8.881784e-16
         dtype: float64
```

The advantage of using statsmodels instead of explicitly constructing NumPy arrays and calling the NumPy least square model is, of course, that much of the process is automated in statsmodels, which makes it possible to add and remove terms in the statistical model without any extra work. Also, when using statsmodels, we have access to a large variety of linear model solvers and statistical tests for analyzing how well the model fits the data. For a summary of the Patsy formula language, see Table 14-1.

Table 14-1. *Simplified Summary of the Patsy Formula Syntax. For a complete specification of the formula syntax, see the Patsy documentation at* `http://patsy.readthedocs.org/en/latest`

Syntax	Example	Description
`lhs ~ rhs`	`y ~ x` (Equivalent to y ~ 1 + x)	The ~ character is used to separate the left-hand side (containing the dependent variables) and the right-hand side (containing the independent variables) of a model equation.
`var * var`	`x1*x2` (Equivalent to x1+x2+x1*x2)	An interaction term that implicitly contains all its lower-order interaction terms.
`var + var + ...`	`x1 + x2 + ...` (Equivalent to y ~ 1 + x1 + x2)	The addition sign is used to denote the union of terms.
`var:var`	`x1:x2`	The colon character denotes a pure interaction term (e.g., $x_1 \cdot x_2$).
`f(expr)`	`np.log(x),` `np.cos(x+y)`	Arbitrary Python functions (often NumPy functions) can be used to transform terms in the expression. The expression for the argument of a function is interpreted as an arithmetic expression rather than the set-like formula operations that are otherwise used in Patsy.
`I(expr)`	`I(x+y)`	I is a Patsy-supplied identity function that can be used to escape arithmetic expression so that they are interpreted as arithmetic operations.
`C(var)`	`C(x), C(x, Poly)`	Treat the variable x as a categorical variable, and expand its values into orthogonal dummy variables.

Now that we have seen how a Patsy formula can be used to construct design matrices, or be used directly with one of the many statistical model classes from statsmodels, we briefly return to the syntax and notational conventions for Patsy formulae before we continue and look in more detail on different statistical models that are available in the statsmodels library. As mentioned in the preceding text, and summarized in Table 14-1, the basic syntax for a model formula has the form "LHS ~ RHS". The ~ character is used to separate the left-hand side (LHS) and the right-hand side (RHS) of the model equation. The LHS specifies the terms that constitute the response variables, and the RHS specifies the terms that constitute the explanatory variables. The terms in the LHS and RHS expressions are separated by + or – signs, but these should not be interpreted as arithmetic operators, but rather as set-union and set-difference operators. For example, a+b means that both *a* and *b* are included in the model, and -a means that the term *a* is *not* included in the model. An expression of the type a*b is automatically expanded to a + b + a:b, where a:b is the pure interaction term *a • b*.

As a concrete example, consider the following formula and the resulting right-hand side terms (which we can extract from the design_info attribute using the term_names attribute):

```
In [31]: from collections import defaultdict
In [32]: data = defaultdict(lambda: np.array([]))
In [33]: patsy.dmatrices("y ~ a", data=data)[1].design_info.term_names
Out[33]: ['Intercept', 'a']
```

Here the two terms are Intercept and a, which correspond to constant and a linear dependence on *a*. By default Patsy always includes the intercept constant, which in the Patsy formula also can be written explicitly using y ~ 1 + a. Including the 1 in the Patsy formula is optional.

```
In [34]: patsy.dmatrices("y ~ 1 + a + b", data=data)[1].design_info.term_names
Out[34]: ['Intercept', 'a', 'b']
```

In this case we have one more explanatory variable (a and b), and here the intercept is explicitly included in the formula. If we do not want to include the intercept in the model, we can use the notation -1 to remove this term:

```
In [35]: patsy.dmatrices("y ~ -1 + a + b", data=data)[1].design_info.term_names
Out[35]: ['a', 'b']
```

Expressions of the type a * b are automatically expanded to include all lower-order interaction terms:

```
In [36]: patsy.dmatrices("y ~ a * b", data=data)[1].design_info.term_names
Out[36]: ['Intercept', 'a', 'b', 'a:b']
```

Higher-order expansions work too:

```
In [37]: patsy.dmatrices("y ~ a * b * c", data=data)[1].design_info.term_names
Out[37]: ['Intercept', 'a', 'b', 'a:b', 'c', 'a:c', 'b:c', 'a:b:c']
```

To remove a specific term from a formula, we can write the term preceded by the minus operator. For example, to remove the pure third-order interaction term a:b:c from the automatic expansion of a*b*c, we can use

```
In [38]: patsy.dmatrices("y ~ a * b * c - a:b:c", data=data)[1].design_
            info.term_names
Out[38]: ['Intercept', 'a', 'b', 'a:b', 'c', 'a:c', 'b:c']
```

In Patsy, the + and - operators are used for set-like operations on sets of terms; if we need to represent the arithmetic operations, we need to wrap the expression in a function call. For convenience, Patsy provides an identity function with the name I that can be used for this purpose. To illustrate this point, consider the following two examples, which show the resulting terms for y ~ a + b and y ~ I(a + b):

```
In [39]: data = {k: np.array([]) for k in ["y", "a", "b", "c"]}
In [40]: patsy.dmatrices("y ~ a + b", data=data)[1].design_info.term_names
Out[40]: ['Intercept', 'a', 'b']
In [41]: patsy.dmatrices("y ~ I(a + b)", data=data)[1].design_info.term_names
Out[41]: ['Intercept', 'I(a + b)']
```

Here the column in the design matrix that corresponds to the term with the name I(a+b) is the arithmetic sum of the arrays for the variables a and b. The same trick must be used if we want to include terms that are expressed as a power of a variable:

```
In [42]: patsy.dmatrices("y ~ a**2", data=data)[1].design_info.term_names
Out[42]: ['Intercept', 'a']
In [43]: patsy.dmatrices("y ~ I(a**2)", data=data)[1].design_info.term_names
Out[43]: ['Intercept', 'I(a ** 2)']
```

The notation I(...) that we used here is an example of a function call notation. We can apply transformations of the input data in a Patsy formula by including arbitrary Python function calls in the formula. In particular, we can transform the input data array using functions from NumPy:

```
In [44]: patsy.dmatrices("y ~ np.log(a) + b", data=data)[1].design_info.
         term_names
Out[44]: ['Intercept', 'np.log(a)', 'b']
```

Or we can even transform variables with arbitrary Python functions:

```
In [45]: z = lambda x1, x2: x1+x2
In [46]: patsy.dmatrices("y ~ z(a, b)", data=data)[1].design_info.term_names
Out[46]: ['Intercept', 'z(a, b)']
```

So far we have considered models with numerical response and explanatory variables. Statistical modeling also frequently includes categorical variables, which can take a discrete set of values that do not have a meaningful numerical order (e.g., "Female" or "Male"; type "A", "B", or "C"; etc.). When using such variables in a linear model, we typically need to recode them by introducing binary dummy variables. In a patsy formula any variable that does not have a numerical data type (float or int) will be interpreted as a categorical variable and automatically encoded accordingly. For numerical variables, we can use the C(x) notation to explicitly request that a variable x should be treated as a categorical variable.

For example, compare the following two examples that show the design matrix for the formula "y ~ - 1 + a" and "y ~ - 1 + C(a)", which corresponds to models where a is a numerical and categorical explanatory variable, respectively:

```
In [48]: data = {"y": [1, 2, 3], "a": [1, 2, 3]}
In [48]: patsy.dmatrices("y ~ - 1 + a", data=data, return_type="dataframe")[1]
Out[48]:
```

	a
0	1
1	2
2	3

For a numerical variable, the corresponding column in the design matrix simply corresponds to the data vector, while for a categorical variable C(a) new binary-valued columns with a mask-like encoding of individual values of the original variable are added to the design matrix:

```
In [49]: patsy.dmatrices("y ~ - 1 + C(a)", data=data, return_
         type="dataframe")[1]
Out[49]:
```

	C(a)[1]	C(a)[2]	C(a)[3]
0	1	0	0
1	0	1	0
2	0	0	1

Variables with nonnumerical values are automatically interpreted and treated as categorical values:

```
In [50]: data = {"y": [1, 2, 3], "a": ["type A", "type B", "type C"]}
In [51]: patsy.dmatrices("y ~ - 1 + a", data=data, return_type="dataframe")[1]
Out[51]:
```

	a[type A]	a[type B]	a[type C]
0	1	0	0
1	0	1	0
2	0	0	1

The default type of encoding of categorical variables into binary-valued treatment fields can be changed and extended by the user. For example, to encode the categorical variables with orthogonal polynomials instead of treatment indicators, we can use C(a, Poly):

```
In [52]: patsy.dmatrices("y ~ - 1 + C(a, Poly)", data=data, return_
type="dataframe")[1]
```

Out[52]:

	C(a, Poly).Constant	C(a, Poly).Linear	C(a, Poly).Quadratic
0	1	-7.071068e-01	0.408248
1	1	-5.551115e-17	-0.816497
2	1	7.071068e-01	0.408248

The automatic encoding of categorical variables by Patsy is a very convenient aspect of Patsy formula, which allows the user to easily add and remove both numerical and categorical variables in a model. This is arguably one of the main advantages of using the Patsy library to define model equations.

Linear Regression

The statsmodels library supports several types of statistical models that are applicable in varying situations, but nearly all follow the same usage pattern, which makes it easy to switch between different models. Statistical models in statsmodels are represented by model classes. These can be initiated given the design matrices for the response and explanatory variables of a linear model or given a Patsy formula and a data frame (or another dictionary-like object). The basic workflow when setting up and analyzing a statistical model with statsmodels includes the following steps:

1. Create an instance of a model class, for example, using `model = sm.MODEL(y, X)` or `model = smf.model(formula, data)`, where MODEL and model are the name of a particular model, such as OLS, GLS, Logit, etc. Here the convention is that uppercase names are used for classes that take design matrices as arguments and lowercase names for classes that take Patsy formulas and data frames as arguments.

2. Creating a model instance does not perform any computations. To fit the model to the data, we must invoke the fit method, `result = model.fit()`, which performs the fit and returns a result object that has methods and attributes for further analysis.

3. Print summary statistics for the result object returned by the fit method. The result object varies in content slightly for each statistical model, but most models implement the method

485

summary, which produces a summary text that describes the result of the fit, including several types of statistics that can be useful for judging if the statistical model successfully explains the data. Viewing the output from the summary method is usually a good starting point when analyzing the result of a fitting process.

4. Post-process the model fit results: in addition to the summary method, the result object also contains methods and attributes for obtaining the fitted parameters (params), the residual for the model and the data (resid), the fitted values (fittedvalues), and a method for predicting the value of the response variables for new independent variables (predict).

5. Finally, it may be useful to visualize the result of the fitting, for example, with the Matplotlib and Seaborn graphics libraries, of using some of the many graphing routines that are directly included in the statsmodels library (see the statsmodels. graphics module).

To demonstrate this workflow with a simple example, in the following we consider fitting a model to generate data whose true value is $y = 1 + 2x_1 + 3x_2 + 4x_1x_2$. We begin with storing the data in a Pandas data frame object:

```
In [53]: N = 100
In [54]: x1 = np.random.randn(N)
In [55]: x2 = np.random.randn(N)
In [56]: data = pd.DataFrame({"x1": x1, "x2": x2})
In [57]: def y_true(x1, x2):
   ...:     return 1 + 2 * x1 + 3 * x2 + 4 * x1 * x2
In [58]: data["y_true"] = y_true(x1, x2)
```

Here we have stored the true value of y in the y_true column in the DataFrame object data. We simulate a noisy observation of y by adding a normal distributed noise to the true values and store the result in the y column:

```
In [59]: e = 0.5 * np.random.randn(N)
In [60]: data["y"] = data["y_true"] + e
```

Now, from the data we know that we have two explanatory variables, x1 and x2, in addition to the response variable y. The simplest possible model we can start with is the linear model $Y = \beta_0 + \beta_1 x_1 + \beta_2 x_2$, which we can define with the Patsy formula "y ~ x1 + x2". Since the response variable is continuous, it is a good starting point to fit the model to the data using ordinary linear squares, for which we can use the smf.ols class.

```
In [61]: model = smf.ols("y ~ x1 + x2", data)
In [62]: result = model.fit()
```

Remember that ordinary least square regression assumes that the residuals of the fitted model and the data are normally distributed. However, before analyzing the data, we might not know if this condition is satisfied or not. Nonetheless, we can start by fitting the data to the model and investigate the distribution of the residual using graphical methods and statistical tests (with the null hypothesis that the residuals are indeed normally distributed). A lot of useful information, including several types of test statistics, can be displayed using the summary method:

```
In [63]: print(result.summary())
                            OLS Regression Results
==============================================================================
Dep. Variable:                      y   R-squared:                       0.380
Model:                            OLS   Adj. R-squared:                  0.367
Method:                 Least Squares   F-statistic:                     29.76
Date:                Wed, 22 Apr 2015   Prob (F-statistic):           8.36e-11
Time:                        22:40:33   Log-Likelihood:                -271.52
No. Observations:                 100   AIC:                             549.0
Df Residuals:                      97   BIC:                             556.9
Df Model:                           2
Covariance Type:            nonrobust
==============================================================================
                 coef    std err          t      P>|t|      [95.0% Conf. Int.]
------------------------------------------------------------------------------
Intercept      0.9868      0.382      2.581      0.011       0.228     1.746
x1             1.0810      0.391      2.766      0.007       0.305     1.857
x2             3.0793      0.432      7.134      0.000       2.223     3.936
==============================================================================
Omnibus:                       19.951   Durbin-Watson:                   1.682
```

Prob(Omnibus):	0.000	Jarque-Bera (JB):	49.964
Skew:	-0.660	Prob(JB):	1.41e-11
Kurtosis:	6.201	Cond. No. 1.32	

==

Warnings: [1] Standard errors assume that the covariance matrix of the errors is correctly specified.

The output produced by the summary method is rather verbose, and a detailed description of all the information provided by this method is beyond the scope of this treatment. Instead, here we only focus on a few key indicators. To begin with, the R-squared value is a statistic that indicates how well the model fits the data. It can take values between 0 and 1, where an R-squared statistic of 1 corresponds to a perfect fit. The R-squared value of 0.380 reported in the preceding summary method is rather poor, and it indicates that we need to refine our model (which is expected, since we left out the interaction term $x_1 \cdot x_2$). We can also explicitly access the R-squared statistic from the result object using the rsquared attribute.

```
In [64]: result.rsquared
Out[64]: 0.38025383255132539
```

Furthermore, the coef column in the middle of the table provides the fitted model parameters. Assuming that the residuals indeed are normally distributed, the std err column provides an estimate of the standard errors for the model coefficients, and the t and P>|t| columns are the t-statistics and the corresponding p-value for the statistical test with the null hypothesis that the corresponding coefficient is zero. Therefore, while keeping in mind that this analysis assumes that the residuals are normally distributed, we can look for the columns with small p-values and judge which explanatory variables have coefficients that are very likely to be different from zero (meaning that they have a significant predictive power).

To investigate whether the assumption of normal distributed errors is justified, we need to look at the residuals of the model fit to the data. The residuals are accessible via the resid attribute of the result object:

```
In [65]: result.resid.head()
Out[65]: 0    -3.370455
         1   -11.153477
         2   -11.721319
         3    -0.948410
```

```
4     0.306215
dtype: float64
```

Using these residuals, we can check for normality using the normaltest function from the SciPy stats module:

```
In [66]: z, p = stats.normaltest(result.fittedvalues.values)
In [67]: p
Out[67]: 4.6524990253009316e-05
```

For this example the resulting p-value is indeed very small, suggesting that we can reject the null hypothesis that the residuals are normally distributed (i.e., we can conclude that the assumption of normal distributed residuals is violated). A graphical method to check for normality of a sample is to use the qqplot from the statsmodels.graphics module. The QQ-plot, which compares the sample quantiles with the theoretical quantiles, should be close to a straight line if the sampled values are indeed normally distributed. The following function call to smg.qqplot produces the QQ-plot shown in Figure 14-1:

```
In [68]: fig, ax = plt.subplots(figsize=(8, 4))
    ...: smg.qqplot(result.resid, ax=ax)
```

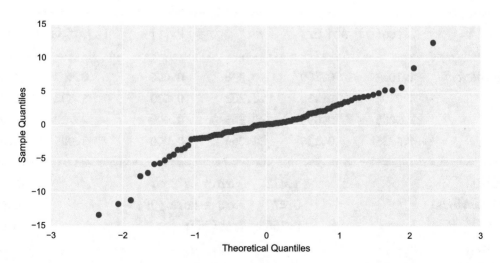

Figure 14-1. *QQ-plot of a linear model with two explanatory variables without any interaction term*

As can be seen in Figure 14-1, the points in the QQ-plot significantly deviate from a linear relation, suggesting that the observed residuals are unlikely to be a sample of a normal distributed random variable. In summary, these indicators provide evidence that the model that we used is not sufficient and that we might need to refine the model. We can include the missing interaction term by adding it to the Patsy formula and repeat the steps from the previous analysis:

```
In [69]: model = smf.ols("y ~ x1 + x2 + x1*x2", data)
In [70]: result = model.fit()
In [71]: print(result.summary())
```

                          OLS Regression Results
==

Dep. Variable:	y	R-squared:	0.963
Model:	OLS	Adj. R-squared:	0.961
Method:	Least Squares	F-statistic:	821.8
Date:	Tue, 21 Apr 2015	Prob (F-statistic):	2.69e-68
Time:	23:52:12	Log-Likelihood:	-138.39
No. Observations:	100	AIC:	284.8
Df Residuals:	96	BIC:	295.2
Df Model:	3		
Covariance Type:	nonrobust		

==

	coef	std err	t	P>|t|	[95.0% Conf. Int.]	
Intercept	1.1023	0.100	10.996	0.000	0.903	1.301
x1	2.0102	0.110	18.262	0.000	1.792	2.229
x2	2.9085	0.095	30.565	0.000	2.720	3.097
x1:x2	4.1715	0.134	31.066	0.000	3.905	4.438

==

Omnibus:	1.472	Durbin-Watson:	1.912
Prob(Omnibus):	0.479	Jarque-Bera (JB):	0.937
Skew:	0.166	Prob(JB):	0.626
Kurtosis:	3.338	Cond. No.	1.54

==

Warnings: [1] Standard errors assume that the covariance matrix of the errors is correctly specified.

In this case we can see that the R-squared statistic is significantly higher, 0.963, indicating a nearly perfect correspondence between the model and the data.

```
In [72]: result.rsquared
Out[72]: 0.96252198253140375
```

Note that we can always increase the R-squared statistic by introducing more variables, but we want to make sure that we do not add variables with low predictive power (small coefficient and high corresponding p-value), since it would make the model susceptible to overfitting, and as usual we require that the residuals be normally distributed. Repeating the normality test and the QQ-plot from the previous analysis with the updated model results in a relatively high *p*-value (0.081) and a relatively linear QQ-plot (see Figure 14-2). This suggests that in this case the residuals could very well be normally distributed (as we know they are, by design, in this example).

```
In [73]: z, p = stats.normaltest(result.fittedvalues.values)
In [74]: p
Out[74]: 0.081352587523644201
In [75]: fig, ax = plt.subplots(figsize=(8, 4))
    ...: smg.qqplot(result.resid, ax=ax)
```

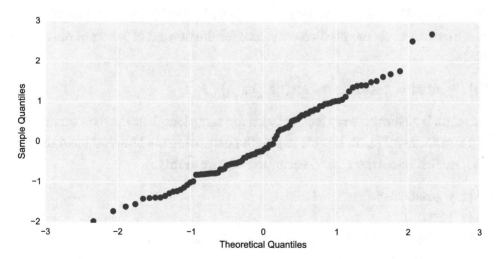

Figure 14-2. *QQ-plot of a linear model with two explanatory variables with an interaction term*

Once we are satisfied with the fit of the model, we can extract the model coefficients from the result object using the `params` attribute.

```
In [76]: result.params
Out[76]: Intercept    1.102297
         x1           2.010154
         x2           2.908453
         x1:x2        4.171501
         dtype: float64
```

Also, we can predict the values of new observations using the `predict` method, which takes as argument a NumPy array or `DataFrame` object with values of the independent variables (x_1 and x_2 in this case). For example, since the current problem has only two independent variables, we can visualize the predictions of the model as a contour plot. To this end, we first construct a DataFrame object with the x_1 and x_2 values for which we want to predict the y-value using the fitted model.

```
In [77]: x = np.linspace(-1, 1, 50)
In [78]: X1, X2 = np.meshgrid(x, x)
In [79]: new_data = pd.DataFrame({"x1": X1.ravel(), "x2": X2.ravel()})
```

Using the `predict` method of the result object obtained from the fitting of the model, we can compute the predicted y values for the new set of values of the response variables.

```
In [80]: y_pred = result.predict(new_data)
```

The result is a NumPy array (vector) with the same length as the data vectors `X1.ravel()` and `X2.ravel()`. To be able to plot the data using the Matplotlib contour function, we first resize the y_pred vector to a square matrix.

```
In [81]: y_pred.shape
Out[81]: (2500,)
In [82]: y_pred = y_pred.values.reshape(50, 50)
```

The contour graphs of the true model and the fitted model are shown in Figure 14-3, which demonstrate that the agreement of the model fitted to the 100 noisy observations of y is sufficient to reproduce the function rather accurately in this example.

```
In [83]: fig, axes = plt.subplots(1, 2, figsize=(12, 5), sharey=True)
    ...: def plot_y_contour(ax, Y, title):
    ...:     c = ax.contourf(X1, X2, Y, 15, cmap=plt.cm.RdBu)
    ...:     ax.set_xlabel(r"$x_1$", fontsize=20)
    ...:     ax.set_ylabel(r"$x_2$", fontsize=20)
    ...:     ax.set_title(title)
    ...:     cb = fig.colorbar(c, ax=ax)
    ...:     cb.set_label(r"$y$", fontsize=20)
    ...:
    ...: plot_y_contour(axes[0], y_true(X1, X2), "true relation")
    ...: plot_y_contour(axes[1], y_pred, "fitted model")
```

Figure 14-3. *The true relation and fit of the correct model to 100 samples from the true relation with normally distributed noise*

In the example we have looked at here, we used the ordinary least square (ols) method to fit the model to the data. Several other options are also available, such as the robust linear model (rlm) that is suitable if there are significant outliers in the observations, and variants of the generalized linear model that is suitable, for example, if the response variable can take only discrete values. This is the topic of the following section. In the following chapter, we will also see examples of regularized regression, where the minimization objective is modified not only to minimize the square of the residuals but also, for example, to penalize large coefficients in the model.

Example Datasets

When working with statistical methods, it is helpful to have example datasets to explore. The statsmodels package provides an interface for loading example datasets from an extensive dataset repository[3] of the R statistical software. The module sm. datasets contains a function get_rdataset that can be used to load datasets listed on the page http://vincentarelbundock.github.io/Rdatasets/datasets.html. The get_rdataset function takes the name of the dataset and optionally also the name of a package (grouping of datasets).

For example, to load a dataset named Icecream from the package Ecdat, we can use

```
In [84]: dataset = sm.datasets.get_rdataset("Icecream", "Ecdat")
```

The result is a data structure with the dataset and metadata describing the dataset. The name of the dataset is given by the title attribute, and the __doc__ attribute contains an explanatory text describing the dataset (too long to display here):

```
In [85]: dataset.title
Out[85]: 'Ice Cream Consumption'
```

The data in the form of a Pandas DataFrame object is accessible via the data attribute:

```
In [86]: dataset.data.info()
<class 'pandas.core.frame.DataFrame'>
Int64Index: 30 entries, 0 to 29
Data columns (total 4 columns):
cons        30 non-null float64
income      30 non-null int64
price       30 non-null float64
temp        30 non-null int64
dtypes: float64(2), int64(2)
memory usage: 1.2 KB
```

From the output given by the DataFrame method info, we can see that the Icecream dataset contains four variables: cons (consumption), income, price, and temp (temperature). Once a dataset is loaded, we can explore it and fit it to statistical models

---

[3]See http://vincentarelbundock.github.io/Rdatasets.

following the usual procedures. For example, to model the consumption as a linear model with price and temperature as independent variables, we can use

```
In [87]: model = smf.ols("cons ~ -1 + price + temp", data=dataset.data)
In [88]: result = model.fit()
```

The result object can be analyzed using descriptive statistics and statistical tests, for example, starting with printing the output from the summary method, as we have seen before. We can also take a graphical approach and plot regression graphs, for example, using the plot_fit function in the smg module (see also the regplot function in the seaborn library):

```
In [89]: fig, (ax1, ax2) = plt.subplots(1, 2, figsize=(12, 4))
    ...: smg.plot_fit(result, 0, ax=ax1)
    ...: smg.plot_fit(result, 1, ax=ax2)
```

Figure 14-4. *Regression plots for the fit of the consumption vs. price and temperature in the Icecream dataset*

From the regression plots shown in Figure 14-4, we can conclude that in this Icecream dataset the consumption seems linearly correlated to the temperature but has no clear dependence on the price (probably because the range of prices is rather small). Graphical tools such as plot_fit can be a useful tool when developing statistical models.

Discrete Regression

Regression with discrete dependent variables (e.g., binary outcomes) requires different techniques than the linear regression model that we have seen so far. The reason is that linear regression requires that the response variable is a normally distributed *continuous* variable, which cannot be used directly for a response variable that has only a few discrete possible outcomes, such as binary variables or variables taking positive integer values. However, using a suitable transformation, it is possible to map a linear predictor to an interval that can be interpreted as a probability of different discrete outcomes. For example, in the case of binary outcomes, one popular transformation is the logistic function $\log(p/(1-p)) = \beta_0 + \beta \cdot x$, or $p = (1 + \exp(-\beta_0 - \beta_1 \cdot x))^{-1}$, which maps $x \in [-\infty, \infty]$ to $p \in [0, 1]$. In other words, the continuous or discrete feature vector x is mapped via the model parameters β_0 and β_1 and the logistic transformation onto a probability p. If $p < 0.5$, it can be taken to predict that $y = 0$, and $p \geq 0.5$ can be taken to predict $y = 1$. This procedure, which is known as logistic regression, is an example of a binary classifier. We will see more about classifiers in Chapter 15 about machine learning.

The statsmodels library provides several methods for discrete regression, including the `Logit` class,[4] the related Probit class (which uses a cumulative distribution function of the normal distribution rather than the logistic function to transform the linear predictor to the [0, 1] interval), the multinomial logistic regression class `MNLogit` (for more than two categories), and the Poisson regression class `Poisson` for Poisson-distributed count variables (positive integers).

Logistic Regression

As an example of how to perform a logistic regression with statsmodels, we first load a classic dataset using the `sm.datasets.get_rdataset` function, which contains sepal and petal lengths and widths for a sample of Iris flowers, together with a classification of the species of the flower. Here we will select a subset of the dataset corresponding to two different species and create a logistic model for predicting the type of species from the values of the petal length and width. The `info` method gives a summary of which variables contained in the dataset:

---

[4]Logistic regression belongs to the class of model that can be viewed as a generalized linear model, with the logistic transformation as link function, so we could alternatively use `sm.GLM` or `smf.glm`.

```
In [90]: df = sm.datasets.get_rdataset("iris").data
In [91]: df.info()
<class 'pandas.core.frame.DataFrame'>
Int64Index: 150 entries, 0 to 149
Data columns (total 5 columns):
Sepal.Length    150 non-null float64
Sepal.Width     150 non-null float64
Petal.Length    150 non-null float64
Petal.Width     150 non-null float64
Species         150 non-null object
dtypes: float64(4), object(1)
memory usage: 7.0+ KB
```

To see how many unique types of species are present in the Species column, we can use the unique method for the Pandas series that is returned when extracting the column from the data frame object:

```
In [92]: df.Species.unique()
Out[92]: array(['setosa', 'versicolor', 'virginica'], dtype=object)
```

This dataset contains three different types of species. To obtain a binary variable that we can use as the response variable in a logistic regression, here we focus only on the data for the two species *versicolor* and *virginica*. For convenience we create a new data frame, df_subset, for the subset of the dataset corresponding to those species:

```
In [93]: df_subset = df[df.Species.isin(["versicolor", "virginica"])].
         copy()
```

To be able to use logistic regression to predict the species using the other variables as independent variables, we first need to create a binary variable that corresponds to the two different species. Using the map method of the Pandas series object, we can map the two species names into binary values 0 and 1.

```
In [94]: df_subset.Species = df_subset.Species.map({"versicolor": 1,
         "virginica": 0})
```

We also need to rename the columns with names that contain period characters to names that are valid symbol names in Python (e.g., by replacing the "." characters with "_"), or else Patsy formulas that include these column names will be interpreted incorrectly. To rename the columns in a `DataFrame` object, we can use the `rename` method and pass a dictionary with name translations as the `columns` argument:

```
In [95]: df_subset.rename(columns={"Sepal.Length": "Sepal_Length",
    ...:                            "Sepal.Width": "Sepal_Width",
    ...:                            "Petal.Length": "Petal_Length",
    ...:                            "Petal.Width": "Petal_Width"},
                           inplace=True)
```

After these transformations we have a `DataFrame` instance that is suitable for use in a logistic regression analysis:

```
In [96]: df_subset.head(3)
Out[96]:
```

	Sepal_Length	Sepal_Width	Petal_Length	Petal_Width	Species
50	7.0	3.2	4.7	1.4	1
51	6.4	3.2	4.5	1.5	1
52	6.9	3.1	4.9	1.5	1

To create a logistic model that attempts to explain the value of the `Species` variable with `Petal_length` and `Petal_Width` as independent variables, we can create an instance of the `smf.logit` class and use the Patsy formula `"Species ~ Petal_Length + Petal_Width"`:

```
In [97]: model = smf.logit("Species ~ Petal_Length + Petal_Width", data=df_
subset)
```

As usual, we need to call the `fit` method of the resulting model instance to actually fit the model to the supplied data. The fit is performed with maximum likelihood optimization.

```
In [98]: result = model.fit()
Optimization terminated successfully.
         Current function value: 0.102818
         Iterations 10
```

As for regular linear regression, we can obtain a summary of the fit of the model to the data by printing the output produced by the `summary` method of the result object. In particular, we can see the fitted model parameters with an estimate for its z-score and the corresponding p-value, which can help us judge whether an explanatory variable is significant or not in the model.

```
In [99]: print(result.summary())
                      Logit Regression Results
==============================================================================
Dep. Variable:               Species   No. Observations:               100
Model:                         Logit   Df Residuals:                    97
Method:                          MLE   Df Model:                         2
Date:               Sun, 26 Apr 2015   Pseudo R-squ.:               0.8517
Time:                       01:41:04   Log-Likelihood:             -10.282
converged:                      True   LL-Null:                    -69.315
LLR p-value:                2.303e-26
==============================================================================
                 coef    std err          z      P>|z|      [95.0% Conf. Int.]
------------------------------------------------------------------------------
Intercept     45.2723     13.612      3.326      0.001      18.594     71.951
Petal_Length  -5.7545      2.306     -2.496      0.013     -10.274     -1.235
Petal_Width  -10.4467      3.756     -2.782      0.005     -17.808     -3.086
==============================================================================
```

The result object for logistic regression also provides the method get_margeff, which returns an object that also implements a summary method that outputs information about the marginal effects of each explanatory variable in the model.

```
In [100]: print(result.get_margeff().summary())
        Logit Marginal Effects
=====================================
Dep. Variable:               Species
Method:                         dydx
At:                          overall
==============================================================================
                dy/dx    std err          z      P>|z|      [95.0% Conf. Int.]
------------------------------------------------------------------------------
Petal_Length  -0.1736      0.052     -3.347      0.001      -0.275     -0.072
Petal_Width   -0.3151      0.068     -4.608      0.000      -0.449     -0.181
==============================================================================
```

When we are satisfied with the fit of the model to the data, we can, for example, use it to predict the value of the response variable for new values of the explanatory variables. For this we can use the `predict` method in the result object produced by the model fitting, and to it we need to pass a data frame object with the new values of the independent variables.

```
In [101]: df_new = pd.DataFrame({"Petal_Length": np.random.randn(20)*0.5 + 5,
     ...:                          "Petal_Width": np.random.randn(20)*0.5 + 1.7})
In [102]: df_new["P-Species"] = result.predict(df_new)
```

The result is an array with probabilities for each observation to correspond to the response $y = 1$, and by comparing this probability to the threshold value 0.5, we can generate predictions for the binary value of the response variable:

```
In [103]: df_new["P-Species"].head(3)
Out[103]: 0     0.995472
          1     0.799899
          2     0.000033
          Name: P-Species, dtype: float64
In [104]: df_new["Species"] = (df_new["P-Species"] > 0.5).astype(int)
```

The intercept and the slope of the line in the plane spanned by the coordinates `Petal_Width` and `Petal_Length` that define the boundary between a point that is classified as $y = 0$ and $y = 1$, respectively, can be computed from the fitted model parameters. The model parameters can be obtained using the `params` attribute of the result object:

```
In [105]: params = result.params
     ...: alpha0 = -params['Intercept']/params['Petal_Width']
     ...: alpha1 = -params['Petal_Length']/params['Petal_Width']
```

Finally, to access the model and its predictions for new data points, we plot a scatter plot of the fitted (squares) and predicted (circles) data where data corresponding to the species *virginica* is coded with blue (dark) color, and the species *versicolor* is coded with green (light) color. The result is shown in Figure 14-10.

```
In [106]: fig, ax = plt.subplots(1, 1, figsize=(8, 4))
     ...:     # species virginica
     ...: ax.plot(df_subset[df_subset.Species == 0].Petal_Length.values,
```

```
...:            df_subset[df_subset.Species == 0].Petal_Width.values,
               's', label='virginica')
...: ax.plot(df_new[df_new.Species == 0].Petal_Length.values,
...:            df_new[df_new.Species == 0].Petal_Width.values,
...:            'o', markersize=10, color="steelblue", label='virginica
               (pred.)')
...:
...: # species versicolor
...: ax.plot(df_subset[df_subset.Species == 1].Petal_Length.values,
...:            df_subset[df_subset.Species == 1].Petal_Width.values,
               's', label='versicolor')
...: ax.plot(df_new[df_new.Species == 1].Petal_Length.values,
...:            df_new[df_new.Species == 1].Petal_Width.values,
...:            'o', markersize=10, color="green", label='versicolor
               (pred.)')
...:
...: # boundary line
...: _x = np.array([4.0, 6.1])
...: ax.plot(_x, alpha0 + alpha1 * _x, 'k')
...: ax.set_xlabel('Petal length')
...: ax.set_ylabel('Petal width')
...: ax.legend()
```

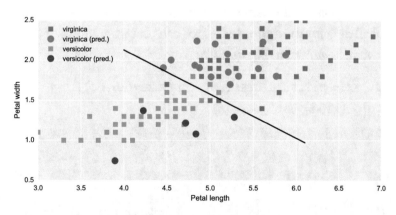

Figure 14-5. *The result of a classification of Iris species using Logit regression with petal length and width and independent variables*

Poisson Model

Another example of discrete regression is the Poisson model, which can describe a process where the response variable is a success count for many attempts that each has a low probability of success. The Poisson model is also an example of a model that can be treated with the generalized linear model, using the natural logarithm as the link function. To see how we can fit data to a Poisson model using the statsmodels library, we will analyze another interesting dataset from the R dataset repository: The discoveries dataset contains counts of the number of great discoveries between 1860 and 1959. Because of the nature of the data, it is reasonable to assume that the counts might be Poisson distributed. To explore this hypothesis, we begin with loading the dataset using the sm.datasets.get_rdataset function and display the first few values to obtain an understanding of the format of the data.

```
In [107]: dataset = sm.datasets.get_rdataset("discoveries")
In [108]: df = dataset.data.set_index("time").rename(columns={"values":
          "discoveries"})
In [109]: df.head(10).T
Out[109]:
```

time	1860	1861	1862	1863	1864	1865	1866	1867	1868	1869
discoveries	5	3	0	2	0	3	2	3	6	1

Here we can see that the dataset contains integer counts in the discoveries series and that the first few years in the series have on average a few great discoveries. To see if this is typical data for the entire series, we can plot a bar graph of the number of discoveries per year, as shown in Figure 14-6.

```
In [109]: fig, ax = plt.subplots(1, 1, figsize=(16, 4))
     ...: df.plot(kind='bar', ax=ax)
```

Figure 14-6. *The number of great discoveries per year*

Judging from Figure 14-6, the number of great discoveries seems to be relatively constant over time, although a slightly declining trend might be noticeable. Nonetheless, the initial hypothesis that the number of discoveries might be Poisson distributed does not look immediately unreasonable. To explore this hypothesis more systematically, we can fit the data to a Poisson process, for example, using the smf.poisson class and the Patsy formula "discoveries ~ 1", which means that we model the discoveries variable with only an intercept coefficient (the Poisson distribution parameter).

```
In [110]: model = smf.poisson("discoveries ~ 1", data=df)
```

As usual we have to call the fit method to actually perform the fit of the model to the supplied data:

```
In [111]: result = model.fit()
Optimization terminated successfully.
         Current function value: 2.168457
         Iterations 7
```

The summary method of the result objects displays a summary of model fit and several fit statistics.

```
In [112]: print(result.summary())
                   Poisson Regression Results
==============================================================================
Dep. Variable:             discoveries   No. Observations:                100
Model:                         Poisson   Df Residuals:                     99
Method:                            MLE   Df Model:                          0
Date:                 Sun, 26 Apr 2015   Pseudo R-squ.:                 0.000
Time:                         14:51:41   Log-Likelihood:              -216.85
converged:                        True   LL-Null:                     -216.85
LLR p-value:                       nan
==============================================================================
                 coef    std err          z      P>|z|      [95.0% Conf. Int.]
------------------------------------------------------------------------------
Intercept      1.1314      0.057     19.920      0.000         1.020      1.243
==============================================================================
```

The model parameters, available via the `params` attribute of the result object, are related to the λ parameter of the Poisson distribution via the exponential function (the inverse of the link function):

```
In [113]: lmbda = np.exp(result.params)
```

Once we have the estimated λ parameter of the Poisson distribution, we can, for example, compare the histogram of the observed count values with the theoretical counts, which we can obtain from a Poisson-distributed random variable from the SciPy stats library.

```
In [114]: X = stats.poisson(lmbda)
```

In addition to the fit parameters, we can also obtain estimated confidence intervals of the parameters using the `conf_int` method:

```
In [115]: result.conf_int()
Out[115]:
```

	0	1
Intercept	1.020084	1.242721

To assess the fit of the data to the Poisson distribution, we also create random variables for the lower and upper bounds of the confidence interval for the model parameter:

```
In [116]: X_ci_l = stats.poisson(np.exp(result.conf_int().values)[0, 0])
In [117]: X_ci_u = stats.poisson(np.exp(result.conf_int().values)[0, 1])
```

Finally, we graph the histogram of the observed counts with the theoretical probability mass functions for the Poisson distributions corresponding to the fitted model parameter and its confidence intervals. The result is shown in Figure 14-7.

```
In [118]: v, k = np.histogram(df.values, bins=12, range=(0, 12),
          normed=True)
In [119]: fig, ax = plt.subplots(1, 1, figsize=(12, 4))
     ...: ax.bar(k[:-1], v, color="steelblue",  align='center',
          label='Discoveries per year')
     ...: ax.bar(k-0.125, X_ci_l.pmf(k), color="red", alpha=0.5,
          align='center', width=0.25,
```

```
...:          label='Poisson fit (CI, lower)')
...: ax.bar(k, X.pmf(k), color="green",  align='center', width=0.5,
     label='Poisson fit')
...: ax.bar(k+0.125, X_ci_u.pmf(k), color="red",  alpha=0.5,
     align='center', width=0.25,
...:          label='Poisson fit (CI, upper)')
...: ax.legend()
```

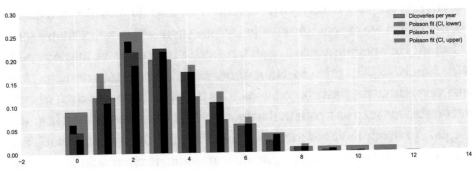

Figure 14-7. *Comparison of histograms of the number of great discoveries per year and the probability mass function for the fitted Poisson model*

The result shown in Figure 14-7 indicates that the dataset of great discoveries is not well described by a Poisson process, since the agreement between Poisson probability mass function and the observed counts deviates significantly. The hypothesis that the number of great discoveries per year is a Poisson process must therefore be rejected. A failure to fit a mode to a given dataset is of course a natural part of the statistical modeling process, and although the dataset turned out not to be Poisson distributed (perhaps because years with a large and small number of great discovers tend to be clustered together), we still have gained insight by the failed attempt to model it as such. Because of the correlations between the number of discoveries at any given year and its recent past, a time-series analysis such as discussed in the following section could be a better approach.

Time Series

Time-series analysis is an important field in statistical modeling that deals with analyzing and forecasting future values of data that is observed as a function of time. Time-series modeling differs in several aspects from the regular regression models that we have looked at so far. Perhaps most importantly, a time series of observations typically cannot be considered as a series of independent random samples from a population. Instead there is often a rather strong component of correlation between observations that are close to each other in time. Also, the independent variables in a time-series model are the past observations of the same series, rather than a set of distinct factors. For example, while a regular regression can describe the demand for a product as a function of its price, in a time-series model it is typical to attempt to predict the future values from the past observations. This is a reasonable approach when there are autocorrelations such as trends in the time series under consideration (e.g., daily or weekly cycles, or steady increasing trends, or inertia in the change of its value). Examples of time series include stock prices, and weather and climate observations, and many other temporal processes in nature and in economics.

An example of a type of statistical model for time series is the autoregressive (AR) model, in which a future value depends linearly on p previous values:

$Y_t = \beta_0 + \sum_{n=1}^{p} \beta_n Y_{t-n} + \varepsilon_t$, where β_0 is a constant and $\beta_n, 1 \le n \le N$ are the coefficients that

define the AR model. The error ε_t is assumed to be white noise without autocorrelation. Within this model, all autocorrelation in the time series should therefore be captured by the linear dependence on the p previous values. A time series which depends linearly on only one previous value (in a suitable unit of time) can be fully modeled with an AR process with $p=1$, denoted as AR(1), and a time series that depends linearly on two previous values can be modeled by an AR(2) process, and so on. The AR model is a special case of the ARMA model, a more general model that also includes a moving average (MA) of q previous residuals of the series: $Y_t = \beta_0 + \sum_{n=1}^{p} \beta_n Y_{t-n} + \sum_{n=1}^{q} \theta_n \varepsilon_{t-n} + \varepsilon_t$,

where the model parameters θ_n are the weight factors for the moving average. This model is known as the ARMA model and is denoted by ARMA(p, q), where p is the number of autoregressive terms and q is the number of moving-average terms. Many other models for time-series model exist, but the AR and ARMA capture the basic ideas that are fundamental to many time-series applications.

The statsmodels library has a submodule dedicated to time-series analysis: `sm.tsa`, which implements several standard models for time-series analysis, as well as graphical and statistical analysis tools for exploring properties of time-series data. For example, let's revisit the time series with outdoor temperature measurements used in Chapter 12 and say that we want to predict the hourly temperature for a few days into the future based on previous observations using an AR model. For concreteness, we will take the temperatures measured during the month of March and predict the hourly temperature for the first 3 days of April. We first load the dataset into a Pandas `DataFrame` object:

```
In [120]: df = pd.read_csv("temperature_outdoor_2014.tsv", header=None,
          delimiter="\t", names=["time", "temp"])
     ...: df.time = pd.to_datetime(df.time, unit="s")
     ...: df = df.set_index("time").resample("H").mean()
```

For convenience we extract the observations for March and April and store them in new `DataFrame` objects, `df_march` and `df_april`, respectively:

```
In [121]: df_march = df[df.index.month == 3]
In [122]: df_april = df[df.index.month == 4]
```

Here we will attempt to model the time series of the temperature observations using the AR model, and an important condition for its applicability is that it is applied to a stationary process, which does not have autocorrelation or trends other than those explained by the terms in the model. The function `plot_acf` in the `smg.tsa` model is a useful graphical tool for visualizing autocorrelation in a time series. It takes an array of time-series observations and graphs the autocorrelation with increasing time delay on the x axis. The optional `lags` argument can be used to determine how many time steps are to be included in the plot, which is useful for long time series and when we only wish to see the autocorrelation for a limited number of time steps. The autocorrelation functions for the temperature observations and its first-, second-, and third-order differences are generated and graphed using the `plot_acf` function in the following code, and the resulting graph is shown in Figure 14-8.

```
In [123]: fig, axes = plt.subplots(1, 4, figsize=(12, 3))
     ...: smg.tsa.plot_acf(df_march.temp, lags=72, ax=axes[0])
     ...: smg.tsa.plot_acf(df_march.temp.diff().dropna(), lags=72,
          ax=axes[1])
```

```
...: smg.tsa.plot_acf(df_march.temp.diff().diff().dropna(), lags=72,
     ax=axes[2])
...: smg.tsa.plot_acf(df_march.temp.diff().diff().diff().dropna(),
     lags=72, ax=axes[3])
```

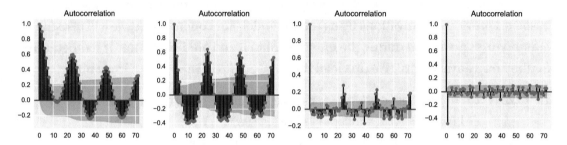

Figure 14-8. *Autocorrelation function for temperature data at increasing order of differentiation, from left to right*

We can see a clear correlation between successive values in the time series in the leftmost graph of Figure 14-8, but for increasing order, differencing of the time series reduces the autocorrelation significantly. Suggesting that while each successive temperature observation is strongly correlated with its preceding value, such correlations are not as strong for the higher-order *changes* between the successive observations. Taking the difference of a time series is often a useful way of detrending it and eliminating correlation. The fact that taking differences diminishes the structural autocorrelation suggests that a sufficiently high-order AR model might be able to model the time series.

To create an AR model for the time series under consideration, we can use the `sm.tsa.AR` class. It can, for example, be initiated with Pandas series that is indexed by `DatetimeIndex` or `PeriodIndex` (see the docstring of AR for alternative ways of pass time-series data to this class):

```
In [124]: model = sm.tsa.AR(df_march.temp)
```

When we fit the model to the time-series data, we need to provide the order of the AR model. Here, since we can see a strong autocorrelation with a lag of 24 periods (24 hours) in Figure 14-8, we must *at least* include terms for 24 previous terms in the model. To be on the safe side, and since we aim to predict the temperature for 3 days, or 72 hours, here we choose to make the order of the AR model correspond to 72 hours as well:

```
In [125]: result = model.fit(72)
```

An important condition for the AR process to be applicable is that the residuals of the series are stationary (no remaining autocorrelation and no trends). The Durbin-Watson statistical test can be used to test for stationary in a time series. It returns a value between 0 and 4, and values close to 2 correspond to time series that do not have remaining autocorrelation. We can also use the plot_acf function to graph the autocorrelation function for the residuals and to verify that the there is no significant autocorrelation.

```
In [126]: sm.stats.durbin_watson(result.resid)
Out[126]: 1.9985623006352975
```

We can also use the plot_acf function to graph the autocorrelation function for the residual and verify that the there is no significant autocorrelation.

```
In [127]: fig, ax = plt.subplots(1, 1, figsize=(8, 3))
     ...: smg.tsa.plot_acf(result.resid, lags=72, ax=ax)
```

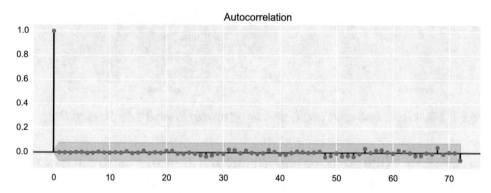

Figure 14-9. *Autocorrelation plot for the residual from the AR(72) model for the temperature observations*

The Durbin-Watson statistic close to 2 and the absence of autocorrelation in Figure 14-9 suggest that the current model successfully explains the fitted data. We can now proceed to forecast the temperature for future dates using the predict method in the result object returned by the model fit method:

```
In [128]: temp_3d_forecast = result.predict("2014-04-01", "2014-04-4")
```

Next we graph the forecast (red) together with the previous 3 days of temperature observations (blue) and the actual outcome (green), for which the result is shown in Figure 14-10:

```
In [129]: fig, ax = plt.subplots(1, 1, figsize=(12, 4))
     ...: ax.plot(df_march.index.values[-72:], df_march.temp.values[-72:],
          label="train data")
     ...: ax.plot(df_april.index.values[:72], df_april.temp.values[:72],
          label="actual outcome")
     ...: ax.plot(pd.date_range("2014-04-01", "2014-04-4", freq="H").values,
          temp_3d_forecast, label="predicted outcome")
     ...:
     ...: ax.legend()
```

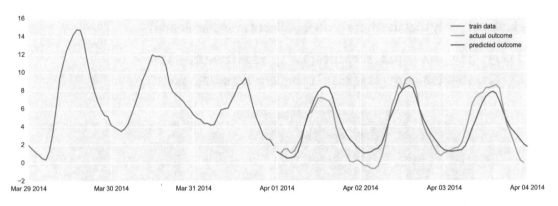

Figure 14-10. *Observed and predicted temperatures as a function of time*

The agreement of the predicted temperature and the actual outcome shown in Figure 14-10 is rather good. However, this will of course not always be the case, as the temperature cannot be forecasted based solely on previous observations. Nonetheless, within a period of a stable weather system, the hourly temperature of a day may be accurately forecasted with an AR model, accounting for the daily variations and other steady trends.

In addition to the basic AR model, statsmodels also provides the ARMA (autoregressive moving average) and ARIMA (autoregressive integrated moving average) models. The usage patterns for these models are similar to that of the AR model we have used here, but there are some differences in the details. Refer to the docstrings for sm. tsa.ARMA and sm.tsa.ARIMA classes and the official statsmodels documentation for further information.

Summary

In this chapter we have briefly surveyed statistical modeling and introduced basic statistical modeling features of the statsmodels library and model specification using Patsy formulas. Statistical modeling is a broad field, and we only scratched the surface of what the statsmodels library can be used for in this chapter. We began with an introduction of how to specify statistical models using the Patsy formula language, which we used in the section on "Linear Regression" for response variables that are continuous (regular linear regression) and discrete (logistic and nominal regression). After having covered linear regression, we briefly looked at time-series analysis, which requires slightly different methods compared to linear regression because of the correlations between successive observations that naturally arise in time series. There are many aspects of statistical modeling that we did not touch upon in this introduction, but the basics of linear regression and time-series modeling that we did cover here should provide a background for further explorations. In Chapter 15 we continue with machine learning, which is a topic that is closely related to statistical modeling in both motivation and methods.

Further Reading

Excellent and thorough introductions to statistical modeling are given in G. James (2013), which is also available for free at `www-bcf.usc.edu/~gareth/ISL/index.html`, and in M. Kuhn (2013). An accessible introduction to time-series analysis is given in R.J. Hyndman (2013), which is also available for free online at `www.otexts.org/fpp`.

References

G. James, D. W. (2013). *An Introduction to Statistical Learning.* New York: Springer-Verlag.

   M. Kuhn, K. J. (2013). *Applied Predictive Modeling.* New York: Springer.

   R.J. Hyndman, G. A. (2013). Forecasting: principles and practice. OTexts.

CHAPTER 15

Machine Learning

In this chapter we explore machine learning. This topic is closely related to statistical modeling, which we considered in Chapter 14, in the sense that both deal with using data to describe and predict outcomes of uncertain or unknown processes. However, while statistical modeling emphasizes the model used in the analysis, machine learning sidesteps the model part and focuses on algorithms that can be trained to predict the outcome of new observations. In other words, the approach taken in statistical modeling emphasizes understanding how the data is generated, by devising models and tuning their parameters by fitting to the data. If the model is found to fit the data well and if it satisfies the relevant model assumptions, then the model gives an overall description of the process, and it can be used to compute statistics with known distributions and for evaluating statistical tests. However, if the actual data is too complex to be explained using available statistical models, this approach has reached its limits. In machine learning, on the other hand, the actual process that generates the data, and potential models thereof, is not central. Instead, the observed data and the explanatory variables are the fundamental starting point of a machine-learning application. Given data, machine-learning methods can be used to find patterns and structure in the data, which can be used to predict the outcome of new observations. Machine learning therefore does not provide an understanding of how data was generated, and because fewer assumptions are made regarding the distribution and statistical properties of the data, we typically cannot compute statistics and perform statistical tests regarding the significance of certain observations. Instead, machine learning puts a strong emphasis on the accuracy with which new observations are predicted.

Although there are significant differences in the fundamental approach taken in statistical modeling and machine learning, many of the mathematical methods that are used are closely related or sometimes even the same. In the course of this chapter, we are going to recognize several methods that we used in Chapter 14 on statistical modeling, but they will be employed with a different mindset and with slightly different goals.

© Robert Johansson 2019
R. Johansson, *Numerical Python*, https://doi.org/10.1007/978-1-4842-4246-9_15

In this chapter we give a brief introduction to basic machine-learning methods, and we survey how such methods can be used in Python. The focus is on machine-learning methods that have broad application in many fields of scientific and technical computing. The most prominent and comprehensive machine-learning library for Python is scikit-learn, although there are several alternative and complementary libraries as well: for example, TensorFlow, Keras, PyTorch, to mention a few. In this chapter we exclusively use the scikit-learn library, which provides implementations of the most common machine-learning algorithm. However, readers that are particularly interested in machine learning are encouraged to also explore the other libraries mentioned in the preceding text as well.

Scikit-learn   The scikit-learn library contains a comprehensive collection of machine-learning-related algorithms, including regression, classification, dimensionality reduction, and clustering. For more information about the project and its documentation, see the project's web page at `http://scikit-learn.org`. At the time of writing, the latest version of scikit-learn is 0.19.2.

Importing Modules

In this chapter we work with the scikit-learn library, which provides the `sklearn` Python module. With the `sklearn` module, here we use the same import strategy as we use with the SciPy library: that is, we explicitly import modules from the library that we need for our work. In this chapter we use the following modules from the `sklearn` library:

```
In [1]: from sklearn import datasets
In [2]: from sklearn import model_selection
In [3]: from sklearn import linear_model
In [4]: from sklearn import metrics
In [5]: from sklearn import tree
In [6]: from sklearn import neighbors
In [7]: from sklearn import svm
In [8]: from sklearn import ensemble
In [9]: from sklearn import cluster
```

For plotting and basic numerics, we also require the Matplotlib and NumPy libraries, which we import in the usual manner:

```
In [10]: import matplotlib.pyplot as plt
In [11]: import numpy as np
```

We also use the Seaborn library for graphics and figure styling:

```
In [12]: import seaborn as sns
```

Brief Review of Machine Learning

Machine learning is a topic in the artificial intelligence field of computer science. Machine learning can be viewed as including all applications where feeding training data into a computer program makes it able to perform a given task. This is a very broad definition, but in practice, machine learning is often associated with a much more specific set of techniques and methods. Here we take a practical approach and explore by example several basic methods and key concepts in machine learning. Before we get started with specific examples, we begin with a brief introduction of the terminology and core concepts.

In machine learning, the process of fitting a model or an algorithm to observed data is known as *training*. Machine-learning applications can often be classified into either of two types: *supervised* and *unsupervised* learning, which differ in the type of data the application is trained with. In *supervised learning*, the data includes feature variables and known response variables. Both feature and response variables can be continuous or discrete. Preparing such data typically requires manual effort and sometimes even expert domain knowledge. The application is thus trained with handcrafted data, and the training can therefore be viewed as supervised machine learning. Examples of applications include regression (prediction of a continuous response variable) and classification (prediction of a discrete response variable), where the value of the response variable is known for the training dataset, but not for new samples.

In contrast, *unsupervised learning* corresponds to situations where machine-learning applications are trained with raw data that is not labeled or otherwise manually prepared. An example of unsupervised learning is clustering of data into groups, or in other words, grouping of data into suitable categories. In contrast to supervised classification, it is typical for unsupervised learning that the final categories are not known in advance, and the training data therefore cannot be labeled accordingly. It may also be the case that the manual labeling of the data is difficult or costly, for example, because the number of samples is too large. It goes without saying that unsupervised machine learning is more difficult and limited in what it can be used for than supervised machine learning, and supervised machine learning therefore should be preferred whenever possible. However, unsupervised machine learning can be a powerful tool when creating labeled training datasets is not possible.

There is naturally much more complexity to machine learning than suggested by the basic types of problems outlined in the preceding text, but these concepts are recurring themes in many machine-learning applications. In this chapter we look at a few examples of basic machine-learning techniques that demonstrate several central concepts of machine learning. Before we do so, we briefly introduce common machine-learning terminology that we will refer to in the following sections:

- *Cross-validation* is the practice of dividing the available data into *training data* and *testing data* (also known as *validation data*), where only the training data is used to train the machine-learning model and where the test data allows the trained application to be tested on previously unseen data. The purpose of this is to measure how well the model predicts new observations and to limit problems with overfitting. There are several approaches to dividing the data into training and testing datasets. For example, one extreme approach is to test all possible ways to divide the data (*exhaustive cross-validation*) and use an aggregate of the result (e.g., average, or the minimum value, depending on the situation). However, for large datasets, the number of possible combinations of train and test data becomes extremely large, making exhaustive cross-validation impractical. Another extreme is to use all but one sample in the training set, and the remaining sample in the training set (*leave-one-out cross-validation*), and to repeat the training-test cycle for all combinations in which one sample is chosen from the available data. A variant of this method is to divide the available data into k groups and perform a leave-one-out cross-validation with the k groups of datasets. This method is known as k-fold cross-validation and is a popular technique that often is used in practice. In the scikit-learn library, the module `sklearn.model_selection` contains functions for working with cross-validation.

- *Feature extraction* is an important step in the preprocessing stage of a machine-learning problem. It involves creating suitable feature variables and the corresponding feature matrices that can be passed to one of many machine-learning algorithms implemented in the scikit-learn library. The scikit-learn module `sklearn.feature_extraction` plays a similar role in many machine-learning applications as the Patsy formula library does in statistical modeling, especially for text- and image-based machine-learning problems. Using methods from the `sklearn.feature_extraction` module, we can automatically assemble feature matrices (design matrices) from various data sources.

- *Dimensionality reduction* and *feature selection* are techniques that are frequently used in machine-learning applications where it is common to have a large number of explanatory variables (features), many of which may not significantly contribute to the predictive power of the application. To reduce the complexity of the model, it is then often desirable to eliminate less useful features and thereby reduce the dimensionality of the problem. This is particularly important when the number of features is comparable to or larger than the number of observations. The scikit-learn modules `sklearn.decomposition` and `sklearn.feature_selection` contain functions for reducing the dimensionality of a machine-learning problem: for example, principal component analysis (PCA) is a popular technique for dimensionality reduction that works by performing a singular-value decomposition of the feature matrix and keeping only dimensions that correspond to the most significant singular vectors.

In the following sections, we look at how scikit-learn can be used to solve examples of machine-learning problems using the techniques discussed in the preceding text. Here we work with generated data and built-in datasets. Like the statsmodels library, scikit-learn comes with a number of built-in datasets that can be used for exploring machine-learning methods. The `datasets` module in `sklearn` provides three groups of functions for loading built-in datasets (with prefix `load_`, e.g., `load_boston`), for fetching external datasets (with prefix `fetch_`, e.g., `fetch_californa_housing`), and finally for generating datasets from random numbers (with prefix `make_`, e.g., `make_regression`).

Regression

Regression is a central part of machine learning and statistical modeling, as we already saw in Chapter 14. In machine learning, we are not so concerned with how well the regression model fits the data, but rather care about how well it predicts new observations. For example, if we have a large number of features and less number of observations, we can typically fit the regression perfectly to the data without it being very useful for predicting new values. This is an example of overfitting: A small residual between the data and regression model is not a guarantee that the model is able to accurately predict future observations. In machine learning, a common method to deal with this problem is to partition the available data into a training dataset and a testing dataset that is used for validating the regression results against previously unseen data.

To see how fitting a training dataset and validating the result against a testing dataset can work out, let's consider a regression problem with 50 samples and 50 features, out of which only 10 features are informative (linearly correlated with the response variable). This simulates a scenario when we have 50 known features, but it turns out that only 10 of those features contribute to the predictive power of the regression model. The make_ regression function in the sklearn.datasets module generates data of kind:

```
In [13]: X_all, y_all = datasets.make_regression(n_samples=50,
            n_features=50, n_informative=10)
```

The result is two arrays, X_all and y_all, of shapes (50, 50) and (50,), corresponding to the design matrices for a regression problem with 50 samples and 50 features. Instead of performing a regression on the entire dataset (and obtain a perfect fit because of the small number of observations), here we split the dataset into two equal size datasets, using the train_test_split function from sklearn.model_selection module. The result is a training dataset X_train, y_train, and a testing dataset X_test, y_test:

```
In [14]: X_train, X_test, y_train, y_test = \
    ...:     model_selection.train_test_split(X_all, y_all, train_size=0.5)
```

In scikit-learn, ordinary linear regression can be carried out using the LinearRegression class from the sklearn.linear_model module, which is comparable with the statsmodels.api.OLS from the statsmodels library. To perform a regression, we first create a LinearRegression instance:

```
In [15]: model = linear_model.LinearRegression()
```

To actually fit the model to the data, we need to invoke the `fit` method, which takes the feature matrix and the response variable vector as the first and second arguments:

```
In [16]: model.fit(X_train, y_train)
Out[16]: LinearRegression(copy_X=True, fit_intercept=True, n_jobs=1,
            normalize=False)
```

Note that compared to the OLS class in statsmodels, the order of the feature matrix and response variable vector is reversed, and in statsmodels the data is specified when the class instance is created instead of when calling the `fit` method. Also, in scikit-learn calling the `fit` method does not return a new result object, but the result is instead stored directly in the model instance. These minor differences are small inconveniences when working interchangeably with the statsmodels and scikit-learn modules and worth taking note of.[1]

Since the regression problem has 50 features and we only trained the model with 25 samples, we can expect complete overfitting that perfectly fits the data. This can be quantified by computing the sum of squared errors (SSEs) between the model and the data. To evaluate the model for a given set of features, we can use the `predict` method, from which we can compute the residuals and the SSE:

```
In [17]: def sse(resid):
   ...:         return np.sum(resid**2)
In [18]: resid_train = y_train - model.predict(X_train)
   ...: sse_train = sse(resid_train)
   ...: sse_train
Out[18]: 8.1172209425431673e-25
```

As expected, for the training dataset, the residuals are all essentially zero, due to the overfitting allowed by having twice as many features as data points. This overfitted model is, however, not at all suitable for predicting unseen data. This can be verified by computing the SSE for our test dataset:

```
In [19]: resid_test = y_test - model.predict(X_test)
   ...: sse_test = sse(resid_test)
   ...: sse_test
Out[19]: 213555.61203039082
```

---

[1]In practice it is common to work with both statsmodels and scikit-learn, as they in many respects complement each other. However, in this chapter we focus solely on scikit-learn.

The result is a very large SSE value, which indicates that the model does not do a good job at predicting new observations. An alternative measure of the fit of a model to a dataset is the R-squared score (see Chapter 14), which we can compute using the score method. It takes a feature matrix and response variable vector as arguments and computes the score. For the training dataset, we obtain, as expected, an r-square score of 1.0, but for the testing dataset, we obtain a low score:

```
In [20]: model.score(X_train, y_train)
Out[20]: 1.0
In [21]: model.score(X_test, y_test)
Out[21]: 0.31407400675201746
```

The big difference between the scores for the training and testing datasets once again indicates that the model is overfitted.

Finally, we can also take a graphical approach and plot the residuals of the training and testing datasets and visually inspect the values of the coefficients and the residuals. From a LinearRegression object, we can extract the fitted parameters using the coef_ attribute. To simplify repeated plotting of the training and testing residuals and the model parameters, here we first create a function plot_residuals_and_coeff for plotting these quantities. We then call the function with the result from the ordinary linear regression model trained and tested on the training and testing datasets, respectively. The result is shown in Figure 15-1, and it is clear that there is a large difference in the magnitude of the residuals for the test and the training datasets, for every sample.

```
In [22]: def plot_residuals_and_coeff(resid_train, resid_test, coeff):
    ...:     fig, axes = plt.subplots(1, 3, figsize=(12, 3))
    ...:     axes[0].bar(np.arange(len(resid_train)), resid_train)
    ...:     axes[0].set_xlabel("sample number")
    ...:     axes[0].set_ylabel("residual")
    ...:     axes[0].set_title("training data")
    ...:     axes[1].bar(np.arange(len(resid_test)), resid_test)
    ...:     axes[1].set_xlabel("sample number")
    ...:     axes[1].set_ylabel("residual")
    ...:     axes[1].set_title("testing data")
    ...:     axes[2].bar(np.arange(len(coeff)), coeff)
    ...:     axes[2].set_xlabel("coefficient number")
```

```
   ...:     axes[2].set_ylabel("coefficient")
   ...:     fig.tight_layout()
   ...:     return fig, axes
In [23]: fig, ax = plot_residuals_and_coeff(resid_train, resid_test,
         model.coef_)
```

Figure 15-1. *The residual between the ordinary linear regression model and the training data (left), the model and the test data (middle), and the values of the coefficients for the 50 features (right)*

The overfitting in this example happens because we have too few samples, and one solution could be to collect more samples until overfitting is no longer a problem. However, this may not always be practical, as collecting observations may be expensive, and because in some applications, we might have a very large number of features. For such situations it is desirable to be able to fit a regression problem in a way that avoids overfitting as much as possible (at the expanse of not fitting the training data perfectly), so that the model can give meaningful predictions for new observations.

Regularized regression is one possible solution to this problem. In the following we look at a few different variations of regularized regression. In ordinary linear regression, the model parameters are chosen such that the sum of squared residuals is minimized. Viewed as an optimization problem, the objective function is therefore $\min_{\beta} \|X\beta - y\|_2^2$, where X is the feature matrix, y is the response variables, and β is the vector of model parameters and where $\|\cdot\|_2$ denotes the L2 norm. In *regularized* regression, we add a *penalty term* in the objective function of the minimization problem. Different types of penalty terms impose different types of regularization on the original regression problem. Two popular regularizations are obtained by adding the L1 or L2 norms of the parameter vector to the minimization objective function, $\min_{\beta} \left\{ \|X\beta - y\|_2^2 + \alpha \|\beta\|_1 \right\}$ and $\min_{\beta} \left\{ \|X\beta - y\|_2^2 + \alpha \|\beta\|_2^2 \right\}$. These are known as LASSO and Ridge regression, respectively.

Here α is a free parameter that determines the strength of the regularization. Adding the L2 norm $\|\beta\|_2^2$ favors model parameter vectors with smaller coefficients, and adding the L1 norm $\|\beta\|_1$ favors a model parameter vectors with as few nonzero elements as possible. Which type of regularization is more suitable depends on the problem at hand: when we wish to eliminate as many features as possible, we can use L1 regularization with LASSO regression, and when we wish to limit the magnitude of the model coefficients, we can use L2 regularization with Ridge regression.

With scikit-learn, we can perform Ridge regression using the Ridge class from the sklearn.linear_model module. The usage of this class is almost the same as the LinearRegression class that we used in the preceding text, but we can also give the value of the α parameter that determines the strength of the regularization as an argument when we initialize the class. Here we chose the value $\alpha = 2.5$. A more systematic approach to choosing α is introduced later in this chapter.

```
In [24]: model = linear_model.Ridge(alpha=2.5)
```

To fit the regression model to the data, we again use the fit method, passing the training feature matrix and response variable as arguments:

```
In [25]: model.fit(X_train, y_train)
Out[25]: Ridge(alpha=2.5, copy_X=True, fit_intercept=True, max_iter=None,
               normalize=False, solver='auto', tol=0.001)
```

Once the model has been fitted to the training data, we can compute the model predictions for the training and testing datasets and compute the corresponding SSE values:

```
In [26]: resid_train = y_train - model.predict(X_train)
    ...: sse_train = sse(resid_train)
    ...: sse_train
Out[26]: 178.50695164950841
In [27]: resid_test = y_test - model.predict(X_test)
    ...: sse_test = sse(resid_test)
    ...: sse_test
Out[27]: 212737.00160105844
```

We note that the SSE of the training data is no longer close to zero, but there is a slight decrease in the SSE for the testing data. For comparison with ordinary regression, we also plot the training and testing residuals and the model parameters using the function plot_residuals_and_coeff that we defined in the preceding text. The result is shown in Figure 15-2.

```
In [28]: fig, ax = plot_residuals_and_coeff(resid_train, resid_test,
            model.coef_)
```

Figure 15-2. *The residual between the Ridge-regularized regression model and the training data (left), the model and the test data (middle), and the values of the coefficients for the 50 features (right)*

Similarly, we can perform the L1-regularized LASSO regression using the Lasso class from the sklearn.linear_model module. It also accepts the value of the α parameter as an argument when the class instance is initialized. Here we choose $\alpha = 1.0$ and perform the fitting of the model to the training data and the computation of the SSE for the training and testing data in the same way as described previously:

```
In [29]: model = linear_model.Lasso(alpha=1.0)
In [30]: model.fit(X_train, y_train)
Out[30]: Lasso(alpha=1.0, copy_X=True, fit_intercept=True, max_iter=1000,
            normalize=False, positive=False, precompute=False, random_
            state=None,
            selection='cyclic', tol=0.0001, warm_start=False)
In [31]: resid_train = y_train - model.predict(X_train)
    ...: sse_train = sse(resid_train)
    ...: sse_train
```

```
Out[31]: 309.74971389531891
In [32]: resid_test = y_test - model.predict(X_test)
    ...: sse_test = sse(resid_test)
    ...: sse_test
Out[32]: 1489.1176065002333
```

Here we note that while the SSE of the training data increased compared to that of the ordinary regression, the SSE for the testing data decreased significantly. Thus, by paying a price in terms of how well the regression model fits the training data, we have obtained a model with significantly improved ability to predict the testing dataset. For comparison with the earlier methods, we graph the residuals and the model parameters once again with the plot_residuals_and_coeff function. The result is shown in Figure 15-3. In the rightmost panel of this figure, we see that the coefficient profile is significantly different from those shown in Figure 15-1 and Figure 15-2, and the coefficient vector produced with the LASSO regression contains mostly zeros. This is a suitable method for the current data because in the beginning, when we generated the dataset, we choose 50 features out of which only 10 are informative. If we suspect that we might have a large number of features that might not contribute much in the regression model, using the L1 regularization of the LASSO regression can thus be a good approach to try.

```
In [33]: fig, ax = plot_residuals_and_coeff(resid_train, resid_test,
            model.coef_)
```

Figure 15-3. *The residual between the LASSO-regularized regression model and the training data (left), the model and the test data (middle), and the values of the coefficients for the 50 features (right)*

The values of α that we used in the two previous examples using Ridge and LASSO regression were chosen arbitrarily. The most suitable value of α is problem dependent, and for every new problem, we need to find a suitable value using trial and error. The scikit-learn library provides methods for assisting this process, as we will see in the following text, but before we explore those methods, it is instructive to look at how the regression model parameters and the SSE for the training and testing datasets depend on the value of α for a specific problem. Here we focus on LASSO regression, since it was seen to work well for the current problem, and we repeatedly solve the same problem using different values for the regularization strength parameter α while storing the values of the coefficients and SSE values in NumPy arrays.

We begin by creating the required NumPy arrays. We use `np.logspace` to create a range of α values that span several orders of magnitude:

```
In [34]: alphas = np.logspace(-4, 2, 100)
In [35]: coeffs = np.zeros((len(alphas), X_train.shape[1]))
In [36]: sse_train = np.zeros_like(alphas)
In [37]: sse_test = np.zeros_like(alphas)
```

Next we loop through the α values and perform the LASSO regression for each value:

```
In [38]: for n, alpha in enumerate(alphas):
    ...:     model = linear_model.Lasso(alpha=alpha)
    ...:     model.fit(X_train, y_train)
    ...:     coeffs[n, :] = model.coef_
    ...:     sse_train[n] = sse(y_train - model.predict(X_train))
    ...:     sse_test[n] = sse(y_test - model.predict(X_test))
```

Finally, we plot the coefficients and the SSE for the training and testing datasets using Matplotlib. The result is shown in Figure 15-4. We can see in the left panel of this figure that a large number coefficients are nonzero for very small values of α. This corresponds to the overfitting regime. We can also see that when α is increased above a certain threshold, many of the coefficients collapse to zero, and only a few coefficients remain nonzero. In the right panel of the figure, we see that while the SSE for the training set is steadily increasing with increasing α, there is also a sharp drop in the SSE for the testing dataset. This is the sought-after effect in LASSO regression. However, for too large values of α, all coefficients converge to zero and the SSEs for both the training and testing datasets become large. Therefore, there is an optimal region of α that prevents overfitting and improves the model's ability to predict unseen data. While these observations are not universally true, a similar pattern can be seen for many problems.

```
In [39]: fig, axes = plt.subplots(1, 2, figsize=(12, 4), sharex=True)
    ...: for n in range(coeffs.shape[1]):
    ...:     axes[0].plot(np.log10(alphas), coeffs[:, n], color='k', lw=0.5)
    ...:
    ...: axes[1].semilogy(np.log10(alphas), sse_train, label="train")
    ...: axes[1].semilogy(np.log10(alphas), sse_test, label="test")
    ...: axes[1].legend(loc=0)
    ...:
    ...: axes[0].set_xlabel(r"${\log_{10}}\alpha$", fontsize=18)
    ...: axes[0].set_ylabel(r"coefficients", fontsize=18)
    ...: axes[1].set_xlabel(r"${\log_{10}}\alpha$", fontsize=18)
    ...: axes[1].set_ylabel(r"sse", fontsize=18)
```

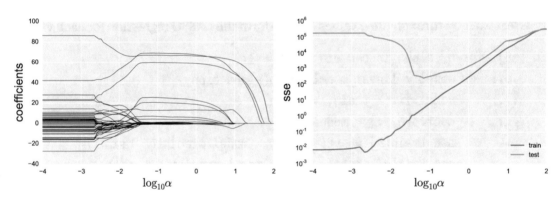

Figure 15-4. *The coefficients (left) and the sum of squared errors (SSEs) for the training and testing datasets (right), for LASSO regression as a function of the logarithm of the regularization strength parameter* α

The process of testing a regularized regression with several values of α can be carried out automatically using, for example, the RidgeCV and LassoCV classes. These variants of the Ridge and LASSO regression internally perform a search for the optimal α using a cross-validation approach. By default, a k-fold cross-validation with $k = 3$ is used, although this can be changed using the cv argument to these classes. Because of the built-in cross-validation, we do not need to explicitly divide the dataset into training and testing datasets, as we have done previously.

To use the LASSO method with an automatically chosen α, we simply create an instance of LassoCV and invoke its fit method:

```
In [40]: model = linear_model.LassoCV()
In [41]: model.fit(X_all, y_all)
Out[41]: LassoCV(alphas=None, copy_X=True, cv=None, eps=0.001,
                 fit_intercept=True, max_iter=1000, n_alphas=100, n_jobs=1,
                 normalize=False, positive=False, precompute='auto',
                 random_state=None, selection='cyclic', tol=0.0001,
                 verbose=False)
```

The value of regularization strength parameter α selected through the cross-validation search is accessible through the `alpha_` attribute:

```
In [42]: model.alpha_
Out[42]: 0.13118477495069433
```

We note that the suggested value of α agrees reasonably well with what we might have guessed from Figure 15-4. For comparison with the previous method, we also compute the SSE for the training and testing datasets (although both were used for training in the call to LassoCV.fit) and graph the SSE values together with the model parameters, as shown in Figure 15-5. By using the cross-validated LASSO method, we obtain a model that predicts both the training and testing datasets with relatively high accuracy, and we are no longer as likely to suffer from the problem of overfitting, in spite of having few samples compared to the number of features[2].

```
In [43]: resid_train = y_train - model.predict(X_train)
    ...: sse_train = sse(resid_train)
    ...: sse_train
Out[43]: 66.900068715063625
In [44]: resid_test = y_test - model.predict(X_test)
    ...: sse_test = sse(resid_test)
    ...: sse_test
Out[44]: 966.39293785448456
In [45]: fig, ax = plot_residuals_and_coeff(resid_train, resid_test,
         model.coef_)
```

---

[2]However, note that we can never be sure that a machine-learning application does not suffer from overfitting before we see how the application performs on new observations, and a repeated reevaluation of the application on a regular basis is a good practice.

Figure 15-5. *The residuals of the LASSO-regularized regression model with cross-validation for the training data (left) and the testing data (middle). The values of the coefficients for the 50 features are also shown (right).*

Finally, yet another type of popular regularized regression, which combines the L1 and L2 regularization of the LASSO and Ridge methods, is known as elastic-net regularization. The minimization objective function for this method is $\min_\beta \left\{ \|X\beta - y\|_2^2 + \alpha\rho\|\beta\|_1 + \alpha(1-\rho)\|\beta\|_2^2 \right\}$, where the parameter ρ (l1_ratio in scikit-learn) determines the relative weight of the L1 and L2 penalties and thus how much the method behaves like the LASSO and Ridge methods. In scikit-learn, we can perform an elastic-net regression using the ElasticNet class, to which we can give explicit values of the α (alpha) and ρ (l1_ratio) parameters, or the cross-validated version ElasticNetCV, which automatically finds suitable values of the α and ρ parameters:

```
In [46]: model = linear_model.ElasticNetCV()
In [47]: model.fit(X_train, y_train)
Out[47]: ElasticNetCV(alphas=None, copy_X=True, cv=None, eps=0.001,
                      fit_intercept=True, l1_ratio=0.5, max_iter=1000,
                      n_alphas=100, n_jobs=1, normalize=False,
                      positive=False, precompute='auto', random_state=None,
                      selection='cyclic', tol=0.0001, verbose=0)
```

The value of regularization parameters α and ρ suggested by the cross-validation search is available through the alpha_ and l1_ratio attributes:

```
In [48]: model.alpha_
Out[48]: 0.13118477495069433
In [49]: model.l1_ratio
Out[49]: 0.5
```

For comparison with the previous method, we once again compute the SSE and plot the model coefficients, as shown in Figure 15-6. As expected with $\rho = 0.5$, the result has characteristics of both LASSO regression (favoring a sparse solution vector with only a few dominating elements) and Ridge regression (suppressing the magnitude of the coefficients).

```
In [50]: resid_train = y_train - model.predict(X_train)
   ...: sse_train = sse(resid_train)
   ...: sse_train
Out[50]: 2183.8391729391255
In [51]: resid_test = y_test - model.predict(X_test)
   ...: sse_test = sse(resid_test)
   ...: sse_test
Out[51]: 2650.0504463382508
In [52]: fig, ax = plot_residuals_and_coeff(resid_train, resid_test,
             model.coef_)
```

Figure 15-6. *The residuals of the elastic-net regularized regression model with cross-validation for the training data (left) and the testing data (middle). The values of the coefficients for the 50 features are also shown (right).*

Classification

Like regression, classification is a central topic in machine learning. In Chapter 14, about statistical modeling, we already saw examples of classification, where we used a logistic regression model to classify observations into discrete categories. Logistic regression is also used in machine learning for the same task, but there are also a wide variety of alternative algorithms for classification, such as nearest neighbor methods, support vector machines (SVM), decision trees, and Random Forest methods. The scikit-learn library provides a convenient unified API that allows all these different methods to be used interchangeably for any given classification problems.

To see how we can train a classification model with a training dataset and test its performance on a testing datasets, let's once again look at the Iris dataset, which provides features for Iris flower samples (sepal and petal width and height), together with the species of each sample (*setosa*, *versicolor*, and *virginica*). The Iris dataset that is included in the scikit-learn library (as well as in the statsmodels library) is a classic dataset that is commonly used for testing and demonstrating machine-learning algorithms and statistical models. Here we revisit the problem of classifying the species of a flower sample given its sepal and petal width and height (see also Chapter 14). First, to load the dataset, we call the `load_iris` function in the `datasets` module. The result is a container object (called a Bunch object in the scikit-learn jargon) that contains the data as well as metadata.

```
In [53]: iris = datasets.load_iris()
In [54]: type(iris)
Out[54]: sklearn.datasets.base.Bunch
```

For example, descriptive names of the features and target classes are available through the `feature_names` and `target_names` attributes:

```
In [55]: iris.target_names
Out[55]: array(['setosa', 'versicolor', 'virginica'], dtype='|S10')
In [56]: iris.feature_names
Out[56]: ['sepal length (cm)',  'sepal width (cm)',  'petal length (cm)',
          'petal width (cm)']
```

and the actual dataset is available through the `data` and `target` attributes:

```
In [57]: iris.data.shape
Out[57]: (150, 4)
In [58]: iris.target.shape
Out[58]: (150,)
```

We begin by splitting the dataset into a training and testing part, using the `train_test_split` function. There we chose to include 70% of the samples in the training set, leaving the remaining 30% for testing and validation:

```
In [59]: X_train, X_test, y_train, y_test = \
    ...:         model_selection.train_test_split(iris.data, iris.target,
            train_size=0.7)
```

The first step in training a classifier and performing classification tasks using scikit-learn is to create a classifier instance. There are, as mentioned in the preceding text and demonstrated in the following, numerous classifiers available. We begin with a logistic regression classifier, which is provided by the LogisticRegression class in the linear_model module:

```
In [60]: classifier = linear_model.LogisticRegression()
```

The training of the classifier is carried out by calling the fit method of the classifier instance. The arguments are the design matrices for the feature and target variables. Here we use the training part of the Iris dataset arrays that was created for us when loading the dataset using the load_iris function. If the design matrices are not already available, we can use the same techniques that we used in Chapter 14: that is, constructing the matrices by hand using NumPy functions or using the Patsy library to automatically construct the appropriate arrays. We can also use the feature extraction utilities in the feature_extraction module in the scikit-learn library.

```
In [61]: classifier.fit(X_train, y_train)
Out[61]: LogisticRegression(C=1.0, class_weight=None, dual=False,
                            fit_intercept=True, intercept_scaling=1,
                            max_iter=100, multi_class='ovr', penalty='l2',
                            random_state=None, solver='liblinear',
                            tol=0.0001, verbose=0)
```

Once the classifier has been trained, we can immediately start using it for predicting the class for new observations using the predict method. Here we apply this method to predict the class for the samples assigned to the testing dataset so that we can compare the predictions with the actual values.

```
In [62]: y_test_pred = classifier.predict(X_test)
```

The sklearn.metrics module contains helper functions for assisting in the analysis of the performance and accuracy of classifiers. For example, the classification_report function, which takes arrays of actual values and the predicted values, returns a tabular summary of the informative classification metrics related to the rate of false negatives and false positives:

```
In [63]: print(metrics.classification_report(y_test, y_test_pred))
             precision    recall  f1-score   support
          0       1.00      1.00      1.00        13
          1       1.00      0.92      0.96        13
          2       0.95      1.00      0.97        19
avg / total       0.98      0.98      0.98        45
```

The so-called confusion matrix, which can be computed using the confusion_
matrix function, also presents useful classification metrics in a compact form: The
diagonals correspond to the number of samples that are correctly classified for each level
of the category variable, and the off-diagonal elements are the number of incorrectly
classified samples. More specifically, the element C_{ij} of the confusion matrix C is the
number of samples of category i that were categorized as j. For the current data, we
obtain the confusion matrix:

```
In [64]: metrics.confusion_matrix(y_test, y_test_pred)
Out[64]: array([[13  0  0]
                [ 0 12  1]
                [ 0  0 19]])
```

This confusion matrix shows that all elements in the first and third class were
classified correctly, but one element of the second class was mistakenly classified as
class three. Note that the elements in each row of the confusion matrix sum up to the
total number of samples for the corresponding category. In this testing sample, we have
13 elements each in the first and second class and 19 elements of the third class, as also
can be seen by counting unique values in the y_test array:

```
In [65]: np.bincount(y_test)
Out[65]: array([13, 13, 19])
```

To perform a classification using a different classifier algorithm, all we need to do is
to create an instance of the corresponding classifier class. For example, to use a decision
tree instead of logistic regression, we can use the DesicisionTreeClassifier class from
the sklearn.tree module. Training the classifier and predicting new observations is
done in exactly the same way for all classifiers:

```
In [66]: classifier = tree.DecisionTreeClassifier()
    ...: classifier.fit(X_train, y_train)
    ...: y_test_pred = classifier.predict(X_test)
```

```
    ...: metrics.confusion_matrix(y_test, y_test_pred)
Out[66]: array([[13,  0,  0],
                [ 0, 12,  1],
                [ 0,  1, 18]])
```

With the decision tree classifier, the resulting confusion matrix is somewhat different, corresponding to one additional misclassification in the testing dataset.

Other popular classifiers that are available in scikit-learn include the nearest neighbor classifier KNeighborsClassifier from the sklearn.neighbors module, the support vector classifier (SVC) from the sklearn.svm module, and the Random Forest classifier RandomForestClassifier from the sklearn.ensemble module. Since they all have the same usage pattern, we can programmatically apply a series of classifiers on the same problem and compare their performance (on this particular problem), for example, as a function of the training and testing sample sizes. To this end, we create a NumPy array with training size ratios, ranging from 10% to 90%:

```
In [67]: train_size_vec = np.linspace(0.1, 0.9, 30)
```

Next we create a list of classifier classes that we wish to apply:

```
In [68]: classifiers = [tree.DecisionTreeClassifier,
    ...:                neighbors.KNeighborsClassifier,
    ...:                svm.SVC,
    ...:                ensemble.RandomForestClassifier]
```

and an array in which we can store the diagonals of the confusion matrix as a function of training size ratio and classifier:

```
In [69]: cm_diags = np.zeros((3, len(train_size_vec), len(classifiers)),
         dtype=float)
```

Finally, we loop over each training size ratio and classifier, and for each combination, we train the classifier, predict the values of the testing data, compute the confusion matrix, and store its diagonal divided by the ideal values in the cm_diags array:

```
In [70]: for n, train_size in enumerate(train_size_vec):
    ...:     X_train, X_test, y_train, y_test = \
    ...:         model_selection.train_test_split(iris.data, iris.target,
    ...:                                          train_size=train_size)
```

```
   ...:        for m, Classifier in enumerate(classifiers):
   ...:            classifier = Classifier()
   ...:            classifier.fit(X_train, y_train)
   ...:            y_test_p = classifier.predict(X_test)
   ...:            cm_diags[:, n, m] = metrics.confusion_matrix(y_test,
                   y_test_p).diagonal()
   ...:            cm_diags[:, n, m] /= np.bincount(y_test)
```

The resulting classification accuracy for each classifier, as a function of training size ratio, is plotted and shown in Figure 15-7.

```
In [71]: fig, axes = plt.subplots(1, len(classifiers), figsize=(12, 3))
   ...: for m, Classifier in enumerate(classifiers):
   ...:     axes[m].plot(train_size_vec, cm_diags[2, :, m], label=iris.
             target_names[2])
   ...:     axes[m].plot(train_size_vec, cm_diags[1, :, m], label=iris.
             target_names[1])
   ...:     axes[m].plot(train_size_vec, cm_diags[0, :, m], label=iris.
             target_names[0])
   ...:     axes[m].set_title(type(Classifier()).__name__)
   ...:     axes[m].set_ylim(0, 1.1)
   ...:     axes[m].set_ylabel("classification accuracy")
   ...:     axes[m].set_xlabel("training size ratio")
   ...:     axes[m].legend(loc=4)
```

Figure 15-7. *Comparison of classification accuracy of four different classifiers*

In Figure 15-7, we see that classification error is different for each model, but for this particular example, they have comparable performance. Which classifier is best depends on the problem at hand, and it is difficult to give any definite answer to which one is more suitable in general. Fortunately, it is easy to switch between different classifiers in scikit-learn and therefore effortless to try a few different classifiers for a given classification problem. In addition to the classification accuracy, another important aspect is the computational performance and scaling to larger problems. For large classification problems, with many features, decision tree methods such as Randomized Forest are often a good starting point.

Clustering

In the two previous sections, we explored regression and classification, which both are examples of supervised learning, since the response variables are given in the dataset. Clustering is a different type of problem that also is an important topic in machine learning. It can be thought of as a classification problem where the classes are unknown, which makes clustering an example of unsupervised learning. The training dataset for a clustering algorithm contains only the feature variables, and the output of the algorithm is an array of integers that assign each sample to a cluster (or class). This output array corresponds to the response variable in a supervised classification problem.

The scikit-learn library implements a large number of clustering algorithms that are suitable for different types of clustering problems and for different types of datasets. Popular general-purpose clustering methods include the *K-means algorithm*, which groups the samples into clusters such that the within-group sum of square deviation from the group center is minimized, and the *mean-shift algorithm*, which clusters the samples by fitting the data to density functions (e.g., Gaussian functions).

In scikit-learn, the `sklearn.cluster` module contains several clustering algorithms, including the K-means algorithm `KMeans` and the mean-shift algorithm `MeanShift`, just to mention a few. To perform a clustering task with one of these methods, we first initialize an instance of the corresponding class and train it with a feature-only dataset using the `fit` method, and we finally obtain the result of the clustering by calling the `predict` method. Many clustering algorithms require the number of clusters as an input parameter, which we can specify using the `n_clusters` parameter when the class instance is created.

As an example of clustering, consider again the Iris dataset that we used in the previous section, but now we will not use the response variable, which was used in supervised classification, but instead we attempt to automatically discover a suitable clustering of the samples using the *K*-means method. We begin by loading the Iris data, as before, and store the feature and target data in the variables X and y, respectively:

```
In [72]: X, y = iris.data, iris.target
```

With the K-means clustering method, we need to specify how many clusters we want in the output. The most suitable number of clusters is not always obvious in advance, and trying clustering with a few different numbers of clusters is often necessary. However, here we know that the data corresponds to three different species of Iris flowers, so we use three clusters. To perform the clustering, we create an instance of Kmeans class, using the n_clusters argument to set the number of clusters.

```
In [73]: n_clusters = 3
In [74]: clustering = cluster.KMeans(n_clusters=n_clusters)
```

To actually perform the computation, we call the fit method with the Iris feature matrix as an argument:

```
In [75]: clustering.fit(X)
Out[75]: KMeans(copy_x=True, init='k-means++', max_iter=300, n_clusters=3,
                n_init=10, n_jobs=1, precompute_distances='auto',
                random_state=None, tol=0.0001, verbose=0)
```

The clustering result is available through the predict method, to which we also pass a feature dataset that optionally can contain features of new samples. However, not all the clustering methods implemented in scikit-learn support predicting clusters for a new sample. In this case, the predict method is not available, and we need to use the fit_predict method instead. Here, we use the predict method with the training feature dataset to obtain the clustering result:

```
In [76]: y_pred = clustering.predict(X)
```

The result is an integer array of the same length and the number of samples in the training dataset. The elements in the array indicate which group (from 0 up to n_samples-1) each sample is assigned to. Since the resulting array y_pred is long, we only display every eighth element in the array using the NumPy stride indexing ::8.

```
In [77]: y_pred[::8]
Out[77]: array([1, 1, 1, 1, 1, 1, 1, 2, 2, 2, 2, 2, 2, 0, 0, 0, 0, 0, 0],
         dtype=int32)
```

We can compare the obtained clustering with the supervised classification of the Iris samples:

```
In [78]: y[::8]
Out[78]: array([0, 0, 0, 0, 0, 0, 0, 1, 1, 1, 1, 1, 1, 2, 2, 2, 2, 2, 2])
```

There seems to be a good correlation between the two, but the output of the clustering has assigned different integer values to the groups than what was used in the target vector in the supervised classification. To be able to compare the two arrays with metrics such as the confusion_matrix function, we first need to rename the elements so that the same integer values are used for the same group. We can do this operation with NumPy array manipulations:

```
In [79]: idx_0, idx_1, idx_2 = (np.where(y_pred == n) for n in range(3))
In [80]: y_pred[idx_0], y_pred[idx_1], y_pred[idx_2] = 2, 0, 1
In [81]: y_pred[::8]
Out[81]: array([0, 0, 0, 0, 0, 0, 0, 1, 1, 1, 1, 1, 1, 2, 2, 2, 2, 2, 2],
         dtype=int32)
```

Now that we represent the corresponding groups with the same integers, we can summarize the overlaps between the supervised and unsupervised classification of the Iris samples using the confusion_matrix function:

```
In [82]: metrics.confusion_matrix(y, y_pred)
Out[82]: array([[50,  0,  0],
                [ 0, 48,  2],
                [ 0, 14, 36]])
```

This confusion matrix indicates that the clustering algorithm was able to correctly identify all samples corresponding to the first species as a group of its own, but due to the overlapping samples in the second and third groups, those could not be completely resolved as different clusters. For example, 2 elements from group one were assigned to group two, and 14 elements from group two were assigned to group one.

The result of the clustering can also be visualized, for example, by plotting scatter plots for each pair of features, as we do in the following. We loop over each pair of features and each cluster and plot a scatter graph for each cluster using different colors (orange, blue, and green, displayed as different shades of gray in Figure 15-8), and we also draw a red square around each sample for which the clustering does not agree with the supervised classification. The result is shown in Figure 15-8.

```
In [83]: N = X.shape[1]
    ...: fig, axes = plt.subplots(N, N, figsize=(12, 12), sharex=True,
         sharey=True)
    ...: colors = ["coral", "blue", "green"]
    ...: markers = ["^", "v", "o"]
    ...: for m in range(N):
    ...:     for n in range(N):
    ...:         for p in range(n_clusters):
    ...:             mask = y_pred == p
    ...:             axes[m, n].scatter(X[:, m][mask], X[:, n][mask], s=30,
    ...:                                marker=markers[p], color=colors[p],
                                        alpha=0.25)
    ...:         for idx in np.where(y != y_pred):
    ...:             axes[m, n].scatter(X[idx, m], X[idx, n], s=30,
    ...:                                marker="s", edgecolor="red",
                                        facecolor=(1,1,1,0))
    ...:     axes[N-1, m].set_xlabel(iris.feature_names[m], fontsize=16)
    ...:     axes[m, 0].set_ylabel(iris.feature_names[m], fontsize=16)
```

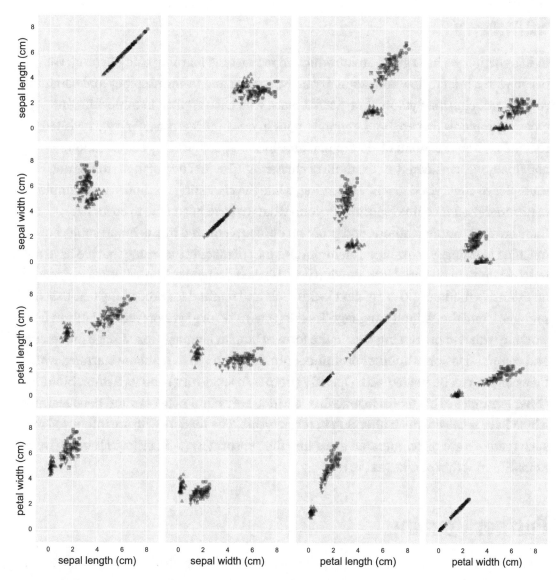

Figure 15-8. *The result of clustering, using the K-means algorithm, of the Iris dataset features*

The result of the clustering of the Iris samples in Figure 15-8 shows that the clustering does a remarkably good job at recognizing which samples belong to distinct groups. Of course, because of the overlap in the features for classes shown in blue (dark gray) and green (medium gray) in the graph, we cannot expect that any unsupervised clustering algorithm can fully resolve the various groups in the dataset, and some deviation from the supervised response variable is therefore expected.

Summary

In this chapter we have given an introduction to machine learning using Python. We began with a brief review and summary of the subject and its terminology and continued with introducing the Python library scikit-learn which we applied in three different types of problems that are fundamental topics in machine learning: First we revisited regression, from the point of view of machine learning, followed by classification, and finally we considered an example of clustering. The first two of these topics are examples of supervised machine learning, while the clustering method is an example of unsupervised machine learning. Beyond what we have been able to cover here, there are many more methods and problem domains covered by the broad subject of machine learning. For example, an important part of machine learning that we have not touched upon in this brief introduction is text-based problems. The scikit-learn contains an extensive module (`sklearn.text`) with tools and method for processing text-based problems, and the Natural Language Toolkit (`www.nltk.org`) is a powerful platform for working with and processing data in the form of human language text. Image processing and computer vision is another prominent problem domain in machine learning, which, for example, can be treated with OpenCV (`http://opencv.org`) and its Python bindings. Other examples of big topics in machine learning are neural networks and deep learning, which have received much attention in recent years. The readers who are interested in such methods are recommended to explore the TensorFlow (`www.tensorflow.org`) and Keras libraries (`http://keras.io`).

Further Reading

Machine learning is a part of the computer science field of artificial intelligence, which is a broad field with numerous techniques, methods, and applications. In this chapter, we have only been able to show examples of a few basic machine-learning methods, which nonetheless can be useful in many practical applications. For a more thorough introduction to machine learning, see T. Hastie (2013), and for introductions to machine learning specific to the Python environment, see, for example, R. Garreta (2013), Hackeling (2014), and L. Pedro Coelho (2015).

References

Hackeling, G. (2014). *Mastering Machine Learning With scikit-learn.* Mumbai: Packt.

L. Pedro Coelho, W. R. (2015). *Building Machine Learning Systems with Python.* Mumbai: Packt.

R. Garreta, G. M. (2013). Learning scikit-learn: Machine Learning in Python. Mumbai: Packt.

T. Hastie, R. T. (2013). The Elements of Statistical Learning: Data Mining, Inference, and Prediction. New York: Springer.

CHAPTER 16

Bayesian Statistics

In this chapter, we explore an alternative interpretation of statistics – Bayesian statistics – and the methods associated with this interpretation. Bayesian statistics, in contrast to the frequentist's statistics that we used in Chapter 13 and Chapter 14, treats probability as a degree of belief rather than as a measure of proportions of observed outcomes. This different point of view gives rise to distinct statistical methods that we can use in problem-solving. While it is generally true that statistical problems can in principle be solved using either frequentist or Bayesian statistics, there are practical differences that make these two approaches to statistics suitable for different types of problems.

Bayesian statistics is based on Bayes' theorem, which relates conditional and unconditional probabilities. Bayes' theorem is a fundamental result in probability theory, and it applies to both the frequentist's and the Bayesian interpretation of statistics. In the context of Bayesian inference, unconditional probabilities are used to describe the prior knowledge of a system, and Bayes' theorem provides a rule for updating this knowledge after making new observations. The updated knowledge is described by a conditional probability, which is conditioned on the observed data. The initial knowledge of a system is described by the *prior probability distribution*, and the updated knowledge, conditioned on the observed data, is the *posterior probability distribution*. In problem-solving with Bayesian statistics, the posterior probability distribution is the unknown quantity that we seek, and from it we can compute expectation values and other statistical quantities for random variables of interest. Although Bayes' theorem describes how to compute the posterior distribution from the prior distribution, for most realistic problems, the calculations involve evaluating high-dimensional integrals that can be prohibitively difficult to compute, both analytically and numerically. This has until recently hindered Bayesian statistics from being widely used in practice. However, with the advent of computational statistics, and the development of efficient simulation methods that allows us to sample directly from the posterior distributions (rather than directly compute it), Bayesian methods

© Robert Johansson 2019
R. Johansson, *Numerical Python*, https://doi.org/10.1007/978-1-4842-4246-9_16

are becoming increasingly popular. The methods that enable us to sample from the posterior distribution are, first and foremost, the so-called Markov chain Monte Carlo (MCMC) methods. Several alternative implementations of MCMC methods are available. For instance, traditional MCMC methods include Gibbs sampling and the Metropolis-Hastings algorithm, and more recent methods include Hamiltonian and No-U-Turn algorithms. In this chapter we explore how to use several of these methods.

Statistical problem-solving with Bayesian inference methods is sometimes known as *probabilistic programming*. The key steps in probabilistic programming are as follows:

> (1) Create a statistical model. (2) Sample from the posterior distribution for the quantity of interest using an MCMC method. (3) Use the obtained posterior distribution to compute properties of interest for the problem at hand and make inference decisions based on the obtained results. In this chapter we explore how to carry out these steps from within the Python environment, with the help of the PyMC library.

PyMc   The PyMC library, currently known as PyMC3, provides a framework for doing probabilistic programming – that is, solving statistical problems using simulation with Bayesian methods. At the time of writing, the latest official release is version 3.4.1. For more information about the project, see the web page at `http://docs.pymc.io`.

Importing Modules

In this chapter we mainly work with the pymc3 library, which we import in the following manner:

```
In [1]: import pymc3 as mc
```

We also require NumPy, Pandas, and Matplotlib for basic numerics, data analytics, and plotting, respectively. These libraries are imported following the usual convention:

```
In [2]: import numpy as np
In [3]: import pandas as pd
In [4]: import matplotlib.pyplot as plt
```

For comparison to non-Bayesian statistics, we also use the `stats` module from SciPy, the `statsmodels` library, and the Seaborn library for visualization:

```
In [5]: from scipy import stats
In [6]: import statsmodels.api as sm
In [7]: import statsmodels.formula.api as smf
In [8]: import seaborn as sns
```

Introduction to Bayesian Statistics

The foundation of Bayesian statistics is Bayes' theorem, which gives a relation between unconditioned and conditional probabilities of two events A and B:

$$P(A|B)P(B) = P(B|A)P(A),$$

where $P(A)$ and $P(B)$ are the unconditional probabilities of event A and B and where $P(A|B)$ is the conditional probability of event A given that event B is true and $P(B|A)$ is the conditional probability of B given that A is true. Both sides of the preceding equation are equal to the probability that both A and B are true: $P(A \cap B)$. In other words, Bayes' rule states the probability that both A and B are equal to the probability of A times the probability of B given that A is true, $P(A)P(B|A)$ or, equivalently, the probability of B times the probability of A given that B is true: $P(B)P(A|B)$.

In the context of Bayesian inference, Bayes' rule is typically employed for the situation when we have a prior belief about the probability of an event A, represented by the unconditional probability $P(A)$, and wish to update this belief after having observed an event B. In this language, the updated belief is represented by the conditional probability of A given the observation B: $P(A|B)$, which we can compute using Bayes' rule:

$$P(A|B) = \frac{P(B|A)P(A)}{P(B)}.$$

Each factor in this expression has a distinct interpretation and a name: $P(A)$ is the *prior* probability of event A, and $P(A|B)$ is the *posterior* probability of A given the observation B. $P(B|A)$ is the *likelihood* of observing B given that A is true, and the probability of observing B regardless of A, $P(B)$, is known as *model evidence* and can be considered as a normalization constant (with respect to A).

In statistical modeling we are typically interested in a set of random variables X that are characterized by probability distributions with certain parameters θ. After collecting data for the process that we are interested in modeling, we wish to infer the values of the model parameters from the data. In the frequentist's statistical approach, we can maximize the likelihood function given the observed data and obtain estimators for the model parameters. The Bayesian approach is to consider the unknown model parameters θ as random variables in their own right and use Bayes' rule to derive probability distributions for the model parameters θ. If we denote the observed data as x, we can express the probability distribution for θ given the observed data x using Bayes' rule as

$$p(\theta|x) = \frac{p(x|\theta)p(\theta)}{p(x)} = \frac{p(x|\theta)p(\theta)}{\int p(x|\theta)p(\theta)d\theta}.$$

The second equality in this equation follows from the law of total probability, $p(x) = \int p(x|\theta)p(\theta)d\theta$. Once we have computed the posterior probability distribution $p(\theta|x)$ for the model parameters, we can, for example, compute expectation values of the model parameters and obtain a result that is similar to the estimators that we can compute in a frequentist's approach. In addition, when we have an estimate of the full probability distribution for $p(\theta|x)$, we can also compute other quantities, such as credibility intervals, and marginal distributions for certain model parameters in the case when θ is multivariate. For example, if we have two model parameters, $\theta = (\theta_1, \theta_2)$, but are interested only in θ_1, we can obtain the marginal posterior probability distribution $p(\theta_1|x)$ by integrating the joint probability distribution $p(\theta_1, \theta_2|x)$ using the expression obtained from Bayes' theorem:

$$p(\theta_1|x) = \int p(\theta_1,\theta_2|x)d\theta_2 = \frac{\int p(x|\theta_1,\theta_2)p(\theta_1,\theta_2)d\theta_2}{\int\int p(x|\theta_1,\theta_2)p(\theta_1,\theta_2)d\theta_1 d\theta_2}.$$

Here note that the final expression contains integrals over the known likelihood function $p(x|\theta_1,\theta_2)$ and the prior distribution $p(\theta_1, \theta_2)$, so we do not need to know the joint probability distribution $p(\theta_1,\theta_2|x)$ to compute the marginal probability distribution $p(\theta_1|x)$. This approach provides a powerful and generic methodology for computing probability distributions for model parameters and successively updating the distributions once new data becomes available. However, directly computing $p(\theta|x)$, or the marginal distributions thereof, requires that we can write down the likelihood function $p(x|\theta)$ and the prior distribution $p(\theta)$ and that we can evaluate the resulting integrals. For many simple but important problems, it is possible to analytically compute these integrals and find the exact closed-form expressions for the posterior distribution. Textbooks, such as Gelman (2013), provide numerous examples of problems that are exactly solvable in this way. However, for more complicated models, with prior distributions and likelihood functions for which the resulting integrals are not easily evaluated, or for multivariate statistical models, for which the resulting integrals can be high-dimensional, both exact and numerical evaluation may be unfeasible.

It is primarily for models that cannot be solved with exact methods that we can benefit from using simulation methods, such as Markov chain Monte Carlo, which allows us to sample the posterior probability distribution for the model parameters and thereby construct an approximation of the joint or marginal posterior distributions, or directly evaluating integrals, such as expectation values. Another important advantage of simulation-based methods is that the modeling process can be automated. Here we exclusively focus on Bayesian statistical modeling using Monte Carlo simulation methods. For a thorough review of the theory, and many examples of analytically solvable problems, see the references given at the end of this chapter. In the remaining part of this chapter, we explore the definition of statistical models and sampling of their posterior distribution with the PyMC library as a probabilistic programming framework.

Before we proceed with computational Bayesian statistics, it is worth taking a moment to summarize the key differences between the Bayesian approach and the classical frequentist's approach that we used in earlier chapters: In both approaches to statistical modeling, we formulate the models in terms of random variables. A key step in the definition of a statistical model is to make assumptions about the probability distributions for the random variables that are defined in the model. In parametric methods, each probability distribution is characterized by a small number of parameters. In the frequentist's approach, those model parameters have some specific true values, and observed data is interpreted as random samples from the true distributions. In other words, the model parameters are assumed to be fixed, and the data is assumed to be stochastic. The Bayesian approach takes the opposite point of view: The data is interpreted as fixed, and the model parameters are described as random variables. Starting from a prior distribution for the model parameters, we can then update the distribution to account for observed data and in the end obtain a probability distribution for the relevant model parameters, conditioned on the observed data.

Model Definition

A statistical model is defined in terms of a set of random variables. The random variables in a given model can be independent or, more interestingly, dependent on each other. The PyMC library provides classes for representing random variables for a large number of probability distributions: For example, an instance of `mc.Normal` can be used to represent a normal distributed random variable. Other examples are `mc.Bernoulli` for representing discrete Bernoulli-distributed random variables, `mc.Uniform` for uniformly distributed random variables, `mc.Gamma` for Gamma-distributed random variables, and so on. For a complete list of available distributions, see `dir(mc.distributions)` and the docstrings for each available distribution for information on how to use them. It is also possible to define custom distributions using the `mc.DensityDist` class, which takes a function that specifies the logarithm of the random variable's probability density function.

In Chapter 13 we saw that the SciPy `stats` module also contains classes for representing random variables. Like the random variable classes in SciPy `stats`, we can use the PyMC distributions to represent random variables with fixed parameters. However, the essential feature of the PyMC random variables is that the distribution parameters, such as the mean μ and variance σ^2 for a random variable following the normal distribution $\mathcal{N}\left(\mu, \sigma^2\right)$, can themselves be random variables. This allows us to chain random variables in a model and to formulate models with a hierarchical structure in the dependencies between random variables that occur in the model.

Let's start with the simplest possible example. In PyMC, models are represented by an instance of the class `mc.Model`, and random variables are added to a model using the Python context syntax: Random variable instances that are created within the body of a model context are automatically added to the model. Say that we are interested in a model consisting of a single random variable that follows the normal distribution with the fixed parameters $\mu = 4$ and $\sigma = 2$. We first define the fixed model parameters and then create an instance of `mc.Model` to represent our model.

```
In [9]: mu = 4.0
In [10]: sigma = 2.0
In [11]: model = mc.Model()
```

Next, we can attach random variables to the model by creating them within the model context. Here, we create a random variable X within the model context, which is activated using a `with model` statement:

```
In [12]: with model:
    ...:        mc.Normal('X', mu, tau=1/sigma**2)
```

All random variable classes in PyMC take as first argument the name of the variable. In the case of mc.Normal, the second argument is the mean of the normal distribution, and the third argument tau is the precision $\tau = 1/\sigma^2$, where σ^2 is the variance. Alternatively, we can use the sd keyword argument to specify the standard deviation rather than the precision: mc.Normal('X', mu, sd=sigma).

We can inspect which random variables exist in a model using the vars attribute. Here we have only one random variable in the model:

```
In [13]: model.vars
Out[13]: [X]
```

To sample from the random variables in the model, we use the mc.sample function, which implements the MCMC algorithm. The mc.sample function accepts many arguments, but at a minimum, we need to provide the number of samples as the first argument and, as the second argument, a step class instance, which implements an MCMC step. Optionally we can also provide a starting point as a dictionary with parameter values from which the sampling is started, using the start keyword argument. For the step method, here we use an instance of the Metropolis class, which implements the Metropolis-Hastings step method for the MCMC sampler[1]. Note that we execute all model-related code within the model context:

```
In [14]: start = dict(X=2)
In [15]: with model:
    ...:        step = mc.Metropolis()
    ...:        trace = mc.sample(10000, start=start, step=step)
[----------------100%----------------] 10000 of 10000 complete in 1.6 sec
```

With these steps, we have sampled 10000 values from the random variable defined within the model, which in this simple case is only a normal distributed random variable. To access the samples, we can use the get_values method of the trace object returned by the mc.sample function:

```
In [16]: X = trace.get_values("X")
```

---

[1] See also the Slice, HamiltonianMC, and NUTS samplers, which can be used more or less interchangeably.

The probability density function (PDF) for a normal distributed is, of course, known analytically. Using SciPy `stats` module, we can access the PDF using the `pdf` method of the `norm` class instance for comparing to the sampled random variable. The sampled values and the true PDF for the present model are shown in Figure 16-1.

```
In [17]: x = np.linspace(-4, 12, 1000)
In [18]: y = stats.norm(mu, sigma).pdf(x)
In [19]: fig, ax = plt.subplots(figsize=(8, 3))
    ...: ax.plot(x, y, 'r', lw=2)
    ...: sns.distplot(X, ax=ax)
    ...: ax.set_xlim(-4, 12)
    ...: ax.set_xlabel("x")
    ...: ax.set_ylabel("Probability distribution")
```

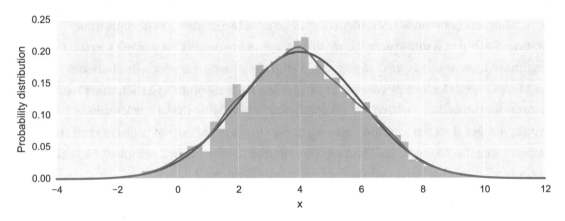

Figure 16-1. *The probability density function for the normal distributed random variable (red/thick line) and a histogram from 10000 MCMC samples*

With the `mc.traceplot` function, we can also visualize the MCMC random walk that generated the samples, as shown in Figure 16-2. The `mc.traceplot` function automatically plots both the kernel-density estimate and the sampling trace for every random variable in the model.

```
In [20]: fig, axes = plt.subplots(1, 2, figsize=(8, 2.5), squeeze=False)
    ...: mc.traceplot(trace, ax=axes)
    ...: axes[0, 0].plot(x, y, 'r', lw=0.5)
```

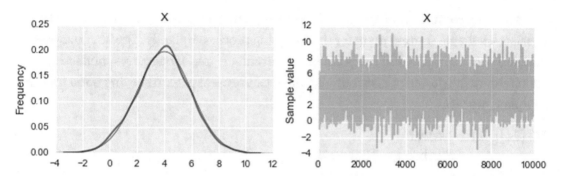

Figure 16-2. *Left panel: the density kernel estimate (blue/thick line) of the sampling trace and the normal probability distribution (red/thin line). Right panel: the MCMC sampling trace*

As a next step in building more complex statistical models, consider again a model with a normal distributed random variable $X \sim \mathcal{N}\left(\mu, \sigma^{2}\right)$, but where parameters μ and σ themselves are random variables. In PyMC, we can easily create dependent variables by passing them as an argument when creating other random variables. For example, with $\mu \sim \mathcal{N}(3, 1)$ and $\sigma \sim |\mathcal{N}(0, 1)|$, we can create the dependent random variable X using the following model specification:

```
In [21]: model = mc.Model()
In [22]: with model:
    ...:     mean = mc.Normal('mean', 3.0)
    ...:     sigma = mc.HalfNormal('sigma', sd=1.0)
    ...:     X = mc.Normal('X', mean, sd=sigma)
```

Here we have used the mc.HalfNormal to represent the random variable $\sigma \sim |\mathcal{N}(0, 1)|$, and the mean and standard deviation arguments to the mc.Normal class for X are random variable instances rather than fixed model parameters. As before we can inspect which random variables a model contains using the vars attribute.

```
In [23]: model.vars
Out[23]: [mean, sigma_log__, X]
```

Note that here the pymc3 library represents the sigma variable with a log-transformed variable sigma_log__, as a mean to handle the half-normal distribution. Nonetheless, we can still directly access the sigma variable from the model, as will be shown in the following text.

When the complexity of the model increases, it may no longer be straightforward to select a suitable starting point for the sampling process explicitly. The `mc.find_MAP` function can be used to find the point in the parameter space that corresponds to the maximum of the posterior distribution, which can serve as a good starting point for the sampling process.

```
In [24]: with model:
    ...:         start = mc.find_MAP()
In [25]: start
Out[25]: {'X': array(3.0), 'mean': array(3.0),
          'sigma': array(0.70710674), 'sigma_log__': array(-0.34657365)}
```

As before, once the model is specified, and a starting point is computed, we can sample from the random variables in the model using the `mc.sample` function, for example, using `mc.Metropolis` as an MCMC sampling step method:

```
In [26]: with model:
    ...:         step = mc.Metropolis()
    ...:         trace = mc.sample(100000, start=start, step=step)
[-----------------100%-----------------] 100000 of 100000 complete in 53.4
sec
```

For example, to obtain the sample trace for the `sigma` variable, we can use `get_values('sigma')`. The result is a NumPy array that contains the sample values, and from it, we can compute further statistics, such as its sample mean and standard deviation:

```
In [27]: trace.get_values('sigma').mean()
Out[27]: 0.80054476153369014
```

The same approach can be used to obtain the samples of X and compute statistics from them:

```
In [28]: X = trace.get_values('X')
In [29]: X.mean()
Out[29]: 2.9993248663922092
In [30]: trace.get_values('X').std()
Out[30]: 1.4065656512676457
```

The trace plot for the current model, created using the mc.traceplot, is shown in Figure 16-3, where we have used the varnames argument to mc.traceplot to explicitly select which random variables to plot.

```
In [31]: fig, axes = plt.subplots(3, 2, figsize=(8, 6), squeeze=False)
    ...: mc.traceplot(trace, varnames=['mean', 'sigma', 'X'], ax=axes)
```

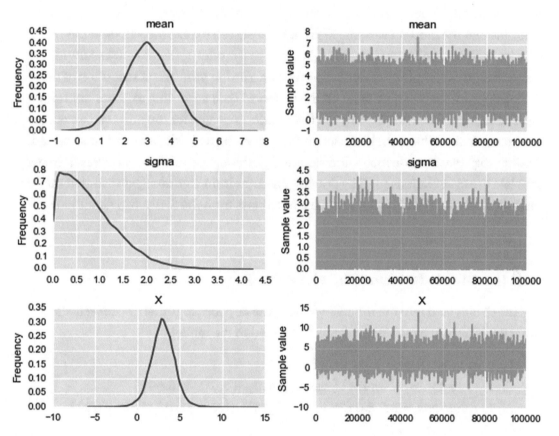

Figure 16-3. *Kernel density estimates (left) and MCMC random sampling trace (right), for the three random variables:* mean, sigma, *and* X

Sampling Posterior Distributions

So far we have defined models and sampled from models that only contain random variables without any references to observed data. In the context of Bayesian models, these types of random variables represent the prior distributions of the unknown model parameters. In the previous examples, we have therefore used the MCMC method

553

to sample from the prior distributions of the model. However, the real application of the MCMC algorithm is to sample from the posterior distribution, which represents the probability distribution for the model variables after having updated the prior distribution to account for the effect of observations.

To condition the model on observed data, all we need to do is to add the data using the observed keyword argument when the corresponding random variable is created within the model: For example, mc.Normal('X', mean, 1/sigma**2, observed=data) indicates that the random variable X has been observed to take the values in the array data. Adding observed random variables to a model automatically results in that subsequent sampling using mc.sample samples the posterior distribution of the model, appropriately conditioned on the observed data according to Bayes' rule and the likelihood function implied by the distribution selected for the observed data. For example, consider the model we used in the preceding text, with a normal distributed random variable X whose mean and standard deviation are random variables. Here we simulate the observations for X by drawing samples from a normally distributed random variable with $\mu = 2.5$ and $\sigma = 1.5$ using the norm class from the SciPy stats module:

```
In [32]: mu = 2.5
In [33]: s = 1.5
In [34]: data = stats.norm(mu, s).rvs(100)
```

The data is feed into the model by setting the keyword argument observed=data when the observed variable is created and added to the model:

```
In [35]: with mc.Model() as model:
    ...:     mean = mc.Normal('mean', 4.0, 1.0) # true 2.5
    ...:     sigma = mc.HalfNormal('sigma', 3.0 * np.sqrt(np.pi/2))
             # true 1.5
    ...:     X = mc.Normal('X', mean, 1/sigma**2, observed=data)
```

A consequence of providing observed data for X is that it is no longer considered as a random variable in the model. This can be seen from inspecting the model using the vars attribute, where X is now absent:

```
In [36]: model.vars
Out[36]: [mean, sigma_log_]
```

Instead, in this case, X is a deterministic variable that is used to construct the likelihood function that relates the priors, represented by mean and sigma, to the posterior distribution for these random variables. Like before, we can find a suitable starting point for the sampling process using the mc.find_MAP function. After creating an MCMC step instance, we can sample the posterior distribution for the model using mc.sample:

```
In [37]: with model:
    ...:        start = mc.find_MAP()
    ...:        step = mc.Metropolis()
    ...:        trace = mc.sample(100000, start=start, step=step)
[----------------100%----------------] 100000 of 100000 complete in 36.1
sec
```

The starting point that was calculated using mc.find_MAP maximizes the likelihood of the posterior given the observed data, and it provides an estimate of the unknown parameters of the prior distribution:

```
In [38]: start
Out[38]: {'mean': array(2.5064940359768246), 'sigma_log':
array(0.394681633456101)}
```

However, to obtain estimates of the distribution of these parameters (which are random variables in their own right), we need to carry out the MCMC sampling using the mc.sample function, as done in the preceding text. The result of the posterior distribution sampling is shown in Figure 16-4. Note that the distributions for the mean and sigma variables are closer to the true parameter values, $\mu = 2.5$ and $\sigma = 1.5$, than to the prior guesses of 4.0 and 3.0, respectively, due to the influence of the data and the corresponding likelihood function.

```
In [38]: fig, axes = plt.subplots(2, 2, figsize=(8, 4), squeeze=False)
    ...: mc.traceplot(trace, varnames=['mean', 'sigma'], ax=axes)
```

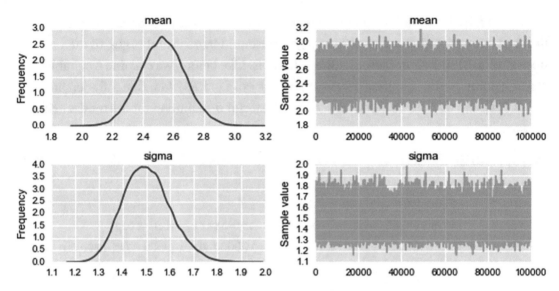

Figure 16-4. *The MCMC sampling trace of the posterior distribution for mean and* sigma

To calculate statistics and estimate quantities using the samples from the posterior distributions, we can access arrays containing the samples using the get_values method, which takes the name of the random variable as argument. For example, in the following code, we compute estimates of the mean of the two random variables in the model and compare to the corresponding true values for the distributions that the data points were drawn from:

```
In [39]: mu, trace.get_values('mean').mean()
Out[39]: (2.5, 2.5290001218008435)
In [40]: s, trace.get_values('sigma').mean()
Out[40]: (1.5, 1.5029047840092264)
```

The PyMC library also provides utilities for analyzing and summarizing the statistics of the marginal posterior distributions obtained from the mc.sample function. For example, the mc.forestplot function visualizes the mean and credibility intervals (i.e., an interval within which the true parameter value is likely to be) for each random variable in a model. The result of visualizing the samples for the current example using the mc.forestplot function is shown in Figure 16-5:

```
In [41]: mc.forestplot(trace, varnames=['mean', 'sigma'])
```

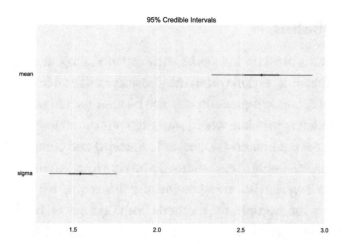

Figure 16-5. *A forest plot for the two parameters, mean and sigma, which show their credibility intervals*

Similar information can also be presented in text form using the mc.summary function, which, for example, includes information such as the mean, standard deviation, and posterior quantiles.

```
In [42]: mc.summary(trace, varnames=['mean', 'sigma'])
mean:
```

Mean	SD	MC Error	95% HPD interval
2.472	0.143	0.001	[2.195, 2.757]

Posterior quantiles:

2.5	25	50	75	97.5
|-------------|==============|==============|-------------|			
2.191	2.375	2.470	2.567	2.754

sigma:

Mean	SD	MC Error	95% HPD interval
1.440	0.097	0.001	[1.256, 1.630]

Posterior quantiles:

2.5	25	50	75	97.5
|-------------|==============|==============|-------------|			
1.265	1.372	1.434	1.501	1.643

Linear Regression

Regression is one of the most basic tools in statistical modeling, and we have already seen examples of linear regression within the classical statistical formalism, for example, in Chapters 14 and 15. Linear regression can also be approached with Bayesian methods and treated as a modeling problem where we assign prior probability distributions to the unknown model parameters (slopes and intercept) and compute the posterior distribution given the available observations. To be able to compare the similarities and differences between Bayesian linear regression and the frequentist's approach to the same problem, using, for example, the methods from Chapter 14, here we begin with a short analysis of a linear regression problem using the statsmodels library. Next, we proceed to analyze the same problem with PyMC.

As example data for performing a linear regression analysis, here we use a dataset that contains the height and weight for 200 men and women, which we can load using the get_rdataset function from the datasets module in the statsmodels library:

```
In [42]: dataset = sm.datasets.get_rdataset("Davis", "carData")
```

For simplicity, to begin with, we work only with the subset of the dataset that corresponds to male subjects, and to avoid having to deal with outliers, we filter out all subjects with a weight that exceeds 110 kg. These operations are readily performed using Pandas methods for filtering data frames using Boolean masks:

```
In [43]: data = dataset.data[dataset.data.sex == 'M']
In [44]: data = data[data.weight < 110]
```

The resulting Pandas data frame object data contains several columns:

```
In [45]: data.head(3)
Out[45]:
```

	sex	weight	height	repwt	repht
0	M	77	182	77	180
3	M	68	177	70	175
5	M	76	170	76	165

Here we focus on a linear regression model for the relationship between the weight and height columns in this dataset. Using the statsmodels library and its model for ordinary least square regression and the Patsy formula language, we create a statistical model for this relationship in a single line of code:

```
In [46]: model = smf.ols("height ~ weight", data=data)
```

To actually perform the fitting of the specified model to the observed data, we use the fit method of the model instance:

```
In [47]: result = model.fit()
```

Once the model has been fitted and the model result object has been created, we can use the predict method to compute the predictions for new observations and for plotting the linear relation between the height and weight, as shown in Figure 16-6.

```
In [48]: x = np.linspace(50, 110, 25)
In [49]: y = result.predict({"weight": x})
In [50]: fig, ax = plt.subplots(1, 1, figsize=(8, 3))
    ...: ax.plot(data.weight, data.height, 'o')
    ...: ax.plot(x, y, color="blue")
    ...: ax.set_xlabel("weight")
    ...: ax.set_ylabel("height")
```

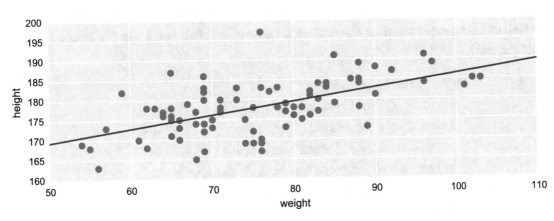

Figure 16-6. *Height vs. weight, with a linear model fitted using ordinary least square*

The linear relation shown in Figure 16-6 summarizes the main result of performing a linear regression on this dataset. It gives the best fitting line, described by specific values of the model parameters (intercept and slope). Within the frequentist's approach to statistics, we can also compute numerous statistics, for example, p-values for various hypotheses, such as the hypotheses that a model parameter is zero (no effect).

The end result of a Bayesian regression analysis is the posterior distribution for the marginal distributions for each model parameter. From such marginal distributions, we can compute the mean estimates for the model parameters, which roughly correspond to the model parameters obtained from a frequentist's analysis. We can also compute other quantities, such as the credibility interval, which characterizes the uncertainty in the estimate. To model the height vs. weight using a Bayesian model, we can use a relation such as $\text{height} \sim \mathcal{N}\left(\text{intercept} + \beta \, \text{weight}, \sigma^2\right)$, where intercept, β, and σ are random variables with unknown distributions and parameters. We also need to give prior distributions to all stochastic variables in the model. Depending on the application, the exact choice of prior can be a sensitive issue, but when there is a lot of data to fit, it is normally sufficient to use reasonable initial guesses. Here we simply start with priors that represent broad distributions for all the model parameters.

To program the model in PyMC, we use the same methodology as earlier in this chapter: First, we create random variables for the stochastic components of the model and assign them to distributions with specific parameters that represent the prior distributions. Next, we create a deterministic variable that are functions of the stochastic variables, but with observed data attached to it using the observed keyword argument, as well as in the expression for the expected value of the distribution of the heights (height_mu).

```
In [51]: with mc.Model() as model:
    ...:     sigma = mc.Uniform('sigma', 0, 10)
    ...:     intercept = mc.Normal('intercept', 125, sd=30)
    ...:     beta = mc.Normal('beta', 0, sd=5)
    ...:     height_mu = intercept + beta * data.weight
    ...:     mc.Normal('height', mu=height_mu, sd=sigma, observed=data.
             height)
    ...:     predict_height = mc.Normal('predict_height', mu=intercept +
             beta * x, sd=sigma, shape=len(x))
```

If we want to use the model for predicting the heights at specific values of weights, we can also add an additional stochastic variable to the model. In the preceding model specification, the predict_height variable is an example of this. Here x is the NumPy array with values between 50 and 110 that was created earlier. Because it is an array, we need to set the shape attribute of the mc.Normal class to the corresponding length of the array. If we inspect the vars attribute of the model, we now see that it contains the two model parameters (intercept and beta), the distribution of the model errors (sigma), and the predict_height variable for predicting the heights at specific values of weight from the x array:

```
In [52]: model.vars
Out[52]: [sigma_interval, intercept, beta, predict_height]
```

Once the model is fully specified, we can turn to the MCMC algorithm to sample the marginal posterior distributions for the model, given the observed data. Like before, we can use mc.find_MAP to find a suitable starting point. Here we use an alternative sampler, mc.NUTS (No-U-Turn Sampler), which is a new and powerful sampler that has been added to version 3 of PyMC.

```
In [53]: with model:
    ...:       start = mc.find_MAP()
    ...:       step = mc.NUTS()
    ...:       trace = mc.sample(10000, step, start=start)
[-----------------100%-----------------] 10000 of 10000 complete in 43.1 sec
```

The result of the sampling is stored in a trace object returned by mc.sample. We can visualize the kernel density estimate of the probability distribution and the MCMC random walk traces that generated the samples using the mc.traceplot function. Here we again use the varnames argument to explicitly select which stochastic variables in the model to show in the trace plot. The result is shown in Figure 16-7.

```
In [54]: fig, axes = plt.subplots(2, 2, figsize=(8, 4), squeeze=False)
    ...: mc.traceplot(trace, varnames=['intercept', 'beta'], ax=axes)
```

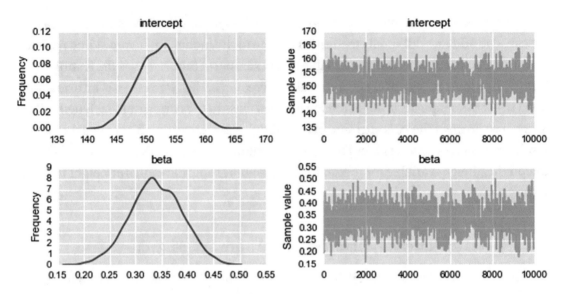

Figure 16-7. *Distribution and sampling trace of the linear model intercept and beta coefficient*

The values of the intercept and coefficient in the linear model that most closely correspond to the results from the statsmodels analysis in the preceding text are obtained by computing the mean of the traces for the stochastic variables in the Bayesian model:

```
In [55]: intercept = trace.get_values("intercept").mean()
In [56]: intercept
Out[56]: 149.97546241676989
In [57]: beta = trace.get_values("beta").mean()
In [58]: beta
Out[58]: 0.37077795098761318
```

The corresponding result from the statsmodels analysis is obtained by accessing the params attribute in the result class returned by the fit method (see the preceding text):

```
In [59]: result.params
Out[59]: Intercept    152.617348
         weight         0.336477
         dtype: float64
```

By comparing these values for the intercepts and the coefficients, we see that the two approaches give similar results for the maximum likelihood estimates of the unknown model parameters. In the statsmodels approach, to predict the expected height for a given weight, say 90 kg, we can use the predict method to get a specific height:

```
In [60]: result.predict({"weight": 90}).values
Out[60]: array([ 182.90030002])
```

The corresponding result in the Bayesian model is obtained by computing the mean for the distribution of the stochastic variable predict_height, for the given weight:

```
In [61]: weight_index = np.where(x == 90)[0][0]
In [62]: trace.get_values("predict_height")[:, weight_index].mean()
Out[62]: 183.33943635274935
```

Again, the results from the two approaches are comparable. In the Bayesian model, however, we have access to an estimate of the full probability distribution of the height at every modeled weight. For example, we can plot a histogram and the kernel density estimate of the probability distribution for the weight 90 kg using the distplot function from the Seaborn library, which results in the graph shown in Figure 16-8:

```
In [63]: fig, ax = plt.subplots(figsize=(8, 3))
    ...: sns.distplot(trace.get_values("predict_height")[:, weight_index],
        ax=ax)
    ...: ax.set_xlim(150, 210)
    ...: ax.set_xlabel("height")
    ...: ax.set_ylabel("Probability distribution")
```

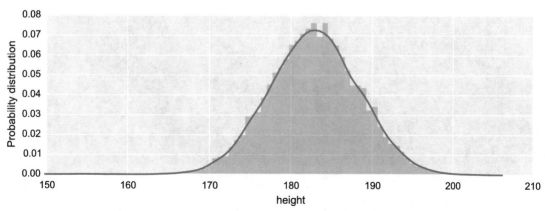

Figure 16-8. *Probability distribution for prediction of the height for the weight 90 kg*

Every sample in the MCMC trace represents a possible value of the intercept and coefficients in the linear model that we wish to fit the observed data. To visualize the uncertainty in the mean intercept and coefficient that we can take as estimates of the final linear model parameters, it is illustrative to plot the lines corresponding to each sample point, along with the data as a scatter plot and the lines that correspond to the mean intercept and slope. This results in a graph like the one shown in Figure 16-9. The spread of the lines represents the uncertainty in the estimate of the height for a given weight. The spread tends to be larger toward the edges where fewer data points are available and tighter in the middle of a cloud of data points.

```
In [64]: fig, ax = plt.subplots(1, 1, figsize=(8, 3))
    ...: for n in range(500, 2000, 1):
    ...:     intercept = trace.get_values("intercept")[n]
    ...:     beta = trace.get_values("beta")[n]
    ...:     ax.plot(x, intercept + beta * x, color='red', lw=0.25,
    ...:         alpha=0.05)
    ...: intercept = trace.get_values("intercept").mean()
    ...: beta = trace.get_values("beta").mean()
    ...: ax.plot(x, intercept + beta * x, color='k', label="Mean Bayesian
    ...:     prediction")
    ...: ax.plot(data.weight, data.height, 'o')
    ...: ax.plot(x, y, '--', color="blue", label="OLS prediction")
    ...: ax.set_xlabel("weight")
    ...: ax.set_ylabel("height")
    ...: ax.legend(loc=0)
```

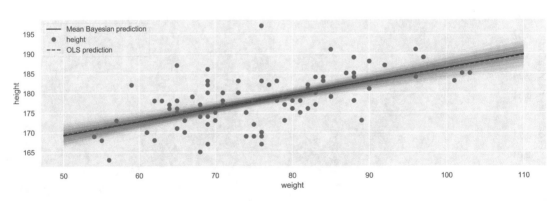

Figure 16-9. *Height vs. weight, with linear fits using OLS and a Bayesian model*

In the linear regression problem we have looked at here, we explicitly defined the statistical model and the stochastic variables included in the model. This illustrates the general steps that are required for analyzing statistical models using the Bayesian approach and the PyMC library. For generalized linear model, however, the PyMC library provides a simplified API that creates the model and the required stochastic variables for us. With the mc.glm.GLM.from_formula function, we can define a generalized linear model using Patsy formula (see Chapter 14) and provide the data using a Pandas data frame. This automatically takes care of setting up the model. With the model setup using mc.glm.glm, we can proceed to sample from the posterior distribution of the model using the same methods as before.

```
In [65]: with mc.Model() as model:
    ...:         mc.glm.GLM.from_formula('height ~ weight', data)
    ...:         step = mc.NUTS()
    ...:         trace = mc.sample(2000, step)
[----------------100%----------------] 2000 of 2000 complete in 99.1 sec
```

The result of the sampling of the GLM model, as visualized by the mc.traceplot function, is shown in Figure 16-10. In these trace plots, sd corresponds to the sigma variable in the explicit model definition used in the preceding text, and it represents the standard error of the residual of the model and the observed data. In the traces, note how the sample requires a few hundred samples before it reaches a steady level. The initial transient period does not contribute samples with the correct distribution, so when using the samples to compute estimates, we should exclude the samples from the initial period.

```
In [66]: fig, axes = plt.subplots(3, 2, figsize=(8, 6), squeeze=False)
    ...: mc.traceplot(trace, varnames=['Intercept', 'weight', 'sd'],
         ax=axes)
```

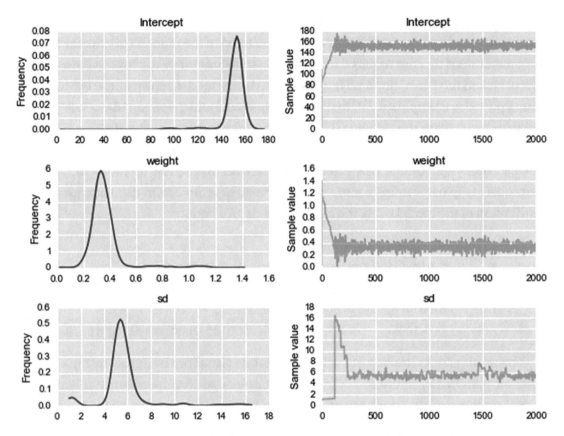

Figure 16-10. *Sample trace plot for a Bayesian GLM model defined using the* `mc.glm` *module*

With the `mc.glm.glm` we can create and analyze linear models using Bayesian statistics in almost the same way as we define and analyze a model using the frequentist's approach with statsmodels. For the simple example studied here, the regression analysis with both statistical approaches gives similar results, and neither methods are much more suitable than the other. However, there are practical differences that depending on the situation can favor one or the other. For example, with the Bayesian approach, we have access to estimates of the full marginal posterior distributions, which can be useful for computing statistical quantities other than the mean. However, performing MCMC on simple models like the one considered here is significantly more computationally demanding than carrying out ordinary least square fitting. The real advantages of the Bayesian methods arise when analyzing complicated models in high dimensions (many unknown model parameters). In such cases, defining appropriate frequentist's models can be difficult and solving the resulting models challenging. The MCMC algorithm has the very attractive property that it scales well to high-dimensional problems and can therefore be highly competitive for complex statistical models. While the model we have considered here all are simple, and can easily be solved using a frequentist's approach, the general methodology used here remains unchanged, and creating more involved models is only a matter of adding more stochastic variables to the model.

As a final example, we illustrate that the same general procedure can be used also when the complexity of the Bayesian model is increased. We return to the height and weight dataset, but instead of selecting only the male subject, here we consider an additional level in the model which accounts for the gender of the subject so that both males and females can be modeled with potentially different slopes and intercepts. In PyMC we can create a multilevel model by using the shape argument to specify the dimension for each stochastic variable that is added to the model, as shown in the following example.

We begin by preparing the dataset. Here we again restrict our analysis to subjects with weight less than 110 kg, to eliminate outliers, and we convert the sex column to a binary variable where 0 represents male and 1 represents female.

```
In [67]: data = dataset.data.copy()
In [68]: data = data[data.weight < 110]
In [69]: data["sex"] = data["sex"].apply(lambda x: 1 if x == "F" else 0)
```

Next, we define the statistical model, which we here take to be height$\sim N$(intercept$_i$ + β_i weight, σ^2), where i is an index that takes the value 0 for male subjects and 1 for female subjects. When creating the stochastic variables for intercept$_i$ and β_i, we indicate the multilevel structure by specifying shape=2 (since in this case, we have two levels: male and female). The only other difference compared to the previous model definition is that we also need to use an index mask when defining the expression for height_mu, so that each value in data.weight is associated with the correct level.

```
In [70]: with mc.Model() as model:
   ...:      intercept_mu, intercept_sigma = 125, 30
   ...:      beta_mu, beta_sigma = 0, 5
   ...:
   ...:      intercept = mc.Normal('intercept', intercept_mu, sd=intercept_
            sigma, shape=2)
   ...:      beta = mc.Normal('beta', beta_mu, sd=beta_sigma, shape=2)
```

```
...:        error = mc.Uniform('error', 0, 10)
...:

...:        sex_idx = data.sex.values
...:        height_mu = intercept[sex_idx] + beta[sex_idx] * data.weight
...:

...:        mc.Normal('height', mu=height_mu, sd=error, observed=data.
       height)
```

Inspecting the model variables using the `vars` attribute object shows that we again have three stochastic variables in the model: `intercept`, `beta`, and `error`. However, in contrast to the earlier model, here `intercept` and `beta` both have two levels.

```
In [71]: model.vars
Out[71]: [intercept, beta, error_interval]
```

The way we invoke the MCMC sampling algorithm is identical to the earlier examples in this chapter. Here we use the NUTS sampler and collect 5000 samples:

```
In [72]: with model:
    ...:        start = mc.find_MAP()
    ...:        step = mc.NUTS()
    ...:        trace = mc.sample(5000, step, start=start)
[----------------100%-----------------] 5000 of 5000 complete in 64.2 sec
```

We can also, like before, use the `mc.traceplot` function to visualize the result of the sampling. This allows us to quickly form an idea of the distribution of the model parameters and to verify that the MCMC sampling has produced sensible results. The trace plot for the current model is shown in Figure 16-11, and unlike earlier examples, here we have multiple curves in the panels for the `intercept` and `beta` variables, reflecting their multilevel nature: the blue (dark) lines show the results for the male subjects, and the green (light) lines show the result for the female subjects.

```
In [73]: mc.traceplot(trace, figsize=(8, 6))
```

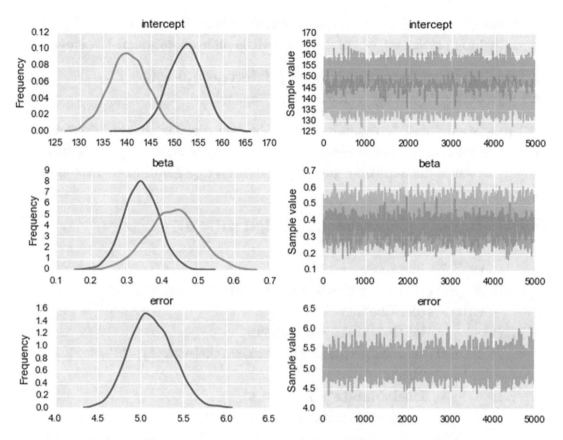

Figure 16-11. *Kernel density estimates of the probability distributions of the model parameters and the MCMC sampling traces for each variable in the multilevel model for height vs. weight*

Using the get_values method of the trace object, we can extract the sampling data for the model variables. Here the sampling data for intercept and beta are two-dimensional arrays with shape (5000, 2): the first dimension represents each sample, and the second dimension represents the level of the variable. Here we are interested in the intercept and the slope for each gender, so we take the mean along the first axis (all samples):

```
In [74]: intercept_m, intercept_f = trace.get_values('intercept').
         mean(axis=0)
In [75]: beta_m, beta_f = trace.get_values('beta').mean(axis=0)
```

569

By averaging over both dimensions, we can also get the intercept and the slope that represent the entire dataset, where male and female subjects are grouped together:

```
In [76]: intercept = trace.get_values('intercept').mean()
In [77]: beta = trace.get_values('beta').mean()
```

Finally, we visualize the results by plotting the data as scatter plots and drawing the lines corresponding to the intercepts and slopes that we obtained for male and female subjects, as well as the result from grouping all subjects together. The result is shown in Figure 16-12.

```
In [78]: fig, ax = plt.subplots(1, 1, figsize=(8, 3))
    ...: mask_m = data.sex == 0
    ...: mask_f = data.sex == 1
    ...: ax.plot(data.weight[mask_m], data.height[mask_m], 'o',
        color="steelblue",
    ...:          label="male", alpha=0.5)
    ...: ax.plot(data.weight[mask_f], data.height[mask_f], 'o',
        color="green",
    ...:          label="female", alpha=0.5)
    ...: x = np.linspace(35, 110, 50)
    ...: ax.plot(x, intercept_m + x * beta_m, color="steelblue",
        label="model male group")
    ...: ax.plot(x, intercept_f + x * beta_f, color="green", label="model
        female group")
    ...: ax.plot(x, intercept + x * beta, color="black", label="model both
        groups")
    ...:
    ...: ax.set_xlabel("weight")
    ...: ax.set_ylabel("height")
    ...: ax.legend(loc=0)
```

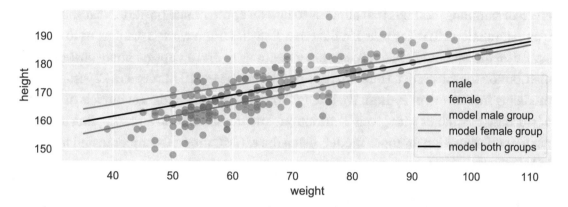

Figure 16-12. *The height vs. weight for male (dark/blue) and female (light/green) subjects*

The regression lines are shown in Figure 16-12, and the distribution plots shown in Figure 16-11 indicate that the model is improved by taking account for different intercepts and slopes for male and female subjects. In a Bayesian model with PyMC, changing the underlying model used in the analysis is only a matter of adding stochastic variables to the model, defining how they are related to each other and assigning a prior distribution for each stochastic variable. The MCMC sampling required to actually solve the model is independent of the model details. This is one of the most attractive aspects of Bayesian statistical modeling. For instance, in the multilevel model considered in the preceding text, instead of specifying the priors for the intercept and slope variables as independent probability distributions, we could relate the distribution parameters of the priors to another stochastic variable and thereby obtain a hierarchical Bayesian model, where the model parameters describing the distribution of the intercept and the slope for each level are drawn from a common distribution. Hierarchical models have many uses and are one of the many applications where Bayesian statistics excel.

Summary

In this chapter, we have explored Bayesian statistics using computational methods provided by the PyMC library. The Bayesian approach to statistics is distinct from classical frequentist's statistics in several fundamental viewpoints. From a practical, computational point of view, Bayesian methods are often very demanding to solve exactly. In fact, computing the posterior distribution for a Bayesian model exactly is often prohibitively expensive. However, what we often can do is to apply powerful and efficient sampling methods that allow us to find an approximate posterior distribution using simulations. The key role of a Bayesian statistics framework is to allow us to define statistical models and then apply sampling methods to find an approximate posterior distribution for the model. Here, we have employed the PyMC library as a Bayesian modeling framework in Python. We briefly explored defining statistical models in terms of stochastic variables with given distributions and the simulation and sampling of the posterior distribution for those models using the MCMC methods implemented in the PyMC library.

Further Reading

For accessible introductions to the theory of Bayesian statistics, see Kruschke (2014) and Downey (2013). A more technical discussion is given in Gelman (2013). A computationally oriented introduction to Bayesian methods with Python is given in *Probabilistic Programming and Bayesian Methods for Hackers*, which is available for free online at `http://camdavidsonpilon.github.io/Probabilistic-Programming-and-Bayesian-Methods-for-Hackers`. An interesting discussion about the differences between the Bayesian and frequentist's approaches to statistics, with examples written in Python, is given in VanderPlas (2014), which is also available at `http://arxiv.org/pdf/1411.5018.pdf`.

References

Downey, A. (2013). *Think Bayes.* Sebastopol: O'Reilly.

Gelman, A. a. (2013). *Bayesian Data Analysis* (3rd ed.). New York: CRC Press.

Kruschke, J. (2014). *Doing Bayesian Data Analysis.* Amsterdam: Academic Press.

VanderPlas, J. (2014). Frequentism and Bayesianism: A Python-driven Primer. *PROC. OF THE 13th PYTHON IN SCIENCE CONF.* Austin: SCIPY.

Signal Processing

In this chapter we explore signal processing, which is a subject with applications in diverse branches of science and engineering. A signal in this context can be a quantity that varies in time (temporal signal) or as a function of space coordinates (spatial signal). For example, an audio signal is a typical example of a temporal signal, while an image is a typical example of a spatial signal in two dimensions. In reality, signals are often continuous functions, but in computational applications, it is common to work with discretized signals, where the original continuous signal is sampled at discrete points with uniform distances. The sampling theorem gives rigorous and quantitative conditions for when a continuous signal can be accurately represented by a discrete sequence of samples.

Computational methods for signal processing play a central role in scientific computing not only because of their widespread application but also because there exist very efficient computational methods for important signal-processing problems. In particular, the fast Fourier transform (FFT) is an important algorithm for many signal-processing problems, and moreover it is perhaps one of the most important numerical algorithms in all of computing. In this chapter we explore how FFTs can be used in spectral analysis, but beyond this basic application, there is also broad usage of FFT both directly and indirectly as a component in other algorithms. Other signal-processing methods, such as convolution and correlation analysis, and linear filters also have widespread applications, in particular in engineering fields such as control theory.

In this chapter we discuss spectral analysis and basic applications of linear filters, using the fftpack and signal modules in the SciPy library.

© Robert Johansson 2019
R. Johansson, *Numerical Python*, https://doi.org/10.1007/978-1-4842-4246-9_17

Importing Modules

In this chapter, we mainly work with the `fftpack` and `signal` modules from the SciPy library. As usual with modules from the SciPy library, we import the modules using the following pattern:

```
In [1]: from scipy import fftpack
In [2]: from scipy import signal
```

We also use the `io.wavefile` module from SciPy to read and write WAV audio files in one of the examples. We import this module in the following way:

```
In [3]: import scipy.io.wavfile
In [4]: from scipy import io
```

For basic numerics and graphics, we also require the NumPy, Pandas, and Matplotlib libraries:

```
In [5]: import numpy as np
In [6]: import pandas as pd
In [7]: import matplotlib.pyplot as plt
In [8]: import matplotlib as mpl
```

Spectral Analysis

We begin this exploration of signal processing by considering spectral analysis. Spectral analysis is a fundamental application of Fourier transforms, which is a mathematical integral transform that allows us to take a signal from the time domain – where it is described as a function of time – to the frequency domain, where it is described as a function of frequency. The frequency-domain representation of a signal is useful for many purposes, for example, extracting features such as dominant frequency components of a signal, applying filters to signals, and solving differential equations (see Chapter 9), just to mention a few.

Fourier Transforms

The mathematical expression for the Fourier transform $F(v)$ of a continuous signal $f(t)$ is[1]

$$F(v) = \int_{-\infty}^{\infty} f(t)e^{-2\pi i v t}dt,$$

and the inverse Fourier transform is given by

$$f(t) = \int_{-\infty}^{\infty} F(v)e^{2\pi i v t}dv.$$

Here $F(v)$ is the complex-valued amplitude spectrum of the signal $f(t)$, and v is the frequency. From $F(v)$ we can compute other types of spectrum, such as the power spectrum $|F(v)|^2$. In this formulation $f(t)$ is a continuous signal with infinite duration. In practical applications we are often more interested in approximating $f(t)$ using a finite number of samples from a finite duration of time. For example, we might sample the function $f(t)$ at N uniformly spaced points in the time interval $t \in [0, T]$, resulting in a sequence of samples that we denote $(x_0, x_1, ..., x_N)$. The continuous Fourier transform shown in the preceding text can be adapted to the discrete case: the discrete Fourier transform (DFT) of a sequence of uniformly spaced samples is

$$X_k = \sum_{n=0}^{N-1} x_n e^{-2\pi i n k / N},$$

and similarly we have the inverse DFT

$$x_n = \frac{1}{N} \sum_{k=0}^{N-1} X_k e^{2\pi i n k / N},$$

where X_k is the discrete Fourier transform of the samples x_n and k is a frequency bin number that can be related to a real frequency. The DFT for a sequence of samples can be computed very efficiently using the algorithm known as fast Fourier transform (FFT).

---

[1]There are several alternative definitions of the Fourier transform, which vary in the coefficient in the exponent and the normalization of the transform integral.

The SciPy fftpack module[2] provides implementations of the FFT algorithm. The fftpack module contains FFT functions for a variety of cases: see Table 17-1 for a summary. Here we focus on demonstrating the usage of the fft and ifft functions and several of the helper functions in the fftpack module. However, the general usage is similar for all FFT functions in Table 17-1.

Table 17-1. *Summary of Selected Functions from the fftpack Module in SciPy. For detailed usage of each function, including their arguments and return values, see their docstrings which are available using, for example,* `help(fftpack.fft)`*.*

Function	Description
fft, ifft	General FFT and inverse FFT of a real- or complex-valued signal. The resulting frequency spectrum is complex valued.
rfft, irfft	FFT and inverse FFT of a real-valued signal.
dct, idct	The discrete cosine transform (DCT) and its inverse.
dst, idst	The discrete sine transform (DST) and its inverse.
fft2, ifft2, fftn, ifftn	The two-dimensional and the N-dimensional FFT for complex-valued signals and their inverses.
fftshift, ifftshift, rfftshift, irfftshift	Shift the frequency bins in the result vector produced by fft and rfft, respectively, so that the spectrum is arranged such that the zero-frequency component is in the middle of the array.
fftfreq	Calculate the frequencies corresponding to the FFT bins in the result returned by fft.

Note that the DFT takes discrete samples as input and outputs a discrete frequency spectrum. To be able to use DFT for processes that are originally continuous, we first must reduce the signals to discrete values using sampling. According to the sampling theorem, a continuous signal with bandwidth B (i.e., the signal does not contain frequencies higher than B) can be completely reconstructed from discrete samples with

[2]There is also an implementation of FFT in the fft module in NumPy. It provides mostly the same functions as scipy.fftpack, which we use here. As a general rule, when SciPy and NumPy provide the same functionality, it is generally preferable to use SciPy if available and fallback to the NumPy implementation when SciPy is not available.

sampling frequency $f_s \geq 2B$. This is a very important result in signal processing because it tells us under what circumstances we can work with discrete instead of continuous signals. It allows us to determine a suitable sampling rate when measuring a continuous process, since it is often possible to know or approximately guess the bandwidth of a process, for example, from physical arguments. While the sampling rate determines the maximum frequency we can describe with a discrete Fourier transform, the spacing of samples in frequency space is determined by the total sampling time T or equivalently from the number of samples points once the sampling frequency is determined, $T = N/f_s$.

As an introductory example, consider a simulated signal with pure sinusoidal components at 1 Hz and at 22 Hz, on top of a normal-distributed noise floor. We begin by defining a function `signal_samples` that generates noisy samples of this signal:

```
In [9]: def signal_samples(t):
   ...:     return (2 * np.sin(2 * np.pi * t) + 3 * np.sin(22 * 2 *
           np.pi * t) + 2 * np.random.randn(*np.shape(t)))
```

We can get a vector of samples by calling this function with an array with sample times as argument. Say that we are interested in computing the frequency spectrum of this signal up to frequencies of 30 Hz. We then need to choose the sampling frequency $f_s = 60$ Hz, and if we want to obtain a frequency spectrum with a resolution of $\Delta f = 0.01$ Hz, we need to collect at least $N = f_s/\Delta f = 6000$ samples, corresponding to a sampling period of $T = N/f_s = 100$ seconds:

```
In [10]: B = 30.0
In [11]: f_s = 2 * B
In [12]: delta_f = 0.01
In [13]: N = int(f_s / delta_f); N
Out[13]: 6000
In [14]: T = N / f_s; T
Out[14]: 100.0
```

Next we sample the signal function at N uniformly spaced points in time by first creating an array t that contains the sample times and then using it to evaluate the `signal_samples` function:

```
In [15]: t = np.linspace(0, T, N)
In [16]: f_t = signal_samples(t)
```

The resulting signal is plotted in Figure 17-1. The signal is rather noisy, both when viewed over the entire sampling time and when viewed for a shorter period of time, and the added random noise mostly masks the pure sinusoidal signals when viewed in the time domain.

```
In [17]: fig, axes = plt.subplots(1, 2, figsize=(8, 3), sharey=True)
    ...: axes[0].plot(t, f_t)
    ...: axes[0].set_xlabel("time (s)")
    ...: axes[0].set_ylabel("signal")
    ...: axes[1].plot(t, f_t)
    ...: axes[1].set_xlim(0, 5)
    ...: axes[1].set_xlabel("time (s)")
```

Figure 17-1. *Simulated signal with random noise. Full signal to the left and zoom into early times on the right*

To reveal the sinusoidal components in the signal, we can use the FFT to compute the spectrum of the signal (or in other words, its frequency-domain representation). We obtain the discrete Fourier transform of the signal by applying the fft function to the array of discrete samples, f_t:

```
In [18]: F = fftpack.fft(f_t)
```

The result is an array F, which contains the frequency components of the spectrum at frequencies that are determined by the sampling rate and the number of samples. When computing these frequencies, it is convenient to use the helper function fftfreq, which takes the number of samples and the time duration between successive samples as parameters and returns an array of the same size as F that contains the frequencies corresponding to each frequency bin.

```
In [19]: f = fftpack.fftfreq(N, 1.0/f_s)
```

The frequency bins for the amplitude values returned by the fft function contain both positive and negative frequencies, up to the frequency that corresponds to half the sampling rate, $f_s/2$. For real-valued signals, the spectrum is symmetric at positive and negative frequencies, and we are for this reason often only interested in the positive-frequency components. Using the frequency array f, we can conveniently create a mask that can be used to extract the part of the spectrum that corresponds to the frequencies we are interested in. Here we create a mask for selecting the positive-frequency components:

```
In [20]: mask = np.where(f >= 0)
```

The spectrum for the positive-frequency components is shown in Figure 17-2. The top panel contains the entire positive-frequency spectrum and is plotted on a log scale to increase the contrast between the signal and the noise. We can see that there are sharp peaks near 1 Hz and 22 Hz, corresponding to the sinusoidal components in the signal. These peaks clearly stand out from the noise floor in the spectrum. In spite of the noise concealing the sinusoidal components in the time-domain signal, we can clearly detect their presence in the frequency-domain representation. The lower two panels in Figure 17-2 show magnifications of the two peaks at 1 Hz and 22 Hz, respectively.

```
In [21]: fig, axes = plt.subplots(3, 1, figsize=(8, 6))
   ...: axes[0].plot(f[mask], np.log(abs(F[mask])), label="real")
   ...: axes[0].plot(B, 0, 'r*', markersize=10)
   ...: axes[0].set_ylabel("$\log(|F|)$", fontsize=14)
   ...: axes[1].plot(f[mask], abs(F[mask])/N, label="real")
   ...: axes[1].set_xlim(0, 2)
   ...: axes[1].set_ylabel("$|F|/N$", fontsize=14)
   ...: axes[2].plot(f[mask], abs(F[mask])/N, label="real")
   ...: axes[2].set_xlim(21, 23)
   ...: axes[2].set_xlabel("frequency (Hz)", fontsize=14)
   ...: axes[2].set_ylabel("$|F|/N$", fontsize=14)
```

Figure 17-2. *Spectrum of the simulated signal with frequency components at 1 Hz and 22 Hz*

Frequency-Domain Filter

Just like we can compute the frequency-domain representation from the time-domain signal using the FFT function fft, we can compute the time-domain signal from the frequency-domain representation using the *inverse* FFT function ifft. For example, applying the ifft function to the F array will reconstruct the f_t array. By modifying the spectrum before we apply the inverse transform, we can realize frequency-domain filters. For example, selecting only frequencies below 2 Hz in the spectrum amounts to applying a 2 Hz low-pass filter, which suppresses high-frequency components in the signal (higher than 2 Hz in this case):

```
In [22]: F_filtered = F * (abs(f) < 2)
In [23]: f_t_filtered = fftpack.ifft(F_filtered)
```

Computing the inverse FFT for the filtered signal results in a time-domain signal where the high-frequency oscillations are absent, as shown in Figure 17-3. This simple example summarizes the essence of many frequency-domain filters. Later in this chapter, we explore in more detail some of the many types of filters that are commonly used in signal-processing analysis.

```
In [24]: fig, ax = plt.subplots(figsize=(8, 3))
    ...: ax.plot(t, f_t, label='original')
    ...: ax.plot(t, f_t_filtered.real, color="red", lw=3, label='filtered')
    ...: ax.set_xlim(0, 10)
    ...: ax.set_xlabel("time (s)")
    ...: ax.set_ylabel("signal")
    ...: ax.legend()
```

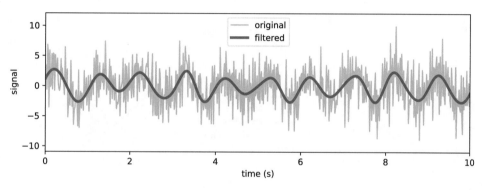

Figure 17-3. *The original time-domain signal and the reconstructed signal after applying a low-pass filter to the frequency-domain representation of the signal*

Windowing

In the previous section, we directly applied the FFT to the signal. This can give acceptable results, but it is often possible to further improve the quality and the contrast of the frequency spectrum by applying a so-called window function to the signal before applying the FFT. A window function is a function that when multiplied with the signal modulates its magnitude so that it approaches zero at the beginning and the end of the sampling duration. There are many possible functions that can be used as a window function, and the SciPy `signal` module provides implementations of many common window functions, including the Blackman function, the Hann function, the Hamming function, Gaussian window functions (with variable standard deviation), and the Kaiser window function.[3] These functions are all plotted in Figure 17-4. This graph shows that while all of these window functions are slightly different, the overall shape is very similar.

---

[3]Several other window functions are also available. See the docstring for the `scipy.signal` module for a complete list.

```
In [25]: fig, ax = plt.subplots(1, 1, figsize=(8, 3))
    ...: N = 100
    ...: ax.plot(signal.blackman(N), label="Blackman")
    ...: ax.plot(signal.hann(N), label="Hann")
    ...: ax.plot(signal.hamming(N), label="Hamming")
    ...: ax.plot(signal.gaussian(N, N/5), label="Gaussian (std=N/5)")
    ...: ax.plot(signal.kaiser(N, 7), label="Kaiser (beta=7)")
    ...: ax.set_xlabel("n")
    ...: ax.legend(loc=0)
```

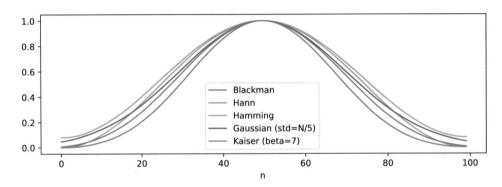

Figure 17-4. *Example of commonly used window functions*

The alternative window functions all have slightly different properties and objectives, but for the most part, they can be used interchangeably. The main purpose of window functions is to reduce spectral leakage between nearby frequency bins, which occur in discrete Fourier transform computation when the signal contains components with periods that are not exactly divisible with the sampling period. Signal components with such frequencies can therefore not fit a full number of cycles in the sampling period, and since discrete Fourier transform assumes that signal is period, the resulting discontinuity at the period boundary can give rise to spectral leakage. Multiplying the signal with a window function reduces this problem. Alternatively, we could also increase the number of sample points (increase the sampling period) to obtain a higher frequency resolution, but this might not always be practical.

To see how we can use a window function before applying the FFT to a time-series signal, let's consider the outdoor temperature measurements that we looked at in Chapter 12. First, we use the Pandas library to load the dataset and resample it to evenly spaced hourly samples, using `fillna` method to aggregate the elements.

```
In [26]: df = pd.read_csv('temperature_outdoor_2014.tsv', delimiter="\t",
    ...:                       names=["time", "temperature"])
In [27]: df.time = (pd.to_datetime(df.time.values, unit="s").
    ...:                   tz_localize('UTC').tz_convert('Europe/Stockholm'))
In [28]: df = df.set_index("time")
In [29]: df = df.resample("H").ffill()
In [30]: df = df[(df.index >= "2014-04-01")*(df.index < "2014-06-01")].
        dropna()
```

Once the Pandas data frame has been created and processed, we exact the underlying NumPy arrays to be able to process the time-series data using the `fftpack` module.

```
In [31]: time = df.index.astype('int64')/1.0e9
In [32]: temperature = df.temperature.values
```

Now we wish to apply a window function to the data in the array `temperature` before we compute the FFT. Here we use the Blackman window function, which is a window function that is suitable for reducing spectral leakage. It is available as the `blackman` function in the `signal` module in SciPy. As the argument to the window function, we need to pass the length of the sample array, and it returns an array of that same length:

```
In [33]: window = signal.blackman(len(temperature))
```

To apply the window function, we simply multiply it with the array containing the time-domain signal and use the result in the subsequent FFT computation. However, before we proceed with the FFT for the windowed temperature signal, we first plot the original temperature time series and the windowed version. The result is shown in Figure 17-5. The result of multiplying the time series with the window function is a signal that approaches zero near the sampling period boundaries, and it can be viewed as a periodic function with smooth transitions between period boundaries, and as such the FFT of the windowed signal has better-behaved properties.

```
In [34]: temperature_windowed = temperature * window
In [35]: fig, ax = plt.subplots(figsize=(8, 3))
    ...: ax.plot(df.index, temperature, label="original")
    ...: ax.plot(df.index, temperature_windowed, label="windowed")
    ...: ax.set_ylabel("temperature", fontsize=14)
    ...: ax.legend(loc=0)
```

Figure 17-5. *Windowed and original temperature time-series signal*

After having prepared the windowed signal, the rest of the spectral analysis proceeds as before: we can use the fft function to compute the spectrum and the fftfreq function to calculate the frequencies corresponding to each frequency bin.

```
In [36]: data_fft_windowed = fftpack.fft(temperature_windowed)
In [37]: f = fftpack.fftfreq(len(temperature), time[1]-time[0])
```

Here we also select the positive frequencies by creating a mask array from the array f and plot the resulting positive-frequency spectrum as shown in Figure 17-6. The spectrum in Figure 17-6 clearly shows peaks at the frequency that corresponds to 1 day (1/86400 Hz) and its higher harmonics (2/86400 Hz, 3/86400 Hz, etc.).

```
In [38]: mask = f > 0
In [39]: fig, ax = plt.subplots(figsize=(8, 3))
    ...: ax.set_xlim(0.000005, 0.00004)
    ...: ax.axvline(1./86400, color='r', lw=0.5)
    ...: ax.axvline(2./86400, color='r', lw=0.5)
```

```
...: ax.axvline(3./86400, color='r', lw=0.5)
...: ax.plot(f[mask], np.log(abs(data_fft_windowed[mask])), lw=2)
...: ax.set_ylabel("$\log|F|$", fontsize=14)
...: ax.set_xlabel("frequency (Hz)", fontsize=14)
```

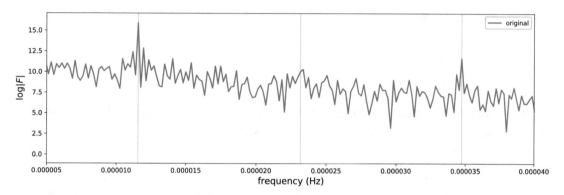

Figure 17-6. *Spectrum of the windowed temperature time series. The dominant peak occurs at the frequency corresponding to a 1-day period and its higher harmonics.*

To get the most accurate spectrum from a given set of samples, it is generally advisable to apply a window function to the time-series signal before applying an FFT. Most of the window functions available in SciPy can be used interchangeably, and the choice of window function is usually not critical. A popular choice is the Blackman window function, which is designed to minimize spectral leakage. For more details about the properties of different window functions, see, for example, Chapter 9 (Smith, 1999).

Spectrogram

As a final example in this section on spectral analysis, here we analyze the spectrum of an audio signal that was sampled from a guitar.[4] First we load sampled data from the guitar.wav file using the io.wavefile.read function from the SciPy library:

```
In [40]: sample_rate, data = io.wavfile.read("guitar.wav")
```

---

[4]The data used in this example was obtained from https://www.freesound.org/people/guitarguy1985/sounds/52047.

The `io.wavefile.read` function returns a tuple containing the sampling rate, `sample_rate`, and a NumPy array containing the audio intensity. For this particular file, we get the sampling rate of 44.1 kHz, and the audio signal was recorded in stereo, which is represented by a data array with two channels. Each channel contains 1181625 samples:

```
In [41]: sample_rate
Out[41]: 44100
In [42]: data.shape
Out[42]: (1181625, 2)
```

Here we will only be concerned with analyzing a single audio channel, so we form the average of the two channels to obtain a mono-channel signal:

```
In [43]: data = data.mean(axis=1)
```

We can calculate the total duration of the audio recording by dividing the number of samples with the sampling rate. The result suggests that the recording is about 26.8 seconds.

```
In [44]: data.shape[0] / sample_rate
Out[44]: 26.79421768707483
```

It is often the case that we like to compute the spectrum of a signal in segments instead of the entire signal at once, for example, if the nature of the signal varies in time on a long timescale but contains nearly periodic components on a short timescale. This is particularly true for music, which can be considered nearly period on short timescales from the point of view of human perception (subsecond timescales) but which varies on longer timescales. In the case of the guitar sample, we would therefore like to apply the FFT on a sliding window in the time-domain signal. The result is a time-dependent spectrum, which is often visualized as an equalizer graph on music equipment and applications. Another approach is to visualize the time-dependent spectrum using a two-dimensional heatmap graph, which in this context is known as a spectrogram. In the following we compute the spectrogram of the guitar sample.

Before we proceed with the spectrogram visualization, we first calculate the spectrum for a small part of the sample. We begin by determining the number of samples to use from the full sample array. If we want to analyze for 0.5 seconds at the time, we can use the sampling rate to compute the number of samples to use:

```
In [45]: N = int(sample_rate/2.0) # half a second -> 22050 samples
```

Next, given the number of samples and the sampling rate, we can compute the frequencies f for the frequency bins for the result of the forthcoming FFT calculation, as well as the sampling times t for each sample in the time-domain signal. We also create a frequency mask for selecting positive frequencies smaller than 1000 Hz, which we will use later on to select a subset of the computed spectrum.

```
In [46]: f = fftpack.fftfreq(N, 1.0/sample_rate)
In [47]: t = np.linspace(0, 0.5, N)
In [48]: mask = (f > 0) * (f < 1000)
```

Next, we exact the first N samples from the full sample array data and apply the fft function on it:

```
In [49]: subdata = data[:N]
In [50]: F = fftpack.fft(subdata)
```

The time- and frequency-domain signals are shown in Figure 17-7. The time-domain signal in the left panel is zero in the beginning before the first guitar string is plucked. The frequency-domain spectrum shows several dominant frequencies that correspond to the different tones produced by the guitar.

```
In [51]: fig, axes = plt.subplots(1, 2, figsize=(12, 3))
    ...: axes[0].plot(t, subdata)
    ...: axes[0].set_ylabel("signal", fontsize=14)
    ...: axes[0].set_xlabel("time (s)", fontsize=14)
    ...: axes[1].plot(f[mask], abs(F[mask]))
    ...: axes[1].set_xlim(0, 1000)
    ...: axes[1].set_ylabel("$|F|$", fontsize=14)
    ...: axes[1].set_xlabel("Frequency (Hz)", fontsize=14)
```

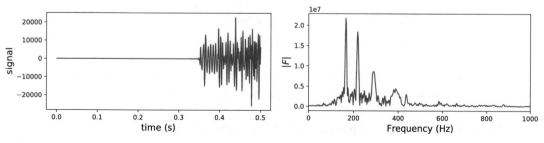

Figure 17-7. *Signal and spectrum for samples half a second duration of a guitar sound*

The next step is to repeat the analysis for successive segments from the full sample array. The time evolution of the spectrum can be visualized as a spectrogram, with frequency on the *x* axis and time on the *y* axis. To be able to plot the spectrogram with the imshow function from Matplotlib, we create a two-dimensional NumPy array spectrogram_data for storing the spectra for the successive sample segments. The shape of the spectrogram_data array is (n_max, f_values), where n_max is the number of segments of length N in the sample array data and f_values are the number of frequency bins with frequencies that match the condition used to compute mask (positive frequencies less than 1000 Hz):

```
In [52]: n_max = int(data.shape[0] / N)
In [53]: f_values = np.sum(mask)
In [54]: spectogram_data = np.zeros((n_max, f_values))
```

To improve the contrast of the resulting spectrogram, we also apply a Blackman window function to each subset of the sample data before we compute the FFT. Here we choose the Blackman window function for its spectral leakage reducing properties, but many other window functions give similar results. The length of the window array must be the same as the length of the subdata array, so we pass its length argument to the Blackman function:

```
In [55]: window = signal.blackman(len(subdata))
```

Finally we can compute the spectrum for each segment in the sample by looping over the array slices of size N, apply the window function, compute the FFT, and store the subset of the result for the frequencies we are interested in in the spectrogram_data array:

```
In [56]: for n in range(0, n_max):
    ...:     subdata = data[(N * n):(N * (n + 1))]
    ...:     F = fftpack.fft(subdata * window)
    ...:     spectogram_data[n, :] = np.log(abs(F[mask]))
```

When the spectrogram_data is computed, we can visualize the spectrogram using the imshow function from Matplotlib. The result is shown in Figure 17-8.

```
In [57]: fig, ax = plt.subplots(1, 1, figsize=(8, 6))
    ...: p = ax.imshow(spectogram_data, origin='lower',
    ...:               extent=(0, 1000, 0, data.shape[0] / sample_rate),
```

```
...:                      aspect='auto',
...:                      cmap=mpl.cm.RdBu_r)
...: cb = fig.colorbar(p, ax=ax)
...: cb.set_label("$\log|F|$", fontsize=14)
...: ax.set_ylabel("time (s)", fontsize=14)
...: ax.set_xlabel("Frequency (Hz)", fontsize=14)
```

The spectrogram in Figure 17-8 contains a lot of information about the sampled signal and how it evolves over time. The narrow vertical stripes correspond to tones produced by the guitar, and those signals slowly decay with increasing time. The broad horizontal bands correspond roughly to periods of time when strings are being plucked on the guitar, which for a short time gives a very broad frequency response. Note, however, that the color axis represents a logarithmic scale, so small variations in the color represent large variation in the actual intensity.

Figure 17-8. *Spectrogram of an audio sampling of a guitar sound*

Signal Filters

One of the main objectives in signal processing is to manipulate and transform temporal or spatial signals to change their characteristics. Typical applications are noise reduction, sound effects in audio signals, and effects such as blurring, sharpening, contrast enhancement, and color balance adjustments in image data. Many common transformations can be implemented as filters that act on the frequency-domain representation of the signal, for example, by suppressing certain frequency components. In the previous section, we saw an example of a low-pass filter, which we implemented by taking the Fourier transform of the signal, removing the high-frequency components, and finally taking the inverse Fourier transform to obtain a new time-domain signal. With this approach, we can implement arbitrary frequency filters, but we cannot necessarily apply them in real time on a streaming signal since they require buffering sufficient samples to be able to perform the discrete Fourier transform. In many applications, it is desirable to apply filters and transform a signal in a continuous fashion, for example, when processing signals in transmission or live audio signals.

Convolution Filters

Certain types of frequency filters can be implemented directly in the time domain using a convolution of the signal with a function that characterizes the filter. An important property of Fourier transformations is that the (inverse) Fourier transform of the product of two functions (e.g., the spectrum of a signal and the filter shape function) is a convolution of the two functions (inverse) Fourier transforms. Therefore, if we want to apply a filter H_k to the spectrum X_k of a signal x_n, we can instead compute the convolution of x_n with h_m, the inverse Fourier transform of the filter function H_k. In general, we can write a filter on convolution form as

$$y_n = \sum_{k=-\infty}^{\infty} x_k h_{n-k},$$

where x_k is the input, y_n is the output, and h_{n-k} is the convolution kernel that characterizes the filter. Note that in this general form, the signal y_n at time step n depends on both earlier and later values of the input x_k. To illustrate this point, let's return to the first example in this chapter, where we applied a low-pass filter to a simulated signal with components at 1 Hz and at 22 Hz. In that example we Fourier transformed the signal and multiplied its spectrum with a step function that suppressed all high-frequency components, and finally we inverse Fourier transformed the signal back into the time domain. The result was a smoothened version of the original noisy signal (Figure 17-3). An alternative approach using convolution is to inverse Fourier transform the frequency response function for the filter H and use the result h as a kernel with which we convolve the original time-domain signal f_t:

```
In [58]: t = np.linspace(0, T, N)
In [59]: f_t = signal_samples(t)
In [60]: H = abs(f) < 2
In [61]: h = fftpack.fftshift(fftpack.ifft(H))
In [62]: f_t_filtered_conv = signal.convolve(f_t, h, mode='same')
```

To carry out the convolution, here we used the convolve function from the signal module in SciPy. It takes as arguments two NumPy arrays containing the signals for which to compute the convolution. Using the optional keyword argument mode, we can set the size of the output array to be the same as the first input (mode='same'), the full convolution output after having zero-padded the arrays to account for transients (mode='full'), or to contain only elements that do not rely on zero-padding (mode='valid'). Here we use mode='same', so we easily can compare and plot the result with the original signal, f_t. The result of applying this convolution filter, f_t_filtered_conv, is shown in Figure 17-9, together with the corresponding result that was computed using fft and ifft with a modified spectrum (f_t_filtered). As expected the two methods give identical results.

```
In [63]: fig = plt.figure(figsize=(8, 6))
    ...: ax = plt.subplot2grid((2,2), (0,0))
    ...: ax.plot(f, H)
    ...: ax.set_xlabel("frequency (Hz)")
    ...: ax.set_ylabel("Frequency filter")
    ...: ax.set_ylim(0, 1.5)
    ...: ax = plt.subplot2grid((2,2), (0,1))
    ...: ax.plot(t - t[-1]/2.0, h.real)
    ...: ax.set_xlabel("time (s)")
    ...: ax.set_ylabel("convolution kernel")
    ...: ax = plt.subplot2grid((2,2), (1,0), colspan=2)
    ...: ax.plot(t, f_t, label='original', alpha=0.25)
```

```
   ...: ax.plot(t, f_t_filtered.real, 'r', lw=2, label='filtered in
        frequency domain')
   ...: ax.plot(t, f_t_filtered_conv.real, 'b--', lw=2, label='filtered
        with convolution')
   ...: ax.set_xlim(0, 10)
   ...: ax.set_xlabel("time (s)")
   ...: ax.set_ylabel("signal")
   ...: ax.legend(loc=2)
```

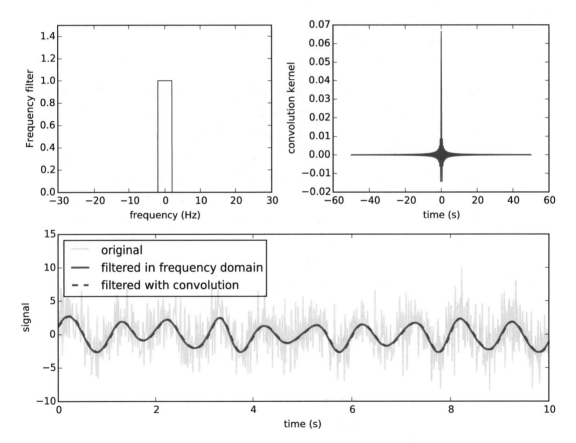

Figure 17-9. *Top left: frequency filter. Top right: convolution kernel corresponding to the frequency filter (its inverse discrete Fourier transform). Bottom: simple low-pass filter applied via convolution*

FIR and IIR Filters

In the example of a convolution filter in the previous section, there is no computational advantage of using a convolution to implement the filter rather than a sequence of a call to `fft`, spectrum modifications, followed by a call to `ifft`. In fact, the convolution here is in general more demanding than the extra FFT transformation, and the SciPy signal module actually provides a function call `fftconvolve`, which implements the convolution using FFT and its inverse. Furthermore, the convolution kernel of the filter has many undesirable properties, such as being noncasual, where the output signal depends on future values of the input (see the upper-right panel in Figure 17-9). However, there are important special cases of convolution-like filters that can be efficiently implemented with both dedicated digital signal processors (DSPs) and general-purpose processors. An important family of such filters is the *finite impulse response* (FIR) filters, which takes the form $y_n = \sum_{k=0}^{M} b_k x_{n-k}$. This time-domain filter is casual because the output y_n only depends on input values at earlier time steps.

Another similar type of filter is the *infinite impulse response* (IIR) filters, which can be written in the form $a_0 y_n = \sum_{k=0}^{M} b_k x_{n-k} - \sum_{k=1}^{N} a_k y_{n-k}$. This is not strictly a convolution since it additionally includes past values of the *output* when computing a new output value (a feedback term), but it is nonetheless in a similar form. Both FIR and IIR filters can be used to evaluate new output values given the recent history of the signal and the output and can therefore be evaluated sequentially in the time domain, if we know the finite sequences of values of b_k and a_k.

Computing the values of b_k and a_k given a set of requirements on filter properties is known as filter design. The SciPy `signal` module provides many functions for this purpose. For example, using the `firwin` function, we can compute the b_k coefficients for an FIR filter given frequencies of the band boundaries, where, for example, the filter transitions from a pass to a stop filter (for a low-pass filter). The `firwin` function takes the number of values in the a_k sequence as the first argument (also known as *taps* in this context). The second argument, `cutoff`, defines the low-pass transition frequency in units of the Nyquist frequency (half the sampling rate). The scale of the Nyquist frequency can optionally be set using the `nyq` argument, which defaults to 1. Finally we can specify the type of window function to use with the `window` argument.

```
In [64]: n = 101
In [65]: f_s = 1 / 3600
In [66]: nyq = f_s/2
In [67]: b = signal.firwin(n, cutoff=nyq/12, nyq=nyq, window="hamming")
```

The result is the sequence of coefficients b_k that defines an FIR filter and which can be used to implement the filter with a time-domain convolution. Given the coefficients b_k, we can evaluate the amplitude and phase response of the filter using the freqz function from the signal module. It returns arrays containing frequencies and the corresponding complex-valued frequency response, which are suitable for plotting purposes, as shown in Figure 17-10.

```
In [68]: f, h = signal.freqz(b)
In [69]: fig, ax = plt.subplots(1, 1, figsize=(12, 3))
    ...: h_ampl = 20 * np.log10(abs(h))
    ...: h_phase = np.unwrap(np.angle(h))
    ...: ax.plot(f/max(f), h_ampl, 'b')
    ...: ax.set_ylim(-150, 5)
    ...: ax.set_ylabel('frequency response (dB)', color="b")
    ...: ax.set_xlabel(r'normalized frequency')
    ...: ax = ax.twinx()
    ...: ax.plot(f/max(f), h_phase, 'r')
    ...: ax.set_ylabel('phase response', color="r")
    ...: ax.axvline(1.0/12, color="black")
```

Figure 17-10. *The amplitude and phase response of a low-pass FIR filter*

The low-pass filter shown in Figure 17-10 is designed to pass through signals with frequencies less than $f_s/24$ (indicated with a vertical line) and suppress higher-frequency signal components. The finite transition region between pass and stop bands and the nonperfect suppression above the cutoff frequency is a price we have to pay to be able to represent the filter in FIR form. The accuracy of the FIR filter can be improved by increasing the number of coefficients b_k, at the expense of higher computational complexity.

The effect of an FIR filter, given the coefficients b_k, and an IIR filter, given the coefficients b_k and a_k, can be evaluated using the lfilter function from the signal module. As first argument, this function expects the array with coefficients b_k, and as second argument, the array with the coefficients a_k in the case of an IIR filter, or the scalar 1 in case of an FIR filter. The third argument to the function is the input signal array, and the return value is the filter output. For example, to apply the FIR filter we created in the preceding text to the array with hourly temperature measurements temperature, we can use

```
In [70]: temperature_filt = signal.lfilter(b, 1, temperature)
```

The effect of applying the low-pass FIR filter to the signal is to smoothen the function by an eliminating the high-frequency oscillations, as shown in Figure 17-11. Another approach to achieve a similar result is to apply a moving average filter, in which the output is a weighted average or median of a few nearby input values. The function medfilt from the signal module applies a median filter of a given input signal, using the number of past nearby values specified with the second argument to the function:

```
In [71]: temperature_median_filt = signal.medfilt(temperature, 25)
```

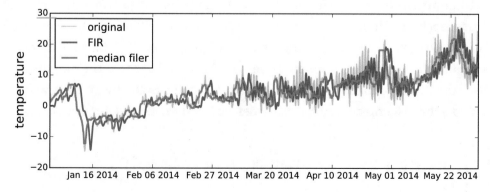

Figure 17-11. *Output of an FIR filter and a median filter*

The result of applying the FIR low-pass filter and the median filter to the hourly temperature measurement dataset is shown in Figure 17-11. Note that the output of the FIR filter is shifted from the original signal by a time delay that corresponds to the number of taps in the FIR filter. The median filter implemented using medfilt does not suffer from this issue because the median is computed with respect to both past and future values, which makes it a noncasual filter that cannot be evaluated on the fly on streaming input data.

```
In [72]: fig, ax = plt.subplots(figsize=(8, 3))
    ...: ax.plot(df.index, temperature, label="original", alpha=0.5)
    ...: ax.plot(df.index, temperature_filt, color="red", lw=2,
        label="FIR")
    ...: ax.plot(df.index, temperature_median_filt, color="green", lw=2,
        label="median filer")
    ...: ax.set_ylabel("temperature", fontsize=14)
    ...: ax.legend(loc=0)
```

To design an IIR filter, we can use the iirdesign function from the signal module or use one of the many predefined IIR filter types, including the Butterworth filter (signal.butter), Chebyshev filters of types I and II (signal.cheby1 and signal.cheby2), and elliptic filter (signal.ellip). For example, to create a Butterworth high-pass filter that allows frequencies above the critical frequency 7/365 Hz to pass, while lower frequencies are suppressed, we can use

```
In [73]: b, a = signal.butter(2, 7/365.0, btype='high')
```

The first argument to this function is the order of the Butterworth filter, and the second argument is the critical frequency of the filter (where it goes from bandstop to bandpass function). The optional argument btype can, for example, be used to specify if the filter is a low-pass filter (low) or high-pass filter (high). More options are described in the function's docstring: see, for example, help(signal.butter). The outputs a and b are the a_k and b_k coefficients that define the IIR filter, respectively. Here we have computed a Butterworth filter of second order, so a and b each has three elements:

```
In [74]: b
Out[74]: array([ 0.95829139, -1.91658277,  0.95829139])
In [75]: a
Out[75]: array([ 1.        , -1.91484241,  0.91832314])
```

Like before we can apply the filter to an input signal (here we again use the hourly temperature dataset as an example):

```
In [76]: temperature_iir = signal.lfilter(b, a, temperature)
```

Alternatively we can apply the filter using the `filtfilt` function, which applies the filter both forward and backward, resulting in a noncasual filter.

```
In [77]: temperature_filtfilt = signal.filtfilt(b, a, temperature)
```

The results of both types of filters are shown in Figure 17-12. Eliminating the low-frequency components detrends the time series and only retains the high-frequency oscillations and fluctuations. The filtered signal can therefore be viewed as measuring the volatility of the original signal. In this example we can see that the daily variations are greater during the spring months of March, April, and May, when compared to the winter month of January and February.

```
In [78]: fig, ax = plt.subplots(figsize=(8, 3))
    ...: ax.plot(df.index, temperature, label="original", alpha=0.5)
    ...: ax.plot(df.index, temperature_iir, color="red", label="IIR
         filter")
    ...: ax.plot(df.index, temperature_filtfilt, color="green",
         label="filtfilt filtered")
    ...: ax.set_ylabel("temperature", fontsize=14)
    ...: ax.legend(loc=0)
```

Figure 17-12. *Output from an IIR high-pass filter and the corresponding filtfilt filter (applied both forward and backward)*

These techniques can be directly applied to audio and image data. For example, to apply a filter to the audio signal of the guitar samples, we can use the use the `lfilter` functions. The coefficients b_k for the FIR filter can sometimes be constructed manually. For example, to apply a naive echo sound effect, we can create an FIR filter that repeats past signals with some time delay: $y_n = x_n + x_{n-N}$, where N is a time delay in units of time steps. The corresponding coefficients b_k are easily constructed and can be applied to the audio signal data.

```
In [79]: b = np.zeros(10000)
    ...: b[0] = b[-1] = 1
    ...: b /= b.sum()
In [80]: data_filt = signal.lfilter(b, 1, data)
```

To be able to listen to the modified audio signal, we can write it to a WAV file using the `write` function from the `io.wavefile` module in SciPy:

```
In [81]: io.wavfile.write("guitar-echo.wav", sample_rate,
    ...:                     np.vstack([data_filt, data_filt]).T.astype(np.
                            int16))
```

Similarly, we can implement many types of image processing filters using the tools from the `signal` module. SciPy also provides a module `ndimage`, which contains many common image manipulation functions and filters that are specially adapted for applying on two-dimensional image data. The Scikit-Image library[5] provides a more advanced framework for working with image processing in Python.

Summary

Signal processing is an extremely broad field with applications in most fields of science and engineering. As such, here we have only been able to cover a few basic applications of signal processing in this chapter, and we have focused on introducing methods for approaching this type of problem with computational methods using Python and the libraries and tools that are available within the Python ecosystem for scientific computing. In particular, we explored spectral analysis of time-dependent signals using fast Fourier transform and the design and application of linear filters to signals using the `signal` module in the SciPy library.

---

[5]See the project's web page at `http://scikit-image.org` for more information.

Further Reading

For a comprehensive review of the theory of signal processing, see Smith (1999), which can also be viewed online at `www.dspguide.com/pdfbook.htm`. For a Python-oriented discussion of signal processing, see Unpingco (2014), from which content is available as IPython notebooks at `http://nbviewer.ipython.org/github/unpingco/Python-for-Signal-Processing`.

References

Smith, S. (1999). The Scientist and Engineer's Guide to Digital Signal Processing. San Diego: Steven W. Smith.

Unpingco, J. (2014). Python for Signal Processing. New York: Springer.

CHAPTER 18

Data Input and Output

In nearly all scientific computing and data analysis applications, there is a need for data input and output. This includes to load datasets and to persistently store results to files on disk or to databases. Getting data in and out of programs is consequently a key step in the computational workflow. There are many standardized formats for storing structured and unstructured data. The benefits of using standardized formats are obvious: You can use existing libraries for reading and writing data, saving yourself both time and effort. In the course of working with scientific and technical computing, it is likely that you will face a variety of data formats through interaction with colleagues and peers or when acquiring data from sources such as equipment and databases. As a computational practitioner, it is important to be able to handle data efficiently and seamlessly, regardless of which format it comes in. This motivates why this entire chapter is devoted to the topic of data input and output.

Python has good support for many file formats. In fact, multiple options exist for dealing with the most common formats. In this chapter we survey data storage formats with applications in computing and discuss typical situations where each format is suitable. We also introduce Python libraries and tools for handling a selection of data formats that are common in computing.

Data can be classified into several categories and types. Important categories are structured and unstructured data, and values can, for example, be categorical (finite set of values), ordinal (values with meaningful ordering), or numerical (continuous or discrete). Values also have types, such as string, integer, floating-point number, etc. A data format for storing or transmitting data should ideally account for these concepts in order to avoid loss of data or metadata, and we frequently need to have fine-grained control of how data is represented.

In computing applications, most of the time we deal with structured data, for example, arrays and tabular data. Examples of unstructured datasets include free-form texts or nested list with nonhomogeneous types. In this chapter we focus on the CSV

© Robert Johansson 2019
R. Johansson, *Numerical Python*, https://doi.org/10.1007/978-1-4842-4246-9_18

family of formats and the HDF5 format for structured data, and toward the end of the chapter, we discuss the JSON format as a lightweight and flexible format that can be used to store both simple and complex datasets, with a bias toward storing lists and dictionaries. This format is well suited for storing unstructured data. We also briefly discuss methods of serializing objects into storable data using the msgpack format and Python's built-in pickle format.

Because of the importance of data input and output in many data-centric computational applications, several Python libraries have emerged with the objective to simplify and assist in handling data in different formats and for moving and converting data. For example, the Blaze library (`http://blaze.pydata.org/en/latest`) provides a high-level interface for accessing data of different formats and from different types of sources. Here we focus mainly on lower-level libraries for reading specific types of file formats that are useful for storing numerical data and unstructured datasets. However, the interested reader is encouraged to also explore higher-level libraries such as Blaze.

Importing Modules

In this chapter we use a number of different libraries for handling different types of data. In particular, we require NumPy and Pandas, which as usual we import as np and pd, respectively:

```
In [1]: import numpy as np
In [2]: import pandas as pd
```

We also use the csv and json modules from the Python standard library:

```
In [3]: import csv
In [4]: import json
```

For working with the HDF5 format for numerical data, we use the h5py and the pytables libraries:

```
In [5]: import h5py
In [6]: import tables
```

Finally, in the context of serializing objects to storable data, we explore the `pickle` and `msgpack` libraries:

```
In [7]: import pickle
In [8]: import msgpack
```

Comma-Separated Values

Comma-separated values (CSV) is an intuitive and loosely defined[1] plain-text file format that is simple yet effective and very prevalent for storing tabular data. In this format each record is stored as a line, and each field of the record is separated with a delimiter character (e.g., a comma). Optionally, each field can be enclosed in quotation marks, to allow for string-valued fields that contain the delimiter character. Also, the first line is sometimes used to store column names, and comment lines are also common. An example of a CSV file is shown in Listing 18-1.

Listing 18-1. Example of a CSV File with a Comment Line, a Header Line, and Mixed Numerical and String-Valued Data Fields (Data source: `www.nhl.com`)

```
# 2013-2014 / Regular Season / All Skaters / Summary / Points
Rank,Player,Team,Pos,GP,G,A,P,+/-,PIM,PPG,PPP,SHG,SHP,GW,OT,S,S%,TOI/
GP,Shift/GP,FO%
1,Sidney Crosby,PIT
,C,80,36,68,104,+18,46,11,38,0,0,5,1,259,13.9,21:58,24.0,52.5
2,Ryan Getzlaf,ANA
,C,77,31,56,87,+28,31,5,23,0,0,7,1,204,15.2,21:17,25.2,49.0
3,Claude Giroux,PHI
,C,82,28,58,86,+7,46,7,37,0,0,7,1,223,12.6,20:26,25.1,52.9
4,Tyler Seguin,DAL
,C,80,37,47,84,+16,18,11,25,0,0,8,0,294,12.6,19:20,23.4,41.5
5,Corey Perry,ANA
,R,81,43,39,82,+32,65,8,18,0,0,9,1,280,15.4,19:28,23.2,36.0
```

---

[1]Although RFC 4180, `http://tools.ietf.org/html/rfc4180`, is sometimes taken as an unofficial specification, in practice there exist many varieties and dialects of CSV.

CSV is occasionally also taken to be an acronym for character-separated value, reflecting the fact that the CSV format commonly refers to a family of formats using different delimiters between the fields. For example, instead of the comma, the Tab character is often used, in which case the format is sometimes called TSV instead of CSV. The term delimiter-separated values (DSV) is also occasionally used to refer to these types of formats.

In Python there are several ways to read and write data in the CSV format, each with different use-cases and advantages. To begin with, the standard Python library contains a module called csv for reading CSV data. To use this module, we can call the csv.reader function with a file handle as argument. It returns a class instance that can be used as an iterator that parses lines from the given CSV file into Python lists of strings. For example, to read the file playerstats-2013-2014.csv (shown in Listing 18-1) into a nested list of strings, we can use

```
In [9]: with open("playerstats-2013-2014.csv") as f:
   ...:         csvreader = csv.reader(f)
   ...:         rows = [fields for fields in csvreader]
In [10]: rows[1][1:6]
Out[10]: ['Player', 'Team', 'Pos', 'GP', 'G']
In [11]: rows[2][1:6]
Out[11]: ['Sidney Crosby', 'PIT', 'C', '80', '36']
```

Note that by default each field in the parsed rows is string-valued, even if the field represents a numerical value, such as 80 (games played) or 36 (goals) in the preceding example. While the csv module provides a flexible way of defining custom CSV reader classes, this module is most convenient for reading CSV files with string-valued fields.

In computational work, it is common to store and load arrays with numerical values, such as vectors and matrices. The NumPy library provides the np.loadtxt and np.savetxt for this purpose. These functions take several arguments to fine-tune the type of CSV format to read or write: for example, with the delimiter argument, we can select which character to use to separate fields, and the header and comments arguments can be used to specify a header row and comment rows that are prepended to the header, respectively.

As an example, consider saving an array with random numbers and of shape $(100, 3)$ to a file data.csv using np.savetxt. To give the data some context, we add a header and a comment line to the file as well, and we explicitly request using the comma character as field delimiter with the argument delimiter="," (the default delimiter is the space character):

```
In [12]: data = np.random.randn(100, 3)
In [13]: np.savetxt("data.csv", data, delimiter=",", header="x,y,z",
    ...:                    comments="# Random x, y, z coordinates\n")
In [14]: !head -n 5 data.csv
# Random x, y, z coordinates
x,y,z
1.652276634254504772e-01,9.522165919962696234e-01,4.659850998659530452e-01
8.699729536125471174e-01,1.187589118344758443e+00,1.788104702180680405e+00
-8.106725710122602013e-01,2.765616277935758482e-01,4.456864674903074919e-01
```

To read data on this format back into a NumPy array, we can use the np.loadtxt function. It takes arguments that are similar to those of np.savetxt: in particular, we again set the delimiter argument to ",", to indicate the fields that are separated by a comma character. We also need to use the skiprows argument to skip over the first two lines in the file (the comment and header line), since they do not contain numerical data:

```
In [15]: data_load = np.loadtxt("data.csv", skiprows=2, delimiter=",")
```

The result is a new NumPy array that is equivalent to the original one written to the data.csv file using np.savetxt:

```
In [16]: (data == data_load).all()
Out[16]: True
```

Note that in contrast to the CSV reader in the csv module in the Python standard library, by default the loadtxt function in NumPy converts all fields into numerical values, and the result is a NumPy with numerical dtype (float64):

```
In [17]: data_load[1,:]
Out[17]: array([ 0.86997295,  1.18758912,  1.7881047 ])
In [18]: data_load.dtype
Out[18]: dtype('float64')
```

To read CSV files that contain nonnumerical data using np.loadtxt – such as the playerstats-2013-2014.csv file that we read using the Python standard library in the preceding text – we must explicitly set the data type of the resulting array using the dtype argument. If we attempt to read a CSV file with nonnumerical values without setting dtype, we get an error:

```
In [19]: np.loadtxt("playerstats-2013-2014.csv", skiprows=2, delimiter=",")
-------------------------------------------------------------------------
ValueError: could not convert string to float: b'Sidney Crosby'
```

Using dtype=bytes (or str or object), we get a NumPy array with unparsed values:

```
In [20]: data = np.loadtxt("playerstats-2013-2014.csv", skiprows=2,
delimiter=",", dtype=bytes)
In [21]: data[0][1:6]
Out[21]: array([b'Sidney Crosby', b'PIT', b'C', b'80', b'36'],
dtype='|S13')
```

Alternatively, if we want to read only columns with numerical types, we can select to read a subset of columns using the usecols argument:

```
In [22]: np.loadtxt("playerstats-2013-2014.csv", skiprows=2, delimiter=",",
usecols=[6,7,8])
Out[22]: array([[  68.,   104.,    18.],
                [  56.,    87.,    28.],
                [  58.,    86.,     7.],
                [  47.,    84.,    16.],
                [  39.,    82.,    32.]])
```

While the NumPy savetxt and loadtxt functions are configurable and flexible CSV writers and readers, they are most convenient for all-numerical data. The Python standard library module csv, on the other hand, is most convenient for CSV files with string-valued data. A third method to read CSV files in the Python is to use the Pandas read_csv function. We have already seen examples of this function in Chapter 12, where we used it to create Pandas data frames from TSV-formatted data files. The read_csv function in Pandas is very handy when reading CSV files with both numerical and string-valued fields, and in most cases, it will automatically determine which type a field has and converts it accordingly. For example, when reading the playerstats-2013-2014. csv file using read_csv, we obtain a Pandas data frame with all the fields parsed into columns with suitable type:

```
In [23]: df = pd.read_csv("playerstats-2013-2014.csv", skiprows=1)
In [24]: df = df.set_index("Rank")
In [25]: df[["Player", "GP", "G", "A", "P"]]
```

Out[25]:

Rank	Player	GP	G	A	P
1	Sidney Crosby	80	36	68	104
2	Ryan Getzlaf	77	31	56	87
3	Claude Giroux	82	28	58	86
4	Tyler Seguin	80	37	47	84
5	Corey Perry	81	43	39	82

Using the info method of the DataFrame instance df, we can see explicitly which type each column has been converted to (here the output is truncated for brevity):

```
In [26]: df.info()
<class 'pandas.core.frame.DataFrame'>
Int64Index: 5 entries, 1 to 5
Data columns (total 20 columns):
Player      5 non-null object
Team        5 non-null object
Pos         5 non-null object
GP          5 non-null int64
G           5 non-null int64
...
S           5 non-null int64
S%          5 non-null float64
TOI/GP      5 non-null object
Shift/GP    5 non-null float64
FO%         5 non-null float64
dtypes: float64(3), int64(13), object(4)
memory usage: 840.0+ bytes
```

Data frames can also be written to CSV files using the to_csv method of the DataFrame object:

```
In [27]: df[["Player", "GP", "G", "A", "P"]].to_csv("playerstats-2013-2014-
subset.csv")
In [28]: !head -n 5 playerstats-2013-2014-subset.csv
Rank,Player,GP,G,A,P
1,Sidney Crosby,80,36,68,104
2,Ryan Getzlaf,77,31,56,87
3,Claude Giroux,82,28,58,86
4,Tyler Seguin,80,37,47,84
```

The combination of the Python standard library, NumPy and Pandas, provides a powerful toolbox for both reading and writing CSV files of various flavors. However, although CSV files are convenient and effective for tabular data, there are obvious shortcomings with the format. For starters, it can only be used to store one- or two-dimensional arrays, and it does not contain metadata that can help interpret the data. Also, it is not very efficient in terms of either storage or reading and writing, and it cannot be used to store more than one array per file, requiring multiple files for multiple arrays even if they are closely related. The use of CSV should therefore be limited to simple datasets. In the following section, we will look at the HDF5 file format, which was designed to store numerical data efficiently and to overcome all the shortcomings of simple data formats such as CSV and related formats.

HDF5

The *Hierarchical Data Format 5* (HDF5) is a format for storing numerical data. It is developed by The HDF Group[2], a nonprofit organization, and it is available under the BSD open source license. The HDF5 format, which was released in 1998, is designed and implemented to efficiently handle large datasets, including support for high-performance parallel I/O. The HDF5 format is therefore suitable for use on distributed high-performance supercomputers and clusters and can be used to store and operate on datasets of terabyte scale, or even larger. However, the beauty of HDF5 is that it is equally suitable for small datasets. As such it is a truly versatile format and an invaluable tool for a computational practitioner.

---

[2]www.hdfgroup.org

The hierarchical aspect of the format allows organizing datasets within a file, using a hierarchical structure that resembles a file system. The terminology used for entities in an HDF5 file is *groups* and *datasets*, which corresponds to directories and files in the file system analogy. Groups in an HDF5 file can be nested to create a tree structure and hence *hierarchical* in the name of the format. A dataset in an HDF5 file is a homogenous array of certain dimensions and elements of a certain type. The HDF5 type system supports all standard basic data types and also allows defining custom compound data types. Both groups and datasets in an HDF5 file can also have *attributes*, which can be used to store metadata about groups and datasets. Attributes can themselves have different types, such as numeric or string-valued.

In addition to the file format itself, The HDF Group also provides a library and a reference implementation of the format. The main library is written in C, and wrappers to its C API are available for many programming languages. The HDF5 library for accessing data from an HDF5 file has sophisticated support for partial read and write operations, which can be used to access a small segment of the entire dataset. This is a powerful feature that enables computations on datasets that are larger than what can be fit a computer's memory[3]. The HDF5 format is a mature file format with widespread support on different platforms and computational environments. This also makes HDF5 a suitable choice for long-term storage of data. As a data storage platform, HDF5 provides a solution to a number of problems: cross-platform storage, efficient I/O, and storage that scales up to very large data files, a metadata system (attributes) that can be used to annotate and describe the groups and datasets in a file to make the data self-describing. Altogether, these features make HDF5 a great tool for computational work.

For Python there are two libraries for using HDF5 files: h5py and PyTables. These two libraries take different approaches to use HDF5, and it is well worth being familiar with both of these libraries. The h5py library provides an API that is relatively close to the basic HDF5 concepts, with a focus on groups and datasets. It provides a NumPy-inspired API for accessing datasets, which makes it very intuitive for someone that is familiar with NumPy.

---

[3]This is also known as out-of-core computing. For another recent project that also provides out-of-core computing capabilities in Python, see the dask library (http://dask.pydata.org/en/latest).

h5py   The h5py library provides a Pythonic interface to the HDF5 file format and a NumPy-like interface to its datasets. For more information about the project, including its official documentation, see its web page at `www.h5py.org`. At the time of writing, the most recent version of the library is 2.7.1.

The PyTables library provides a higher-level data abstraction based on the HDF5 format, providing database-like features, such as tables with easily customizable data types. It also allows querying datasets as a database and the use of advanced indexing features.

PyTables   The PyTables library provides a database-like data model on top of HDF5. For more information about the project and its documentation, see the web page at `http://pytables.github.io`. At the time of writing, the latest version of PyTables is 3.4.3.

In the following two sections, we explore in more detail how the h5py and PyTables libraries can be used to read and write numerical data with HDF5 files.

h5py

We begin with a tour of the h5py library. The API for h5py is surprisingly simple and pleasant to work with, yet at the same time full-featured. This is accomplished through the thoughtful use of Pythonic idioms such as dictionary and NumPy's array semantics. A summary of basic objects and methods in the h5py library is shown in Table 18-1. In the following, we explore how to use these methods through a series of examples.

Table 18-1. *Summary of the Main Objects and Methods in the h5py API*

Object	Method/Attribute	Description
h5py.File	__init__(name, mode, ...)	Open an existing HDF5, or create a new one, with filename name. Depending on the value of the mode argument, the file can be opened in read-only or read-write mode (see main text).
	flush()	Write buffers to file.
	close()	Close an open HDF5 file.
h5py.File, h5py.Group	create_group(name)	Create a new group with name name (can be a path) within the current group.
	create_dataset(name, data=..., shape=..., dtype=..., ...)	Create a new dataset.
	[] dictionary syntax	Access items (groups and datasets) within a group.
h5py.Dataset	dtype	Data type.
	shape	Shape (dimensions) of the dataset.
	value	The full array of the underlying data of the dataset.
	[] array syntax	Access elements or subsets of the data in a dataset.
h5py.File, h5py.Group, h5py.Dataset	name	Name (path) of the object in the HDF5 file hierarchy.
	attrs	Dictionary-like attribute access.

Files

We begin by looking at how to open existing and create new HDF5 files using the
h5py.File object. The initializer for this object only takes a filename as a required
argument, but we will typically also need to specify the mode argument, with which
we can choose to open a file in read-only or read-write mode and if a file should be
truncated or not when opened. The mode argument takes string values similar to
the built-in Python function open: "r" is used for read-only (file must exist), "r+" for
read-write (file must exist), "w" for creating a new file (truncate if file exists), "w-" for
creating a new file (error if file exists), and "a" for read-write (if file exists, otherwise
create). To create a new file in read-write mode, we can therefore use

```
In [29]: f = h5py.File("data.h5", mode="w")
```

The result is a file handle, here assigned to the variable f, which we can use to access
and add content to the file. Given a file handle, we can see which mode it is opened in
using the mode attribute:

```
In [30]: f.mode
Out[30]: 'r+'
```

Note that even though we opened the file in mode "w", once the file has been opened it is
either read-only ("r") or read-write ("r+"). Other file-level operations that can be performed
using the HDF5 file object are flushing buffers containing data that has not yet been written
to the file using the flush method and closing the file using the close method:

```
In [31]: f.flush()
In [32]: f.close()
```

Groups

At the same time as representing an HDF5 file handle, the File object also represents the
HDF5 group object known as the *root group*. The name of a group is accessible through
the name attribute of the group object. The name takes the form of a path, similar to a
path in a file system, which specifies where in the hierarchical structure of the file the
group is stored. The name of the root group is "/":

```
In [33]: f = h5py.File("data.h5", "w")
In [34]: f.name
Out[34]: '/'
```

A group object has the method create_group for creating a new group within an existing group. A new group created with this method becomes a subgroup of the group instance for which the create_group method is invoked:

```
In [35]: grp1 = f.create_group("experiment1")
In [36]: grp1.name
Out[36]: '/experiment1'
```

Here the group experiment1 is a subgroup of root group, and its name and path in the hierarchical structure is therefore /experiment1. When creating a new group, its immediate parent group does not necessarily have to exist beforehand. For example, to create a new group /experiment2/measurement, we can directly use the create_group method of the root group *without* first creating the experiment2 group explicitly. Intermediate groups are created automatically.

```
In [37]: grp2_meas = f.create_group("experiment2/measurement")
In [38]: grp2_meas.name
Out[38]: '/experiment2/measurement'
In [39]: grp2_sim = f.create_group("experiment2/simulation")
In [40]: grp2_sim.name
Out[40]: '/experiment2/simulation'
```

The group hierarchy of an HDF5 file can be explored using a dictionary-style interface. To retrieve a group with a given path name, we can perform a dictionary-like lookup from one of its ancestor groups (typically the root node):

```
In [41]: f["/experiment1"]
Out[41]: <HDF5 group "/experiment1" (0 members)>
In [42]: f["/experiment2/simulation"]
Out[42]: <HDF5 group "/experiment2/simulation" (0 members)>
```

The same type of dictionary lookup works for subgroups too (not only the root node):

```
In [43]: grp_experiment2 = f["/experiment2"]
In [44]: grp_experiment2['simulation']
Out[44]: <HDF5 group "/experiment2/simulation" (0 members)>
```

The keys method returns an iterator over the names of subgroups and datasets within a group, and the items method returns an iterator over (name, value) tuples for each entity in the group. These can be used to traverse the hierarchy of groups programmatically.

```
In [45]: list(f.keys())
Out[45]: ['experiment1', 'experiment2']
In [46]: list(f.items())
Out[46]: [('experiment1', <HDF5 group "/experiment1" (0 members)>),
          ('experiment2', <HDF5 group "/experiment2" (2 members)>)]
```

To traverse the hierarchy of groups in an HDF5 file, we can also use the method visit, which takes a function as argument and calls that function with the name for each entity in the file hierarchy:

```
In [47]: f.visit(lambda x: print(x))
experiment1
experiment2
experiment2/measurement
experiment2/simulation
```

or the visititems method which does the same thing except that it calls the function with both the item name and the item itself as arguments:

```
In [48]: f.visititems(lambda name, item: print(name, item))
experiment1 <HDF5 group "/experiment1" (0 members)>
experiment2 <HDF5 group "/experiment2" (2 members)>
experiment2/experiment <HDF5 group "/experiment2/measurement" (0 members)>
experiment2/simulation <HDF5 group "/experiment2/simulation" (0 members)>
```

In keeping with the semantics of Python dictionaries, we can also operate on Group objects using the set membership testing with the in Python keyword:

```
In [49]: "experiment1" in f
Out[49]: True
In [50]: "simulation" in f["experiment2"]
Out[50]: True
In [51]: "experiment3" in f
Out[51]: False
```

Using the `visit` and `visititems` methods, together with the dictionary-style methods `keys` and `items`, we can easily explore the structure and content of an HDF5 file, even if we have no prior information on what it contains and how the data is organized within it. The ability to conveniently explore HDF5 is an important aspect of the usability of the format. There are also external non-Python tools for exploring the content of HDF5 files that often are useful when working with this type of files. In particular, the `h5ls` command-line tool is handy for quickly inspecting the content of an HDF5 file:

```
In [52]: f.flush()
In [53]: !h5ls -r data.h5
/                          Group
/experiment1               Group
/experiment2               Group
/experiment2/measurement   Group
/experiment2/simulation    Group
```

Here we used the `-r` flag to the `h5ls` program to recursively show all items in the file. The `h5ls` program is part of a series of HDF5 utility programs provided by a package called `hdf5-tools` (see also `h5stat`, `h5copy`, `h5diff`, etc.). Even though these are not Python tools, they are very useful when working with HDF5 files in general, also from within Python.

Datasets

Now that we have explored how to create and access groups within an HDF5 file, it is time to look at how to store datasets. Storing numerical data is after all the main purpose of the HDF5 format. There are two main methods to create a dataset in an HDF5 file using h5py. The easiest way to create a dataset is to simply assign a NumPy array to an item within an HDF5 group, using the dictionary index syntax. The second method is to create an empty dataset using the `create_dataset` method, as we will see examples later in this section.

For example, to store two NumPy arrays, array1 and meas1, into the root group and the experiment2/measurement groups, respectively, we can use

```
In [54]: array1 = np.arange(10)
In [55]: meas1 = np.random.randn(100, 100)
In [56]: f["array1"] = array1
In [57]: f["/experiment2/measurement/meas1"] = meas1
```

To verify that the datasets for the assigned NumPy arrays were added to the file, let's traverse through the file hierarchy using the visititems method:

```
In [58]: f.visititems(lambda name, value: print(name, value))
array1 <HDF5 dataset "array1": shape (10,), type "<i8">
experiment1 <HDF5 group "/experiment1" (0 members)>
experiment2 <HDF5 group "/experiment2" (2 members)>
experiment2/measurement <HDF5 group "/experiment2/measurement" (1 members)>
experiment2/measurement/meas1 <HDF5 dataset "meas1": shape (100, 100),
type "<f8">
experiment2/simulation <HDF5 group "/experiment2/simulation" (0 members)>
```

We see that, indeed, the array1 and meas1 datasets are now added to the file. Note that the paths used as dictionary keys in the assignments determine the locations of the datasets within the file. To retrieve a dataset, we can use the same dictionary-like syntax as we used to retrieve a group. For example, to retrieve the array1 dataset, which is stored in the root group, we can use f["array1"]:

```
In [59]: ds = f["array1"]
In [60]: ds
Out[60]: <HDF5 dataset "array1": shape (10,), type "<i8">
```

The result is a Dataset object, not a NumPy array like the one that we assigned to the array1 item. The Dataset object is a proxy for the underlying data within the HDF5. Like a NumPy array, a Dataset object has several attributes that describe the dataset, including name, dtype, and shape. It also has the method len that returns the length of the dataset:

```
In [61]: ds.name
Out[61]: '/array1'
In [62]: ds.dtype
```

```
Out[62]: dtype('int64')
In [63]: ds.shape
Out[63]: (10,)
In [64]: ds.len()
Out[64]: 10
```

The actual data for the dataset can be accessed, for example, using the value attribute. This returns the entire dataset as a NumPy array, which here is equivalent to the array that we assigned to the array1 dataset.

```
In [65]: ds.value
Out[65]: array([0, 1, 2, 3, 4, 5, 6, 7, 8, 9])
```

To access a dataset deeper down the group hierarchy, we can use a file system-like path name. For example, to retrieve the meas1 dataset in the group experiment2/measurement, we can use

```
In [66]: ds = f["experiment2/measurement/meas1"]
In [67]: ds
Out[67]: <HDF5 dataset "meas1": shape (100, 100), type "<f8">
```

Again we get a Dataset object, whose basic properties can be inspected using the object attributes we introduced earlier:

```
In [68]: ds.dtype
Out[68]: dtype('float64')
In [69]: ds.shape
Out[69]: (100, 100)
```

Note that the data type of this dataset is float64, while for the dataset array1, the data type is int64. This type information was derived from the NumPy arrays that were assigned to the two datasets. Here again we could use the value attribute to retrieve the array as a NumPy array. An alternative syntax for the same operation is to use bracket indexing with the ellipsis notation: ds[...].

```
In [70]: data_full = ds[...]
In [71]: type(data_full)
Out[71]: numpy.ndarray
In [72]: data_full.shape
Out[72]: (100, 100)
```

This is an example of NumPy-like array indexing. The `Dataset` object supports most of the indexing and slicing types used in NumPy, and this provides a powerful and flexible method for partially reading data from a file. For example, to retrieve only the first column from the `meas1` dataset, we can use

```
In [73]: data_col = ds[:, 0]
In [74]: data_col.shape
Out[74]: (100,)
```

The result is a 100-element array corresponding the first column in the dataset. Note that this slicing is performed within the HDF5 library, and not in NumPy, so in this example only 100 elements were read from the file and stored in the resulting NumPy array, without every fully loading the dataset into memory. This is an important feature when working with large datasets that do not fit in memory.

For example, the `Dataset` object also supports strided indexing:

```
In [75]: ds[10:20:3, 10:20:3] # 3 stride
Out[75]: array([[-0.22321057, -0.61989199,  0.78215645,  0.73774187],
                [-1.03331515,  2.54190817, -0.24812478, -2.49677693],
                [ 0.17010011,  1.88589248,  1.91401249, -0.63430569],
                [ 0.4600099 , -1.3242449 ,  0.41821078,  1.47514922]])
```

as well as "fancy indexing", where a list of indices is given for one of the dimensions of the array (does not work for more than one index):

```
In [76]: ds[[1,2,3], :].shape
Out[76]: (3, 100)
```

We can also use Boolean indexing, where a Boolean-valued NumPy array is used to index a `Dataset`. For example, to single out the first five columns (index `:5` on the second axis) for each row whose value in the first column (`ds[:, 0]`) is larger than 2, we can index the dataset with the Boolean mask `ds[:, 0] > 2`:

```
In [77]: mask = ds[:, 0] > 2
In [78]: mask.shape, mask.dtype
Out[78]: ((100,), dtype('bool'))
In [79]: ds[mask, :5]
```

```
Out[79]: array([[ 2.1224865 ,  0.70447132, -1.71659513,  1.43759445,
                  -0.61080907],
                [ 2.11780508, -0.2100993 ,  1.06262836,
                  -0.46637199,  0.02769476],
                [ 2.41192679, -0.30818179, -0.31518842, -1.78274309,
                  -0.80931757],
                [ 2.10030227,  0.14629889,  0.78511191,
                  -0.19338282,  0.28372485]])
```

Since the Dataset object uses the NumPy's indexing and slicing syntax to select subsets of the underlying data, working with large HDF5 datasets in Python using h5py comes very naturally to someone who is familiar with NumPy. Also remember that for large files, there is a big difference in index slicing on the Dataset object rather than on the NumPy array that can be accessed through the value attribute, since the former avoids loading the entire dataset into memory.

So far we have seen how to create datasets in an HDF5 file by explicitly assigning data into an item in a group object. We can also create datasets explicitly using the create_ dataset method. It takes the name of the new dataset as the first argument, and we can either set the data for the new dataset using the data argument or create an empty array by setting the shape argument. For example, instead of the assignment f["array2"] = np.random.randint(10, size=10), we can also use the create_dataset method:

```
In [80]: ds = f.create_dataset("array2", data=np.random.randint(10,
         size=10))
In [81]: ds
Out[81]: <HDF5 dataset "array2": shape (10,), type "<i8">
In [82]: ds.value
Out[82]: array([2, 2, 3, 3, 6, 6, 4, 8, 0, 0])
```

When explicitly calling the create_dataset method, we have a finer level of control of the properties of the resulting dataset. For example, if we can explicitly set the data type for the dataset using the dtype argument, we can choose a compression method using the compress argument, setting the chunk size using the chunks argument and setting the maximum allowed array size for resizable datasets using the maxsize argument. There are also many other advanced features related to the Dataset object. See the docstring for create_dataset for details.

When creating an empty array by specifying the shape argument instead of providing an array for initializing a dataset, we can also use the fillvalue argument to set the default value for the dataset. For example, to create an empty dataset of shape (5, 5) and default value -1, we can use

```
In [83]: ds = f.create_dataset("/experiment2/simulation/data1",
            shape=(5, 5), fillvalue=-1)
In [84]: ds
Out[84]: <HDF5 dataset "data1": shape (5, 5), type "<f4">
In [85]: ds.value
Out[85]: array([[-1., -1., -1., -1., -1.],
                [-1., -1., -1., -1., -1.],
                [-1., -1., -1., -1., -1.],
                [-1., -1., -1., -1., -1.],
                [-1., -1., -1., -1., -1.]], dtype=float32)
```

HDF5 is clever about disk usage for an empty dataset and will not store more data than necessary, in particular if we select a compression method using the compression argument. There are several compression methods available, for example, 'gzip'. Using dataset compression we can create a very large dataset and gradually fill them with data, for example, when measurement results or results of computations become available, without initially wasting a lot of storage space. For example, let's create a large dataset with shape (5000, 5000, 5000) with the data1 in the group experiment1/simulation:

```
In [86]: ds = f.create_dataset("/experiment1/simulation/data1",
            shape=(5000, 5000, 5000), fillvalue=0, compression='gzip')
In [87]: ds
Out[87]: <HDF5 dataset "data1": shape (5000, 5000, 5000), type "<f4">
```

To begin with this dataset uses neither memory nor disk space, until we start filling it with data. To assign values to the dataset, we can again use the NumPy-like indexing syntax and assign values to specific elements in the dataset or to subsets selected using slicing syntax:

```
In [88]: ds[:, 0, 0] = np.random.rand(5000)
In [89]: ds[1, :, 0] += np.random.rand(5000)
In [90]: ds[:2, :5, 0]
```

```
Out[90]: array([[ 0.67240328, 0.         , 0.         , 0.         , 0.          ],
               [ 0.99613971, 0.48227152, 0.48904559, 0.78807044, 0.62100351]],
              dtype=float32)
```

Note that the elements that have not been assigned values are set to the value of fillvalue that was specified when the array was created. If we do not know what fill value a dataset has, we can find out by looking at the fillvalue attribute of the Dataset object:

```
In [91]: ds.fillvalue
Out[91]: 0.0
```

To see that the newly created dataset is indeed stored in the group where we intended to assign it, we can again use the visititems method to list the content of the experiment1 group:

```
In [92]: f["experiment1"].visititems(lambda name, value: print(name,
         value))
simulation <HDF5 group "/experiment1/simulation" (1 members)>
simulation/data1 <HDF5 dataset "data1": shape (5000, 5000, 5000), type
"<f4">
```

Although the dataset experiment1/simulation/data1 is very large (4×5000^3 bytes ~ 465 Gb), since we have not yet filled it with much data, the HDF5 file still does not take a lot of disk space (only about 357 Kb):

```
In [93]: f.flush()
In [94]: f.filename
Out[94]: 'data.h5'
In [95]: !ls -lh data.h5
-rw-r--r--@ 1 rob   staff    357K Apr  5 18:48 data.h5
```

So far we have seen how to create groups and datasets within an HDF5 file. It is of course sometimes also necessary to delete items from a file. With h5py we can delete items from a group using the Python del keyword, again complying with the semantics of Python dictionaries:

```
In [96]: del f["/experiment1/simulation/data1"]
In [97]: f["experiment1"].visititems(lambda name, value: print(name, value))
simulation <HDF5 group "/experiment1/simulation" (0 members)>
```

Attributes

Attributes are a component of the HDF5 format that makes it a great format for annotating data and providing self-describing data through the use of metadata. For example, when storing experimental data, there are often external parameters and conditions that should be recorded together with the observed data. Likewise, in a computer simulation, it is usually necessary to store additional model or simulation parameters together with the generated simulation results. In all these cases, the best solution is to make sure that the required additional parameters are stored as metadata together with the main datasets.

The HDF5 format supports this type of metadata through the use of attributes. An arbitrary number of attributes can be attached to each group and dataset within an HDF5 file. With the h5py library, attributes are accessed using a dictionary-like interface, just like groups are. The Python attribute `attrs` of `Group` and `Dataset` objects are used to access the HDF5 attributes:

```
In [98]: f.attrs
Out[98]: <Attributes of HDF5 object at 4462179384>
```

To create an attribute, we simply assign to the `attrs` dictionary for the target object. For example, to create an attribute `description` for the root group, we can use

```
In [99]: f.attrs["description"] = "Result sets for experiments and
simulations"
```

Similarly, to add `date` attributes to the `experiment1` and `experiment2` groups:

```
In [100]: f["experiment1"].attrs["date"] = "2015-1-1"
In [101]: f["experiment2"].attrs["date"] = "2015-1-2"
```

We can also add attributes directly to datasets (not only groups):

```
In [102]: f["experiment2/simulation/data1"].attrs["k"] = 1.5
In [103]: f["experiment2/simulation/data1"].attrs["T"] = 1000
```

Like for groups, we can use the keys and items method of the Attribute object to retrieve iterators over the attributes it contains:

```
In [104]: list(f["experiment1"].attrs.keys())
Out[104]: ['date']
In [105]: list(f["experiment2/simulation/data1"].attrs.items())
Out[105]: [('k', 1.5), ('T', 1000)]
```

The existence of an attribute can be tested with the Python in operator, in keeping with the Python dictionary semantics:

```
In [106]: "T" in f["experiment2/simulation/data1"].attrs
Out[106]: True
```

To delete existing attributes, we can use the del keyword:

```
In [107]: del f["experiment2/simulation/data1"].attrs["T"]
In [108]: "T" in f["experiment2/simulation"].attrs
Out[108]: False
```

The attributes of HDF5 groups and datasets are suitable for storing metadata together with the actual datasets. Using attributes generously can help to provide context to the data, which often must be available for the data to be useful.

PyTables

The PyTables library offers an alternative interface to HDF5 for Python. The focus on this library is higher-level table-based data model implemented using the HDF5 format, although PyTables can also be used to create and read generic HDF5 groups and datasets, like the h5py library. Here we focus on the table data model, as it complements the h5py library that we discussed in the previous section. We demonstrate the use of PyTables table objects using the NHL player statistics dataset that we used earlier in this chapter and where we construct a PyTables table from a Pandas data frame for that dataset. We therefore begin with reading in the dataset into a DataFrame object using the read_csv function:

```
In [109]: df = pd.read_csv("playerstats-2013-2014.csv", skiprows=1)
     ...: df = df.set_index("Rank")
```

Next we proceed to create a new PyTables HDF5 file handle by using the `tables.open_file` function[4]. This function takes a filename as the first argument and the file mode as an optional second argument. The result is a PyTables HDF5 file handle (here assigned to the variable f):

```
In [110]: f = tables.open_file("playerstats-2013-2014.h5", mode="w")
```

Like with the h5py library, we can create HDF5 groups with the method `create_group` of the file handle object. It takes the path to the parent group as the first argument, the group name as the second argument, and optionally also the argument `title`, with which a descriptive HDF5 attribute can be set on the group.

```
In [111]: grp = f.create_group("/", "season_2013_2014",
    ...:                        title="NHL player statistics for the
                               2013/2014 season")
In [112]: grp
Out[112]: /season_2013_2014 (Group) 'NHL player statistics for the
2013/2014 season'
            children := []
```

Unlike the h5py library, the file handle object in PyTables does not represent the root group in the HDF5 file. To access the root node, we must use the `root` attribute of the file handle object:

```
In [113]: f.root
Out[113]: / (RootGroup) "
            children := ['season_2013_2014' (Group)]
```

A nice feature of the PyTables library is that it is easy to create tables with mixed column types, using the struct-like compound data type of HDF5. The simplest way to define such a table data structure with PyTables is to create a class that inherits from the `tables.IsDescription` class. It should contain fields composed of data-type representations from the `tables` library. For example, to create a specification of the table structure for the player statistics dataset, we can use

---

[4]Note that the Python module provided by the PyTables library is named `tables`. Therefore, `tables.open_file` refers to `open_file` function in the `tables` module provided by the PyTables library.

```
In [114]: class PlayerStat(tables.IsDescription):
     ...:       player = tables.StringCol(20, dflt="")
     ...:       position = tables.StringCol(1, dflt="C")
     ...:       games_played = tables.UInt8Col(dflt=0)
     ...:       points = tables.UInt16Col(dflt=0)
     ...:       goals = tables.UInt16Col(dflt=0)
     ...:       assists = tables.UInt16Col(dflt=0)
     ...:       shooting_percentage = tables.Float64Col(dflt=0.0)
     ...:       shifts_per_game_played = tables.Float64Col(dflt=0.0)
```

Here the class PlayerStat represents the table structure of a table with eight
columns, where the first two columns are fixed-length strings (tables.StringCol),
where the following four columns are unsigned integers (tables.UInt8Col and tables.
UInt16Col, of 8- and 16-bit size), and where the last two columns have floating-point
type (tables.Float64Col). The optional dflt argument to data-type objects specifies
the fields' default value. Once the table structure is defined using a class on this form,
we can create the actual table in the HDF5 file using the create_table method. It takes
a group object or the path to the parent node as the first argument, the table name as
the second argument, the table specification class as the third argument, and optionally
a table title as the fourth argument (stored as an HDF5 attribute for the corresponding
dataset):

```
In [115]: top30_table = f.create_table(grp, 'top30', PlayerStat, "Top 30
          point leaders")
```

To insert data into the table, we can use the row attribute of the table object to
retrieve a Row accessor class that can be used as a dictionary to populate the row with
values. When the row object is fully initialized, we can use the append method to actually
insert the row into the table:

```
In [116]: playerstat = top30_table.row
In [117]: for index, row_series in df.iterrows():
     ...:       playerstat["player"] = row_series["Player"]
     ...:       playerstat["position"] = row_series["Pos"]
     ...:       playerstat["games_played"] = row_series["GP"]
     ...:       playerstat["points"] = row_series["P"]
     ...:       playerstat["goals"] = row_series["G"]
```

```
    ...:        playerstat["assists"] = row_series["A"]
    ...:        playerstat["shooting_percentage"] = row_series["S%"]
    ...:        playerstat["shifts_per_game_played"] = row_series["Shift/GP"]
    ...:        playerstat.append()
```

The flush method forces a write of the table data to the file:

```
In [118]: top30_table.flush()
```

To access data from the table, we can use the cols attribute to retrieve columns as NumPy arrays:

```
In [119]: top30_table.cols.player[:5]
Out[119]: array([b'Sidney Crosby', b'Ryan Getzlaf', b'Claude Giroux',
                  b'Tyler Seguin', b'Corey Perry'], dtype='|S20')
In [120]: top30_table.cols.points[:5]
Out[120]: array([104,  87,  86,  84,  82], dtype=uint16)
```

To access data in a row-wise fashion, we can use the iterrows method to create an iterator over all the rows in the table. Here we use this approach to loop through all the rows and print them to the standard output (here the output is truncated for brevity):

```
In [121]: def print_playerstat(row):
    ...:        print("%20s\t%s\t%s\t%s" %
    ...:             (row["player"].decode('UTF-8'), row["points"],
    ...:              row["goals"], row["assists"]))
In [122]: for row in top30_table.iterrows():
    ...:        print_playerstat(row)
  Sidney Crosby     104     36      68
Ryan Getzlaf         87     31      56
Claude Giroux        86     28      58
Tyler Seguin         84     37      47
...
Jaromir Jagr         67     24      43
John Tavares         66     24      42
Jason Spezza         66     23      43
Jordan Eberle        65     28      37
```

One of the most powerful features of the PyTables table interface is the ability to selectively extract rows from the underlying HDF5 using queries. For example, the where method allows us to pass an expression in terms of the table columns as a string that is used by PyTables to filter rows:

```
In [123]: for row in top30_table.where("(points > 75) & (points <= 80)"):
     ...:         print_playerstat(row)
Phil Kessel         80     37     43
Taylor Hall         80     27     53
Alex Ovechkin       79     51     28
Joe Pavelski        79     41     38
Jamie Benn          79     34     45
Nicklas Backstrom   79     18     61
Patrick Sharp       78     34     44
Joe Thornton        76     11     65
```

With the where method, we can also define conditions in terms of multiple columns:

```
In [124]: for row in top30_table.where("(goals > 40) & (points < 80)"):
     ...:         print_playerstat(row)
Alex Ovechkin       79     51     28
Joe Pavelski        79     41     38
```

This feature allows us to query a table in a database-like fashion. Although for a small dataset, like the current one, we could just as well perform these kinds of operations directly in memory using a Pandas data frame, but remember that HDF5 files are stored on disk, and the efficient use of I/O in the PyTables library enables us to work with very large datasets that do not fit in memory, which would prevent us from using, for example, NumPy or Pandas on the entire dataset.

Before we conclude this section, let us inspect the structure of the resulting HDF5 file that contains the PyTables table that we have just created:

```
In [125]: f
Out[125]: File(filename=playerstats-2013-2014.h5, title='', mode='w',
             root_uep='/', filters=Filters(complevel=0, shuffle=False,
             fletcher32=False, least_significant_digit=None))
```

```
/ (RootGroup) " /season_2013_2014 (Group) 'NHL player stats for
    the 2013/2014 season'
/season_2013_2014/top30 (Table(30,)) 'Top 30 point leaders'
    description := {
    "assists": UInt16Col(shape=(), dflt=0, pos=0),
    "games_played": UInt8Col(shape=(), dflt=0, pos=1),
    "goals": UInt16Col(shape=(), dflt=0, pos=2),
    "player": StringCol(itemsize=20, shape=(), dflt=b", pos=3),
    "points": UInt16Col(shape=(), dflt=0, pos=4),
    "position": StringCol(itemsize=1, shape=(), dflt=b'C',
    pos=5),
    "shifts_per_game_played": Float64Col(shape=(), dflt=0.0,
    pos=6),
    "shooting_percentage": Float64Col(shape=(), dflt=0.0, pos=7)}
    byteorder := 'little'
    chunkshape := (1489,)
```

From the string representation of the PyTables file handle, and the HDF5 file hierarchy that it contains, we can see that the PyTables library has created a dataset / season_2013_2014/top30 that uses an involved compound data type that was created according to the specification in the PlayerStat object that we created earlier. Finally, when we are finished modifying a dataset in a file, we can flush its buffers and force a write to the file using the flush method, and when we are finished working with a file, we can close it using the close method:

```
In [126]: f.flush()
In [127]: f.close()
```

Although we do not cover other types of datasets here, such as regular homogenous arrays, it is worth mentioning that the PyTables library also supports these types of data structures. For example, we can use the create_array, create_carray, and create_earray to construct fixed-size arrays, chunked arrays, and enlargeable arrays, respectively. For more information on how to use these data structures, see the corresponding docstrings.

Pandas HDFStore

A third way to store data in HDF5 files using Python is to use the HDFStore object in Pandas. It can be used to persistently store data frames and other Pandas objects in an HDF5 file. To use this feature in Pandas, the PyTables library must be installed. We can create an HDFStore object by passing a filename to its initializer. The result is an HDFStore object that can be used as a dictionary to which we can assign Pandas DataFrame instances to have them stored into the HDF5 file:

```
In [128]: store = pd.HDFStore('store.h5')
In [129]: df = pd.DataFrame(np.random.rand(5,5))
In [130]: store["df1"] = df
In [131]: df = pd.read_csv("playerstats-2013-2014-top30.csv", skiprows=1)
In [132]: store["df2"] = df
```

The HDFStore object behaves as a regular Python dictionary, and we can, for example, see what objects it contains by calling the keys method:

```
In [133]: store.keys()
Out[133]: ['/df1', '/df2']
```

and we can test for the existence of an object with a given key using the Python in keyword:

```
In [134]: 'df2' in store
Out[134]: True
```

To retrieve an object from the store, we again use the dictionary-like semantic and index the object with its corresponding key:

```
In [135]: df = store["df1"]
```

From the HDFStore object, we can also access the underlying HDF5 handle using the root attribute. This is actually nothing but a PyTables file handle:

```
In [136]: store.root
Out[136]: / (RootGroup) "    children := ['df1' (Group), 'df2' (Group)]
```

Once we are finished with an HDFStore object, we should close it using the close method, to ensure that all data associated with it is written to the file.

```
In [137]: store.close()
```

Since HDF5 is a standard file format, there is, of course, nothing that prevents us from opening an HDF5 file created with Pandas HDFStore or PyTables with any other HDF5 compatible software, such as the h5py library. If we open the file produced with HDFStore with h5py, we can easily inspect its content and see how the HDFStore object arranges the data of the DataFrame objects that we assigned to it:

```
In [138]: f = h5py.File("store.h5")
In [139]: f.visititems(lambda x, y: print(x, "\t" * int(3 -
          len(str(x))//8), y))
df1                    <HDF5 group "/df1" (4 members)>
df1/axis0              <HDF5 dataset "axis0": shape (5,), type "<i8">
df1/axis1              <HDF5 dataset "axis1": shape (5,), type "<i8">
df1/block0_items       <HDF5 dataset "block0_items": shape (5,), type "<i8">
df1/block0_values      <HDF5 dataset "block0_values": shape (5, 5), type "<f8">
df2                    <HDF5 group "/df2" (8 members)>
df2/axis0              <HDF5 dataset "axis0": shape (21,), type "|S8">
df2/axis1              <HDF5 dataset "axis1": shape (30,), type "<i8">
df2/block0_items       <HDF5 dataset "block0_items": shape (3,), type "|S8">
df2/block0_values      <HDF5 dataset "block0_values": shape (30, 3), type "<f8">
df2/block1_items       <HDF5 dataset "block1_items": shape (14,), type "|S4">
df2/block1_values      <HDF5 dataset "block1_values": shape (30, 14), type "<i8">
df2/block2_items       <HDF5 dataset "block2_items": shape (4,), type "|S6">
df2/block2_values      <HDF5 dataset "block2_values": shape (1,), type "|O8">
```

We can see that the HDFStore object stores each DataFrame object in a group of its own and that it has split each data frame into several heterogeneous HDF5 datasets (blocks) where the columns are grouped by their data type. Furthermore, the column names and values are stored in separate HDF5 datasets.

```
In [140]: f["/df2/block0_items"].value
Out[140]: array([b'S%', b'Shift/GP', b'FO%'], dtype='|S8')
In [141]: f["/df2/block0_values"][:3]
```

```
Out[141]: array([[ 13.9,  24. ,  52.5],
                  [ 15.2,  25.2,  49. ],
                  [ 12.6,  25.1,  52.9]])
In [142]: f["/df2/block1_values"][:3, :5]
Out[142]: array([[  1,  80,  36,  68, 104],
                  [  2,  77,  31,  56,  87],
                  [  3,  82,  28,  58,  86]])
```

JSON

The JSON[5] (JavaScript Object Notation) is a human-readable, lightweight plain-text format that is suitable for storing datasets made up from lists and dictionaries. The values of such lists and dictionaries can themselves be lists or dictionaries or must be of the following basic data types: string, integer, float, and Boolean, or the value null (like the None value in Python). This data model allows storing complex and versatile datasets, without structural limitations such as the tabular form required by formats such as CSV. A JSON document can, for example, be used as a key-value store, where the values for different keys can have different structure and data types.

The JSON format was primarily designed to be used as a data interchange format for passing information between web services and JavaScript applications. In fact, JSON is a subset of JavaScript language and, as such, a valid JavaScript code. However, the JSON format itself is a language-independent data format that can be readily parsed and generated from essentially every language and environment, including Python. The JSON syntax is also almost valid Python code, making it familiar and intuitive to work with from Python as well.

We have already seen an example of a JSON dataset in Chapter 10, where we looked at the graph of the Tokyo Metro network. Before we revisit that dataset, we begin with a brief overview of JSON basics and how to read and write JSON in Python. The Python standard library provides the module json for working with JSON-formatted data. Specifically, this module contains functions for generating JSON data from a Python data structure (list or dictionary), json.dump and json.dumps, and for parsing JSON data into a Python data structure: json.load and json.loads. The functions loads and dumps take Python strings as input and output, while the load and dump operate on a file handle and read and write data to a file.

---

[5]For more information about JSON, see http://json.org.

For example, we can generate the JSON string of a Python list by calling the `json.dumps` function. The return value is a JSON string representation of the given Python list that closely resembles the Python code that could be used to create the list. However, a notable exception is the Python value None, which is represented as the value `null` in JSON:

```
In [143]: data = ["string", 1.0, 2, None]
In [144]: data_json = json.dumps(data)
In [145]: data_json
Out[145]: '["string", 1.0, 2, null]'
```

To convert the JSON string back into a Python object, we can use `json.loads`:

```
In [146]: data = json.loads(data_json)
In [147]: data
Out[147]: ['string', 1.0, 2, None]
In [148]: data[0]
Out[148]: 'string'
```

We can use exactly the same method to store Python dictionaries as JSON strings. Again, the resulting JSON string is essentially identical to the Python code for defining the dictionary:

```
In [149]: data = {"one": 1, "two": 2.0, "three": "three"}
In [150]: data_json = json.dumps(data)
In [151]: data_json
Out[151]: '{"two": 2.0, "three": "three", "one": 1}'
```

To parse the JSON string and convert it back into a Python object, we again use `json.loads`:

```
In [152]: data = json.loads(data_json)
In [153]: data["two"]
Out[153]: 2.0
In [154]: data["three"]
Out[154]: 'three'
```

The combination of lists and dictionaries makes a versatile data structure. For example, we can store lists or dictionaries of lists with a variable number of elements. This type of data would be difficult to store directly as a tabular array, and further levels of nested lists and dictionaries would make it very impractical. When generating JSON data with the json.dump and json.dumps functions, we can optionally give the argument indent=True, to obtain indented JSON code that can be easier to read:

```
In [155]: data = {"one": [1],
    ...:         "two": [1, 2],
    ...:         "three": [1, 2, 3]}
In [156]: data_json = json.dumps(data, indent=True)
In [157]: data_json
Out[157]: {
             "two": [
              1,
              2
             ],
             "three": [
              1,
              2,
              3
             ],
             "one": [
              1
             ]
           }
```

As an example of a more complex data structure, consider a dictionary containing a list, a dictionary, a list of tuples, and a text string. We could use the same method as in the preceding text to generate a JSON representation of the data structure using json.dumps, but instead here we write the content to a file using the json.dump function. Compared to json.dumps, it additionally takes a file handle as a second argument, which we need to create beforehand:

```
In [158]: data = {"one": [1],
    ...:         "two": {"one": 1, "two": 2},
    ...:         "three": [(1,), (1, 2), (1, 2, 3)],
```

```
    ...:           "four": "a text string"}
In [159]: with open("data.json", "w") as f:
    ...:     json.dump(data, f)
```

The result is that the JSON representation of the Python data structure is written to the file data.json:

```
In [160]: !cat data.json
{"four": "a text string", "two": {"two": 2, "one": 1}, "three": [[1],
       [1, 2], [1, 2, 3]],
 "one": [1]}
```

To read and parse a JSON-formatted file into a Python data structure, we can use json.load, to which we need to pass a handle to an open file:

```
In [161]: with open("data.json", "r") as f:
    ...:     data_from_file = json.load(f)
In [162]: data_from_file["two"]
Out[162]: [1, 2]
In [163]: data_from_file["three"]
Out[163]: [[1], [1, 2], [1, 2, 3]]
```

The data structure returned by json.load is not always identical to the one stored with json.dump. In particular, JSON is stored as Unicode, so strings in the data structure returned by json.load are always Unicode strings. Also, as we can see from the preceding example, JSON does not distinguish between tuples and lists, and the json.load always produces lists rather than tuples, and the order in which keys for a dictionary are displayed is not guaranteed, unless using the sorted_keys=True argument to the dumps and dump functions.

Now that we have seen how Python lists and dictionaries can be converted to and from JSON representation using the json module, it is worthwhile to revisit the Tokyo Metro dataset in Chapter 10. This is a more realistic dataset and an example of a data structure that mixes dictionaries, lists of variable lengths, and string values. The first 20 lines of the JSON file are shown here:

```
In [164]: !head -n 20 tokyo-metro.json
{
    "C": {
        "color": "#149848",
```

```
"transfers": [
    [
        "C3",
        "F15"
    ],
    [
        "C4",
        "Z2"
    ],
    [
        "C4",
        "G2"
    ],
    [
        "C7",
        "M14"
    ],
```

To load the JSON data into a Python data structure, we use `json.load` in the same way as before:

```
In [165]: with open("tokyo-metro.json", "r") as f:
     ...:     data = json.load(f)
```

The result is a dictionary with a key for each metro line:

```
In [166]: data.keys()
Out[166]: ['N', 'M', 'Z', 'T', 'H', 'C', 'G', 'F', 'Y']
```

The dictionary value for each metro line is again a dictionary that contains line color, lists of transfer points, and the travel times between stations on the line:

```
In [167]: data["C"].keys()
Out[167]: ['color', 'transfers', 'travel_times']
In [168]: data["C"]["color"]
Out[168]: '#149848'
In [169]: data["C"]["transfers"]
```

```
Out[169]: [['C3', 'F15'],  ['C4', 'Z2'],  ['C4', 'G2'],  ['C7', 'M14'],
           ['C7', 'N6'],
           ['C7', 'G6'],  ['C8', 'M15'],  ['C8', 'H6'],  ['C9', 'H7'],
           ['C9', 'Y18'],
           ['C11', 'T9'],  ['C11', 'M18'],  ['C11', 'Z8'],  ['C12', 'M19'],
           ['C18', 'H21']]
```

With the dataset loaded as a nested structure of Python dictionaries and lists, we can iterate over and filter items from the data structure with ease, for example, using Python's list comprehension syntax. The following example demonstrates how to select the set of connected nodes in the graph on the C line which has a travel time of 1 minute:

```
In [170]: [(s, e, tt) for s, e, tt in data["C"]["travel_times"] if tt == 1]
Out[170]: [('C3', 'C4', 1), ('C7', 'C8', 1), ('C9', 'C10', 1)]
```

The hierarchy of dictionaries and the variable length of the lists stored in the dictionaries make this a good example of a dataset that does not have a strict structure and which therefore is suitable to store in a versatile format such as JSON.

Serialization

In the previous section, we used the JSON format to generate a representation of in-memory Python objects, such as lists and dictionaries. This process is called serialization, which in this case resulted in a JSON plain-text representation of the objects. An advantage of the JSON format is that it is language independent and can easily be read by other software. Its disadvantages are that JSON files are not space efficient, and they can only be used to serialize a limited type of objects (list, dictionaries, basic types, as discussed in the previous section). There are many alternative serialization techniques that address these issues. Here we briefly will look at two alternatives that address the space efficiency issue and the types of objects that can be serialized, respectively: the msgpack library and the Python `pickle` module.

We begin with msgpack, which is a binary protocol for storing JSON-like data efficiently. The msgpack software is available for many languages and environments. For more information about the library and its Python bindings, see the project's web page at `http://msgpack.org`. In analogy to the JSON module, the msgpack library provides two sets of functions that operate on byte lists (`msgpack.packb` and `msgpack.unpackb`) and file handles (`msgpack.pack` and `msgpack.unpack`), respectively. The pack and packb functions convert a Python data structure into a binary representation, and the unpack and unpackb functions perform the reverse operation. For example, the JSON file for the Tokyo Metro dataset is relatively large and takes about 27 Kb on disk:

```
In [171]: !ls -lh tokyo-metro.json
-rw-r--r--@ 1 rob   staff    27K Apr  7 23:18 tokyo-metro.json
```

Packing the data structure with msgpack rather than JSON results in a considerably smaller file, at around 3 Kb:

```
In [172]: data_pack = msgpack.packb(data)
In [173]: type(data_pack)
Out[173]: bytes
In [174]: len(data_pack)
Out[174]: 3021
In [175]: with open("tokyo-metro.msgpack", "wb") as f:
     ...:     f.write(data_pack)
In [176]: !ls -lh tokyo-metro.msgpack
-rw-r--r--@ 1 rob   staff   3.0K Apr  8 00:40 tokyo-metro.msgpack
```

More precisely, the byte list representation of the dataset uses only 3021 bytes. In applications where storage space or bandwidth is essential, this is can be a significant improvement. However, the price we have paid for this increased storage efficiency is that we must use the msgpack library to unpack the data, and it uses a binary format and therefore is not human-readable. Whether this is an acceptable trade-off or not will depend on the application at hand. To unpack a binary `msgpack` byte list, we can use the `msgpack.unpackb` function, which recovers the original data structure:

```
In [177]: del data
In [178]: with open("tokyo-metro.msgpack", "rb") as f:
     ...:     data_msgpack = f.read()
     ...:     data = msgpack.unpackb(data_msgpack)
In [179]: list(data.keys())
Out[179]: ['T', 'M', 'Z', 'H', 'F', 'C', 'G', 'N', 'Y']
```

The other issue with JSON serialization is that only certain type of Python objects can be stored as JSON. The Python `pickle` module[6] can create a binary representation of nearly any kind of Python object, including class instances and function. Using the `pickle` module follows the exact same use pattern as the `json` module: we have the dump and dumps functions for serializing an object to a byte array and a file handle, respectively, and the load and loads for deserializing a pickled object.

```
In [180]: with open("tokyo-metro.pickle", "wb") as f:
   ...:       pickle.dump(data, f)
In [181]: del data
In [182]: !ls -lh tokyo-metro.pickle
-rw-r--r--@ 1 rob  staff   8.5K Apr  8 00:40 tokyo-metro.pickle
```

The size of the pickled object is considerably smaller than the JSON serialization but larger than the serialization produced by msgpack. We can recover a pickled object using the `pickle.load` function, which expects a file handle as argument:

```
In [183]: with open("tokyo-metro.pickle", "rb") as f:
   ...:       data = pickle.load(f)
In [184]: data.keys()
Out[184]: dict_keys(['T', 'M', 'Z', 'H', 'F', 'C', 'G', 'N', 'Y'])
```

The main advantage with pickle is that almost any type of Python object can be serialized. However, Python pickles cannot be read by software not written in Python, and it is also not a recommended format for long-term storage, because compatibility between Python versions and with different versions of libraries that defines the objects that are pickled cannot always be guaranteed. If possible, using JSON for serializing list- and dictionary-based data structures is generally a better approach, and if the file size is an issue, the msgpack library provides a popular and easily accessible alternative to JSON.

---

[6]An alternative to the `pickle` module is the `cPickle` module, which is a more efficient reimplementation that is also available in the Python standard library. See also the `dill` library at https://pypi.org/project/dill.

Summary

In this chapter we have reviewed common data formats for reading and writing numerical data to files on disk, and we introduced a selection of Python libraries that are available for working with these formats. We first looked at the ubiquitous CSV file format, which is a simple and transparent format that is suitable for small and simple datasets. The main advantage of this format is that it is human-readable plain text, which makes it intuitively understandable. However, it lacks many features that are desirable when working with numerical data, such as metadata describing the data and support for multiple datasets. The HDF5 format naturally takes over as the go-to format for numerical data when the size and complexity of the data grow beyond what is easily handled using a CSV format. HDF5 is a binary file format, so it is not a human-readable format like CSV, but there are good tools for exploring the content in an HDF5 file, both programmatically and using command-line and GUI-based user interfaces. In fact, due to the possibility of storing metadata in attributes, HDF5 is a great format for self-describing data. It is also a very efficient file format for numerical data, both in terms of I/O and storage, and it can even be used as a data model for computing with very large datasets that do not fit in the memory of the computer. Overall, HDF5 is a fantastic tool for numerical computing that anyone working with computing should benefit greatly from being familiar with. Toward the end of the chapter, we also briefly reviewed JSON, msgpack, and Python pickles for serializing data into text and binary format.

Further Reading

An informal specification of the CSV file is given in RFC 4180, `http://tools.ietf.org/html/rfc4180`. It outlines many of the commonly used features of the CSV format, although not all CSV readers and writers comply with every aspect of this document. An accessible and informative introduction to the HDF5 format and the h5py library is given by the creator of h5py in Collette (2013). It is also worth reading about the NetCDF (Network Common Data Format), `www.unidata.ucar.edu/software/netcdf`, which is another widely used format for numerical data. The Pandas library also provides I/O functions beyond what we have discussed here, such as the ability to read Excel files (`pandas.io.excel.read_excel`) and the fixed-width format (`read_fwf`). Regarding the JSON format, a concise but complete specification of the format is available at the web site `http://json.org`. With the increasingly important role of data in computing, there has been a rapid diversification of formats and data storage technologies in recent years. As a computational practitioner, reading data from databases, such as SQL and NoSQL databases, is now also an important task. Python provides a common database API for standardizing database access from Python applications, as described by PEP 249 (`www.python.org/dev/peps/pep-0249`). Another notable project for reading databases from Python is SQLAlchemy (`www.sqlalchemy.org`).

Reference

Collette, A. (2013). *Python and HDF5*. Sebastopol: O'Reilly.

CHAPTER 19

Code Optimization

In this book we have explored various topics of scientific and technical computing using Python and its ecosystem of libraries. As touched upon in the very first chapter of this book, the Python environment for scientific computing generally strikes a good balance between a high-level environment that is suitable for exploratory computing and rapid prototyping – which minimizes development efforts – and high-performance numerics, which minimize application runtimes. High-performance numerics is achieved not through the Python language itself, but rather through leveraging libraries that contain or use externally compiled code, typically written in C or in Fortran. Because of this, in computing applications that rely heavily on libraries such as NumPy and SciPy, most of the number crunching is performed by compiled code, and the performance is therefore vastly better than if the same computation were to be implemented purely in Python.

The key to high-performance Python programs is therefore to efficiently utilize libraries such as NumPy and SciPy for array-based computations. The vast majority of scientific and technical computations can be expressed in terms of common array operations and fundamental computational routines. Much of this book has been dedicated to exploring this style of scientific computing with Python, by introducing the main Python libraries for different fields of scientific computing. However, occasionally there is a need for computations that cannot easily be formulated as array expressions or do not fit existing computing patterns. In such cases it may be necessary to implement the computation from the ground up, for example, using pure Python code. However, pure Python code tends to be slow compared to the equivalent code written in a compiled language, and if the performance overhead of pure Python is too large, it can be necessary to explore alternatives. The traditional solution is to write an external library in, for example, C or Fortran, which performs the time-consuming computations, and interface it to Python code using an extension module.

© Robert Johansson 2019
R. Johansson, *Numerical Python*, https://doi.org/10.1007/978-1-4842-4246-9_19

There are several methods to create extension modules for Python. The most fundamental method is to use Python's C API to build an extension module with functions implemented in C that can be called from Python. This is typically very tedious and requires a significant effort. The Python standard library itself provides the module ctypes to simplify the interfacing between Python and C. Other alternatives include the CFFI (C foreign function interface) library[1] for interfacing Python with C and the F2PY[2] program for generating interfaces between Python and Fortran. These are all effective tools for interfacing Python with compiled code, and they all play an important role in making Python suitable for scientific computing. However, using these tools requires programming skills and efforts in other languages than Python, and they are the most useful when working with a code base that is already written in, say, C or Fortran.

For new development there are alternatives closer to Python that are worth considering before embarking on a complete implementation of a problem directly in a compiled language. In this chapter we explore two such methods: Numba and Cython. These offer a middle ground between Python and low-level languages that retains many of the advantages of a high-level language while achieving performance comparable to compiled code.

Numba is a just-in-time (JIT) compiler for Python code using NumPy that produces machine code that can be executed more efficiently than the original Python code. To achieve this, Numba leverages the LLVM compiler suite (http://llvm.org), which is a compiler toolchain that has become very popular in recent years for its modular and reusable design and interface, enabling, for example, applications such as Numba. Numba is a relatively new project and is not yet widely used in a lot of scientific computing libraries, but it is a promising project with strong backing by Continuum Analytics Inc.,[3] and it is likely to have a bright future in scientific computing with Python.

---

[1]http://cffi.readthedocs.org

[2]http://docs.scipy.org/doc/numpy-dev/f2py/index.html

[3]The producers of the Anaconda Python environment, see Chapter 1 and Appendix.

Numba   The Numba library provides a just-in-time compiler for Python and NumPy code that is based on the LLVM compiler. The main advantage of Numba is that it can generate machine code with minimal or no changes to the original Python code. For more information about the project and its documentation, see the project's web page at `http://numba.pydata.org`. At the time of writing, the latest version of the library is 0.39.0. Numba is an open source project created by Continuum Analytics Inc., which also offers a commercial extended version of Numba called NumbaPro (for more information, see `http://docs.continuum.io/numbapro/index`).

Cython is a superset of the Python language that can be automatically translated into C or C++ and compiled into a machine code, which can run much faster than Python code. Cython is widely used in computationally oriented Python projects for speeding up time-critical parts of a code base that is otherwise written in Python. Several of the libraries that we have used earlier in this book heavily rely on Cython. These include NumPy, SciPy, Pandas, and scikit-learn, just to mention a few.

Cython   The Cython library provides a translation of Python code, or decorated Python code, into C or C++, which can be compiled into a binary extension module. For more information about the project and its documentation, see the project's web page at `http://cython.org`. At the time of writing, the latest version of Cython is 0.28.4.

In this chapter we explore how Numba and Cython can be used to speed up code originally written in Python. These methods can be tried when a Python implementation turns out to be unacceptably slow. However, before trying to optimize anything that is written in Python, it is advisable to first profile the code, for example, using the `cProfile` module or IPython's profiling utilities (see Chapter 1) and identifying exactly which parts of a code are the bottlenecks. If clear bottlenecks can be identified, they may be good candidates for optimization efforts. The first optimization attempt should be to use already existing libraries, such as NumPy as SciPy, in the most efficient way to solve the problem at hand, as well as use the Python language itself in the most efficient

643

manner possible.[4] Only when existing libraries do not already provide functions and methods that allow us to implement a computation in an efficient way should we consider optimizing our code with Numba or Cython. Code optimization should only be used as a last resort, since premature optimization is often fruitless and results in less maintainable code: "premature optimization is the root of all evil" (Donald Knuth).

Importing Modules

In this chapter we will work with Numba and Cython. Numba is used as a regular Python module, and here we assume that this library is imported in its entirety using

```
In [1]: import numba
```

Cython can be used in several different ways, as we will see later in this chapter. Typically we are not required to explicitly import the Cython library when using Cython code from Python, but instead, we import the pyximport library provided by Cython and register an import hook using pyximport.install():

```
In [2]: import pyximport
```

This will alter the way Python modules are imported, and in particular it will allow us to import Cython files with the file-ending *pyx* directly as if they were pure Python modules. Occasionally it is also useful to explicitly import the Cython library, in which case we assume that it is imported in the following manner:

```
In [3]: import cython
```

For basic numerics and plotting, we also require the NumPy and Matplotlib libraries:

```
In [4]: import numpy as np
In [5]: import matplotlib.pyplot as plt
```

Numba

One of the most attractive aspects of the Numba library is that it can often be used to speed up Python code that uses NumPy without changing the target code. The only thing that we need to do is decorating a function with the @numba.jit decorator, which results in the

---

[4]For example, carefully consider which data structures to use, and make good use of iterators to avoid unnecessary memory copy operations.

function being just-in-time (JIT) compiled into code that can be significantly faster than the pure Python code, by as much as a factor of several hundred or more. The speedup is obtained mainly for functions that use NumPy arrays, for which Numba can automatically perform type interference and generate optimized code for the required type signatures.

To get started using Numba, consider the following simple problem: compute the sum of all elements in an array. A function that performs this computation is simple to implement in Python using for loops:

```
In [6]: def py_sum(data):
   ...:     s = 0
   ...:     for d in data:
   ...:         s += d
   ...:     return s
```

Although this function is nearly trivial, it nicely illustrates the potential and power of Numba. For loops in Python are notoriously slow, due to Python's flexibility and dynamic typing. To quantify this statement and benchmark the py_sum function, we generate an array with 50000 random numbers and use the %timeit IPython command to measure the typical computation time:

```
In [7]: data = np.random.randn(50000)
In [8]: %timeit py_sum(data)
100 loops, best of 3: 8.43 ms per loop
```

The result suggests that summing the 50000 elements in the data array using the py_sum function typically takes 8.43 milliseconds on this particular system. Compared to other methods that we explore below, this is not a good performance. The usual solution is to use array operations, such as those provided by NumPy, instead of iterating over the arrays manually. Indeed, NumPy provides the sum function that does exactly what we want to do here. To verify that the py_sum function defined in the preceding text produces the same results as the NumPy sum function, we first issue an assert statement to this effect:

```
In [9]: assert abs(py_sum(data) - np.sum(data)) < 1e-10
```

Since assert does not raise an error, we conclude that the two functions produce the same result. Next we benchmark the NumPy sum function using %timeit in the same way we used in the preceding example:

```
In [10]: %timeit np.sum(data)
10000 loops, best of 3: 29.8 µs per loop
```

The NumPy sum function is several hundred times faster than the py_sum function, demonstrating that vectorized expressions and operations using, for example, NumPy are the key to good performance in Python. We see the same phenomena for other functions that use for loops. For example, consider also the accumulative sum, py_cumsum, which takes an array as input and produces an array as output:

```
In [11]: def py_cumsum(data):
    ...:     out = np.zeros_like(data)
    ...:     s = 0
    ...:     for n in range(len(data)):
    ...:         s += data[n]
    ...:         out[n] = s
    ...:     return out
```

Benchmarking this function also gives a result that is much slower than the corresponding array-based NumPy function:

```
In [12]: %timeit py_cumsum(data)
100 loops, best of 3: 14.4 ms per loop
In [13]: %timeit np.cumsum(data)
10000 loops, best of 3: 147 µs per loop
```

Now let's see how Numba can be used to speed up the slow py_sum and py_cumsum functions. To activate JIT compilation of a function, we simply apply the decorator @numba.jit:

```
In [14]: @numba.jit
    ...: def jit_sum(data):
    ...:     s = 0
    ...:     for d in data:
    ...:         s += d
    ...:     return s
```

Next we verify that the JIT-compiled function produces the same result as the NumPy sum function and benchmark it using the %timeit function.

```
In [15]: assert abs(jit_sum(data) - np.sum(data)) < 1e-10
In [16]: %timeit jit_sum(data)
10000 loops, best of 3: 47.7 µs per loop
```

Compared to the pure Python function, the jit_sum function is about 300 times faster and reaches performance that is comparable to the NumPy sum function, in spite of being written in pure Python.

In addition to JIT compiling a function by applying the numba.jit decorator when the function is defined, we can apply the decorator after the fact. For example, to JIT compile the py_cumsum function that we defined earlier, we can use

```
In [17]: jit_cumsum = numba.jit()(py_cumsum)
```

We verify that the resulting function jit_cumsum indeed produces the same result as the corresponding NumPy function and benchmark it using %timeit:

```
In [18]: assert np.allclose(np.cumsum(data), jit_cumsum(data))
In [19]: %timeit jit_cumsum(data)
10000 loops, best of 3: 66.6 µs per loop
```

In this case the jit_cumsum function outperforms the NumPy cumsum function by a factor of two. The NumPy function cumsum is more versatile than the jit_cumsum function, so the comparison is not entirely fair, but it is remarkable that we can reach performance that is comparable to compiled code by JIT compiling Python code with a single function decorator. This allows us to use loop-based computations in Python without performance degradation, which is particularly useful for algorithms that are not easily written in vectorized form.

An example of such an algorithm is the computation of the Julia fractal, which requires a variable number of iterations for each element of a matrix with coordinate points in the complex plane: A point z in the complex plane belongs to the Julia set if the iteration formula $z \leftarrow z^2 + c$ does not diverge after a large number of iterations. To generate a Julia fractal graph, we can therefore loop over a set of coordinate points and iterate $z \leftarrow z^2 + c$ and store the number of iterations required to diverge beyond some predetermined bound (absolute value larger than 2.0 in the following implementation):

```
In [20]: def py_julia_fractal(z_re, z_im, j):
    ...:     for m in range(len(z_re)):
    ...:         for n in range(len(z_im)):
    ...:             z = z_re[m] + 1j * z_im[n]
    ...:             for t in range(256):
    ...:                 z = z ** 2 - 0.05 + 0.68j
    ...:                 if np.abs(z) > 2.0:
```

```
...:                          j[m, n] = t
...:                          break
```

This implementation is very simple and straightforward when using explicit loops, but in pure Python these three nested loops are inhibitively slow, as we will see in the following text. However, with JIT compilation using Numba, we can obtain a great speedup.

By default Numba gracefully falls back on the standard Python interpreter in cases when it fails to produce optimized code. An exception to this rule is when the nopython=True argument to numba.jit is given, in which case the JIT compilation will fail if Numba is unable to generate statically typed code. When automatic type interference fails, the resulting JIT-compiled code generated by Numba typically does not provide any speedup, so it is often advisable to use the nopython=True argument to the jit decorator so that we fail quickly when the produced JIT-compiled code is unlikely to result in a speedup. To assist Numba in the code generation, it is sometimes useful to explicitly define types of variables that occur in a function body, which we can do using the locals keyword argument to the jit decorator that can be assigned to a dictionary that maps symbol names to explicit types: for example, locals=dict(z=numba.complex) specifies that the variable z is a complex number. However, with the current example, we do not need to specify the types of local variables explicitly, since they can all be inferred from the data types of the NumPy arrays that are passed to the function. We can verify that this is the case by using the nopython=True argument to numba.jit when decorating the py_julia_fractal function:

```
In [21]: jit_julia_fractal = numba.jit(nopython=True)(py_julia_fractal)
```

Next we call the resulting jit_julia_fractal function to compute the Julia set. Note that here we have written the function such that all the involved NumPy arrays are defined outside the function. This helps Numba recognizing which types are involved in the calculation and allows it to generate efficient code in the JIT compilation:

```
In [22]: N = 1024
In [23]: j = np.zeros((N, N), np.int64)
In [24]: z_real = np.linspace(-1.5, 1.5, N)
In [25]: z_imag = np.linspace(-1.5, 1.5, N)
In [26]: jit_julia_fractal(z_real, z_imag, j)
```

After the call to the jit_julia_fractal function, the result of the computation is stored in the j array. To visualize the result, we can plot the j array using the Matplotlib imshow function. The result is shown in Figure 19-1:

```
In [27]: fig, ax = plt.subplots(figsize=(8, 8))
    ...: ax.imshow(j, cmap=plt.cm.RdBu_r, extent=[-1.5, 1.5, -1.5, 1.5])
    ...: ax.set_xlabel("$\mathrm{Re}(z)$", fontsize=18)
    ...: ax.set_ylabel("$\mathrm{Im}(z)$", fontsize=18)
```

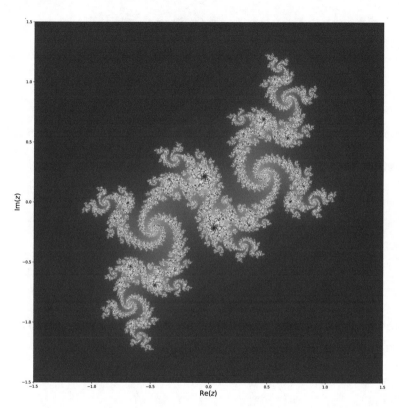

Figure 19-1. *The Julia fractal generated by a JIT-compiled Python function using Numba*

We can compare the speed of the pure Python function py_julia_fractal and the corresponding JIT-compiled function jit_julia_fractal using the %timeit command:

```
In [28]: %timeit py_julia_fractal(z_real, z_imag, j)
1 loops, best of 3: 60 s per loop
In [29]: %timeit jit_julia_fractal(z_real, z_imag, j)
10 loops, best of 3: 140 ms per loop
```

The speedup in this particular case is a remarkable 430 times, again by simply adding a decorator to the Python function. With this type of speedup, for loops in Pythons do not really need to be avoided after all.

Another useful decorator in the Numba library is `numba.vectorize`. It generates and JIT compiles a vectorized function from a kernel function written for scalar input and output, much like the NumPy `vectorize` function. Consider, for example, the Heaviside step function:

$$\Theta(x) = \begin{cases} 0, & x < 0 \\ \dfrac{1}{2}, & x = 0. \\ 1, & x > 0 \end{cases}$$

If we wanted to implement this function for scalar input x, we could use

```
In [30]: def py_Heaviside(x):
    ...:     if x == 0.0:
    ...:         return 0.5
    ...:     if x < 0.0:
    ...:         return 0.0
    ...:     else:
    ...:         return 1.0
```

This function only works for scalar input, and if we want to apply it to an array or list, we have to explicitly iterate over the array and apply it to each element:

```
In [31]: x = np.linspace(-2, 2, 50001)
In [32]: %timeit [py_Heaviside(xx) for xx in x]
100 loops, best of 3: 16.7 ms per loop
```

This is inconvenient and slow. The NumPy vectorize function solves the inconvenience problem, by automatically wrapping the scalar kernel function into a NumPy-array aware function:

```
In [33]: np_vec_Heaviside = np.vectorize(py_Heaviside)
In [34]: np_vec_Heaviside(x)
Out[34]: array([ 0.,  0.,  0., ...,  1.,  1.,  1.])
```

However, the NumPy vectorize function does not solve the performance problem. As we see from benchmarking the np_vec_Heaviside function with %timeit, its performance is comparable to explicitly looping over the array and consecutively calls the py_Heaviside function for each element:

```
In [35]: %timeit np_vec_Heaviside(x)
100 loops, best of 3: 13.6 ms per loop
```

Better performance can be achieved by using NumPy array expressions instead of using NumPy vectorize on a scalar kernel written in Python:

```
In [36]: def np_Heaviside(x):
   ...:     return (x > 0.0) + (x == 0.0)/2.0
In [37]: %timeit np_Heaviside(x)
1000 loops, best of 3: 268 µs per loop
```

However, even better performance can be achieved using Numba and the vectorize decorator, which takes a list of function signatures for which to generate JIT-compiled code. Here we choose to generate vectorized functions for two signatures – one that takes arrays of 32-bit floating-point numbers as input and output, defined as numba.float32(numba.float32), and one that takes arrays of 64-bit floating-point numbers as input and output, defined as numba.float64(numba.float64):

```
In [38]: @numba.vectorize([numba.float32(numba.float32),
   ...:                     numba.float64(numba.float64)])
   ...: def jit_Heaviside(x):
   ...:     if x == 0.0:
   ...:         return 0.5
   ...:     if x < 0:
   ...:         return 0.0
   ...:     else:
   ...:         return 1.0
```

Benchmarking the resulting jit_Heaviside function shows the best performance of the methods we have looked at:

```
In [39]: %timeit jit_Heaviside(x)
10000 loops, best of 3: 58.5 µs per loop
```

and the jit_Heaviside function can be used as any NumPy universal function, including support for broadcasting and other NumPy features. To demonstrate that the function indeed implements the desired function, we can test it on a simple list of input values:

```
In [40]: jit_Heaviside([-1, -0.5, 0.0, 0.5, 1.0])
Out[40]: array([ 0. ,  0. ,  0.5,  1. ,  1. ])
```

In this section we have explored speeding up Python code using JIT compilation with the Numba library. We looked at four examples: two simple examples for demonstrating the basic usage of Numba, the summation and accumulative summation of an array. For a more realistic use-case of Numba that is not so easily defined in terms of vector expressions, we looked at the computation of the Julia set. Finally, we explored the vectorization of scalar kernel with the implementation of the Heaviside step function. These examples demonstrate the common use patterns for Numba, but there is also much more to explore in the Numba library, such as code generation for GPUs. For more information about this and other topics, see the official Numba documentation at `http://numba.pydata.org/doc.html`.

Cython

Like Numba, Cython is a solution for speeding up Python code, although Cython takes a completely different approach to this problem. Whereas Numba is a Python library that converts pure Python code to LLVM code that is JIT-compiled into machine code, Cython is a programming language that is a superset of the Python programming language: Cython extends Python with C-like properties. Most notably, Cython allows us to use explicit and static type declarations. The purpose of the extensions to Python introduced in Cython is to make it possible to translate the code into efficient C or C++ code, which can be compiled into a Python extension module that can be imported and used from regular Python code.

There are two main usages of Cython:speeding up Python code and generating wrappers for interfacing with compiled libraries. When using Cython, we need to modify the targeted Python code, so compared to using Numba, there is a little bit more work involved, and we need to learn the syntax and behavior of Cython in order to use it to speed up Python code. However, as we will see in this section, Cython provides more fine-grained control of how the Python code is processed, and Cython also has features that are out of scope for Numba, such as generating interfaces between Python and external libraries and speeding up Python code that does not use NumPy arrays.

While Numba uses transparent just-in-time compilation, Cython is mainly designed for using traditional ahead-of-time compilation. There are several ways to compile Cython code into a Python extension module, each with different use-cases. We begin with reviewing options for compiling Cython code and then proceed to introduce features of Cython that are useful for speeding up computations written in Python.

Throughout this section we will work with mostly the same examples that we looked at in the previous section using Numba, so that we can easily compare both the methods and the results. We begin by looking at how to speed up the py_sum and py_cumsum functions defined in the previous section.

To use Cython code from Python, it has to pass through the Cython compilation pipeline: first the Cython code must be translated into C or C++ code, after which it has to be compiled into machine code using a C or C++ compiler. The translation from Cython code to C or C++ can be done using the cython command-line tool. It takes a file with Cython code, which we typically store in files using the *pyx* file extension, and produces a C or C++ file. For example, consider the file cy_sum.pyx, with the content shown in Listing 19-1. To generate a C file from this Cython file, we can run the command cython cy_sum.pyx. The result is the file cy_sum.c, which we can compile using a standard C compiler into a Python extension module. This compilation step is platform dependent and requires using the right compiler flags and options as to produce a proper Python extension.

Listing 19-1. Content of the Cython File cy_sum.pyx

```
def cy_sum(data):
    s = 0.0
    for d in data:
        s += d
    return s
```

To avoid the complications related to platform-specific compilation options for C and C++ code, we can use the distutils and Cython libraries to automate the translation of Cython code into a useful Python extension module. This requires creating a setup.py script that calls the setup function from distutils.core (which knows how to compile C code into a Python extension) and the cythonize function from Cython. Build (which knows how to translate Cython code into C code), as shown in Listing 19-2. When the setup.py file is prepared, we can compile the Cython module using the command python setup.py build_ext --inplace, which instructs distutils to build the extension module and place it in the same directory as the source code.

Listing 19-2. A setup.py Script That Can Be Used to Automatically Compile a Cython File into a Python Extension Module

```
from distutils.core import setup
from Cython.Build import cythonize
import numpy as np

setup(ext_modules=cythonize('cy_sum.pyx'),
      include_dirs=[np.get_include()],
      requires=['Cython', 'numpy'])
```

Once the Cython code has been compiled into a Python extension module, whether by hand or using `distutils` library, we can import it and use it as a regular module in Python:

```
In [41]: from cy_sum import cy_sum
In [42]: cy_sum(data)
Out[42]: -189.70046227549025
In [43]: %timeit cy_sum(data)
100 loops, best of 3: 5.56 ms per loop
In [44]: %timeit py_sum(data)
100 loops, best of 3: 8.08 ms per loop
```

Here we see that for this example, compiling the pure Python code in Listing 19-1 using Cython directly gives a speedup of about 30%. This is a nice speedup, but arguably not worth the trouble of going through the Cython compilation pipeline. We will see later how to improve on this speedup using more features of Cython.

Listing 19-3. Content of the Cython File `cy_cumsum.pyx`

```
cimport numpy
import numpy

def cy_cumsum(data):
    out = numpy.zeros_like(data)
    s = 0
    for n in range(len(data)):
        s += data[n]
        out[n] = s
    return out
```

The explicit compilation of Cython code into a Python extension module shown in the preceding code is useful for distributing prebuilt modules written in Cython, as the end result does not require Cython to be installed to use the extension module. An alternative way to implicitly invoke the Cython compilation pipeline automatically during the import of a module is provided by the `pyximport` library, which is distributed with Cython. To seamlessly import a Cython file directly from Python, we can first invoke the `install` function from the `pyximport` library:

```
In [45]: pyximport.install(setup_args=dict(include_dirs=np.get_include()))
```

This will modify the behavior of the Python `import` statement and add support for Cython *pyx* files. When a Cython module is imported, it will first be compiled in C or C++ and then to machine code in the format of a Python extension module that the Python interpreter can import. These implicit steps sometimes require additional configuration, which we can pass to the `pyximport.install` function via arguments. For example, to be able to import Cython code that uses NumPy-related features, we need the resulting C code to be compiled against the NumPy C header files. We can configure this by setting the `include_dirs` to the value given by `np.get_include()` in the `setup_args` argument to the `install` function, as shown in the preceding code. Several other options are also available, and we can also give custom compilation and linking arguments. See the docstring for `pyximport.install` for details. Once `pyximport.install` has been called, we can use a standard Python `import` statement to import a function from a Cython module:

```
In [46]: from cy_cumsum import cy_cumsum
In [47]: %timeit cy_cumsum(data)
100 loops, best of 3: 5.91 ms per loop
In [48]: %timeit py_cumsum(data)
100 loops, best of 3: 13.8 ms per loop
```

In this example too, we see a welcome but not very impressive speedup of a factor of two for the Python code that has been passed through the Cython compilation pipeline.

Before we get into the detailed usage of Cython that allow us to improve upon this speedup factor, we quickly introduce yet another way of compiling and import Cython code. When using IPython, and especially the Jupyter Notebook, we can use the convenient %%cython command, which automatically compiles and loads Cython code in a code cell as Python extension and makes it available in the IPython session. To be able to use this command, we first have to activate it using the %load_ext cython command:

```
In [49]: %load_ext cython
```

With the %%cython command activated, we can write and load Cython code interactively in an IPython session:

```
In [50]: %%cython
    ...: def cy_sum(data):
    ...:     s = 0.0
    ...:     for d in data:
    ...:         s += d
    ...:     return s
In [51]: %timeit cy_sum(data)
100 loops, best of 3: 5.21 ms per loop
In [52]: %timeit py_sum(data)
100 loops, best of 3: 8.6 ms per loop
```

As before, see a direct speedup by simply adding the %%cython at the first line of the IPython code cell. This is reminiscent of adding the @numba.jit decorator to a function, but the underlying mechanics of these two methods are rather different. In the rest of this section, we will use this method for compiling and loading Cython code. When using the %%cython IPython command, it is also useful to add the -a argument. This results in Cython annotation output to be displayed as the output of the code cell, as shown in Figure 19-2. The annotation shows each code line in a shade of yellow, where bright yellow indicates that the line of code is translated to C code with strong dependencies on the Python C/API, and a where a while line of code is directly translated into pure C code. When working on optimizing Cython code, we generally need to strive for Cython code that gets translated into as pure C code as possible, so it is extremely useful to inspect the annotation output and look for yellow lines, which typically represent the bottlenecks in the code. As an added bonus, clicking a line of code in the annotation output toggles between the Cython code that we provided and the C code that it is being translated into.

```
In [88]:  %%cython -a
          def cy_sum(data):
              s = 0.0
              for d in data:
                  s += d
              return s

Out[88]:
          Generated by Cython 0.22
          +1:  def cy_sum(data):
          +2:      s = 0.0
          +3:      for d in data:
          +4:          s += d
          +5:      return s
```

Figure 19-2. *Annotation generated by Cython using the %%cython IPython command with the -a argument*

In the rest of the section, we explore ways of speeding up Cython code using language features that are introduced by Cython that are particularly useful for computational problems. We first revisit the implementation of the cy_sum given in the preceding code. In our first attempt to speed up this function, we simply used the pure Python and passed it through the Cython compilation pipeline, and as a result, we saw a speedup of about 30%. The key step to see much larger speedups is to add type declarations for all the variables and arguments of the function. By explicitly declaring the types of variables, the Cython compiler will be able to generate more efficient C code. To specify a type of a variable, we need to use the Cython keyword cdef, which we can use with any standard C type. For example, to declare the variable n of integer type, we can use cdef int n. We can also use type definitions from the NumPy library: for example, cdef numpy.float64_t s declares the variable s to be a 64-bit floating-point number. NumPy arrays can be declared using the type specification in the format numpy.ndarray[numpy.float64_t, ndim=1] data, which declares data to be an array with 64-bit floating-point number elements, with one dimension (a vector) and of unspecified length. Adding type declarations of this style to the previous cy_sum function results in the following code:

```
In [53]: %%cython
    ...: cimport numpy
    ...: cimport cython
    ...:
    ...: @cython.boundscheck(False)
    ...: @cython.wraparound(False)
```

```
...: def cy_sum(numpy.ndarray[numpy.float64_t, ndim=1] data):
...:     cdef numpy.float64_t s = 0.0
...:     cdef int n, N = len(data)
...:     for n in range(N):
...:         s += data[n]
...:     return s
```

In this implementation of the cy_sum function, we have also applied the two decorators @cython.boundscheck(False) and @cython.wraparound(False), which disable time-consuming bound checks on the indexing of NumPy arrays. This results in less safe code, but if we are confident that the NumPy arrays in this function will not be index outside of their valid ranges, we can obtain additional speedup by disabling such checks. Now that we have explicitly declared the type of all variables and arguments of the function, Cython is able to generate efficient C code that when compiled into a Python module provides performance that is comparable to the JIT-compiled code using Numba and not far from the built-in sum function from NumPy (which also is implemented in C):

```
In [54]: %timeit cy_sum(data)
10000 loops, best of 3: 49.2 µs per loop
In [55]: %timeit jit_sum(data)
10000 loops, best of 3: 47.6 µs per loop
In [56]: %timeit np.sum(data)
10000 loops, best of 3: 29.7 µs per loop
```

Next let's turn our attention to the cy_cumsum function. Like the cy_sum function, this function too will benefit from explicit type declarations. To simplify the declarations of NumPy array types, here we use the ctypedef keyword to create an alias for numpy.float64_t to the shorter FTYPE_t. Note also that in Cython code, there are two different import statements: cimport and import. The import statement can be used to import any Python module, but it will result in C code that calls back into the Python interpreter and can therefore be slow. The cimport statement works like a regular import, but is used for importing other Cython modules. Here cimport numpy imports a Cython module named numpy that provides Cython extensions to NumPy, mostly type and function declarations. In particular, the C-like types such as numpy.float64_t are declared in this Cython module. However, the function call numpy.zeros in the function defined in the following code results in a call to the function zeros in the NumPy module, and for it, we need to include the Python module numpy using import numpy.

Adding these type declarations to the previously defined cy_cumsum function results in the implementation given in the following:

```
In [57]: %%cython
    ...: cimport numpy
    ...: import numpy
    ...: cimport cython
    ...:
    ...: ctypedef numpy.float64_t FTYPE_t
    ...:
    ...: @cython.boundscheck(False)
    ...: @cython.wraparound(False)
    ...: def cy_cumsum(numpy.ndarray[FTYPE_t, ndim=1] data):
    ...:     cdef int n, N = data.size
    ...:     cdef numpy.ndarray[FTYPE_t, ndim=1] out = numpy.zeros
              (N, dtype=data.dtype)
    ...:     cdef numpy.float64_t s = 0.0
    ...:     for n in range(N):
    ...:         s += data[n]
    ...:         out[n] = s
    ...:     return out
```

Like for cy_sum, we see a significant speedup after having declared the types of all variables in the function, and the performance of cy_cumsum is now comparable to the JIT-compiled Numba function jit_cumsum and faster than the built-in cumsum function in NumPy (which on the other hand is more versatile):

```
In [58]: %timeit cy_cumsum(data)
10000 loops, best of 3: 69.7 µs per loop
In [59]: %timeit jit_cumsum(data)
10000 loops, best of 3: 70 µs per loop
In [60]: %timeit np.cumsum(data)
10000 loops, best of 3: 148 µs per loop
```

When adding explicit type declarations, we gain performance when compiling the function with Cython, but we lose generality as the function is now unable to take any other type of arguments. For example, the original py_sum function, as well as the NumPy sum function, accepts a much wider variety of input types. We can sum Python lists and NumPy arrays of both floating-point numbers and integers:

```
In [61]: py_sum([1.0, 2.0, 3.0, 4.0, 5.0])
Out[61]: 15.0
In [62]: py_sum([1, 2, 3, 4, 5])
Out[62]: 15
```

The Cython-compiled version with explicit type declaration, on the other hand, only works for exactly the type we declared it:

```
In [63]: cy_sum(np.array([1.0, 2.0, 3.0, 4.0, 5.0]))
Out[63]: 15.0
In [64]: cy_sum(np.array([1, 2, 3, 4, 5]))
---------------------------------------------------------------------------
ValueError: Buffer dtype mismatch, expected 'float64_t' but got 'long'
```

Often it is desirable to support more than one type of input, such as providing the ability to sum arrays of both floating-point numbers and integers with the same function. Cython provides a solution to this problem through its ctypedef fused keyword, with which we can define new types that are one out of several provided types. For example, consider the modification to the py_sum function given in py_fused_sum here:

```
In [65]: %%cython
    ...: cimport numpy
    ...: cimport cython
    ...:
    ...: ctypedef fused I_OR_F_t:
    ...:     numpy.int64_t
    ...:     numpy.float64_t
    ...:
    ...: @cython.boundscheck(False)
    ...: @cython.wraparound(False)
    ...: def cy_fused_sum(numpy.ndarray[I_OR_F_t, ndim=1] data):
    ...:     cdef I_OR_F_t s = 0
```

```
...:      cdef int n, N = len(data)
...:      for n in range(N):
...:          s += data[n]
...:      return s
```

Here the function is defined in terms of the type I_OR_F_t, which is defined using ctypedef fused to be either numpy.int64_t or numpy.float64_t. Cython will automatically generate the necessary code for both types of functions, so that we can use the function on both floating-point and integer arrays (at the price of a small decrease in performance):

```
In [66]: cy_fused_sum(np.array([1.0, 2.0, 3.0, 4.0, 5.0]))
Out[66]: 15.0
In [67]: cy_fused_sum(np.array([1, 2, 3, 4, 5]))
Out[67]: 15
```

As a final example of how to speed up Python code with Cython, consider again the Python code for generating the Julia set that we looked at in the previous section. To implement a Cython version of this function, we simply take the original Python code and explicitly declare the types of all the variables used in the function, following the procedure we used in the preceding text. We also add the decorators for disabling index bound checks and wraparound. Here we have both NumPy integer arrays and floating-point arrays as input, so we define the arguments as types numpy.ndarray[numpy.float64_t, ndim=1] and numpy.ndarray[numpy.int64_t, ndim=2], respectively.

The implementation of cy_julia_fractal given in the following code also includes a Cython implementation of the square of the absolute value of a complex number. This function is declared as inline using the inline keyword, which means that the compiler will put the body of the function at every place it is called, rather than creating a function that is called from those locations. This will result in large code, but avoid the overhead of an additional function call. We also define this function using cdef rather than the usual def keyword. In Cython, def defines a function that can be called from Python, while cdef defines a function that can be called from C. Using the cpdef keyword, we can simultaneously define a function that is callable both from C and from Python. As it is written here, using cdef, we cannot call the abs2 function from the IPython session after executing this code cell, but if we change cdef to cpdef, we can.

```
In [68]: %%cython
    ...: cimport numpy
    ...: cimport cython
    ...:
    ...: cdef inline double abs2(double complex z):
    ...:     return z.real * z.real + z.imag * z.imag
    ...:
    ...: @cython.boundscheck(False)
    ...: @cython.wraparound(False)
    ...: def cy_julia_fractal(numpy.ndarray[numpy.float64_t, ndim=1] z_re,
    ...:                      numpy.ndarray[numpy.float64_t, ndim=1] z_im,
    ...:                      numpy.ndarray[numpy.int64_t, ndim=2] j):
    ...:     cdef int m, n, t, M = z_re.size, N = z_im.size
    ...:     cdef double complex z
    ...:     for m in range(M):
    ...:         for n in range(N):
    ...:             z = z_re[m] + 1.0j * z_im[n]
    ...:             for t in range(256):
    ...:                 z = z ** 2 - 0.05 + 0.68j
    ...:                 if abs2(z) > 4.0:
    ...:                     j[m, n] = t
    ...:                     break
```

If we call the cy_julia_fractal function with the same arguments as we previously called the Python implementation that was JIT-compiled using Numba, we see that the two implementations have comparable performance.

```
In [69]: N = 1024
In [70]: j = np.zeros((N, N), dtype=np.int64)
In [71]: z_real = np.linspace(-1.5, 1.5, N)
In [72]: z_imag = np.linspace(-1.5, 1.5, N)
In [73]: %timeit cy_julia_fractal(z_real, z_imag, j)
10 loops, best of 3: 113 ms per loop
In [74]: %timeit jit_julia_fractal(z_real, z_imag, j)
10 loops, best of 3: 141 ms per loop
```

The slight edge to the `cy_julia_fractal` implementation is mainly due to the inline definition of the innermost loop call to the `abs2` function and the fact that `abs2` avoids computing the square root. Making a similar change in `jit_julia_fractal` improves its performance and approximately accounts for the difference shown here.

So far we have explored Cython as a method to speed up Python code by compiling it into machine code that is made available as Python extension modules. There is another important use-case of Cython, which is at least as important to its widespread use in the Python scientific computing community: Cython can also be used to easily create wrappers to compiled C and C++ libraries. We will not explore this in depth here, but will give a simple example that illustrates that using Cython we can call out to arbitrary C libraries in just a few lines of code. As an example, consider the math library from the C standard library. It provides mathematical functions, similar to those defined in the Python standard library with the same name: `math`. To use these functions in a C program, we would include the `math.h` header file to obtain their declarations and compile and link the program against the `libm` library. From Cython we can obtain function declarations using the `cdef extern from` keywords, after which we need to give the name of the C header file and list the declarations of the function we want to use in the following code block. For example, to make the `acos` function from `libm` available in Cython, we can use the following code:

```
In [75]: %%cython
    ...: cdef extern from "math.h":
    ...:         double acos(double)
    ...:
    ...: def cy_acos1(double x):
    ...:     return acos(x)
```

Here we also defined the Python function `cy_acos1`, which we can call from Python:

```
In [76]: %timeit cy_acos1(0.5)
10000000 loops, best of 3: 83.2 ns per loop
```

Using this method we can wrap arbitrary C functions into functions that are callable from regular Python code. This is a very useful feature for scientific computing applications since it makes existing code written in C and C++ easily available from Python. For the standard libraries, Cython already provides type declarations via the `libc` module, so we do not need to explicitly define the functions using `cdef extern` from. For the `acos` example, we could therefore instead directly import the function from `libc.math` using the `cimport` statement:

```
In [77]: %%cython
    ...: from libc.math cimport acos
    ...:
    ...: def cy_acos2(double x):
    ...:     return acos(x)
In [78]: %timeit cy_acos2(0.5)
10000000 loops, best of 3: 85.6 ns per loop
```

The resulting function `cy_acos2` is identical to `cy_acos1` that was explicitly imported from `math.h` earlier. It is instructive to compare the performance of these C math library functions to the corresponding functions defined in NumPy and the Python standard math library:

```
In [79]: from numpy import arccos
In [80]: %timeit arccos(0.5)
1000000 loops, best of 3: 1.07 µs per loop
In [81]: from math import acos
In [82]: %timeit acos(0.5)
10000000 loops, best of 3: 95.9 ns per loop
```

The NumPy version is about ten times slower than the Python `math` function and Cython wrappers to the C standard library function, because of the overhead related to NumPy array data structures.

Summary

In this chapter we have explored methods for speeding up Python code using Numba, which produces optimized machine code using just-in-time compilation, and Cython, which produces C code that can be compiled into machine code using ahead-of-time compilation. Numba works with pure Python code, but heavily relies on type interference using NumPy arrays, while Cython works with an extension to the Python language that allows explicit type declarations. The advantages of these methods are that we can achieve performance that is comparable to compiled machine code while staying in a Python or Python-like programming environment. The key to speeding up Python code is the use of typed variables, either by using type interference from NumPy arrays, as in Numba, or by explicitly declaring the types of variables, as in Cython. Explicitly typed code can be translated into much more efficient code than the dynamically typed code in pure Python and can avoid much of the overhead involved in type lookups in Python. Both Numba and Cython are convenient ways to obtain impressive speedups of Python code, and they often produce code with similar performance. Cython also provides an easy-to-use method for creating interfaces to external libraries so that they can be accessed from Python. In both Numba and Cython, the common theme is use type information (from NumPy arrays or from explicit declarations) to generate more efficient typed machine code. Within the Python community, there has also recently been a movement toward adding support for optional type hints to the Python language itself. For more details about type hints, see PEP 484 (`www.python.org/dev/peps/pep-0484`), which has been included in Python as of version 3.5. While type hints in Python are not likely to be widely available in the near future, it is certainly an interesting development to follow.

Further Reading

Thorough guides to using Cython are given in Smith (2015) and Herron (2013). For more information about Numba, see its official documentation at `http://numba.pydata.org/numba-doc`. For a detailed discussion of high-performance computing with Python, see also M. Gorelick (2014).

References

Herron, P. (2013). *Learning Cython Programming.* Mumbai: Packt.

M. Gorelick, I. O. (2014). High Performance Python: Practical Performant Programming for Humans. Sebastopol: O'Reilly.

Smith, K. (2015). *Cython A Guide for Python Programmers.* Sebastopol: O'Reilly.

APPENDIX

Installation

This appendix covers the installation and setup of a Python environment for scientific computing on commonly used platforms. As discussed in Chapter 1, the scientific computing environment for Python is not a single product, but rather a diverse ecosystem of packages and libraries, and there are numerous possible ways to install and configure a Python environment on any given platform. Python itself is rather easy to install,[1] and on many operating systems, it is even preinstalled. All pure Python libraries that are hosted on the *Python Package Index*[2] are also easily installed, for example, using `pip` and a command such as `pip install PACKAGE`, where `PACKAGE` is the name of the package to install. The *pip* software then searches for the package on the Python Package Index, and downloads and installs it, if it is found. For example, to install IPython, we can use

```
$ pip install ipython
```

and to upgrade an already installed package, we simply add the `--upgrade` flag to the `pip` command:

```
$ pip install --upgrade ipython
```

However, many libraries for computing with Python are not pure Python libraries, and they frequently have dependencies on system libraries written in other languages, such as C and Fortran. These dependencies cannot be handled by `pip` and the Python Package Index, and to build such libraries from source requires C and Fortran compilers to be installed. In other words, installing a full scientific computing software stack for Python manually can be difficult or at least time-consuming and tedious. To solve this problem, there have emerged a number of prepackaged Python environments with automated installers. The most popular

---

[1]Installers for all major platforms are available for download at `http://www.python.org/downloads`

[2]`http://pypi.python.org`

R. Johansson, *Numerical Python*, https://doi.org/10.1007/978-1-4842-4246-9

environments are Continuum Analytics' *Anaconda*[3] and Enthought's *Canopy*,[4] which are both sponsored by corporations with close connections to the open source scientific Python community, and *Python(x,y)*,[5] which is a community-maintained environment that targets Microsoft's operating systems. These environments all have in common that they bundle the Python interpreter, the required system libraries and tools, and a large number of scientific-computing-oriented Python libraries in an easy-to-install distribution. Any of these environments can readily be used to set up the software required to run the code discussed in this book, but in the following we use the Anaconda environment from Continuum Analytics. In particular, we discuss Miniconda – a lightweight version of Anaconda – and the package manager conda.

Miniconda and Conda

The Anaconda environment, which comes bundled with a large number of libraries, is a convenient way to get a scientific computing environment for Python up and running quickly. However, for clarity, here we start with a Miniconda environment and explicitly install the packages that we need. In this way, we control exactly which packages are included in the environment we set up. Miniconda is a minimal version of Anaconda, which only includes the most basic components: a Python interpreter, a few fundamental libraries, and the conda package manager. The download page for the Miniconda project (`http://conda.pydata.org/miniconda.html`) contains installers for Linux, Mac OS X, and Windows.[6] Download and run the installer, and follow the on-screen instructions. When the installation has finished, you should have a directory named miniconda in your home directory, and if you choose to add it to your PATH variable during the installation, you should now be able to invoke the conda package manager by running conda at the command prompt.

---

[3]`http://continuum.io/downloads`

[4]`http://www.enthought.com/products/canopy`

[5]`http://code.google.com/p/pythonxy`

[6]Miniconda is available in both 32- and 64-bit versions. Generally the 64-bit version is recommended for modern computers, but on Windows a 64-bit compiler might not always be readily available, so staying with the 32-bit version might be better in some cases on this platform.

Conda[7] is a cross-platform package manager that can handle dependencies on Python packages as well as system tools and libraries. This is essential for installing scientific computing software, which by nature uses a diverse set of tools and libraries. Conda packages are prebuilt binaries for the target platform and are therefore fast and convenient to install. To verify that conda is available on your system, you can try

```
$ conda --version
conda 4.5.11
```

In this case, the output tells us that conda is installed and that the version of conda is 4.5.11. To update to the latest version of conda, we can use the conda package manager itself:

```
$ conda update conda
```

and to update all packages installed in a particular conda environment, we can use

```
$ conda update --all
```

Once conda is installed, we can use it to install Python interpreters and libraries. When doing so we can optionally specify precise versions of the packages we want to install. The Python software ecosystem consists of a large number of independent projects, each with their own release cycles and development goals, and there are constantly new versions of different libraries being released. This is both exciting – because there is steady progress and new features are frequently made available – but unfortunately not all new releases of all libraries are backward compatible. This presents a dilemma for a user that requires a stable and reproducible environment over the long term and for users that simultaneously work on projects with different versions of dependencies.

The best solution in the Python ecosystem for this problem is to use a package manager such as conda to set up virtual Python environments for different projects, in which different versions of the required dependencies are installed. With this approach, it is easy to maintain multiple environments with different configurations, such as separate Python 2 and Python 3 environments, or environments with stable versions and development versions of relevant packages. I strongly recommend using virtual Python environments rather than using the default system-wide Python environments for the reasons given in the preceding text.

---

[7]http://conda.pydata.org/docs/index.html

With conda, new environments are created with the conda create command, to which we need to provide a name for the new environment using -n NAME or alternatively a path to where the environment is to be stored using -p PATH. When providing a name, the environment is by default stored in the miniconda/envs/ NAME directory. When creating a new environment, we can also give a list of packages to install. At least one package must be specified. For example, to create two new environments based on Python 2.7 and Python 3.6, we can use the commands

```
$ conda create -n py2.7 python=2.7
$ conda create -n py3.6 python=3.6
```

where we have given the Python 2 and Python 3 environments the names py2.7 and py3.6, respectively. To use one of these environments, we need to *activate* it using the command source activate py2.7 or source activate py3.6, respectively, and to *deactivate* an environment, we use source deactivate.[8] With this method it is easy to switch between different environments, as illustrated in the following sequence of commands:

```
$ source activate py2.7
discarding /Users/rob/miniconda/bin from PATH
prepending /Users/rob/miniconda/envs/py2.7/bin to PATH

(py2.7)$ python --version
Python 2.7.14 :: Continuum Analytics, Inc.

(py2.7)$ source activate py3.6
discarding /Users/rob/miniconda/envs/py2.7/bin from PATH
prepending /Users/rob/miniconda/envs/py3.6/bin to PATH

(py3.6)$ python --version
Python 3.6.5 :: Continuum Analytics, Inc.

(py3.6)$ source deactivate
discarding /Users/rob/miniconda/envs/py3.6/bin from PATH
$
```

---

[8]On Windows, leave out source from these commands.

To manage environments, the conda env, conda info, and conda list commands are helpful tools. The conda info command can be used to list available environments (same as conda env list):

```
$ conda info --envs
# conda environments:
#
base                     *   /Users/rob/miniconda
py2.7                        /Users/rob/miniconda/envs/py2.7
py3.6                        /Users/rob/miniconda/envs/py3.6
```

and the conda list command can be used to list installed packages and their versions, in a given environment:

```
$ conda list -n py3.6
# packages in environment at /Users/rob/miniconda/envs/py3.6:
#
# Name                    Version                   Build  Channel
ca-certificates           2017.08.26            ha1e5d58_0
certifi                   2018.1.18                py36_0
libcxx                    4.0.1                 h579ed51_0
libcxxabi                 4.0.1                 hebd6815_0
libedit                   3.1                   hb4e282d_0
libffi                    3.2.1                 h475c297_4
ncurses                   6.0                   hd04f020_2
openssl                   1.0.2o                h26aff7b_0
pip                       9.0.3                    py36_0
python                    3.6.5                 hc167b69_0
readline                  7.0                   hc1231fa_4
setuptools                39.0.1                   py36_0
sqlite                    3.22.0                h3efe00b_0
tk                        8.6.7                 h35a86e2_3
wheel                     0.30.0           py36h5eb2c71_1
xz                        5.2.3                 h0278029_2
zlib                      1.2.11                hf3cbc9b_2
```

Similar information in YAML format[9] is produced by the `conda env export` command:

```
(py3.6)$ conda env export
name: py3.6
channels:
  - defaults
dependencies:
  - ca-certificates=2017.08.26=ha1e5d58_0
  - certifi=2018.1.18=py36_0
  - libcxx=4.0.1=h579ed51_0
  - libcxxabi=4.0.1=hebd6815_0
  - libedit=3.1=hb4e282d_0
  - libffi=3.2.1=h475c297_4
  - ncurses=6.0=hd04f020_2
  - openssl=1.0.2o=h26aff7b_0
  - pip=9.0.3=py36_0
  - python=3.6.5=hc167b69_0
  - readline=7.0=hc1231fa_4
  - setuptools=39.0.1=py36_0
  - sqlite=3.22.0=h3efe00b_0
  - tk=8.6.7=h35a86e2_3
  - wheel=0.30.0=py36h5eb2c71_1
  - xz=5.2.3=h0278029_2
  - zlib=1.2.11=hf3cbc9b_2
prefix: /Users/rob/miniconda/envs/py3.6
```

To install additional packages in an environment, we can either specify a list of packages when the environment is created or activate the environment and use `conda install` or use the `conda install` command with the `-n` flag to specify a target environment for the installation. For example, to create a Python 3.6 environment with NumPy version 1.14, we could use

```
$ conda create -n py3.6-np1.14 python=3.6 numpy=1.14
```

---

[9]http://yaml.org

672

To verify that the new environment py3.6-np1.14 indeed contains NumPy of the specified version, we can use the conda list command again:

```
$ conda list -n py3.6-np1.14
# packages in environment at /Users/rob/miniconda/envs/py3.6-np1.14:
#
# Name                    Version                   Build  Channel
ca-certificates           2017.08.26                ha1e5d58_0
certifi                   2018.1.18                    py36_0
intel-openmp              2018.0.0                          8
libcxx                    4.0.1                     h579ed51_0
libcxxabi                 4.0.1                     hebd6815_0
libedit                   3.1                       hb4e282d_0
libffi                    3.2.1                     h475c297_4
libgfortran               3.0.1                     h93005f0_2
mkl                       2018.0.2                          1
mkl_fft                   1.0.1                  py36h917ab60_0
mkl_random                1.0.1                  py36h78cc56f_0
ncurses                   6.0                       hd04f020_2
numpy                     1.14.2                 py36ha9ae307_1
openssl                   1.0.2o                    h26aff7b_0
pip                       9.0.3                        py36_0
python                    3.6.5                     hc167b69_0
readline                  7.0                       hc1231fa_4
setuptools                39.0.1                       py36_0
sqlite                    3.22.0                    h3efe00b_0
tk                        8.6.7                     h35a86e2_3
wheel                     0.30.0                 py36h5eb2c71_1
xz                        5.2.3                     h0278029_2
zlib                      1.2.11                    hf3cbc9b_2
```

Here we see that NumPy is indeed installed, and the precise version of the library is 1.8.14. If we do not explicitly specify the version of a library, the latest stable release is used.

To use the second method – to install additional packages in an already existing environment – we first activate the environment

```
$ source activate py3.6
```

and then use `conda install PACKAGE` to install the package with name PACKAGE. Here we can also give a list of package names. For example, to install the NumPy, SciPy, and Matplotlib libraries, we can use

```
(py3.6)$ conda install numpy scipy matplotlib
```

or, equivalently,

```
$ conda install -n py3.6 numpy scipy matplotlib
```

When installing packages using conda, all required dependencies are also installed automatically, and the preceding command actually also installed the packages dateutil, freetype, libpng, pyparsing, pytz, and six packages, among others, which are dependencies for the matplotlib package:

```
(py3.6)$ conda list
# packages in environment at /Users/rob/miniconda/envs/py3.6:
#
# Name                    Version                   Build  Channel
ca-certificates           2017.08.26                ha1e5d58_0
certifi                   2018.1.18                     py36_0
cycler                    0.10.0                py36hfc81398_0
freetype                  2.8                     h12048fb_1
intel-openmp              2018.0.0                           8
kiwisolver                1.0.1                 py36h792292d_0
libcxx                    4.0.1                   h579ed51_0
libcxxabi                 4.0.1                   hebd6815_0
libedit                   3.1                     hb4e282d_0
libffi                    3.2.1                   h475c297_4
libgfortran               3.0.1                   h93005f0_2
libpng                    1.6.34                  he12f830_0
matplotlib                2.2.2                 py36ha7267d0_0
mkl                       2018.0.2                           1
mkl_fft                   1.0.1                 py36h917ab60_0
mkl_random                1.0.1                 py36h78cc56f_0
ncurses                   6.0                     hd04f020_2
numpy                     1.14.2                py36ha9ae307_1
openssl                   1.0.2o                  h26aff7b_0
```

pip	9.0.3	py36_0
pyparsing	2.2.0	py36hb281f35_0
python	3.6.5	hc167b69_0
python-dateutil	2.7.2	py36_0
pytz	2018.3	py36_0
readline	7.0	hc1231fa_4
scipy	1.0.1	py36hcaad992_0
setuptools	39.0.1	py36_0
six	1.11.0	py36h0e22d5e_1
sqlite	3.22.0	h3efe00b_0
tk	8.6.7	h35a86e2_3
tornado	5.0.1	py36_1
wheel	0.30.0	py36h5eb2c71_1
xz	5.2.3	h0278029_2
zlib	1.2.11	hf3cbc9b_2

Note that not all of the packages installed in this environment are Python libraries. For example, libpng and freetype are system libraries, but conda is able to handle them and install them automatically as dependencies. This is one of the strengths of conda compared to, for example, the Python-centric package manager pip.

To update selected packages in an environment, we can use the conda update command. For example, to update NumPy and SciPy in the currently active environment, we can use

```
(py3.4)$ conda update numpy scipy
```

To remove a package, we can use conda remove PACKAGE, and to completely remove an environment, we can use conda remove -n NAME --all. For example, to remove the environment py2.7-np1.8, we can use

```
$ conda remove -n py2.7-np1.8 --all
```

Conda locally caches packages that have once been installed. This makes it fast to reinstall a package in a new environment and also quick and easy to tear down and set up new environments for testing and trying out different things, without any risk of breaking environments used for other projects. To re-create a conda environment, all we need to do is to keep track of the installed packages. Using the -e flag with the conda list command gives a list of packages and their versions, in a format that is also compatible with the `pip` software. This list can be used to replicate a conda environment, for example, on another system or at a later point in time:

```
$ conda list -e > requirements.txt
```

With the file `requirements.txt`, we can now update an existing conda environment in the following manner:

```
$ conda install --file requirements.txt
```

or create a new environment that is a replication of the environment that was used to create the `requirement.txt` file:

```
$ conda create -n NAME --file requirements.txt
```

Alternatively, we can use the YAML format dump of an environment produced by `conda env export`:

```
$ conda env export -n NAME > env.yml
```

and in this case we can reproduce the environment using

```
$ conda env create --file env.yml
```

Note that here we do not need to specify the environment name since the `env.yml` file also contains this information. Using this method also has the advantage that packages installed using `pip` are installed when the environment is replicated or restored.

A Complete Environment

Now that we have explored the conda package manager, and seen how it can be used to set up environments and install packages, next we cover the procedures for setting up a complete environment with all the dependencies that are required for the material in this book. In the following we use the `py3.6` environment, which was previously created using the command:

```
$ conda create -n py3.6 python=3.6
```

This environment can be activate using

```
$ source activate py3.6
```

Once the target environment is activated, we can install the libraries that we use in this book with the following commands:

```
conda install ipython jupyter jupyterlab spyder pylint pyflakes pep8
conda install numpy scipy sympy matplotlib networkx pandas seaborn
conda install patsy statsmodels scikit-learn pymc3
conda install h5py pytables msgpack-python cython numba cvxopt
conda install -c conda-forge fenics mshr
conda install -c conda-forge pygraphviz

pip install scikit-monaco
pip install version_information
```

The FEniCS libraries have many intricate dependencies, which can make it difficult to install using this standard approach on some platforms.[10] For this reason, if the FEniCS installation using conda fails, it is most easily installed using the prebuilt environments available from the project's web site: `http://fenicsproject.org/download`. Another good solution for obtaining a complete FEniCS environment can be to use a Docker[11] container with FEniCS preinstalled. See, for example, `https://registry.hub.docker.com/repos/fenicsproject` for more information about this method.

Table A-1 presents a breakdown of the installation commands for the dependencies, on a chapter-by-chapter basis.

---

[10]There are recent efforts to create conda packages for the FEniCS libraries and their dependencies: `http://fenicsproject.org/download/`. However, this method is currently only available for Linux and MacOS.

[11]For more information about software container solution Docker, see `https://www.docker.com`.

Table A-1. *Installation Instructions for Dependencies for Each Chapter*

Chapter	Used Libraries	Installation

1	IPython, Spyder, Jupyter	`conda install ipython jupyter jupyterlab`
		`conda install spyder pylint pyflakes pep8`
		Here `pylint`, `pyflakes`, and `pep8` are code analysis tools that can be used by Spyder.
		For converting IPython notebooks to PDF, you also need a working LaTeX installation.
		To book-keep which versions of libraries that were used to execute the IPython notebooks that accompany this book, we have used IPython extension command `%version_`
		`information`, which is available in the `version_`
		`information` package that can be installed with `pip`:
		`pip install version_information`

2	NumPy	`conda install numpy`
3	NumPy, SymPy	`conda install numpy sympy`
4	NumPy, Matplotlib	`conda install numpy matplotlib`
5	NumPy, SymPy, SciPy, Matplotlib	`conda install numpy sympy scipy matplotlib`
6	NumPy, SymPy, SciPy, Matplotlib, cvxopt	`conda install numpy sympy scipy matplotlib cvxopt`
7	NumPy, SciPy, Matplotlib	`conda install numpy scipy matplotlib`

8	NumPy, SymPy, SciPy, Matplotlib, Scikit-Monaco	`conda install numpy sympy scipy matplotlib`
		There is no conda package for scikit-monaco, so we need to install this library using `pip`:
		`pip install scikit-monaco`

9	NumPy, SymPy, SciPy, Matplotlib	`conda install numpy sympy scipy matplotlib`

(continued)

Table A-1. (*continued*)

Chapter	Used Libraries	Installation

10	NumPy, SciPy, Matplotlib, NetworkX	`conda install numpy scipy matplotlib networkx`
		To visualize NetworkX graphs, we also need the Graphviz library (see `www.graphviz.org`) and its Python bindings in the `pygraphviz` library:
		`conda install -c conda-forge pygraphviz`

11	NumPy, SciPy, Matplotlib, and FEniCS	`conda install numpy scipy matplotlib` `conda install -c conda-forge fenics mshr`
12	NumPy, Pandas, Matplotlib, Seaborn	`conda install numpy pandas matplotlib seaborn`
13	NumPy, SciPy, Matplotlib, Seaborn	`conda install numpy scipy matplotlib seaborn`
14	NumPy, Pandas, Matplotlib, Seaborn, Patsy, Statsmodels	`conda install numpy pandas matplotlib seaborn patsy statsmodels`
15	NumPy, Matplotlib, Seaborn, scikit-learn	`conda install numpy matplotlib seaborn scikit-learn`
16	NumPy, Matplotlib, PyMC3	`conda install numpy matplotlib pymc3`
17	NumPy, SciPy, Matplotlib	`conda install numpy scipy matplotlib`

18	NumPy, Pandas, h5py, PyTables, msgpack	`conda install numpy pandas h5py pytables msgpack-python`
		At the time of writing, the `msgpack-python` conda package is not available for all platforms. When conda packages are not available, the msgpack library needs to be installed manually, and its python bindings can be installed using `pip`:
		`pip install msgpack-python`

19	NumPy, Matplotlib, Cython, Numba	`conda install numpy matplotlib cython numba`

A list of the packages and their exact versions that were used to run the code included in this book is also available in the `requirements.txt` file that is available for download together with the code listing. With this file we can directly create an environment with all the required dependencies with a single command:

```
$ conda create -n py3.6 --file requirements.txt
```

Alternatively, we can re-create the `py2.7` and `py3.6` environments using the exports `py2.7-env.yml` and `py3.6-env.yml`. These files are also available together with the source code listings.

```
$ conda env create --file py2.7-env.yml
$ conda env create --file py3.6-env.yml
```

Summary

In this appendix we have reviewed the installation of the various Python libraries used in this book. The Python environment for scientific computing is not a monolithic environment, but rather consists of an ecosystem of diverse libraries that are maintained and developed by different groups of people, following different release cycles and development paces. As a consequence, it can be difficult to collect all the necessary pieces of a productive setup from scratch. In response to this problem, several solutions addressing this situation have appeared, typically in the form of prepackaged Python distributions. In the Python scientific computing community, Anaconda and Canopy are two popular examples of such environments. Here we focused on the conda package manager from the Anaconda Python distribution, which in addition to being a package manager also allows to create and to manage virtual installation environments.

Further Reading

If you are interested in creating Python source packages for your own projects, see, for example, `http://packaging.python.org/en/latest/index.html`. In particular, study the `setuptools` library and its documentation at `http://pythonhosted.org/setuptools`. Using `setuptools`, we can create installable and distributable Python source packages. Once a source package has been created using `setuptools`, it is usually straightforward to create binary conda packages for distribution. For information on creating and distributing conda packages, see `http://conda.pydata.org/docs/build_tutorials/pkgs.html`. See also the `conda-recipes` repository at `github.com`, which contains many examples of conda packages: `http://github.com/conda/conda-recipes`. Finally, `http://www.anaconda.org` is a conda package hosting service with many public channels (repositories) where custom-built conda packages can be published and installed directly using the conda package manager. Many packages that are not available in the standard Anaconda channel can be found on user-contributed channels on `anaconda.org`. In particular, many packages are available in the `conda-forge` channel, built from conda recipes available from `conda-forge.org`.

Index

A

Adams-Bashforth methods, 316
Adams-Moulton methods, 316
add_axes method, 143, 146
add_subplot function, 177
Analytical value, 287
Application programming
        interfaces (APIs), 136
Arbitrary-precision integration, 285
Array creation, 50–51
    constant values, 52–54
    incremental sequence, 54
    list, 52
    logarithmic sequence, 54
    matrix array, 57–58
    uninitialized array, 56
Autoregressive (AR) model, 506
    ARIMA, 510
    ARMA, 506, 510
    Durbin-Watson statistical test, 509
    high-order, 508
    sm.tsa.AR class, 508
    temperature observations, 507, 509
Autoregressive integrated moving
        average (ARIMA) model, 510
Autoregressive moving average (ARMA)
        model, 506, 510
autoscale method, 157
AutoSubDomain instance, 400
ax.annotation, 154–155

Axes, 138, 144–146
    legend method, 152–153
    line properties, 147–151
    plot types, 146–147
    properties (*see* Properties, axis)
    text annotations, 153
Axes layouts
    GridSpec, 173
    insets, 168–169
    subplot2grid, 172
    subplots, 170–171
Axes.text method, 154
Axis ticks, 158
ax.legend method, 140
ax.set_xlabel method, 140
ax.set_ylabel method, 140
ax.text, 154–155

B

Bayes' theorem, 545–546
bbox_to_anchor argument, 152
BFGS method, 225–226, 228
Bisection method, 201–203, 217–218
Bivariate interpolation, 264
Blackman
        function, 581, 583, 588
Boolean-valued array, 64
Bracketing methods, 217
brent function, 218

Printed in the United States
By Bookmasters